As You Examine HOLT PRE-ALGEE

Please Notice HOLT PRE-ALGEBRA provides a com_____
to introduce students to algebraic concepts while mai_____ _____ _____ _____.
All of the concepts needed to prepare students for a future, full-year algebra course are
found in **HOLT PRE-ALGEBRA**.

Format HOLT PRE-ALGEBRA's five-part lesson format facilitates students' learning
by actively involving them in the learning process. ● Lesson objectives are on the pupil
page ● Recall items review prerequisite skills needed for each lesson ● Examples develop
concepts in a clear, orderly way ● Practice exercises check students' understanding of
the examples ● Exercises reinforce the lesson objectives

1. The course begins with an introduction to basic algebraic concepts which are developed and maintained throughout the book. (see pp. 1, 58, 142)

2. Basic mathematics skills are maintained and developed throughout the book. (see pp. 26, 82, 288)

3. Suggestions, hints, comments, and summaries guide students through lessons and clarify concepts taught. (see pp. 38, 316)

4. Practice exercises appear after the examples so students can check their understanding of the examples. (see pp. 79, 133)

5. Oral Exercises appear before written exercises so teachers can quickly test students' understanding and readiness. (see pp. 7, 130)

6. Exercises are plentiful and provide practice of lesson concepts. They are graded from easiest to most challenging. (see pp. 7, 290)

7. Reading instruction helps students analyze and better understand mathematical language. (see pp. 12, 322)

8. Each chapter ends with a Chapter Review and Chapter Test. The Review is diagnostic and keys items to lesson pages. (see pp. 23, 50)

9. Extra Practice exercises provide further reinforcement and practice of lesson concepts. (see pp. 436–454)

10. A four-part problem solving strategy guides students through problems and allows them to apply their skills to real-world applications and careers. (see pp. 13, 68, 98)

11. Non-Routine Problems challenge students to use logical reasoning to solve problems. (see pp. 135, 235)

12. Math Aptitude Tests provide experience in taking standardized tests. (see pp. 48, 363)

13. Computer Activities appear in every chapter, demonstrating how computers can be used to enhance students' understanding of mathematics concepts. A computer section in the back of the book introduces students to the fundamentals of programming. (see pp. 75, 420)

14. Answers to practice exercises and odd-numbered answers to extra practice exercises are provided in the back of the pupil edition. (see pp. 455, 466)

15. A Teacher's Edition with teaching suggestions, lesson commentaries, enrichment activities, projects, and a complete set of answers makes teaching **HOLT PRE-ALGEBRA** easy.

16. Assignments for reinforcement and maintenance allows teachers to make assignments on two levels of difficulty. Assignments are designed to provide practice of the current lesson and previous lessons.

Supplementary Materials Teacher's Resource Package provides *Skillmasters* with additional exercises and cumulative reviews. *Testmasters*, which test students' understanding of the course objectives and basic mathematics concepts, and *Lesson Aids*, which help teachers individualize the course, are all in blackline master format. *Skillmasters* and *Testmasters* are also available as duplicating masters.

HOLT
PRE-ALGEBRA
Teacher's Edition

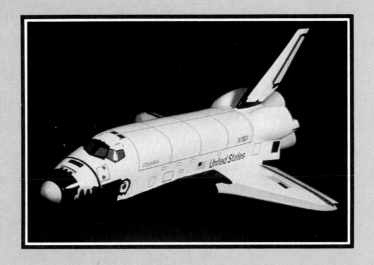

Eugene D. Nichols • Mervine L. Edwards
Sylvia A. Hoffman • Albert Mamary

HOLT, RINEHART AND WINSTON, PUBLISHERS
New York • Toronto • Mexico City • London • Sydney • Tokyo

Staff Credits

Editorial Development	Everett T. Draper, Earl D. Paisley, Tony Y. Maksoud, Eugene R. McCormick
Product Manager	Daniel M. Loch
Field Advisory Board	Douglas A. Nash, Sam Sherwood, Robert Wolff, Wendell Anthony, Gary Crump, Jack M. Custer, Roy Eliason, Kenneth C. Scupp, Dennis Spurgeon, Jeffra Ann Nicholson
Marketing Research	Erica S. Felman, Linda A. DeLora
Editorial Processing	Margaret M. Byrne, Pamela K. Caugherty
Art and Design	Carol Steinberg, Amy Newberg
Production	Bev Silver, Joan McNeil, Heidi J. Henney
Photo Resources	Linda Sykes, Rita Longabucco

ISBN 0-03-001859-5

901 040 65

Teacher's Edition Contents

Benefits

HOLT PRE-ALGEBRA is designed to help students obtain the skills necessary for success in a future full-year algebra course. The book gives students a feel for some of the key areas of introductory algebra while maintaining basic mathematics skills.

Algebra Strand Algebraic concepts are introduced beginning with the first chapter, and a strong algebra-maintenance feature occurs throughout the book.

Five-Part Lesson Development A consistent five-part lesson format enhances the gradual development of skills and concepts.

Problem-Solving Strand A clear, logical problem-solving strategy and special problem-solving application lessons including non-routine problems, provide a meaningful approach to problem solving.

Enrichment Calculator activities, problem formulation, computer lessons, special Challenge sections, plus enrichment sections provide a wide selection of activity-oriented experiences for the student.

Testing and Review Extra-practice exercises, chapter reviews, chapter tests, and cumulative reviews geared to the material taught in each chapter keep students and teachers aware of student progress.

Teacher's Annotated Edition The Teacher's Edition includes full-sized, annotated pupil's pages, chapter objectives, cumulative reviews, activities, and common errors, as well as detailed teaching suggestions.

Teacher's Resource Package contains Blackline Masters for review, extra practice, diagnosing areas of difficulty, and supplementary testing, as well as a complete testing program.

Supplementary Materials Skillmasters (duplicating masters) for review, extra practice, diagnosing areas of difficulty, and supplementary testing; Testmasters (duplicating masters) for supplementary testing, as well as a complete testing program containing chapter tests, cumulative and final tests

Five-Part Lesson Format

HOLT PRE-ALGEBRA uses a carefully structured five-part lesson format to actively involve students in the learning process.

Recall reviews the pre-requisite skills the students need to understand the lesson.

Objectives tell the students what they will learn in each lesson.

Examples develop the mathematical concepts being taught, in a series of carefully-structured steps.

Introduction to Variables

— ◇ **OBJECTIVES** ◇ —   — ◇ **RECALL** ◇ —

To identify the variables, constants, and terms in an expression
To evaluate algebraic expressions for given values of variables

Compute $18 - 5 + 6$.
$$18 - 5 + 6$$
$$13 \quad + 6$$
$$19$$

Think of a number. Add 2 to the number.

Number $+ 2$
$\downarrow$ $\downarrow$

Use x to represent the number. $x \quad + 2$
$\uparrow$
Variable

A *variable* is a letter that may be replaced by different values.

Example 1
To evaluate means to find the value.

Evaluate $y + 7$ if $y = 8$.
$$y + 7$$
$$\downarrow$$
Substitute 8 for y. $8 + 7$
$$15$$
So, the value of $y + 7$ is 15 if $y = 8$.

practice ▷ Evaluate if $x = 2$, $m = 3$, $y = 5$, and $a = 2$.

1. $x + 5$ 2. $7 + m$ 3. $6 - y$ 4. $a + 9$

Example 2

Evaluate $c - d$ if $c = 90$ and $d = 23$.
$$c - d$$
$$90 - 23$$
$$67$$
So, the value of $c - d$ is 67 if $c = 90$ and $d = 23$.

practice ▷ Evaluate if $x = 40$, $y = 26$, $a = 56$, $b = 39$, and $d = 13$.

5. $x - y$ 6. $a + b$ 7. $a - b$ 8. $b + d$

INTRODUCTION TO VARIABLES   1

Example 3

Evaluate $a + b - c + d$ if $a = 17, b = 48, c = 19$, and $d = 8$.

$$\begin{array}{r} {\scriptstyle 1} \\ 17 \\ +48 \\ \hline 65 \end{array}$$

$$\begin{array}{r} {\scriptstyle 5\ 15} \\ 65 \qquad \not{6}\ \not{5} \\ -19 \qquad -\ 1\ 9 \\ \hline 4\ 6 \end{array}$$

$$a + b \ - \ c + d$$
$$17 + 48 \ - \ 19 + 8$$
$$\overline{65 \ - \ 19} + 8$$
$$46 \quad + 8$$
$$54$$

So, the value of $a + b - c + d$ is 54.

practice ▷ Evaluate if $x = 13, y = 59, z = 32, w = 6, a = 48, b = 44, c = 17$, and $d = 5$.

9. $x + y - z + w$ **10.** $a + b - c + d$ **11.** $y - a + z - d$

The expression $x + y - 6$ has three *terms:*
x, y, and 6.
Terms are added or subtracted.
The *variables* are x and y; 6 is a *constant*.

Example 4

Name the terms, variables, and constants of $x - 7 + y$.
The terms are x, 7, and y.
The variables are x and y. The constant is 7.

practice ▷ Name the terms, variables, and constants of each expression.

12. $11 - p$ **13.** $t + 4 - u$ **14.** $x + 8 - y$ **15.** $x - 9 + y - 4$

◇ EXERCISES ◇

Evaluate the following expressions if $r = 7, s = 9, t = 4$, and $w = 5$.

1. $r + 5$	**2.** $9 + t$	**3.** $11 - w$	**4.** $s + 6$	**5.** $17 - t$
6. $r + t$	**7.** $s - w$	**8.** $w + r$	**9.** $s + 22$	**10.** $19 - w$

Evaluate the following expressions if $a = 70, b = 42, c = 73$, and $d = 49$.

11. $b + 19$	**12.** $c - 19$	**13.** $a + b$	**14.** $c - d$
15. $a + b - 13$	**16.** $a - b + d$	**17.** $d - b + a$	**18.** $39 + a - d$
19. $a + c - b + d$	**20.** $c - d + a - b$	**21.** $d - b + c - a$	

Name the terms, variables, and constants of each expression.

22. $a - 4$ **23.** $p - 12 + t$ **24.** $5 + a + 7 - n$

Special Features

A variety of features guide students through activities designed to stimulate thinking.

Oral Exercises provide quick review of skills to be used in the Exercises. The teacher can quickly check the students' understanding of skills and concepts before they move on to the written exercises.

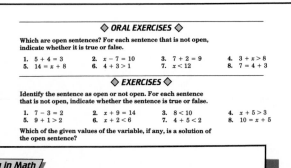

◇ **ORAL EXERCISES** ◇

Which are open sentences? For each sentence that is not open, indicate whether it is true or false.

1. $5 + 4 = 3$ 2. $x - 7 = 10$ 3. $7 + 2 = 9$ 4. $3 + x > 8$
5. $14 = x + 8$ 6. $4 + 3 > 1$ 7. $x < 12$ 8. $7 = 4 + 3$

◇ **EXERCISES** ◇

Identify the sentence as open or not open. For each sentence that is not open, indicate whether the sentence is true or false.

1. $7 - 3 = 2$ 2. $x + 9 = 14$ 3. $8 < 10$ 4. $x + 5 > 3$
5. $9 + 1 > 2$ 6. $x + 2 < 6$ 7. $4 + 5 < 2$ 8. $10 = x + 5$

Which of the given values of the variable, if any, is a solution of the open sentence?

Reading in Math helps students develop the skills needed to read, interpret, and use the language of mathematics effectively.

/ Reading in Math /

Match each English phrase with its mathematical form.

1. x increased by 10   **a.** $x - 15$
2. 10 less than x **b.** $x + 10$
3. the sum of x and 15   **c.** $x + 15$
4. 10 more than x **d.** $x - 10$
5. x decreased by 15   **e.** $15 - x$
     f. $10 - x$

◇ **EXERCISES** ◇

Write in mathematical form.

Chapter Openers provide opportunities to practice problem formulation, estimation, and application.

Equations–Adding and Subtracting

1

Up to 62,000 people can be seated in the Houston Astrodome, the first domed stadium.

Enrichment Materials

HOLT PRE-ALGEBRA provides many resources for enriching the curriculum, depending on the students' readiness and learning aptitude.

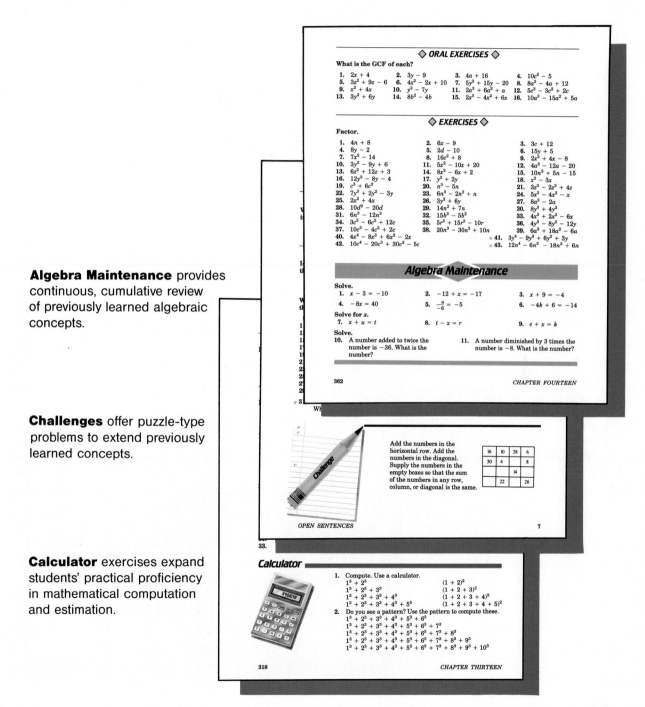

Algebra Maintenance provides continuous, cumulative review of previously learned algebraic concepts.

Challenges offer puzzle-type problems to extend previously learned concepts.

Calculator exercises expand students' practical proficiency in mathematical computation and estimation.

◇ ORAL EXERCISES ◇

What is the GCF of each?

1. $2x + 4$ 2. $3y - 9$ 3. $4a + 16$ 4. $10c^2 - 5$
5. $3z^2 + 9z - 6$ 6. $4x^2 - 2x + 10$ 7. $5y^2 + 15y - 20$ 8. $8a^2 - 4a + 12$
9. $x^2 + 4x$ 10. $y^2 - 7y$ 11. $2a^3 + 6a^2 + a$ 12. $5c^3 - 3c^2 + 2c$
13. $3y^2 + 6y$ 14. $8b^2 - 4b$ 15. $2x^3 - 4x^2 + 6x$ 16. $10a^3 - 15a^2 + 5a$

◇ EXERCISES ◇

Factor.

1. $4n + 8$ 2. $6x - 9$ 3. $3c + 12$
4. $8y - 2$ 5. $2d - 10$ 6. $15y + 5$
7. $7x^2 - 14$ 8. $16c^2 + 8$ 9. $2x^2 + 4x - 8$
10. $3y^2 - 9y + 6$ 11. $5x^2 - 10x + 20$ 12. $4a^2 - 12a - 20$
13. $6z^2 + 12z + 3$ 14. $8x^2 - 6x + 2$ 15. $10n^2 + 5n - 15$
16. $12y^2 - 8y - 4$ 17. $y^2 + 2y$ 18. $x^2 - 3x$
19. $c^3 + 6c^2$ 20. $n^3 - 5n$ 21. $3z^3 - 2z^2 + 4z$
22. $7y^3 + 2y^2 - 3y$ 23. $6n^3 - 2n^2 + n$ 24. $5x^3 - 4x^2 - x$
25. $2x^2 + 4x$ 26. $3y^2 + 6y$ 27. $8a^2 - 2a$
28. $10d^2 - 20d$ 29. $14n^2 + 7n$ 30. $8y^3 + 4y^2$
31. $6n^3 - 12n^2$ 32. $15b^3 - 5b^2$ 33. $4x^3 + 2x^2 - 6x$
34. $3c^3 - 6c^2 + 12c$ 35. $5r^3 + 15r^2 - 10r$ 36. $4y^3 - 8y^2 - 12y$
37. $10c^3 - 4c^2 + 2c$ 38. $20n^3 - 30n^2 + 10n$ 39. $6a^3 + 18a^2 - 6a$
40. $4x^4 - 8x^3 + 6x^2 - 2x$ ★ 41. $3y^4 - 9y^3 + 6y^2 + 3y$
42. $10c^4 - 20c^3 + 30c^2 - 5c$ ★ 43. $12n^4 - 6n^3 - 18n^2 + 6n$

Algebra Maintenance

Solve.
1. $x - 3 = -10$ 2. $-12 + x = -17$ 3. $x + 9 = -4$
4. $-8x = 40$ 5. $-\frac{u}{6} = -5$ 6. $-4k + 6 = -14$

Solve for x.
7. $x + u = t$ 8. $t - x = r$ 9. $e + x = k$

Solve.
10. A number added to twice the number is -36. What is the number?
11. A number diminished by 3 times the number is -8. What is the number?

362   CHAPTER FOURTEEN

Add the numbers in the horizontal row. Add the numbers in the diagonal. Supply the numbers in the empty boxes so that the sum of the numbers in any row, column, or diagonal is the same.

16	10	28	6
30	4		8
		14	
	22		26

OPEN SENTENCES   7

Calculator

1. Compute. Use a calculator.
$1^3 + 2^3$ $(1 + 2)^2$
$1^3 + 2^3 + 3^3$ $(1 + 2 + 3)^2$
$1^3 + 2^3 + 3^3 + 4^3$ $(1 + 2 + 3 + 4)^2$
$1^3 + 2^3 + 3^3 + 4^3 + 5^3$ $(1 + 2 + 3 + 4 + 5)^2$
2. Do you see a pattern? Use the pattern to compute these.
$1^3 + 2^3 + 3^3 + 4^3 + 5^3 + 6^3$
$1^3 + 2^3 + 3^3 + 4^3 + 5^3 + 6^3 + 7^3$
$1^3 + 2^3 + 3^3 + 4^3 + 5^3 + 6^3 + 7^3 + 8^3$
$1^3 + 2^3 + 3^3 + 4^3 + 5^3 + 6^3 + 7^3 + 8^3 + 9^3$
$1^3 + 2^3 + 3^3 + 4^3 + 5^3 + 6^3 + 7^3 + 8^3 + 9^3 + 10^3$

318   CHAPTER THIRTEEN

Computer Strand

A computer program is a set of written instructions for a computer. If you have access to a computer that uses the language of **BASIC**, you may use the computer to type in the program in the Example below.

Example Write a program to find S for S = 219 − 43 + 124
Then type the program into a computer and **RUN** it. (At the end of each line, press the **RETURN** key to get to the next line.)

Write a formula telling the computer how to find S. This uses the LET command.	`10  LET S = 219 - 43 + 124`
The PRINT command tells the computer to print S.	`20  PRINT S`
END tells the computer that the program is finished.	
The computer will do nothing until you tell it to **RUN** the program. Type **RUN**	
The computer now displays the value of S.	

Notice that the lines of the program are numbered back and insert more lines in between any two li present program from the computer's memory:

Press **RETURN**

Type **NEW**

Press **RETURN**

Exercises

Find the value of each by first doing your own cor **RUN** a program to check that the computer's resu

1. Find A. A = 76 + 49 − 18
3. Find A. A = 69 + 49
5. Find P. P = 89 − 62 + 43 − 26

22

Every chapter has a computer lesson, which includes computer applications with examples and hands-on programming exercises.

COMPUTER SECTION

INT

OBJECTIVE
To use **INT** to determine if one number is a factor of another

The INTeger part of the number 8.679543 is 8.
In **BASIC**, the instruction **PRINT** the INTeger part of 8.679543 is written as **PRINT INT(8.679543)**.

Example 1 Find each: **INT(3.45649)**   **INT(3.87569)**   **INT(3)**

                     INT(3.45649) = 3   **INT(3.87569)** = 3   **INT(3.0)** = 3

Example 2 Use **INT** to determine whether 4 is a factor of 15.

Does 4 divide into 15 evenly?

$\dfrac{3.75}{4)15.00}$

Does 15/4 = **INT(15/4)**?

Does 3.75 = **INT(3.75)**?

No. 3.75 ≠ 3

So, 4 is not a factor of 15.

Example 3 Write a program to determine whether a number is a factor of 528.

```
10  INPUT "TYPE IN A POSSIBLE FACTOR OF 528 ";F
20  IF 528 / F = INT (528 / F) THEN 50
30  PRINT F" IS NOT A FACTOR OF 528."
40  GOTO 60
50  PRINT F" IS A FACTOR OF 528."
60  END
```

Exercises

1. Find each of the following: **INT**(32.18976); **INT**(10 ∗ 0.3455654 + 1).

2. Modify the program above with a loop so that you can test whether any 5 different numbers are factors of 528. Then **RUN** the program to determine which of the following numbers are factors of 528: 4, 17, 24, 96, 176.

434   *COMPUTER SECTION*

The back of the book contains a 16-page computer section designed to give students more instruction and practice in using and understanding statements in BASIC.

Problem Solving Strand

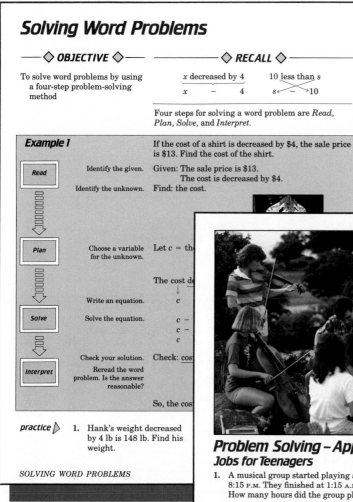

Solving Word Problems

—◇ **OBJECTIVE** ◇—   ◇ **RECALL** ◇—

To solve word problems by using a four-step problem-solving method

x decreased by 4

$$x \quad - \quad 4$$

10 less than s

$$s \quad - \quad 10$$

Four steps for solving a word problem are *Read, Plan, Solve,* and *Interpret.*

Example 1

If the cost of a shirt is decreased by $4, the sale price is $13. Find the cost of the shirt.

Read

Identify the given.   Given: The sale price is $13.
                                   The cost is decreased by $4.

Identify the unknown.   Find: the cost.

Plan

Choose a variable for the unknown.   Let c = th

The cost d

Write an equation. c

Solve

Solve the equation. c —
                                    c —
                                    c

Interpret

Check your solution.   Check: cos

Reread the word problem. Is the answer reasonable?

So, the cos

practice ▷   1. Hank's weight decreased by 4 lb is 148 lb. Find his weight.

SOLVING WORD PROBLEMS

A four-part problem-solving strategy, easily identified by its logo, encourages students to approach problem solving in a logical manner.

Special *Problem-Solving Applications* include Jobs for Teenagers and Careers; two areas that offer abundant opportunities to practice problem-solving strategies.

Problem Solving – Applications
Jobs for Teenagers

1. A musical group started playing at 8:15 P.M. They finished at 1:15 A.M. How many hours did the group play?

2. One evening the Keynotes played for three hours. Each song lasted an average of 4 minutes. How many songs did they play?

3. A band earns an average of $450 a weekend. They wanted to buy new equipment that costs $2,700. How many weekends must they work in order to buy the new equipment?

4. A three-piece band played from 9:00 P.M. to midnight on Friday, from 8:00 P.M. to 2:00 A.M. on Saturday, and from 9:00 P.M. to 11:45 P.M. on Sunday. How many hours did they play altogether on that weekend?

5. A rock group charges a flat fee of $620. They played for 5 hours. How much did they make per hour?

6. The sophomore class hires a band that charges $350. Students are charged $3.00 a ticket. 275 students attend the dance. Find the class profit.

7. A piano player earned $20 per hour for the first three hours of playing and $24 per hour for each hour over 3 hours. How much did she earn for a 4-hour engagement?

8. A drummer earns $19.50 per hour for playing until midnight. She is then paid $22.00 per hour for playing after midnight. She played from 8:00 P.M. to 2:00 A.M. How much did she earn?

CHAPTER THREE

Reinforcing Skills

Summary provides a quick review of the key concepts and procedures taught in the lesson.

Non-Routine Problems provide students with the opportunity to solve unusual problems using non-standard approaches or strategies.

Mathematics Aptitude Tests give students early practice and experience in taking multiple-choice tests similar to the standardized tests they will encounter in the future.

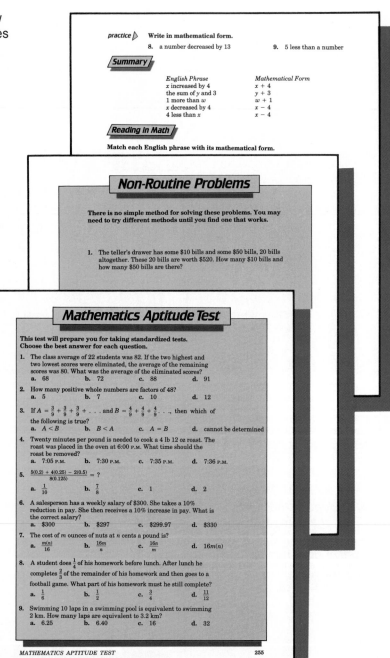

practice ▷ **Write in mathematical form.**

8. a number decreased by 13   **9.** 5 less than a number

Summary

English Phrase	Mathematical Form
x increased by 4	$x + 4$
the sum of y and 3	$y + 3$
1 more than w	$w + 1$
x decreased by 4	$x - 4$
4 less than x	$x - 4$

Reading in Math

Match each English phrase with its mathematical form.

Non-Routine Problems

There is no simple method for solving these problems. You may need to try different methods until you find one that works.

1. The teller's drawer has some $10 bills and some $50 bills, 20 bills altogether. These 20 bills are worth $520. How many $10 bills and how many $50 bills are there?

Mathematics Aptitude Test

This test will prepare you for taking standardized tests. Choose the best answer for each question.

1. The class average of 22 students was 82. If the two highest and two lowest scores were eliminated, the average of the remaining scores was 80. What was the average of the eliminated scores?
   a. 68   **b.** 72   **c.** 88   **d.** 91

2. How many positive whole numbers are factors of 48?
   a. 5   **b.** 7   **c.** 10   **d.** 12

3. If $A = \frac{3}{9} + \frac{3}{9} + \frac{3}{9} + \ldots$ and $B = \frac{4}{9} + \frac{4}{9} + \frac{4}{9} \ldots$, then which of the following is true?
   a. $A < B$ **b.** $B < A$ **c.** $A = B$ **d.** cannot be determined

4. Twenty minutes per pound is needed to cook a 4 lb 12 oz roast. The roast was placed in the oven at 6:00 P.M. What time should the roast be removed?
   a. 7:05 P.M.   **b.** 7:30 P.M.   **c.** 7:35 P.M.   **d.** 7:36 P.M.

5. $\frac{5(0.2) + 4(0.25) - 2(0.5)}{8(0.125)} = ?$
   a. $\frac{1}{10}$ **b.** $\frac{7}{8}$ **c.** 1   **d.** 2

6. A salesperson has a weekly salary of $300. She takes a 10% reduction in pay. She then receives a 10% increase in pay. What is the correct salary?
   a. $300   **b.** $297   **c.** $299.97   **d.** $330

7. The cost of m ounces of nuts at n cents a pound is?
   a. $\frac{m(n)}{16}$ **b.** $\frac{16m}{n}$ **c.** $\frac{16n}{m}$ **d.** $16m(n)$

8. A student does $\frac{1}{4}$ of his homework before lunch. After lunch he completes $\frac{2}{3}$ of the remainder of his homework and then goes to a football game. What part of his homework must he still complete?
   a. $\frac{1}{6}$ **b.** $\frac{1}{2}$ **c.** $\frac{3}{4}$ **d.** $\frac{11}{12}$

9. Swimming 10 laps in a swimming pool is equivalent to swimming 2 km. How many laps are equivalent to 3.2 km?
   a. 6.25   **b.** 6.40   **c.** 16   **d.** 32

MATHEMATICS APTITUDE TEST   **255**

Testing and Review

Tests in **HOLT PRE-ALGEBRA** assist the teacher in measuring the students' understanding of the major skills and concepts taught.

Chapter Tests are designed to measure the students' mastery of the chapter objectives.

Chapter Reviews contain exercise sets which are lesson-coded for easy reference to corresponding lessons when review is needed.

Cumulative Reviews test the students' understanding of algebraic concepts and mathematical skills that have been previously taught.

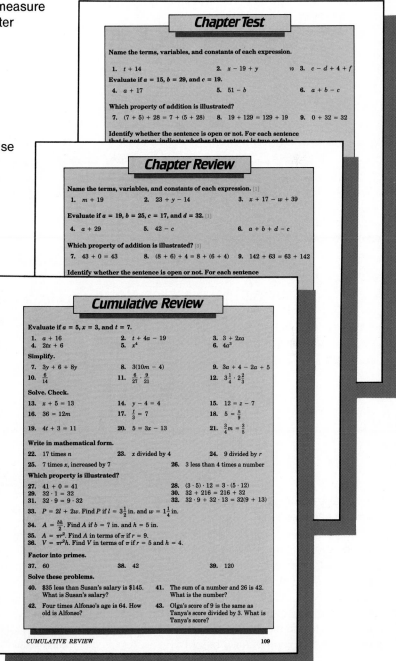

Chapter Test

Name the terms, variables, and constants of each expression.

1. $t + 14$ 2. $x - 19 + y$ 19   3. $c - d + 4 + f$

Evaluate if $a = 15, b = 29,$ and $c = 19.$

4. $a + 17$ 5. $51 - b$ 6. $a + b - c$

Which property of addition is illustrated?

7. $(7 + 5) + 28 = 7 + (5 + 28)$ 8. $19 + 129 = 129 + 19$ 9. $0 + 32 = 32$

Identify whether the sentence is open or not. For each sentence that is not open, indicate whether the sentence is true or false.

Chapter Review

Name the terms, variables, and constants of each expression. [1]

1. $m + 19$ 2. $23 + y - 14$ 3. $x + 17 - w + 39$

Evaluate if $a = 19, b = 25, c = 17,$ and $d = 32.$ [1]

4. $a + 29$ 5. $42 - c$ 6. $a + b + d - c$

Which property of addition is illustrated? [3]

7. $43 + 0 = 43$ 8. $(8 + 6) + 4 = 8 + (6 + 4)$ 9. $142 + 63 = 63 + 142$

Identify whether the sentence is open or not. For each sentence

Cumulative Review

Evaluate if $a = 5, x = 3,$ and $t = 7.$

1. $a + 16$ 2. $t + 4a - 19$ 3. $3 + 2xa$
4. $2tx + 6$ 5. x^4 6. $4a^3$

Simplify.

7. $3y + 6 + 8y$ 8. $3(10m - 4)$ 9. $3a + 4 - 2a + 5$
10. $\frac{6}{14}$ 11. $\frac{6}{27} \cdot \frac{9}{21}$ 12. $3\frac{1}{4} \cdot 2\frac{2}{3}$

Solve. Check.

13. $x + 5 = 13$ 14. $y - 4 = 4$ 15. $12 = z - 7$
16. $36 = 12m$ 17. $\frac{t}{3} = 7$ 18. $5 = \frac{n}{9}$
19. $4t + 3 = 11$ 20. $5 = 3x - 13$ 21. $\frac{3}{4}m = \frac{3}{5}$

Write in mathematical form.

22. 17 times n 23. x divided by 4   24. 9 divided by r
25. 7 times x, increased by 7   26. 3 less than 4 times a number

Which property is illustrated?

27. $41 + 0 = 41$ 28. $(3 \cdot 5) \cdot 12 = 3 \cdot (5 \cdot 12)$
29. $32 \cdot 1 = 32$ 30. $32 + 216 = 216 + 32$
31. $32 \cdot 9 = 9 \cdot 32$ 32. $32 \cdot 9 + 32 \cdot 13 = 32(9 + 13)$
33. $P = 2l + 2w.$ Find P if $l = 3\frac{1}{2}$ in. and $w = 1\frac{1}{4}$ in.
34. $A = \frac{bh}{2}.$ Find A if $b = 7$ in. and $h = 5$ in.
35. $A = \pi r^2.$ Find A in terms of π if $r = 9.$
36. $V = \pi r^2 h.$ Find V in terms of π if $r = 5$ and $h = 4.$

Factor into primes.

37. 60   38. 42   39. 120

Solve these problems.

40. $35 less than Susan's salary is $145. What is Susan's salary?
41. The sum of a number and 26 is 42. What is the number?
42. Four times Alfonso's age is 64. How old is Alfonso?
43. Olga's score of 9 is the same as Tanya's score divided by 3. What is Tanya's score?

CUMULATIVE REVIEW   109

Teacher's Edition

The Teacher's Edition of **HOLT PRE-ALGEBRA** provides an annotated, full-size, four-color reproduction of the pupil's page and includes detailed chapter and lesson commentaries.

Provides practice in problem formulation, estimation and applications

Tells how to use the computer page in the lesson

Focuses on the key topics to be taught

Provides challenge for better students

States the mathematical content of the chapter

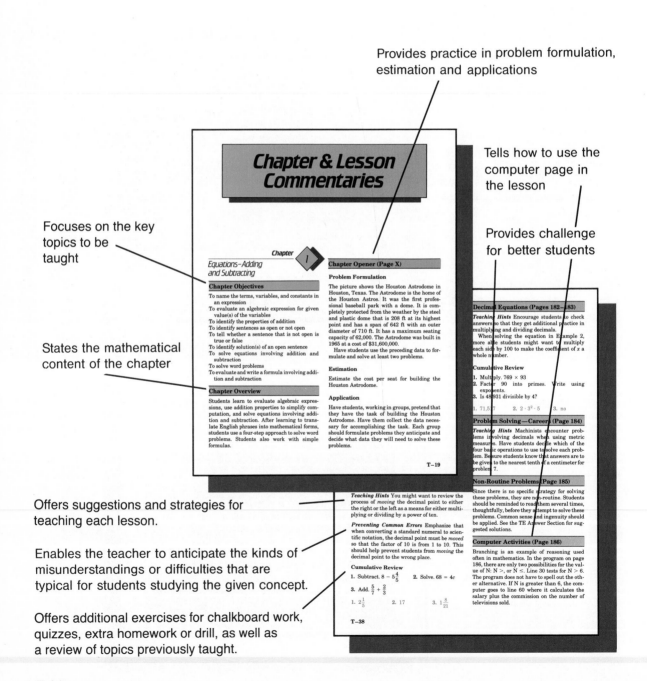

Chapter & Lesson Commentaries

Chapter 1

Equations–Adding and Subtracting

Chapter Objectives

To name the terms, variables, and constants in an expression
To evaluate an algebraic expression for given value(s) of the variables
To identify the properties of addition
To identify sentences as open or not open
To tell whether a sentence that is not open is true or false
To identify solution(s) of an open sentence
To solve equations involving addition and subtraction
To solve word problems
To evaluate and write a formula involving addition and subtraction

Chapter Overview

Students learn to evaluate algebraic expressions, use addition properties to simplify computation, and solve equations involving addition and subtraction. After learning to translate English phrases into mathematical forms, students use a four-step approach to solve word problems. Students also work with simple formulas.

Chapter Opener (Page X)

Problem Formulation

The picture shows the Houston Astrodome in Houston, Texas. The Astrodome is the home of the Houston Astros. It was the first professional baseball park with a dome. It is completely protected from the weather by the steel and plastic dome that is 208 ft at its highest point and has a span of 642 ft with an outer diameter of 710 ft. It has a maximum seating capacity of 62,000. The Astrodome was built in 1965 at a cost of $31,600,000.
Have students use the preceding data to formulate and solve at least two problems.

Estimation

Estimate the cost per seat for building the Houston Astrodome.

Application

Have students, working in groups, pretend that they have the task of building the Houston Astrodome. Have them collect the data necessary for accomplishing the task. Each group should formulate problems they anticipate and decide what data they will need to solve these problems.

T–19

Decimal Equations (Pages 182–183)

Teaching Hints Encourage students to check answers so that they get additional practice in multiplying and dividing decimals.
When solving the equation in Example 2, more able students might want to multiply each side by 100 to make the coefficient of x a whole number.

Cumulative Review

1. Multiply. 769×93
2. Factor 90 into primes. Write using exponents.
3. Is 48,931 divisible by 4?

1. 71,517   2. $2 \cdot 3^2 \cdot 5$ 3. no

Problem Solving—Careers (Page 184)

Teaching Hints Machinists encounter problems involving decimals when using metric measures. Have students decide which of the four basic operations to use to solve each problem. Be sure students know that answers are to be given to the nearest tenth of a centimeter for problem 7.

Non-Routine Problems (Page 185)

Since there is no specific strategy for solving these problems, they are non-routine. Students should be reminded to read them several times, thoughtfully, before they attempt to solve these problems. Common sense and ingenuity should be applied. See the TE Answer Section for suggested solutions.

Computer Activities (Page 186)

Branching is an example of reasoning used often in mathematics. In the program on page 186, there are only two possibilities for the value of N: N >, or N ≤. Line 30 tests for N > 6. The program does not have to spell out the other alternative. If N is greater than 6, the computer goes to line 60 where it calculates the salary plus the commission on the number of televisions sold.

Offers suggestions and strategies for teaching each lesson.

Enables the teacher to anticipate the kinds of misunderstandings or difficulties that are typical for students studying the given concept.

Offers additional exercises for chalkboard work, quizzes, extra homework or drill, as well as a review of topics previously taught.

Teaching Hints You might want to review the process of *moving* the decimal point to either the right or the left as a means for either multiplying or dividing by a power of ten.

Preventing Common Errors Emphasize that when converting a standard numeral to scientific notation, the decimal point must be *moved* so that the factor of 10 is from 1 to 10. This should help prevent students from *moving* the decimal point to the wrong place.

Cumulative Review

1. Subtract. $8 - 5\frac{4}{5}$ 2. Solve. $68 = 4c$

3. Add. $\frac{5}{7} + \frac{2}{3}$

1. $2\frac{1}{5}$ 2. 17   3. $1\frac{8}{21}$

T–38

Supplementary Materials

HOLT PRE-ALGEBRA supplementary materials include Testmasters, Skillmasters, and a Teacher's Resource Package with Blackline Masters, so that the teacher can instruct each student according to his or her needs and level of skill.

Skillmasters
contain two parallel sections for every lesson, which can be used as pre- or posttests for diagnosis, extra practice, quizzes, and assignments. There is also a cumulative review for every two lessons.

Testmasters
supply chapter tests, cumulative tests, and final tests.

Teacher's Resource Package
contains Skillmasters and Testmasters in blackline form, Competency Tests, and Lesson Aids. The Competency Tests contain the type of material found on standard competency tests used in many states. The Lesson Aids provide additional material that can be used for activities, projects, and extension.

About Holt Pre-Algebra

Organization of the Course   HOLT PRE-ALGEBRA bridges the conceptual and operational gap between arithmetic and algebra. It provides a sound course of mathematical study that builds on previously learned skills, while introducing algebraic concepts of increasing difficulty.

Starting with the first chapter, where equation properties and the concept of the variable are introduced, to the final chapter on coordinate geometry, an algebra strand that includes topics such as graphs, statistics, probability, exponents and real numbers is maintained throughout. Integrated with this strand is the mathematical development of the book, which has both variety and continuity. It moves in a logical progression from a quick review of the basic arithmetical operations in the first chapter, to more sophisticated concepts.

Teaching Math to Teenagers   The authors of HOLT PRE-ALGEBRA place a great premium on motivation and have provided a number of features that will bring the study of mathematics closer to the students' daily lives and interests. The full-color chapter openers on subjects of interest to teenagers can provide exciting and stimulating lessons in problem formulation, estimation, and application. Problem-solving applications to jobs and future careers for teenagers bring home the practical need for good mathematical skills and problem-solving ability, while providing an integration of skills and concepts into word problems.

Individualizing the Course   HOLT PRE-ALGEBRA is designed so that each student can proceed at his or her own pace, while making steady progress in the subject matter. The five-part lesson development is structured for step-by-step ease of comprehension and checking to ensure that learning has occurred.

The Assignments for Reinforcement and Maintenance section at the end of the Lesson Commentaries provides guidelines for the teacher wishing to allow extra time for reinforcement/recycling (level 1) or to provide enrichment (level 2) for one or more students.

The Skillmasters and Testmasters are valuable tools for assessing the individual student's level of comprehension and skill in a given lesson or topic. The short cumulative review section for each lesson, found in the Teacher's Lesson Commentary, is another valuable tool for spot-checking students' skill and comprehension levels.

Students with Handicapping Conditions   Whenever possible, students with handicapping conditions should participate either partially or totally in regular school programs and classes, as part of the requirement of Federal law PL94-142, which states that handicapped children be educated in the "least restrictive environment" possible.

By using the unit, chapter, and lesson objectives in the book, the teacher can select appropriate learning goals for each student and communicate them easily to parents or special education teachers who may be assisting the handicapped student.

Success of Holt Pre-Algebra   HOLT PRE-ALGEBRA has been used successfully for a variety of students and courses, ranging from specific pre-algebra courses to general mathematics courses. It has been found to be stimulating, highly teachable and motivational with a good balance of skill, theory, and application.

This new edition of HOLT PRE-ALGEBRA, with its expanded and new features, is based on suggestions from users of the previous edition.

Course Objectives

PART I: COMPUTATION AND CONCEPTS

1. To evaluate expressions for the given values of variables
2. To use properties to make computations easier
3. To identify open and not open sentences
4. To solve equations and inequalities
5. To write and evaluate formulas
6. To simplify algebraic expressions
7. To factor a number into primes
8. To find the GCF of several numbers
9. To add, subtract, multiply, and divide whole numbers, fractions, mixed numbers and decimals
10. To use scientific notation
11. To find the LCM of two numbers
12. To find place value and round whole numbers and decimals
13. To change percents to decimals, decimals to percents, and fractions to percents
14. To find a percent of a number
15. To find the percent one number is of another
16. To find a number, given the percent of the number
17. To find the lengths of the sides of similar triangles
18. To compute tangent, sine, and cosine of an acute angle in a right triangle
19. To compare integers
20. To add, subtract, multiply, and divide integers
21. To divide rational numbers
22. To evaluate algebraic expressions
23. To classify polynomials
24. To simplify polynomials
25. To multiply a monomial by a polynomial
26. To factor polynomials
27. To classify angles
28. To tell whether two angles are complementary or supplementary
29. To name pairs of vertical angles, alternate interior angles, and corresponding angles
30. To determine measures of the angles created by the intersection of two parallel lines by a transversal, given the measure of one of the angles
31. To classify triangles according to their sides and their angles
32. To identify perfect squares and approximate the square root of a whole number
33. To find the third side of a right triangle, given the other two sides
34. To graph inequalities on a number line
35. To graph equations in a plane
36. To find the slope of a line, given two points on the line
37. To find the slope and y-intercept of a line, given its equation

PART II: APPLICATIONS

1. To translate between English and mathematical expressions
2. To read and use vertical, horizontal, bar, and line graphs, pictographs and tables
3. To change between metric units of length, area, capacity, and mass
4. To find perimeters, areas, surface areas, and volumes of geometric figures, and to find the circumference of a circle
5. To tell which unit of measurement is best for measuring an object
6. To change between Celsius and Fahrenheit units
7. To measure indirectly by applying geometric concepts
8. To find total wages based on commission and salary

Course Objectives Continued on page **T-18**

Course Objectives Continued from page **T-17**

9. To find the cost of an item if a discount is given
10. To determine the percent of increase and decrease
11. To find the range, mean, median, and mode for a given set of data
12. To find the mean of grouped data
13. To find the probability of simple, dependent and independent events
14. To apply the counting principle in finding the number of possible arrangements
15. To identify, solve, and apply proportions and direct and inverse variations
16. To read and use circle graphs
17. To find distances by using a scale
18. To solve word problems

Reference Chart for Skillmasters

LESSON PAGES	SKILLMASTER NUMBER	LESSON PAGES	SKILLMASTER NUMBER
2, 4	1	210, 218	32
7, 10	2	221, 223	33
12, 15	3	225, 228	34
19, 27	4	231, 242	35
29, 32	5	245, 246	36
34, 38	6	249, 251	37
41, 44	7	253, 262	38
45, 47	8	265, 268	39
55, 57	9	269, 272	40
59, 62	10	276, 279	41
64, 67	11	282, 290	42
71, 73	12	293, 297	43
81, 84	13	300, 302	44
87, 89	14	304, 306	45
91, 94	15	313, 315	46
97, 100	16	318, 322	47
101, 103	17	324, 327	48
105, 113	18	329, 336	49
115, 118	19	338, 341	50
121, 124	20	346, 349	51
128, 130	21	352, 355	52
134, 142	22	357, 359	53
144, 149	23	362, 369	54
152, 156	24	371, 374	55
159, 167	25	376, 378	56
170, 172	26	380, 384	57
174, 179	27	387, 395	58
181, 183	28	398, 403	59
191, 194	29	405, 408	60
197, 201	30	412, 414	61
204, 207	31		

Chapter & Lesson Commentaries

Chapter

Equations–Adding and Subtracting

Chapter Objectives

To name the terms, variables, and constants in an expression

To evaluate an algebraic expression for given value(s) of the variables

To identify the properties of addition

To identify sentences as open or not open

To tell whether a sentence that is not open is true or false

To identify solution(s) of an open sentence

To solve equations involving addition and subtraction

To solve word problems

To evaluate and write a formula involving addition and subtraction

Chapter Overview

Students learn to evaluate algebraic expressions, use addition properties to simplify computation, and solve equations involving addition and subtraction. After learning to translate English phrases into mathematical forms, students use a four-step approach to solve word problems. Students also work with simple formulas.

Chapter Opener (Page X)

Problem Formulation

The picture shows the Houston Astrodome in Houston, Texas. The Astrodome is the home of the Houston Astros. It was the first professional baseball park with a dome. It is completely protected from the weather by the steel and plastic dome that is 208 ft at its highest point and has a span of 642 ft with an outer diameter of 710 ft. It has a maximum seating capacity of 62,000. The Astrodome was built in 1965 at a cost of $31,600,000.

Have students use the preceding data to formulate and solve at least two problems.

Estimation

Estimate the cost per seat for building the Houston Astrodome.

Application

Have students, working in groups, pretend that they have the task of building the Houston Astrodome. Have them collect the data necessary for accomplishing the task. Each group should formulate problems they anticipate and decide what data they will need to solve these problems.

Introduction to Variables (Pages 1–2)

Teaching Hints As you discuss the new vocabulary introduced in this lesson, remind students that for a *variable,* the value *varies,* but for a *constant,* the value remains the *same.* Point out that *terms* are separated by addition or subtraction signs.

Cumulative Review

Compute.

1. $5 + 7$ **2.** 9×8 **3.** $49 \div 7$

1. 12   2. 72   3. 7

Properties of Addition (Pages 3–4)

Teaching Hints To help students remember the names of the properties, point out that when *commuting,* a person goes from a to b and then from b to a. For the associative property, you may *associate* a with b first or b with c first. The additive *identity* is 0 because when you add 0 to any number, the sum is *identical* to that number.

Enrichment After students do Exercise 27, ask them if the commutative property holds for subtraction. Have students give an example to show that the commutative property does not hold for subtraction of whole numbers.

$$12 - 5 \neq 5 - 12$$

Cumulative Review

Evaluate if $a = 2, b = 6, c = 15,$ and $d = 36$.

1. $c - a$ **2.** $a + d - c$
3. $a + c - b + d$

1. 13   2. 23   3. 47

Open Sentences (Pages 5–7)

Teaching Hints The chart that follows may help to summarize the concepts of this lesson.

SENTENCES

	True	False
Not open	$5 + 6 = 11$	$6 - 4 < 1$
Open	$x - 5 = 8$ if $x = 13$	$7 + y = 9$ if $y = 3$

Preventing Common Errors For students who confuse $>$ and $<$, have them observe that the symbol resembles an arrowhead pointing to the smaller number.

Cumulative Review

Which property is shown?

1. $(5 + 6) + 2 = 5 + (6 + 2)$
2. $0 + 8 = 8$

3. Compute the easiest way. $17 + 56 + 3$

1. associative property of addition
2. property of additive identity
3. $(17 + 3) + 56 = 20 + 56 = 76$

Solving Equations (Pages 8–10)

Teaching Hints As students work with the addition and subtraction properties to solve equations, stress that addition and subtraction are *inverse operations.*

Cumulative Review

1. Is $5 + 2 < 6$ open or not open? If not open, indicate whether the sentence is true or false.
2. Evaluate $x + 2 - y$ if $x = 8$ and $y = 5$.
3. Is 6 a solution of $x + 3 > 9$?

1. not open; false   2. 5
3. no

Algebraic Phrases (Pages 11–12)

Teaching Hints Point out to students that when we write an English phrase in mathematical terms, we are translating the English into Algebra. The process is similar to that of translating English into another language.

Preventing Common Errors Students should note that when translating the phrases "increased by" or "decreased by," the numbers should be put down in the order in which they occur. However, when translating the phrases "less than" or "more than," the numbers should be reversed.

Cumulative Review

Choose the solution of each open sentence.

1. $x + 15 = 30$; 0, 2, 15, 20
2. $b - 3 < 10$; 3, 7, 10, 13
3. $y + 2 > 6$; 2, 4, 6, 8

1. 15   2. 3, 7, 10   3. 6, 8

Solving Word Problems (Pages 13–15)

Teaching Hints In this lesson we introduce a four-step method for solving problems. This method will be used throughout the book, so it is important to go over it step by step.

Cumulative Review

Write in mathematical form.
1. x increased by 12   **2.** 7 less than y
3. a decreased by 5

1. $x + 12$ 2. $y - 7$ 3. $a - 5$

Problem Solving—Applications (Pages 16–17)

Teaching Hints This is the first lesson reflecting the theme of teenagers at work. Key phrases that indicate addition or subtraction in word problems are introduced, and the steps, *Read, Plan, Solve,* and *Interpret,* are used to help students solve problems.

Formulas (Pages 18–19)

Teaching Hints Example 1 illustrates the use of a simple formula. Some students will need to be reminded that a perimeter means the distance around something.

Enrichment Have students interpret this formula.

$$\text{selling price} = \text{cost} + \text{profit}$$
$$s = c + p$$

Challenge them to write a formula for c in terms of s and p and a formula for p in terms of s and c. Then have them do the same for the formula in Example 1.

$$(P = a + b + c)$$

Cumulative Review

Solve and check.

1. $x + 18 = 29$ **2.** $x - 15 = 32$
3. $93 - x = 42$

1. 11   2. 47   3. 51

Problem Solving—Careers (Page 20)

Teaching Hints This is the first lesson reflecting careers in problem solving. Students solve a variety of problems representing real-life situations encountered in the career highlighted.

Non-Routine Problems (Page 21)

Since there is no specific strategy for solving these problems, they are non-routine. Students should be reminded to read them several times, thoughtfully, before they attempt to solve these problems. Common sense and ingenuity should be applied. See TE Answer Section for suggested solutions.

Computer Activities (Page 22)

The first lesson on the computer provides an introduction to the format of a single program. You may notice that the word **LET** is not necessary for the program to work. However, retain it at the early stages of the course to reinforce the idea that the student is assigning a value to an expression. Point out that there is no line number for **RUN**. Note that on some computers the **ENTER** key serves the same function as **RETURN**.

Chapter

Equations–Multiplying and Dividing

Chapter Objectives

To evaluate an algebraic expression for given value(s) of the variables

To identify the properties of multiplication

To compute arithmetic expressions and evaluate algebraic expressions using the rules for the order of operations

To solve equations involving multiplication and division

To write English phrases in mathematical form

To solve word problems

To evaluate geometric formulas

Chapter Overview

Students evaluate algebraic expressions and use the properties of multiplication to simplify computation. They apply the rules for the order of operations and solve equations and word problems involving multiplication and division. Mixed types of equations and word problems are introduced. Students also evaluate geometric formulas.

Chapter Opener (Page 25)

Problem Formulation

The picture shows the Space Needle in Seattle, Washington. The Space Needle was built for the 1962 "Century 21 Exposition," world's fair. The exposition emphasized technological and scientific achievements of the spage age. The Space Needle is 607 ft high. It is topped by an observation tower and revolving restaurant. The Space Needle ranks with the Eiffel Tower as one of the world's most widely recognized visual symbols for a city. The needle's success is aided by fine elevator service to the observation deck. Rates are as follows: $3.00 adults, $1.50 children age 6–12, and $2.00 each for groups of 20 or more. Children 5 and under ride free.

Have students use the preceding data to formulate and solve at least two problems.

Estimation

Have students estimate the cost of going to the observation deck for a family, a school group of 25 students and 3 teachers, and a group of 15 teenagers chaperoned by 4 adults.

Application

Have students, working in groups, determine what problems might be encountered in operating the Space Needle.

Multiplication and Division (Pages 26–27)

Teaching Hints As you go over Example 3, some students may want to substitute 4 for a and then simplify as follows:

$$\frac{5 \cdot \overset{2}{\cancel{4}}}{\underset{1}{\cancel{2}}} = 10$$

Since this method is mathematically sound, do not discourage students from using it. It shortens the computation process.

Preventing Common Errors Warn students to be careful when they write raised dots for multiplication so that the dots are not mistaken for decimal points.

Cumulative Review

Evaluate.

1. $n + 3$ if $n = 7$ **2.** $a - 15$ if $a = 18$
3. $45 - b$ if $b = 25$

1. 10 2. 3 3. 20

Properties of Multiplication (Pages 28–29)

Teaching Hints Point out that the three properties of multiplication given here are the same as those given for addition on pages 3–4. Mention that 0 is the identity element for addition, while 1 is the identity element for multiplication.

Enrichment After students do Exercise 22, ask them if the commutative property holds for division. It does not.

Cumulative Review

Evaluate.
1. $n \cdot 7$ if $n = 8$
2. $9x$ if $x = 5$
3. $a + 8$ if $a = 34$

1. 56   2. 45   3. 42

Order of Operations (Pages 30–32)

Teaching Hints Stress the reason for making an agreement for the order of operations. It is made so that each mathematical expression involving constants and the four basic operations will have a *unique* value; that is, such an expression can name only one number.

Preventing Common Errors Students must exercise care in applying the order of operations agreement as they work with expressions, such as those in Examples 2 and 3. Multiplications and divisions must be done *first,* even though additions or subtractions may occur first.

Cumulative Review

Solve.

1. $a - 6 = 9$ 2. $16 = x + 5$ 3. $12 = y - 5$

1. 15   2. 11   3. 17

Order of Operations: Variables (Pages 33–34)

Teaching Hints As students go over Example 3, some may want to compute as follows:

$$4 \cdot 2 \cdot 5 - 10 = 4 \cdot 10 - 10 = 40 - 10 = 30.$$

This is acceptable because of the associative property of multiplication.

Preventing Common Errors As students go over Example 1, some may be tempted to add 4 to 6 first. Point out that this would violate the order of operations.

Cumulative Review

Evaluate for $x = 7, y = 4$.

1. $3x - 4y$ 2. $\dfrac{7y}{x}$ 3. $2x - y + 5$

1. 5   2. 4   3. 15

Problem Solving—Careers (Page 35)

Teaching Hints Students may be interested in more information about the career side of computer science. Indicate that a two-year community college or technical school program will prepare them for jobs as technicians and as programmers. Systems analysts require at least a B.A. in computer science.

Solving Equations (Pages 36–38)

Teaching Hints As students work to solve equations involving multiplication and division, point out that just as addition and subtraction are called *inverse operations,* so are multiplication and division.

Cumulative Review

Compute.
1. $8 + 2 \cdot 2$
2. $6 \cdot 4 - 2 \cdot 5 + 6 \cdot 5$
3. Which property is shown?
   $(5 \cdot 4) \cdot 3 = 5 \cdot (4 \cdot 3)$

1. 12   2. 44
3. associative property of multiplication

Solving Word Problems (Pages 39–41)

Teaching Hints Once again point out the importance of carefully translating the English into a mathematical form. Be sure that students identify the variable, and specify what the variable stands for.

Cumulative Review

Solve.

1. $3x = 24$ 2. $\dfrac{x}{5} = 1$ 3. $4 = \dfrac{x}{6}$

1. 8   2. 5   3. 24

Problem Solving—Applications (Pages 42–43)

Teaching Hints Problems that may be encountered by teenagers in fund-raising activities are presented. Key phrases that indicate multiplication or division are introduced.

Mixed Types of Equations (Page 44)

Teaching Hints As you go over the four equations in the examples, ask students to tell which of the four equation-solving strategies they would use to solve each equation.

Preventing Common Errors Caution students to watch for equations in which the variable is on the right side.

Cumulative Review

Compute.

1. $14 + 6 \cdot 5$ **2.** $46 - 25 + 3$
3. $7 + 2(9 - 5)$

1. 44   2. 24   3. 15

Problem Solving: Mixed Types (Page 45)

Teaching Hints Start the lesson with a warm-up or a quiz covering the key phrases.

Cumulative Review

Solve.

1. $x + 9 = 19$ **2.** $12 = 4y$ **3.** $\dfrac{w}{4} = 12$

1. 10   2. 3   3. 48

Evaluating Geometric Formulas (Pages 46–47)

Teaching Hints After students substitute the given value for the variable in the formula in Example 2, they may want to shorten their computation as follows:

$$A = \frac{\overset{7}{\cancel{14}} \cdot 7}{\underset{1}{\cancel{2}}} = 49$$

The perimeter formula given in Example 3 is often written in this form: $P = 2(l + w)$. Have students find the perimeter given in Example 3 using this formula and compare the results to those shown on page 47.

Preventing Common Errors Students often confuse area and perimeter. Use a rectangle divided into unit squares, and have students count the square units in the area and then count the linear units in the perimeter.

Cumulative Review

What property is illustrated?

1. $5 \cdot 8 = 8 \cdot 5$ **2.** $6 \cdot 1 = 6$
3. $(7 \cdot 9) \cdot 3 = 7 \cdot (9 \cdot 3)$

1. commutative property of multiplication
2. property of multiplicative identity
3. associative property of multiplication

Mathematics Aptitude Test (Page 48)

The problems on this page are structured to resemble those found on standardized tests. The format is multiple choice and can provide students with practice in transferring their answers to a master sheet. Have them use the sheet provided in the *Teacher Resource Package*.

Computer Activities (Page 49)

Stress the significance of the proper use of parentheses. Ask the class to find the average of 4 math grades: 80, 60, 100, and 88. The average is 82. Now ask them to find A when A = 80 + 60 + 100 + 88/4. By the order of operations, do division first. A = 262. This is not the average. Parentheses can be used to break the order of operations. Thus,

$$A = (80 + 60 + 100 + 88)/4$$

Have students write a program to find the average of the math scores, type it, and **RUN** it.

Chapter 3

Expressions and Formulas

Chapter Objectives

To rewrite expressions using the distributive property

To simplify and evaluate algebraic expressions for given values of the variables

To solve equations using more than one property of equations

To write English phrases in mathematical forms and solve word problems involving two operations

To find the value of expressions with exponents

To evaluate algebraic expressions and formulas involving exponents

Chapter Overview

Students begin by working with the distributive property. They use the property to combine like terms. After they learn to solve two-step equations, they work with two-step word problems. They also evaluate geometric formulas. Factors and exponents are introduced.

Chapter Opener (Page 54)

Problem Formulation

The picture shows a covered bridge in Warner, New Hampshire. The barn-like covering of the timber-truss bridge was needed to protect the structural members against the elements. The following information is on three covered bridges around Warner, New Hampshire. The Warner-Dalton Bridge, built around 1800, is 86 ft long. The Bradford-Bement Bridge, built in 1854, is 71 ft long. Its original construction cost was $500. Its repair in 1970 cost $20,000. The Werner-Waterloo Bridge, built in the 1840's, is 76 ft long. The longest covered bridge in the world at Hartland, New Brunswick, spans 1,282 ft.

Have students use the preceding data to formulate and solve at least two problems.

Estimation

Estimate how many times greater was the repair cost of the bridge compared to the original cost of construction.

Application

Have students, working in groups, pretend that they have the task of repairing a covered bridge. Have them collect the data they will need to accomplish the task.

The Distributive Property (Pages 53–55)

Teaching Hints As you introduce the distributive property, you might want to show a concrete geometric model using the area of a rectangle.

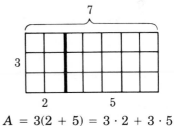

$$A = 3(2 + 5) = 3 \cdot 2 + 3 \cdot 5$$

Preventing Common Errors Give special attention to applying the distributive property from the right. This direction tends to give students more difficulty than working from the left, and it will provide the basis for combining like terms in the next lesson.

Cumulative Review

Solve.

1. $n \cdot 3 = 27$ **2.** $\frac{x}{4} = 7$ **3.** $y + 25 = 75$

1. 9   2. 28   3. 50

Combining Like Terms (Pages 56–57)

Teaching Hints You might want to have some students use the vertical form for combining like terms. Students could write Example 3 as follows:

$$
\begin{array}{r}
3x + 5y + 7 \\
+\ 6x - 2y + 4 \\
\hline
9x + 3y + 11
\end{array}
$$

Cumulative Review

1. Find the area of the rectangle.
   $l = 15$ cm, $w = 9$ cm
2. Find the area of the triangle.
   $b = 16$ cm, $h = 7$ cm
3. Find the area of the rectangle.
   $l = 12$ cm, $w = 5$ cm

1. 135 cm^2 2. 56 cm^2 3. 60 cm^2

Simplifying and Evaluating Expressions (Pages 58–59)

Teaching Hints For Example 1, you may want to have students first substitute 2 for a and evaluate. Then have students simplify, substitute 2 for a, and evaluate again. The result should be 20 in each case.

Preventing Common Errors Students often forget the order of operations. Caution them to multiply first when evaluating expressions such as $6 - 4 \cdot 8$.

Cumulative Review

Rewrite using the distributive property or simplify.

1. $3(x + 5)$ 2. $8x - 6x$ 3. $(y + 8)6$

1. $3x + 15$ 2. $2x$ 3. $6y + 48$

Problem Solving—Applications (Page 60)

Teaching Hints The jobs for teenagers strand is reflected in problems associated with musical groups.

Solving Equations: Using Two Properties (Pages 61–62)

Teaching Hints As students begin working through the three examples of solving two-step equations, remind them that their goal is to isolate the variable on one side of the equation. First they must transfer the term that does not contain the variable to the other side of the equation. Then they can isolate the variable by "undoing" the operation.

Cumulative Review

Evaluate if $x = 4$, $y = 7$, $z = 3$.

1. $5x + 8$ 2. $3xy - z$ 3. $\frac{x}{2} + 5$

1. 28   2. 81   3. 7

English Phrases to Algebra (Pages 63–64)

Teaching Hints Point out the need for commas in Examples 1 and 2. Without commas, Example 1 could be interpreted as $3(n + 5)$.

Cumulative Review

Simplify if possible.

1. $7 \cdot 4y$ 2. $3(5x - 6)$ 3. $3a + 2b$

1. $28y$ 2. $15x - 18$ 3. not possible

Word Problems: Two Operations (Pages 65–67)

Teaching Hints This lesson extends the work done in previous lessons. Students identify what is given in a word problem and what is to be found. They assign a variable to the unknown, write an equation for the word problem, and solve. Go over each example carefully.

Cumulative Review

Write in mathematical form.

1. 5 more than twice a number
2. 3 times a number, decreased by 7
3. 10 more than a number, divided by 4

1. $2x + 5$ 2. $3x - 7$ 3. $\frac{x}{4} + 10$

Problem Solving—Careers (Page 68)

Teaching Hints Students solve a variety of problems representing real-life situations encountered by dieticians in their work.

Factors and Exponents (Pages 69–71)

Teaching Hints Practice the reading of powers with students. 5^3 is read five to the third power, five cubed, the third power of five, or five raised to the third power. Point out the idea that an exponent tells how many times the base is used as a factor.

Enrichment Have students complete the pattern.

$$2^4 = 2 \cdot 2 \cdot 2 \cdot 2 = 16$$
$$2^3 =$$
$$2^2 =$$
$$2^1 =$$
$$2^0 =$$

While this pattern does not prove $2^0 = 1$, it shows that 2^0 should be defined to be 1 if the pattern is to continue.

Cumulative Review

Solve.

1. $x + 17 = 24$ **2.** $\dfrac{x}{8} = 9$
3. $3x + 2 = 17$

1. 7   **2.** 72   **3.** 5

Exponents in Formulas (Pages 72–73)

Teaching Hints Review the differences in finding area and volume. Students should not be required to memorize the formulas. Review the formulas and have students use them. Answers may be left in terms of π.

Enrichment Have students discover what happens to the area of a square and the volume of a cube if the length of a side is doubled or tripled.

Cumulative Review

Write using exponents.

1. $6 \cdot 6 \cdot 6 \cdot 6$ **2.** $x \cdot x \cdot x \cdot y \cdot y$ **3.** $5a \cdot b \cdot b$

1. 6^4 **2.** $x^3 \cdot y^2$ **3.** $5ab^2$

Non-Routine Problems (Page 74)

Since there is no specific strategy for solving these problems, they are non-routine. Students should be reminded to read them several times, thoughtfully, before they attempt to solve these problems. Common sense and ingenuity should be applied. See TE Answer Section for suggested solutions.

Computer Activities (Page 75)

Students are now ready to learn more about the techniques of programming. If students have already learned **BASIC,** they need not study the section at the end of the book. This section is a minicourse in **BASIC** and can be handled in one of several ways:

1. Teach the entire computer section at the end of the text. It will take several class periods of instruction.
2. Form a minicourse during a study period or after school for those who need the instruction.
3. Allow students to study the end-of-the-text material on their own during their "free time." Check the availability of a computer.

Students can **RUN** the program in each computer lesson without having a background in computer literacy. Their computer experience, however, will be enriched if they have a working knowledge of **BASIC.**

Chapter

4

Multiplying and Dividing Fractions

Chapter Objectives

To find numbers corresponding to points on a number line

To determine whether numbers are prime or composite, or neither

To factor numbers into primes

To find the Greatest Common Factor (GCF) of several numbers

To simplify fractions

To multiply and divide fractions and mixed numbers

To solve equations and evaluate formulas involving multiplication of fractions

Chapter Overview

The number line and shaded regions are used as tools for introducing fractions. Students are shown the algorithm for multiplying fractions. After using prime factorization to find the GCF of two numbers, they learn to simplify fractions, products, and quotients. The chapter closes with lessons in algebra.

Chapter Opener (Page 78)

Problem Formulation

The picture shows Bryce Canyon National Park in southern Utah. Named for Ebenezer Bryce, an early settler in the region, the canyon is similar to nearby Grand Canyon. The national park was established in 1928 and covers approximately 35,835 acres. Five years earlier part of the park, 7,040 acres, had been set aside as a national monument. The formations in the canyon are oddly shaped, beautifully-colored limestone and sandstone. A system of trails goes from the rim to wind among the formations.

Estimation

Estimate how many times greater the national park is than the national monument.

Application

Have students, working in groups, pretend they are exploring the canyon. Have them decide what they will need to spend a week on the canyon trail. Each group should formulate problems they may encounter.

Introduction to Fractions (Pages 79–81)

Teaching Hints The first lesson of the chapter paves the way for multiplication of a fraction by a whole number. Once a student understands that $3 \cdot \frac{1}{5}$ can be thought of as $\frac{3 \cdot 1}{5}$ or $\frac{3}{5}$ it is easier to think of $2 \cdot \frac{3}{5}$ as $\frac{2 \cdot 3}{5}$ or $\frac{6}{5}$.

Preventing Common Errors Many students do not know how to count on the number line.

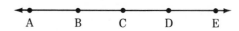

They erroneously begin by assigning 1 to point A. Stress that we count intervals and not points. Show that we begin by assigning 0 to point A, then 1 to point B, and so on.

Cumulative Review

1. Solve. $x - 8 = 3$ **2.** Divide. $21\overline{)672}$
3. Factor 18 into primes. Use exponents.

1. 11   2. 32   3. $2 \cdot 3^2$

Multiplying by a Whole Number (Pages 82–84)

Teaching Hints In this lesson, we make $8 \cdot \frac{3}{4}$ a direct application of the results of the first lesson. Students learned that $3 \cdot \frac{1}{4} = \frac{3}{4}$. So it logically follows that $8 \cdot \frac{3}{4} = \frac{8 \cdot 3}{4}$ or $\frac{24}{4} = 6$.

Enrichment Have students show why the denominator of a fraction cannot be 0. $\frac{5}{9}$ means $0\overline{)5}$. Check division by multiplication. $0 \cdot x = 5$. 0 times any number is 0, not 5. Thus the denominator of a fraction cannot be 0.

Cumulative Review

1. Multiply. $719 \cdot 24$
2. Evaluate. $18 - x$ if $x = 13$
3. Divide. $21\overline{)1,365}$

1. 17,256   2. 5   3. 65

Multiplying by a Fraction (Pages 85–87)

Teaching Hints Make certain that practice on this lesson contains only relatively prime numerators and denominators. Rewriting fractions in simplest form occurs on page 92.

Cumulative Review

1. Subtract. $72.32 − $49.95
2. Multiply. $6 \cdot \dfrac{1}{6}$ 3. Solve. $7 + x = 10$

1. $22.37   2. 1   3. 3

Prime Factorization (Pages 88–89)

Teaching Hints As you go over Example 2, mention that it can be proved that every composite number can be factored into prime factors in only one way except for the order of the factors. This important fact is called the *fundamental theorem of arithmetic*.

Cumulative Review

Write using exponents.

1. $9 \cdot 9 \cdot 9 \cdot 9 \cdot 9$
2. $4 \cdot 4 \cdot 7 \cdot 7 \cdot 7 \cdot 7$
3. $a \cdot a \cdot a \cdot b \cdot b \cdot c \cdot c \cdot c$

1. 9^5 2. $4^2 \cdot 7^4$ 3. $a^3 b^2 c^3$

Greatest Common Factor (GCF) (Pages 90–91)

Teaching Hints The approach to finding the GCF of two numbers, such as 20 and 30, can be summarized as follows: Factor each number into primes.

$$20 = 2^2 \cdot 5 \qquad 30 = 2 \cdot 3 \cdot 5$$

Multiply the common factors, $2 \cdot 5 = 10$. The GCF of 20 and 30 is 10.

Enrichment Relatively prime numbers are numbers that are not necessarily prime that have the number 1 as their only common factor. 26 and 27 are relatively prime numbers (GCF = 1), but neither is prime. Have students find relatively prime pairs of numbers.

Cumulative Review

Evaluate.

1. x^3 if $x = 5$ 2. $9a^4$ if $a = 3$
3. $A = \pi r^2$. Find A in terms of π if $r = 9$.

1. 125   2. 729   3. 81π

Simplest Form (Pages 92–94)

Teaching Hints Begin the lesson with a brief oral drill on prime factorization.

Cumulative Review

1. Simplify. $5 \cdot \dfrac{2}{3}$ 2. Multiply. $\dfrac{2}{3} \cdot \dfrac{5}{7}$
3. Evaluate. $17a$ if $a = 4$

1. $3\dfrac{1}{3}$ 2. $\dfrac{10}{21}$ 3. 68

Simplifying Products (Pages 95–97)

Teaching Hints As students divide out like factors in the numerator and the denominator, remind them that they are actually removing factors of 1. Show several cases, such as $\dfrac{5}{5} = 1$.

Preventing Common Errors It is important to emphasize that, in simplifying, you are *dividing out* like factors. The term *cancelling out* can lead to errors like this:

$$\frac{2 + \cancel{3}}{\cancel{3}} = 2$$

Cumulative Review

1. Write $\dfrac{18}{24}$ in simplest form.
2. Solve. $7x + 1 = 57$
3. Multiply. $319 \cdot 108$

1. $\dfrac{3}{4}$ 2. 8   3. 34,452

Problem Solving—Applications (Page 98)

Teaching Hints For the summer, teenagers can find jobs in local or county parks. These word problems are typical problems many teenagers encounter as park attendants.

Dividing Fractions (Pages 99–100)

Teaching Hints Show the advantage of rewriting $2\frac{1}{4} \div 6$ as $2\frac{1}{4} \div \frac{6}{1}$. Students frequently do not think of 6 as a fraction and therefore do not know how to rewrite $2\frac{1}{4} \div 6$ as $2\frac{1}{4} \cdot \frac{1}{6}$.

Cumulative Review

1. Multiply. $3\frac{1}{2} \cdot \frac{1}{14}$
2. Divide. $43\overline{)9{,}288}$
3. Solve. $6x - 9 = 33$

1. $\frac{1}{4}$
2. 216
3. 7

Evaluating Expressions (Page 101)

Teaching Hints As you go over the Example, point out that the numerator and denominator can be factored into primes so that like factors can be divided out.

Enrichment Allow more able students to try problems of this type: Evaluate $\frac{3}{4}x^3$ if $x = \frac{2}{3}$.

Equations (Pages 102–103)

Teaching Hints Multiplying by the reciprocal provides a simple one-step method for solving equations such as $\frac{3}{5}x = 9$. Another method is to multiply each side by 5 and then divide each side by 3.

Cumulative Review

1. Factor 24 into primes.
2. Multiply. $7\frac{1}{2} \cdot \frac{2}{5}$
3. Theater tickets cost $3.75 each. What is the cost of 7 tickets?

1. $2^3 \cdot 3$
2. 3
3. $26.25

Evaluating Formulas (Pages 104–105)

Teaching Hints Remind students that mixed numbers and whole numbers should be rewritten as fractions in order to multiply or square them.

Enrichment Challenge students to solve problems like this: $A = \frac{1}{2}bh$. Find h, if $A = 32$ and $b = 16$. Make up a few problems like this, but be sure the answers are whole numbers.

Cumulative Review

Solve.

1. $\frac{1}{3}x = 2$
2. $\frac{3}{4}x = 9$
3. $\frac{2}{3}x = \frac{3}{4}$

1. 6
2. 12
3. $\frac{9}{8}$

Computer Activities (Page 106)

The program is for simplifying any fraction of the form $\frac{A}{B}$, $B \neq 0$ by finding the GCF and dividing the numerator and denominator by the GCF. First the computer decides which is smaller, the numerator or the denominator. This is the function of line 130. If $A - B < 0$, then A must be the smaller number. Then the computer divides the numerator and denominator by the smaller number, in this case, 12. This is the trial divisor. Since 12 is not a common divisor of both 28 and 12, the computer divides numerator and denominator by 11, 10, 9, and so on until it finds a common divisor, 4. The computer then prints the result of dividing both 28 and 12 by 4, $\frac{7}{3}$.

Chapter

5

Adding and Subtracting Fractions

Chapter Objectives

To find the Least Common Multiple (LCM) of two numbers

To add and subtract fractions and mixed numbers

To evaluate algebraic expressions involving fractions

To solve fractional equations and word problems

Chapter Overview

Students learn to add and subtract fractions and mixed numbers with like and unlike denominators. They learn to evaluate algebraic expressions and solve equations involving fractions. In every lesson students see problems stated horizontally as well as vertically.

A unique development is finding common denominators by factoring denominators into primes. For example, to add $\frac{2}{3} + \frac{1}{8} + \frac{1}{12}$ many students are not able to see that 24 is the LCD. We tell students to begin by factoring each denominator into primes.

Chapter Opener (Page 110)

Problem Formulation

The picture shows the Gateway Arch in St. Louis, Missouri. The arch was designed by the American architect Eero Saarinen, who was born in Finland in 1910. The Saarinen family of four came to the United States in 1923. Saarinen designed the Gateway Arch as a dominant part of the Jefferson National Expansion Memorial that occupies 82 acres. It was designed in 1948 and completed in 1964.

The arch is made of stainless steel and has a span and height of 630 ft or 190 m. Visitors are carried to the top of the arch by small cars inside the monument.

Have students use the preceding data to formulate and solve at least two problems.

Estimation

Have students estimate the area of the 82 acre Jefferson National Expansion Memorial in ft^2 to the nearest 1,000 ft^2.

Application

Have students, working in groups, pretend that they have the task of designing an arch. Have them collect the data necessary for accomplishing the task. Each group should formulate problems they might encounter.

Fractions with the Same Denominator (Pages 111–113)

Teaching Hints You might wish to use the number line to teach these concepts rather than use the distributive property. Sketch a number line on the board. Divide the length from 0 to 1 into sevenths; label $0, \frac{1}{7}, \frac{2}{7}$, and so on, to $\frac{7}{7}$ or 1. Show by counting that $\frac{4}{7} + \frac{2}{7}$ is $\frac{6}{7}$.

Preventing Common Errors It is essential to stress that only the numerators are to be added, not the denominators.

Cumulative Review

1. Solve. $\frac{5}{3}x = \frac{10}{7}$
2. Write using exponents. $a \cdot a \cdot a \cdot b \cdot b$
3. Factor completely. Write using exponents. 100

1. $\frac{6}{7}$ 2. a^3b^2 3. $2^2 \cdot 5^2$

Least Common Multiple (LCM) (Pages 114–115)

Teaching Hints The approach to finding the LCM of two numbers, such as 20 and 30, can be summarized as follows:

$$20 = 2^2 \cdot 5 \qquad 30 = 2 \cdot 3 \cdot 5$$

Multiply the greatest power of each prime factor. $2^2 \cdot 3 \cdot 5 = 60$. The LCM of 20 and 30 is 60.

Preventing Common Errors Students tend to confuse the concepts of GCF and LCM. Ask them to find both the GCF and the LCM of several pairs of numbers. Ask them to compare the two procedures.

Cumulative Review

Find the number.
1. 6 more than a number is 20.
2. A number divided by 8 is 3.
3. 5 less than 3 times a number is 16.

1. 14   2. 24   3. 7

Fractions with Unlike Denominators (Pages 116–118)

Teaching Hints Emphasize the mathematical reason for why the process of multiplying numerator and denominator by the same number works. In Example 1 we can multiply both numerator and denominator by 3 because $\frac{3}{3} = 1$ and 1 is the multiplicative identity. Multiplying by 1 does not change the number.

Preventing Common Errors In Example 1, students might write:

$$\frac{1}{5 \cdot 3} + \frac{2}{5 \cdot 3}$$

Remind students that they need to multiply both the numerator and denominator of $\frac{1}{5}$ by 3.

Cumulative Review

1. Find the value of n if $n = 98 + 79 + 68$.
2. Round 18,976 to the nearest thousand.
3. Jean practiced from 10:20 A.M. to 1:15 P.M. How long did she practice?

1. 245   2. 19,000   3. 2 h, 55 min

Addition and Subtraction (Pages 119–121)

Teaching Hints Put practice exercises on the board or on a spirit master. Ask students what factors are needed to make each denominator the same.

1. $\frac{7}{5 \cdot 2} + \frac{2}{5 \cdot 3}$ 2. $\frac{1}{2} + \frac{1}{2 \cdot 2} + \frac{9}{2 \cdot 2 \cdot 5}$

    3   2   2, 5   5

3. $\frac{5}{3 \cdot 2 \cdot 2} + \frac{3}{2 \cdot 3 \cdot 3} + \frac{7}{3 \cdot 2 \cdot 2 \cdot 2}$

   2, 3   2, 2   3

Cumulative Review

1. Evaluate. $\frac{28}{n}$ if $n = 4$

2. Multiply.   508   3. Solve. $\frac{x}{15} = 4$
            $\times\,648$

1. 7   2. 329,184   3. 60

Mixed Numbers (Pages 122–124)

Teaching Hints Students tend to have trouble with simplifying results like $9\frac{10}{6}$. Be sure the student changes $\frac{10}{6}$ to $1\frac{2}{3}$ and remembers to add the $1\frac{2}{3}$ to the 9 to get $10\frac{2}{3}$.

Cumulative Review

1. Find the area of the rectangle.
   $l = 74$ cm, $w = 22$ cm
2. Find the area of the right triangle.
   $h = 18$ cm, $b = 14$ cm
3. Find the area of the square. $s = 9$ cm

1. 1,628 cm^2 2. 126 cm^2 3. 81 cm^2

Problem Solving—Applications (Page 125)

Teaching Hints The word problems reflect real-life problems involving fractions one might encounter as an athletic coach.

Renaming in Subtraction (Pages 126–128)

Teaching Hints The lesson begins with rewriting fractions like $6\frac{3}{10}$ as $5\frac{13}{10}$ to prepare the way for the concept of renaming. The lesson closes with subtracting a mixed number from a whole number. Pay particular attention to the Oral Exercises that test understanding of the renaming technique.

Cumulative Review

Add or subtract. Simplify if possible.

1. $8\frac{3}{4} + 1\frac{2}{3}$ **2.** $7\frac{1}{5} - 3$ **3.** $6\frac{2}{5} - 5\frac{3}{7}$

1. $10\frac{5}{12}$ 2. $4\frac{1}{5}$ 3. $\frac{34}{35}$

Evaluating Expressions (Pages 129–130)

Teaching Hints Remind students of the rule for the order of operations: multiply and divide from left to right then add and subtract.

Cumulative Review

1. Multiply. $4\frac{1}{5} \cdot 3$ **2.** Simplify. $2\frac{2}{3} \cdot \frac{9}{16}$

3. Solve. $\frac{3}{4}x = \frac{5}{8}$

1. $12\frac{3}{5}$ 2. $1\frac{1}{2}$ 3. $\frac{5}{6}$

Problem Solving—Careers (Page 131)

Teaching Hints Bookkeeping is one of many careers that uses mathematics. These problems reflect situations that may be encountered by bookkeepers in their work.

Equations (Pages 132–134)

Teaching Hints For the Oral Exercises, you may wish to ask students whether they will subtract or add to solve the equations.

Cumulative Review

1. Write using exponents. $x \cdot x \cdot x \cdot x \cdot x$
2. Divide if possible. $0 \div 97$
3. Divide if possible. $100 \div 0$

1. x^5 2. 0   3. not possible

Non-Routine Problems (Page 135)

Since there is no specific strategy for solving these problems, they are non-routine. Students should be reminded to read them several times, thoughtfully, before they attempt to solve these problems. Common sense and ingenuity should be applied. See the TE Answer Section for suggested solutions.

Computer Activities (Page 136)

The most important outcome of this lesson is not the writing of the program itself but the generalization of the mathematical process involved in adding two fractions. Students must be able to generalize the process before they can instruct the computer on the procedure. Students may be interested in deriving the general formula for adding any two fractions before they see the solution on page 136. Provide a hint by adding two fractions with prime denominators at the chalkboard. Then ask students to parallel each step with the generalization as shown below:

$$\frac{2}{3} + \frac{1}{5} \qquad\qquad \frac{a}{b} + \frac{c}{d}$$

$$\frac{2 \cdot 5}{3 \cdot 5} + \frac{1 \cdot 3}{5 \cdot 3} \qquad\qquad \frac{a \cdot d}{b \cdot d} + \frac{c \cdot b}{d \cdot b}$$

$$\frac{10 + 3}{15} \text{ or } \frac{13}{15} \qquad\qquad \frac{a \cdot d + c \cdot b}{b \cdot d}$$

Chapter

6

Organizing Data

Chapter Objectives

To read a table
To read and make pictographs, bar graphs, and line graphs
To solve problems using graphs
To use equations to find the missing side in perimeter problems

Chapter Overview

Students learn to read tables and read and construct a variety of graphs including pictographs, horizontal and vertical bar graphs, and line graphs. They use graphs to solve problems. They use equations to find the missing side of a rectangle or triangle given the perimeter and the other sides.

Chapter Opener (Page 139)

Problem Formulation

The picture shows Pennsylvania Avenue and the Capitol in Washington, D.C. Its cornerstone was laid in September, 1793, and the government moved to Washington in 1800. The Capitol is 751 ft long and 350 ft deep. Its dome is 287 ft high and is cast iron. Atop the dome is the bronze Statue of Freedom sculptured by Thomas Crawford. The statue is almost 20 ft high and weighs more than 14,000 lbs. In the south wing of the Capitol is the Chamber of the House of Representatives and in its north wing is the Senate Chamber. The rotunda of the Capitol is 96 ft in diameter and more than 180 ft from the floor to the dome.

Have students use the preceding data to formulate and solve at least two problems.

Estimation

Have students estimate the maximum number of football fields (100 yd by 50 yd) that can be placed in the space occupied by the Capitol.

Application

Have students draw the football fields to scale to show the number of fields that can be placed in the area occupied by the Capitol.

Reading Tables (Pages 140–142)

Teaching Hints Much of this lesson can be done orally with the entire class. Many students may never have seen a tax table, let alone read one. You may wish to bring tax tables to class and give students additional practice in reading them. Have students look for tables, train, bus, or airline schedules in newspapers, magazines, or other sources. Have students bring in their tables and schedules, explain them to the class, and use them to make a bulletin board display.

Enrichment Have students gather data about student activities and make a table. Fund raising, class test scores, and team scores are some data students can gather.

Cumulative Review

Evaluate.

1. $\dfrac{42}{x}$ if $x = 7$
2. $7 + 2x + y$ if $x = 5$ and $y = 3$
3. $2x^2 + 2$ if $x = 3$

1. 6   2. 20   3. 20

Pictographs (Pages 143–144)

Any number of different pictures can be used to represent data; the choice is arbitrary. Students should be encouraged to create their own pictures even for the examples shown. Have students give a justification for the pictures they chose. Every pictograph should have a value assigned to the picture. A large enough value must be used to avoid having too many pictures in one graph. One limitation of the pictograph is that it is often difficult to tell the value of a fraction of a picture.

Enrichment Have students gather data, such as birthdays and months of the year, athletic game scores, and school attendance. Have them make pictographs and display them.

Cumulative Review

1. Subtract. $\dfrac{7}{8} - \dfrac{2}{8}$ 2. Add. $\dfrac{5}{7} + \dfrac{1}{3}$
3. Solve. $52 - x = 12$

1. $\dfrac{5}{8}$ 2. $1\dfrac{1}{21}$ 3. 40

Problem Solving—Applications (Pages 145–146)

Teaching Hints In this lesson word problems involving more than one operation are presented. Go over Examples 1 and 2 with the class. Be sure students understand what operations to use for the problems.

Bar Graphs (Pages 147–149)

Teaching Hints As you go over Examples 1 and 2, point out the difference between vertical and horizontal bar graphs.

Preventing Common Errors Bar graphs indicate, at a glance, a comparison of data. But bar graphs can also be used to distort or misrepresent data. Therefore it is essential to use appropriate widths of bars, heights of bars, and space between them.

Cumulative Review

1. Evaluate. $9n$ if $n = 8$
2. Multiply. 685,433
$\underline{\times\ 96}$
3. Divide $97,094 \div 86$

1. 72   2. 65,801,568   3. 1,129

Line Graphs (Pages 150–152)

Teaching Hints Discuss with students the advantages and disadvantages of line graphs, bar graphs, and pictographs. Line graphs are frequently used to show trends or patterns while bar graphs are used to show comparisons. Pictographs are easy to interpret.

One way to motivate students in their classwork is to have them make a line graph of their quiz or test scores. Students can see their progress over a period of time.

Enrichment Have students make line graphs using information that is meaningful to them. Test scores, sports scores, sales, or climate characteristics are a few ideas. Have students find examples of line graphs and display them.

Cumulative Review

1. Evaluate. $4x^3$ if $x = 3$
2. Factor completely. 81
3. Evaluate. Leave answers in terms of π. $V = \pi r^2 h$ if $r = 6$ and $h = 3$

1. 108   2. 3^4 3. 108π

Problem Solving—Careers (Page 153)

Teaching Hints Students apply the skills they have learned to real-life problem-solving situations encountered by appliance repair specialists.

Using Graphs (Pages 154–156)

Teaching Hints Students are provided with varied uses of graphs in applied situations. You may have to spend some time on terminology that is not familiar to students.

Cumulative Review

1. Multiply. $4\frac{1}{3} \cdot \frac{1}{4}$ 2. Solve. $7 = \frac{y}{9}$

3. Divide. $9\frac{1}{3} \div \frac{4}{3}$

1. $1\frac{1}{12}$ 2. 63   3. 7

Applying Equations (Pages 157–159)

Teaching Hints Some students may enjoy working with the formula $\frac{P}{2} = l + w$ for the perimeter of a rectangle. It works nicely for the problems in this lesson.

Cumulative Review

1. Subtract. $\frac{5}{6} - \frac{1}{3}$ 2. Add. $6\frac{1}{5} + 3\frac{1}{2}$

3. Solve. $x - 2\frac{1}{3} = 4\frac{1}{3}$

1. $\frac{1}{2}$ 2. $9\frac{7}{10}$ 3. $6\frac{2}{3}$

Mathematics Aptitude Test (Page 160)

The problems on this page are structured to resemble those found on standardized tests. The format is multiple choice and can provide students with practice in transferring their answers to a master sheet. Have them use the sheet provided in the blackline copymaster package.

Computer Activities (Page 161)

Students might need clarification of how the **READ** and **DATA** statements work in conjunction with the loop. The **READ** statement is similar to **INPUT.** The main difference is that the **INPUT** statement allows you to enter new data each time the program is **RUN,** whereas the **READ** statement makes the data an actual part of the written program. Line 30, **READ W, H,** tells the computer that the data occur in pairs, the first member of each pair being W and the second member being H. In the program on page 161 the set of **DATA** is:

5.75, 35, 6.70, 40, 5.75, 39, 7, 35

The computer will read the first set of data with the understanding that W = 5.75 and H = 35, when K = 1. Then when K = 2, the computer will read the second set of data as W = 6.70 and H = 40, and so on until all 4 sets of data have been read.

Chapter 7

Decimals

Chapter Objectives

To write decimals for word names and word names for decimals
To change a fraction to a decimal
To round a decimal to the nearest tenth or hundredth
To add, subtract, multiply, and divide decimals
To evaluate algebraic expressions and solve equations and word problems involving decimals
To write a number in scientific notation

Chapter Overview

Decimals are introduced through place value. Students learn to add, subtract, multiply, and divide decimals. An example of evaluating algebraic expressions in which the variable is replaced by a decimal is provided in each lesson. Scientific notation for large numbers is introduced and students solve equations involving decimals.

Chapter Opener (Page 164)

Problem Formulation

The picture shows Niagara Falls. The falls are located in the Niagara River, which forms a boundary between Ontario, Canada, and New York State. At its midway point the river falls into two streams. The larger of the streams drops on the Canadian side and forms the Horseshoe Falls. The smaller one drops to the east and forms the American Falls. The Horseshoe Falls are 173 ft high and 2,600 ft wide at their widest point. The American Falls are 182 ft high, and 1,000 ft wide.

During the tourist season the minimum daytime flow is about 100,000 ft^3/sec. At all other times the minimum flow is about 50,000 ft^3/sec.

Have students use the preceding data to formulate and solve at least two problems.

Estimation

Ask students to estimate the daytime flow of Niagara Falls during the tourist season.

Application

Have students do a comparison between Horseshoe Falls and American Falls.

Introduction to Decimals (Pages 165–167)

Teaching Hints In Examples 2 and 3 when changing decimal fractions to decimals, stress the method of *moving* the decimal point as many places to the left as there are zeros in the denominator.

When rounding decimals to the nearest tenth or hundredth, remind students to look at the digit to the *right* of the place to which they are rounding.

Preventing Common Errors Students frequently have trouble changing fractions like $\frac{4}{100}$ to a decimal. The common error is to write 0.40 instead of 0.04. Contrast $\frac{4}{10}$ and $\frac{4}{100}$ and show how each is written as a decimal.

Cumulative Review

1. Add. 48,756 + 62,431 + 104,103
2. Find the value of n if $n = 71 + 79 + 16$.
3. Find the perimeter of a rectangle whose length is 73 cm and whose width is 49 cm.

1. 215,290   2. 166   3. 244 cm

Changing Fractions to Decimals (Pages 168–170)

Teaching Hints Stress the need to continue to write 0's when dividing until there is either no remainder or until a pattern of repetition becomes obvious.

Enrichment Show that $\frac{4}{7}$ is a repeating decimal.

Cumulative Review

1. Multiply.   1,003
                $\times$ 7,096
2. Compute. $6 \cdot 2 - 2 + 3 \cdot 4$
3. Solve. $7x = 84$

1. 7,117,288   2. 22   3. 12

Adding and Subtracting Decimals (Pages 171–172)

Teaching Hints Students need to be reminded of the importance of the proper use of zeros. For example, to subtract 3.45 from 20 emphasize that the number of 0's we write after 20 matches the number of decimal places after 3 in 3.45, which is two.

For the Calculator activity, a memory function comes in handy. Show those students whose calculators have this function how to use the memory to subtract two sums.

Preventing Common Errors Students sometimes have difficulty in aligning decimal points when rewriting horizontal sums in the vertical form. Drawing a vertical line through the first decimal point helps students to line up the remaining decimals.

Cumulative Review

1. Divide. $\frac{3}{4} \div \frac{8}{12}$ 2. Solve. $\frac{3}{5}x = \frac{2}{15}$
3. Mary made $19.00 cutting grass. It took her $3\frac{1}{6}$ hours. How much did she make an hour?

1. $1\frac{1}{8}$ 2. $\frac{2}{9}$ 3. $6.00

Multiplying Decimals (Pages 173–174)

Teaching Hints When multiplying a whole number by a decimal it might be helpful to write a decimal point after the whole number. This helps the student see that there are 0 places after the whole number.

Preventing Common Errors Students need to be reminded to place additional zeros to the left of the product when necessary. For example:

$$\begin{array}{r} 7.6 \\ \times\ 0.001 \\ \hline 0.0076 \end{array}$$

Some students will forget to write the two 0's and will place the decimal point in front of the 7 in the product.

Cumulative Review

1. Add. $\frac{5}{6} + \frac{1}{3} + \frac{1}{4}$ 2. Subtract. $6\frac{1}{5}$
                                                                      $-4\frac{1}{3}$

3. Solve. $x - 1\frac{2}{3} = 2\frac{1}{6}$

1. $1\frac{5}{12}$ 2. $1\frac{13}{15}$ 3. $3\frac{5}{6}$

Problem Solving—Applications (Pages 175–176)

Teaching Hints Real-life problems that may be encountered by teenagers working as clerks in a delicatessen form the framework for problems involving decimals.

Dividing Decimals (Pages 177–179)

Teaching Hints Two-digit divisors, as in Example 4, are difficult for some students. Encourage them to read the left-hand column comments on how to approximate quotients.

Preventing Common Errors Students tend to forget to write 0 in quotients like this:

$$
\begin{array}{r}
0.70 \\
48)\overline{33.750} \\
\underline{33\ 6} \\
15
\end{array}
\qquad
\text{Display: } 48)\overline{15.}
$$

Cumulative Review

1. Write using exponents. $x \cdot x \cdot x \cdot x \cdot y \cdot y$
2. Divide if possible. $0 \div 130$
3. Evaluate. $3a^5$ if $a = 1$

1. $x^4 y^2$ 2. 0   3. 3

Scientific Notation and Large Numbers (Pages 180–181)

Teaching Hints You might want to review the process of *moving* the decimal point to either the right or the left as a means for either multiplying or dividing by a power of ten.

Preventing Common Errors Emphasize that when converting a standard numeral to scientific notation, the decimal point must be *moved* so that the factor of 10 is from 1 to 10. This should help prevent students from *moving* the decimal point to the wrong place.

Cumulative Review

1. Subtract. $8 - 5\frac{4}{5}$ 2. Solve. $68 = 4c$

3. Add. $\frac{5}{7} + \frac{2}{3}$

1. $2\frac{1}{5}$ 2. 17   3. $1\frac{8}{21}$

Decimal Equations (Pages 182–183)

Teaching Hints Encourage students to check answers so that they get additional practice in multiplying and dividing decimals.

When solving the equation in Example 2, more able students might want to multiply each side by 100 to make the coefficient of x a whole number.

Cumulative Review

1. Multiply. 769×93
2. Factor 90 into primes. Write using exponents.
3. Is 48,931 divisible by 4?

1. 71,517   2. $2 \cdot 3^2 \cdot 5$ 3. no

Problem Solving—Careers (Page 184)

Teaching Hints Machinists encounter problems involving decimals when using metric measures. Have students decide which of the four basic operations to use to solve each problem. Be sure students know that answers are to be given to the nearest tenth of a centimeter for problem 7.

Non-Routine Problems (Page 185)

Since there is no specific strategy for solving these problems, they are non-routine. Students should be reminded to read them several times, thoughtfully, before they attempt to solve these problems. Common sense and ingenuity should be applied. See the TE Answer Section for suggested solutions.

Computer Activities (Page 186)

Branching is an example of reasoning used often in mathematics. In the program on page 186, there are only two possibilities for the value of N: N > 6, or N ≤ 6. Line 30 tests for N > 6. The program does not have to spell out the other alternative. If N is greater than 6, the computer goes to line 60 where it calculates the salary plus the commission on the number of televisions sold.

Project Ask students to obtain a federal tax schedule. Discuss graduated income tax. Stress that the rate increases as a person's income increases. Discuss how a program could use branching to compute taxes.

Chapter

Measurement

Chapter Objectives

To measure segments to the nearest centimeter and millimeter

To change among metric units of length, area, capacity, and mass

To find areas of geometric figures

To tell which unit is best for measuring an object

To change between Celsius and Fahrenheit scales

To solve problems involving measurement

Chapter Overview

In this chapter students work with metric units of length, area, capacity, weight, and temperature. It is suggested that students begin each of these topics by performing actual measurement activities.

Chapter Opener (Page 189)

Problem Formulation

The picture shows the *Columbia* space shuttle at the Kennedy Space Center. Astronauts Thomas K. Mattingly II and Henry W. Hartsfield, Jr. piloted the *Columbia* space shuttle for 112 orbits between June 27 and July 4, 1982. They carried the first military and commercial payloads. The overall shuttle length is 184.2 ft and the height is 76.6 ft. Its cargo bay is 60 ft long and 15 ft wide. The weight of the shuttle at lift-off was $4\frac{1}{2}$ million lb and the weight of its orbitor was 212,000 lb. The maximum payload a shuttle can carry is 65,000 lb. It has 3 main liquid rocket engines, each with approximately 470,000 lb of thrust at lift-off.

Have students use the preceding data to formulate and solve at least two problems.

Estimation

Have students round the overall length and height of the *Columbia* space shuttle to the nearest 10 ft and estimate its volume, assuming that it is of cylindrical shape.

Application

Have students work in groups to explore some of the major problems that have to be solved in order to launch a space shuttle. Have them indicate the kind of data needed.

Centimeters and Millimeters (Pages 190–191)

Teaching Hints Encourage students to estimate lengths before they actually measure the objects. Point out that in real-life situations, there will be more occasions when they will estimate lengths than when they will actually measure them.

When students are measuring in millimeters, quite often their measurements will vary by a millimeter or two. This is a good time to point out that every measurement is an approximation. No matter how small our unit is, the best we can do is to measure to the nearest unit.

Cumulative Review

1. Multiply. $2\frac{1}{5} \cdot 3\frac{1}{3}$ 2. Simplify. $\frac{2}{3}$ of 12

3. Twice a number is 16. Find the number.

1. $7\frac{1}{3}$ 2. 8   3. 8

Units of Length (Pages 192–194)

Teaching Hints Since the metric system, like our decimal numeration system, is based on ten, converting from one metric unit to another is easy to learn. As students are introduced to the chart of metric units of length, point out that the most commonly used units are the millimeter, the centimeter, the meter, and the kilometer.

Do not attempt to teach the customary system of inches, feet, yards, and miles at the time students are working with the metric system. However, you might want to point out that a meter is a little longer than a yard, a kilometer is about five-eighths of a mile, and so on.

Cumulative Review

1. Divide. $4\frac{2}{5} \div 11$ 2. Divide. $12 \div 1\frac{1}{3}$

3. Solve. $\frac{4}{5}x = \frac{8}{15}$

1. $\frac{2}{5}$ 2. 9   3. $\frac{2}{3}$

Changing Units of Length (Pages 195–197)

Teaching Hints As an introduction to this lesson, review the method of *moving* the decimal point when multiplying or dividing a number by a power of 10. The two examples in the recall section illustrate this procedure.

Preventing Common Errors Note that the charts used in Examples 1–4 are arranged with the metric units progressing from *largest to smallest* as we move from *left to right*. Help students remember this arrangement by pointing out that in our place value system of numeration, the place values also progress from largest to smallest as we move from left to right. For example, consider the number 645.317.

hundreds	tens	ones		tenths	hundredths	thousandths
6	4	5	.	3	1	7

These place values correspond to those in the metric system.

Cumulative Review

1. Solve. $5 = \frac{x}{7}$ 2. Add. $\frac{7}{9} + \frac{2}{3}$

3. Subtract. $7\frac{2}{3} - 4\frac{1}{6}$

1. 35   2. $1\frac{4}{9}$ 3. $3\frac{1}{2}$

Problem Solving — Applications (Pages 198–199)

Teaching Hints Real-life problems encountered in servicing a car and traveling by car are dealt with on these pages. Students use a formula to calculate the mpg. The four-step problem-solving strategy is reviewed.

Area (Pages 200–201)

Teaching Hints An alternate approach to what is shown in the Examples in this lesson would be to convert the linear measures before multiplying to find the area. Thus, for Example 2, students would change 5.8 m to 580 cm and then multiply. Point out that the unit *are* is pronounced är.

Activity Have students take measurements, compute areas, and change area units. Some items to measure might be their math books, a desk top, or the floor of the classroom.

Cumulative Review

1. Estimate to the nearest thousand. 76,394 + 7,380

2. Solve. $0.3x = 36$

3. Solve. $\frac{3}{5}x + 1 = 31$

1. 83,000   2. 120   3. 50

Capacity (Pages 202–204)

Teaching Hints When introducing the liter, it is helpful to show objects of different shapes that have the capacity of 1 liter. For example, a cube 1 decimeter on an edge, some soft drink bottles, and some small coffee cans all have the capacity of about 1 liter. In fact, many coffee cans make good liter measuring devices since they are often divided into fourths (250 mL) by creases.

Activity It is beneficial to offer students the opportunity to estimate and measure capacities in liters and in milliliters. Collect several containers with a capacity of 1 liter or greater. Ask students to arrange them in order from the greatest to the smallest capacity. Then have them estimate the capacity of each container.

Cumulative Review

1. Change to a decimal. $\frac{73}{100}$

2. Change to a decimal rounded to the nearest tenth and the nearest hundredth. $\frac{5}{12}$

3. Add. $48.35 + 0.007 + 3.657$

1. 0.73   2. 0.4, 0.42   3. 52.014

Weight (Pages 205–207)

Teaching Hints When introducing the gram and the kilogram, it is a good idea to allow students to feel the weight of objects that weigh 1 gram and 1 kilogram. The weight of a standard paper clip is close to 1 gram. For a kilogram, use a liter container filled nearly to the top with water. Point out that a liter of water without the container weighs 1 kilogram.

Be sure to point out the relationship among the different types of metric measurement. 1 cm³ of water equals 1 mL and weighs 1 g. 1,000 cm³ of water equals 1 L and weighs 1 kg.

A milligram is such a small unit of weight that it is difficult to visualize. Ask students to think of the weight of a cubic millimeter of water. That weight is 1 milligram.

Activity Students will benefit from the opportunity to estimate and measure the weight of various small objects in grams. A simple balance scale can be made from rod and disk construction sets and plastic cups.

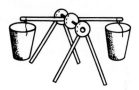

Small paper clips can be used as gram weights to balance the objects. Nickels can serve as 5-gram weights.

Cumulative Review

1. Subtract. $18.643 - 8.797$

2. Multiply. $(93.4)(4.7)$

3. Divide $0.03\overline{)0.41375}$ to the nearest tenth.

1. 9.846   2. 438.98   3. 13.8

Problem Solving—Careers (Page 208)

Teaching Hints Problems encountered by stonemasons include area problems, earnings, and the weight of the stone. Remind students to label their answers with the correct unit of measure.

Temperature: Celsius and Fahrenheit (Pages 209–210)

Teaching Hints It would be beneficial to demonstrate this lesson with actual Fahrenheit and Celsius thermometers. Place them in a glass of water to get equivalent readings on the two scales. Use several glasses with water at different temperatures.

Enrichment Ask students who have worked with negative integers to find the temperature at which the Fahrenheit and Celsius readings are the same. The answer is $-40°$.

Cumulative Review

1. Solve. $3x - \dfrac{1}{2} = 11\dfrac{1}{2}$ 2. Solve. $5y = 0.2965$
3. Factor 1,250 into primes. Write using exponents.

   1. 4   2. 0.0593   3. $2 \cdot 5^4$

Computer Activities (Page 211)

You need not write **STEP** for sequential input that involves data that increase by 1. If the data information of Example 2 is changed from 25, 30, 35, 40, 45, 50 to 25, 26, 27, 28, 29, 30, 31, 32, 33, line 10 would be:

10 FOR K = 25 TO 33

The computer understands that the data are being *stepped* up by 1 at a time. It would not be incorrect to write:

10 FOR K = 25 TO 33 STEP 1

However, the **STEP** 1 is not necessary.

Enrichment Ask students which of the three methods of supplying data to a computer would be most appropriate for each of the following.

1. A = 3X   Find A for X = 7, 2, 19, 4.
2. Y = 7X − 4   Find Y for X = 3, 7, 11, 15, 19.
3. M = 3B + 2   Find M for any 6 values of B.

   1. READ-DATA
   2. Sequential INPUT   3. INPUT

To find the percent one number is of another
To find a number given a percent of the number
To solve problems involving commission, discount, and percent increase and decrease

Chapter Overview

The chapter opens with a brief introduction to the meaning of percent. The three classic cases of percent are developed algebraically, and percents are then applied to business and consumer situations.

Chapter Opener (Page 215)

Problem Formulation

The picture shows the Chicago skyline and Merchandise Mart as viewed from Lake Michigan. Three of the world's tallest skyscrapers—the Sears Tower at 1,454 ft or 110 stories, the Standard Oil Building at 1,136 ft or 80 stories, and the John Hancock Center at 1,127 ft or 100 stories—are in Chicago.

The Merchandise Mart was built in 1930. It is the largest commercial structure in the world, enclosing about 4,000,000 sq ft. It contains 95 acres of floor space.

Have students use the preceding data to formulate and solve at least two problems.

Estimation

Have students estimate the floor space in the Merchandise Mart in square feet to the nearest 1,000 ft^2.

Application

Have students make a bar graph to show a comparison among the Sears Tower, the Standard Oil Building, and the John Hancock Center.

Chapter

9

Percents

Chapter Objectives

To change percents to decimals, decimals to percents, and fractions to percents
To find a percent of a number

Introduction to Percent (Pages 216–218)

Teaching Hints Example 3 may prove difficult for many students. Students tend to write 6% as 0.60. A good analogy to help students distinguish between 6% and 60% is through money. For example, 6 cents is $0.06 but 60 cents is $0.60.

In this lesson avoid interpreting a fractional percent, such as $3\frac{1}{4}\%$, as a decimal, 3.25%. This concept is developed in the next lesson.

Cumulative Review

1. Multiply. $5\frac{1}{3} \cdot \frac{15}{4}$
2. Simplify. $2\frac{3}{4} \cdot 5\frac{1}{3}$
3. Divide. $5\frac{1}{4} \div \frac{3}{4}$

1. 20   2. $14\frac{2}{3}$ 3. 7

Percent of a Number (Pages 219–221)

Teaching Hints The next two lessons use an algebraic-equation approach. However, do not compel students to write an equation for finding a percent of a number.

Cumulative Review

1. Write in simplest form. $\frac{24}{16}$
2. Simplify. $\frac{3}{4}$ of 15
3. Solve. $\frac{6}{7}x = \frac{14}{3}$

1. $1\frac{1}{2}$ 2. $11\frac{1}{4}$ 3. $5\frac{4}{9}$

Finding Percents (Pages 222–223)

Teaching Hints Point out that writing the equation is a direct, easy translation from the English to the Algebraic.

Preventing Common Errors Students tend to forget to write the percent symbol in their answers. This needs to be emphasized.

Cumulative Review

Add or subtract. Simplify if possible.

1. $\frac{4}{5}$
   $+\frac{2}{3}$

2. $\frac{8}{9}$
   $-\frac{1}{6}$

3. $7\frac{1}{5}$
   $-2\frac{1}{4}$

1. $1\frac{7}{15}$ 2. $\frac{13}{18}$ 3. $4\frac{19}{20}$

Finding the Number (Pages 224–225)

Teaching Hints Emphasize the ease of translating the English sentence into an algebraic equation. Stress that *of* means times and *is* means equals.

Cumulative Review

1. Add. $6\frac{3}{4} + 5\frac{1}{2} + 7\frac{1}{3}$
2. Subtract. $8\frac{1}{5} - 5\frac{2}{3}$
3. Solve. $x - \frac{1}{4} = \frac{1}{5}$

1. $19\frac{7}{12}$ 2. $2\frac{8}{15}$ 3. $\frac{9}{20}$

Commission (Pages 226–228)

Teaching Hints You may have to define commission before beginning the lesson. Point out that amounts are rounded down to the lower penny. Example 3 involves the use of several operations to solve one problem.

Cumulative Review

1. Add. 347.6 + 1.056 + 18.49
2. Subtract. 4.15 from 37
3. Evaluate. $0.03s^2$ if $s = 0.03$

1. 367.146   2. 32.85   3. 0.000027

Discounts (Pages 229–231)

Teaching Hints Rounding is done in favor of the merchant. You may have to point out that after the discount is computed, the problem is still not completed. The discount must then be subtracted from the original price to find the actual amount the customer pays.

Enrichment Students might be interested in a shorter, different approach, for example:

$$\textit{Think: } 100\% - 20\% = 80\%$$

$$\begin{array}{r} \$40.00 \\ \times\ 0.80 \\ \hline \$32.00 \end{array}$$

Cumulative Review

1. Change to a decimal. 56.3%
2. Change to a percent. 0.357
3. Find the percent. 52% of 138

1. 0.563   2. 35.7%   3. 71.76

Problem Solving—Applications (Pages 232–233)

Teaching Hints Real-life problem situations involving work in a beauty salon include sales tax, total cost of goods and services, wages, and discounts. Remind students that the sales tax is on purchases but not on services.

Problem Solving—Careers (Page 234)

Teaching Hints Problems using percents encountered by sales representatives in their work are presented.

Non-Routine Problems (Page 235)

Since there is no specific strategy for solving these problems, they are non-routine. Students should be reminded to read them several times, thoughtfully, before they attempt to solve these problems. Common sense and ingenuity should be applied. See the TE Answer Section for suggested solutions.

Computer Activities (Page 236)

This computer lesson lays the foundation for understanding direct variation in algebra. In Exercise 3, doubling the deposit will double the amount. But in Exercise 4, doubling the rate will provide almost 10 times the amount. This is due to the exponent.

This can be seen more easily by looking at areas of squares. A square 6 units on a side has an area of 36 square units. Doubling the side to 12 units results in an area of 144 square units. This is not double the original but is 4 times as large.

Project Write a program to compare areas of squares when lengths of sides are doubled. Use **READ-DATA** and the following lengths of sides and their doubles: 6, 12; 8, 16; 10, 20.

```
10 PRINT "SIDE", "AREA", "TWICE
SIDE", "NEW AREA"
20 FOR K = 1 TO 3
30 READ S1, S2    S1 is length of side;
                  S2 is length doubled.
40 LET A1 = S1 ∧ 2
50 LET A2 = S2 ∧ 2
60 PRINT S1, A1, S2, A2
70 NEXT K
80 END
```

Have students adjust the program to compare areas of squares when lengths of sides are tripled. Encourage them to use their own data.

Chapter

10

Statistics

Chapter Objectives

To find the range, mean, median, and mode for a set of data

To make a frequency table

To find the mean of grouped data

To collect, display, and analyze data

To find the probability of simple, independent, and dependent events

To apply the counting principle in finding the number of possible arrangements

Chapter Overview

Basic statistics are introduced: range, mean, median, and mode. Students group data to make a frequency table and use bar graphs as a tool for analyzing data. The probability of simple and independent events is developed. Students use the counting principle to find the number of possible arrangements.

Chapter Opener (Page 239)

Problem Formulation

The picture shows one of the most impressive earthwork mounds, the Great Serpent Mound, located near Sunking Spring in southern Ohio. Prehistoric Hopewell Indians are credited with its construction. It is about 4 ft high, about $\frac{1}{4}$ mi long, and is shaped like a coiled serpent.

Human–made mounds of earth used as prehistoric Indian burial grounds are frequently found from the Great Lakes to Mexico. It is estimated that the first mounds were built about 1000 B.C. The most recent type of mound built was the flat-topped pyramid, similar to the Mexican pyramids. One of the largest is Monk's Mound in Illinois. This mound is 1,080 ft long, 710 ft wide, and 100 ft high.

Have students use the preceding data to formulate and solve at least two problems.

Estimation

Have students estimate the volume of the Monk's Mound.

Application

Have them decide what data should be gathered and what measurements should be made at a prehistoric site.

Range, Mean, Median, Mode (Pages 240–242)

Teaching Hints Extremes can affect the mean and the range so as to be misleading. To compensate for extremes in ranges, it is sometimes appropriate to delete extreme scores on both ends before calculating averages.

Enrichment Some students might like using the algebraic formula in finding the mean.

Cumulative Review

1. Multiply. $3\frac{1}{7} \cdot \frac{3}{12}$ 2. Divide. $5\frac{1}{4} \div \frac{7}{2}$

3. Solve. $\frac{3}{5}x = \frac{9}{10}$

1. $\frac{11}{14}$ 2. $1\frac{1}{2}$ 3. $1\frac{1}{2}$

Mean of Grouped Data Pages 243–245)

Teaching Hints This lesson is an extension of the previous lesson, but because of the greater number of numbers in the data it is best to make a frequency table to calculate the mean.

Cumulative Review

1. Add. $\frac{5}{6}$
            $+\frac{3}{5}$

2. Subtract. $6\frac{3}{8}$
                 $-4\frac{1}{2}$

3. Anne swims $14\frac{1}{2}$ laps, rests, then swims $8\frac{1}{4}$ laps more. How many laps did she swim altogether?

1. $1\frac{13}{30}$ 2. $1\frac{7}{8}$ 3. $22\frac{3}{4}$

Analyzing Data (Page 246)

Teaching Hints For Exercise 3, have students use scores for whatever sport is in season. They could use calculators to check their work.

Preventing Common Errors For Exercises 1 and 2, students may want to divide by 8 since 8 scores are given. Remind them that there are 4 scores for each team, so they should divide by 4.

Cumulative Review

1. Find the percent. 37% of 18.3
2. 30 is what percent of 40?
3. 50 is 5% of what number?

1. 6.771   2. 75%   3. 1,000

Problem Solving—Applications (Page 247)

Teaching Hints Real-life problem situations that may be encountered in a school store provide the background for reviewing skills taught thus far.

Simple Events (Pages 248–249)

Teaching Hints You may wish to make a spinner and conduct an experiment. Compare actual results against theoretical results. Rarely are actual results the same as the predicted results. It might be interesting to combine all the tallies of the class and check to see how close the actual results are to the mathematical probability.

Enrichment Have students discuss why all of the sectors on a spinner must be the same size. Introduce the concept of the ratio of the area of a sector to the total area of the circle as the probability of an event. Have students perform experiments suggested in the lesson and compare the likelihood of the event with the actual experimental results.

Cumulative Review

1. Add. 13.75 + 6.389 + 104.2
2. Multiply.   38.3
              $\times$ 0.045

3. Divide. Round to the nearest tenth.

   $0.07\overline{)0.2437}$

1. 124.339   2. 1.7235   3. 3.5

Independent and Dependent Events (Pages 250–251)

Teaching Hints On page 250, students may need help in reading $P(A \text{ and } B) = P(A) \cdot P(B)$. $P(A \text{ and } B)$ means the probability of *both* A and B occurring. $P(A) \cdot P(B)$ means the probability of A times the probability of B.

Enrichment Challenge students to find the probability of 3 or more independent events.

Cumulative Review

Solve.

1. $x - 8 = 9$ 2. $y + 4 = 13$
3. $4x - 3 = 17$

1. 17   2. 9   3. 5

Counting Principle (Pages 252–253)

Teaching Hints Some students might be fascinated in seeing the large number of possible arrangements suggested by the counting principle.

   The concept of repeating is introduced in starred Exercises 7 and 8 and could be made more real by using physical or board activities.

Enrichment Have students determine how many different telephone numbers can be formed with and without using area codes. Note that the middle digit in an area code is always either 0 or 1 and that neither 0 nor 1 is used for the first two numbers. This will make a good discussion problem.

Cumulative Review

1. Change to a decimal. 4%
2. Change to a decimal. 73%
3. Change to a percent. 0.653

1. 0.04   2. 0.73   3. 65.3%

Problem Solving—Careers (Page 254)

Teaching Hints Real-life problems that may be encountered by manufacturing inspectors are used to review skills taught thus far.

Mathematics Aptitude Test (Page 255)

The problems on this page are structured to resemble those found on standardized tests. The format is multiple choice and can provide students with practice in transferring their answers to a master sheet.

Computer Activities (Page 256)

The use of the counter C in the sample program of this lesson illustrates the use of an **Accumulator,** which will be studied again in the computer activity on page 307 of Chapter 12. The **Accumulator** is a device of **BASIC** that allows you to add results as you go without actually using the "+" symbol. It is essential to *initialize* C = 0 at the beginning. This clears all previous numbers from the memory so that they will not be added to new numbers. Notice that the **Accumulator** can be part of the **IF . . . THEN** statement on line 50. Students can save themselves the trouble of typing **RUN** for each new trial by using another loop; add a new line, 5. **5 FOR W = 1 TO 5.** Add line 90. **90 NEXT W.** You can now **RUN** the program 5 times. This illustrates the use of nested loops. The K loop is nested within the W loop. The computer will select 240 random numbers from 1 through 6. It will do this again 5 times.

Chapter

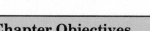

Ratio and Proportion

11

Chapter Objectives

To write ratios
To identify, solve, and apply proportions
To read and make circle graphs
To find distances by using a scale
To write trigonometric ratios

Chapter Overview

Ratios, proportions, and problems using these concepts are developed. Similar triangles and an optional lesson trigonometry follow naturally from the discussion of ratio and proportion.

Chapter Opener (Page 259)

Problem Formulation

The picture shows the Empire State Building in New York City. The Empire State Building was erected in 1 year and 45 days by 3,500 workers and was completed in 1931. Ten million bricks, 750 tons of aluminum and stainless steel for the outer walls, 1,670,000 ft^3 of concrete, 60 mi of plumbing pipe, and 15,000,000 ft of telephone wire were used in the construction of this magnificent building. Its height is 1,250 ft or 378 m and comprises 102 stories. The elevator speed in the upper levels is 1,200 ft/min or 365 m/min. Its television antenna is 222 ft, which brings the total height of the building to 1,472 ft.

The World Trade Center was dedicated in 1973 and comprises two twin towers, each 110 stories high. Its height is 1,377 ft, and it contains nearly 9,000,000 ft^2 of office space.

Have students use the preceding data to formulate and solve at least two problems.

Estimation

Have students estimate how far the telephone wire used in the Empire State Building would stretch from New York City to Los Angeles.

Application

Have students determine costs of various parts of the construction if the Empire State Building were to be built today. How much would 10 million bricks cost today?

Ratios (Pages 260–262)

Teaching Hints The idea of ratio is used in everyday language, such as the ratio of boys to girls, height to weight, and length to width. Ratio is used quite frequently in sports, such as the ratio of hits to times at bat, shots made to shots attempted, and wins to losses.

For computation, it is convenient to express ratios using fractions. Students should also realize that fractions are ratios.

Cumulative Review

1. Simplify. $2\frac{2}{3} \cdot \frac{5}{14}$ 2. Find. $\frac{3}{4}$ of 15

3. Divide. $\frac{2}{3} \div \frac{5}{6}$

1. $\frac{20}{21}$ 2. $11\frac{1}{4}$ 3. $\frac{4}{5}$

Proportions (Pages 263–265)

Teaching Hints You may want to encourage some students to use the method given in the Calculator activity when solving proportions.

Enrichment Ask students to reverse the cross-multiplication process and form several proportions from one product, $3 \times 4 = 2 \times 6$.

$$\frac{3}{2} = \frac{6}{4} \qquad \frac{2}{3} = \frac{4}{6} \qquad \frac{2}{4} = \frac{3}{6}$$

$$\frac{6}{4} = \frac{3}{2} \qquad \frac{4}{6} = \frac{2}{3} \qquad \frac{3}{6} = \frac{2}{4}$$

Cumulative Review

Solve and simplify.

1. $x - \frac{1}{3} = \frac{1}{5}$ 2. $x + \frac{2}{3} = 10\frac{1}{5}$

3. Jack weighed $188\frac{1}{3}$ lb. He gained $5\frac{1}{5}$ lb. How much does he weigh now?

1. $\frac{8}{15}$ 2. $9\frac{8}{15}$ 3. $193\frac{8}{15}$ lb

Applying Proportions (Pages 266–268)

Teaching Hints Have students study the tables to see that both values either increase or decrease. Students may be interested in finding a general formula for direct variation, $y = cx$, where c is a constant. After students have learned how to solve problems using proportions, you may want to discuss unit pricing, how to determine it, and how to solve problems involving it.

Cumulative Review

1. Add. $48.36 + 0.015 + 6.593$
2. Subtract. $\begin{array}{r} 19.073 \\ -\ 11.187 \end{array}$ 3. Multiply. $\begin{array}{r} 497.3 \\ \times\ 6.03 \end{array}$

1. 54.968   2. 7.886   3. 2,998.719

Inverse Variation (Page 269)

Teaching Hints Have students study the tables to see that as one value increases, the other decreases. Have students locate and solve problems in newspapers that involve indirect variation.

Circle Graphs (Pages 270–271)

Teaching Hints Circle graphs usually indicate percent or fractional parts of the whole. Circle graphs also show the relationships among numerical information by the size of the sectors and the relationship between each item and the whole.

Preventing Common Errors For Exercises 3 and 4 on page 272, students often have difficulty in forming fractions to determine the percentages for dividing circles. Guide them in determining these percentages.

Cumulative Review

1. 12% of what number is 18?
2. 14 is 70% of what number?
3. 30 is what percent of 50?

1. 150   2. 20   3. 60%

Problem Solving—Careers (Page 273)

Teaching Hints Real-life problems that may be encountered by service people include service charge, hourly labor rate, cost of parts, and travel charges.

Scale Drawing (Pages 274–276)

Teaching Hints The ideas presented are applicable to most scale drawings. You may want to have students bring in copies of maps and practice reading them. You may want to give students some practical experience in making scale drawings of known situations, such as the classroom, the school, or the schoolyard.

Enrichment Have students contact architects, builders, interior designers, or others who use scale drawings and if possible borrow copies of scale drawings from them. Have students interpret the drawings for the class.

Cumulative Review

1. C.B. Regular price: $90.50. Discount: 20%. Cost?
2. Winter coat. Regular price: $120.00. Discount: $\frac{1}{3}$ off. Actual cost?
3. Find the total actual cost of the following purchases at 30% discount: $32.00 sweater, $15.75 shirt, $27.95 shoes.

1. $72.40   2. $80.00   3. $52.99

Similar Triangles (Pages 277–279)

Teaching Hints The fact that similar triangles have corresponding sides in proportion is fundamental in developing right-triangle trigonometry. All right triangles with the same acute angle are similar and therefore have their corresponding sides in proportion. Discuss differences between similar triangles and congruent triangles.

Cumulative Review

Change as indicated.

1. 28 mm to cm   2. 3.45 km to m
3. 20 m to km

1. 2.8 cm   2. 3,450 m   3. 0.02 km

Trigonometric Ratios (Pages 280–282)

Teaching Hints Students should memorize the trigonometric ratios of sine, cosine, and tangent. They may ask about the ratios of the other sides, giving rise to cotangent, secant, and cosecant. You may wish to discuss these three functions, depending upon the ability of your students.

You may want to show students a table of trig values and have them read the table. You also may want to introduce these students to a calculator that has these functions.

Preventing Common Errors Students often confuse the trigonometric ratios given here. A mnemonic device is often used to aid students in memorizing these functions.

SOH   CAH   TOA or
$$\text{Sin} = \frac{\text{opp}}{\text{hyp}}, \text{Cos} = \frac{\text{adj}}{\text{hyp}}, \text{Tan} = \frac{\text{opp}}{\text{adj}}$$

Cumulative Review

1. Find the range. Find the mean.
   15, 7, 8, 12, 3
2. Find the median.
   93, 78, 65, 91, 87
3. Find the mode(s).
   7, 9, 10, 12, 16, 10, 6, 8

1. 12, 9   2. 87   3. 10

Non-Routine Problems (Page 283)

Since there is no specific strategy for solving these problems, they are non-routine. Students should be reminded to read them several times, thoughtfully, before they attempt to solve these problems. Common sense and ingenuity should be applied. See TE Answer Section for suggested solutions.

Before a program for determining whether a particular equation is a proportion can be written, a general formula for the process must be derived. Students can derive this formula for themselves before reading the solution.

Project Assign a volunteer to write a program to solve a proportion if the first three terms are given and the fourth term is unknown, for example, solve $\frac{3}{5} = \frac{21}{Y}$.

The student must first write the general problem, then state the general solution. The general proportion is $\frac{A}{B} = \frac{X}{Y}$. The solution is $Y = \frac{BX}{A}$.

```
10 INPUT "WHAT IS B? "; B
20 INPUT "WHAT IS X? "; X
30 INPUT "WHAT IS A? "; A
40 LET Y = B * X/A
50 PRINT "THE SOLUTION IS " Y
60 END
RUN
```

Chapter

Adding and Subtracting Integers

Chapter Objectives

To compare integers using > or <
To add and subtract integers
To solve problems involving integers
To solve equations involving addition and subtraction of integers

Chapter Overview

The number line model is used for representing integers, as well as for adding and subtracting integers. Students learn to subtract an integer by adding its opposite, to solve problems using integers, and to solve equations where solutions are integers.

Problem Formulation

The picture shows the Golden Gate Bridge in San Francisco, California. The bridge was completed in 1937 and cost over 35 million dollars. The bridge hangs from 2 steel cables 36.5 in. in diameter. The cables are held up by the two towers. Each tower is 746 ft high and stands about 1,125 ft from each shore. The section between the towers is 4,200 ft long, one of the largest spans in the world.

Have students use the preceding data to formulate and solve at least two problems.

Estimation

Have students estimate the total length of the Golden Gate Bridge.

Application

Have students compare the Golden Gate Bridge with some other well-known bridge.

Integers on a Number Line (Pages 288–290)

Teaching Hints Point out that each integer has exactly one point corresponding to it on the number line, but only some points correspond to integers. Students will see later that, given real numbers, there is a one-to-one correspondence between all real numbers and all points on a number line.

Activities Ask students to write down all situations they can think of in which integers might be of use.

Cumulative Review

1. Multiply. $5\frac{1}{7} \cdot \frac{4}{5}$ 2. Divide. $3\frac{5}{6} \div 1\frac{2}{3}$

3. Solve. $\frac{1}{4}x = 4$

1. $4\frac{4}{35}$ 2. $2\frac{3}{10}$ 3. 16

Adding Integers (Pages 291–293)

Teaching Hints Adding two positive integers and two negative integers is easy. It is just like adding whole numbers, but when adding negative integers one must remember that the sum is a negative integer.

The most complex case of addition of integers is that of adding a positive and a negative integer. Drawing trips on a number line to represent this addition is helpful.

Enrichment Explain the meaning of absolute value and give some examples. Suggest that students develop a rule for adding a positive and a negative integer in terms of the absolute value. (The sum can be found by subtracting the smaller of the absolute values of the numbers from the greater. The sign of the sum is that of the integer with the greater absolute value.)

Cumulative Review

1. Add. $\frac{1}{3} + \frac{4}{5} + \frac{3}{4}$ 2. Subtract. $6\frac{5}{8} - 1\frac{3}{7}$

3. Subtract. $9 - 6\frac{1}{4}$

1. $1\frac{53}{60}$ 2. $5\frac{11}{56}$ 3. $2\frac{3}{4}$

Problem Solving—Applications (Page 294)

Teaching Hints The real-life problems involving costs and profit that teenagers encounter as newspaper carriers are dealt with on this page.

Subtracting Integers (Pages 295–297)

Teaching Hints Example 2 develops the case of subtracting a larger positive number from a smaller positive number. Some students will observe that this is like the preceding case (subtracting the smaller number from the larger), but the difference is negative.

Preventing Common Errors The cases illustrated in Examples 4 and 5 are somewhat more complex than the previous cases. Let students use the number line if they have difficulty with cases of this type.

Cumulative Review

Solve. Simplify, if possible.

1. $x - 3\frac{1}{5} = 5\frac{1}{3}$ 2. $x + 5\frac{1}{8} = 7$

3. $x + 1\frac{1}{8} = 4\frac{5}{6}$

1. $8\frac{8}{15}$ 2. $1\frac{7}{8}$ 3. $3\frac{17}{24}$

Problem Solving—Careers (Page 298)

Teaching Hints Real-life problems involving time, cost, labor, and percent decrease, typical problems encountered by mechanics, are dealt with in this lesson.

Opposites (Pages 299–300)

Teaching Hints Students see that opposites are the same distance from 0 on a number line but in opposite directions. They learn that every integer has a unique opposite integer and that the sum of the two is 0. Finally, a practical situation dealing with a gain and a loss of money is used to illustrate opposites.

Cumulative Review

1. Subtract. 4.35 from 18.67
2. Multiply.   4.003
            $\times\ 0.076$

3. Evaluate. $0.07x^2$ if $x = 0.03$

1. 14.32   2. 0.304228   3. 0.000063

Subtracting by Adding (Pages 301–302)

Teaching Hints Students are shown that the number-line methods yield the same answer for $7 - 4$ and $7 + (-4)$. Thus, we can conclude that $7 - 4 = 7 + (-4)$, or subtracting positive 4 is the same as adding negative 4.

Have students go through Examples 1–5 and observe that the same pattern holds in each case. That is, subtracting an integer is the same as adding its opposite.

Cumulative Review

Solve. Round to the nearest tenth.

1. $0.07x = 9.3$ **2.** $0.3x = 0.784$
3. $0.45x = 15$

1. 132.9   2. 2.6   3. 33.3

Using Integers (Pages 303–304)

Teaching Hints This lesson illustrates the many situations in daily life in which positive and negative numbers are used.

Cumulative Review

1. Change to a decimal. 43.9%

2. Change to a percent. $\dfrac{4}{9}$

3. Change to a percent. 0.65

1. 0.439   2. 44.44%   3. 65%

Equations (Pages 305–306)

Teaching Hints Point out that the techniques for solving the equations given here are the same as in previous lessons. The only difference here is that the equations contain integers and the solutions are integers.

Cumulative Review

A baseball team won 20 games and lost 3. Find each ratio.

1. Wins to losses   **2.** Wins to total
3. Losses to total games   games

1. 20:3   2. 3:23   3. 20:23

Computer Activities (Page 307)

Students are sometimes confused by the nature of the **Accumulator** line 50. How can A = A + X? This does not mean A = A + X. The expression is read as "A is now assigned to A + X," or the previous value of A is added to the newly entered value of X and this sum is now designated as the new value of A. A good application of this is to find the average of a set of grades. If you wanted to average 10 grades without the **Accumulator** technique, you would need 10 input statements.

Project Ask students to write a program that will find the average of any number of grades.

```
10 INPUT "HOW MANY GRADES? "; N
20 S = 0
30 FOR J = 1 TO N
40 INPUT "ENTER A GRADE "; G
50 LET S = S + G
60 NEXT J
70 PRINT "TOTAL IS " S
80 LET A = S/N
90 PRINT "AVERAGE " A
100 END
```

Chapter

Multiplying and Dividing Integers

Chapter Objectives

To multiply and divide integers
To use the distributive property
To divide rational numbers
To simplify algebraic expressions involving integers
To solve equations involving multiplication and division of integers

Chapter Overview

The main purpose of this chapter is to teach students multiplication and division of integers. Rationals are introduced and division of rational numbers is developed. Students use their skills to simplify algebraic expressions.

Chapter Opener (Page 311)

Problem Formulation

The picture shows Okefenokee Swamp Park in Georgia. Most of the swamp was purchased by the government in 1937 and designated as a wildlife refuge. It is the home for many alligators, bears, birds, deer, opossums, otters, raccoons, and wildcats and about 50 species of fish. The swamp has many small islands called hammocks. Plants include cypress trees, Spanish moss, orchids, and white and golden lilies. The swamp is about 40 mi long and 30 mi wide at its widest point and covers an area of about 700 mi^2, or 293,826 acres.

Have students use the preceding data to formulate and solve at least two problems.

Estimation

Have students estimate the area of the swamp if it were shaped like a rectangle. How much greater would the area be than it actually is?

Application

Have students plan a camping trip to Okefenokee Swamp Park. Have them determine the approximate cost of the trip. Have them formulate and solve problems they may encounter on the trip.

Multiplying Integers (Pages 312–313)

Teaching Hints Have students start with two different positive integers and, following the idea of Examples 2 and 3, built their own sequences to demonstrate that a continuation of the pattern leads to the conclusion that the product of a positive and a negative integer is a negative integer.

Cumulative Review

1. Multiply. $\dfrac{3}{4} \cdot \dfrac{5}{4}$ **2.** Simplify. $2\dfrac{1}{2} \cdot \dfrac{16}{5}$

3. Divide. $\dfrac{5}{9} \div 3\dfrac{1}{3}$

1. $\dfrac{15}{16}$ 2. 8 3. $\dfrac{1}{6}$

Multiplying Negative Integers (Pages 314–315)

Teaching Hints After students study Examples 1 and 2, have them build patterns of their own, starting with two different negative integers. They will see that maintaining the pattern leads to the conclusion that the product of two negative integers is a positive integer.

Enrichment Suggest that students make up some messages of their own like the one in the Challenge.

Cumulative Review

1. Add. $487.63 + 0.005 + 42.1$
2. Subtract. 4.15 from 20.8
3. Divide. Round to the nearest tenth.

$0.05\overline{)9.07}$

1. 529.735 2. 16.65 3. 181.4

Properties of Multiplication (Pages 316–318)

Teaching Hints Students become aware of the power of algebraic notation. For example, the brief statement $n \cdot 1 = n$ covers an infinite number of cases. Ask students to suggest other statements that are universally true. It might be an interesting activity to have the class compile a list of universally-true statements.

Enrichment Ask students to use their imaginations in making up examples that would look very complicated at first but would be simple to evaluate because multiplication by 0 and addition and subtraction patterns are involved. For example,

$$-369 \cdot [1 + (-1)] - (543 - 789) \cdot 0$$

Cumulative Review

1. Change to a decimal. 18.7%
2. Change to a percent. $0.40\frac{1}{4}$
3. 15 is what percent of 50?

1. 0.187   2. $40\frac{1}{4}\%$ 3. 30%

Problem Solving—Applications (Page 319)

Teaching Hints One of the jobs open for teenagers is that of camp counselor. Real-life problems involving the activities of a camp counselor are used to review skills taught thus far.

Dividing Integers (Pages 320–322)

Teaching Hints Examples 1–4 cover all possible combinations of signs for divisors and dividends. You might ask students to state generalizations as they study these examples.

Guide students to conclude from Example 5 that any integer divided by 1 is that integer, from Example 6 that any integer divided by −1 is the opposite of that integer, from Example 7 that 0 divided by any non-zero integer is 0, and from Example 8 that division by 0 does not yield any answer.

Enrichment Suggest that some students develop a careful analysis and differentiate between the case of dividing a non-zero integer by 0 and dividing 0 by 0. They should discover that in the former case there is no answer at all and in the latter case any number could be an answer.

Cumulative Review

1. Kathy earns 15% commission. She brought in $450.75 in sales. Find her commission.
2. Maria earns $240.00 per week plus 9% commission on sales. Her sales last week were $874.00. Find her total earnings.
3. Shoes: $45.75. Discount: 20%. Actual cost?

1. $67.61   2. $318.68   3. $36.60

From Integers to Rationals (Pages 323–324)

Teaching Hints Students should observe that the only difference between dividing fractions and dividing rational numbers is that in the latter case they must decide whether the quotient is positive or negative.

Cumulative Review

1. A man drove 180 km in 3 hours. How far did he drive in 5 hours?
2. Two cups of blueberries make 15 pancakes. How many cups of blueberries make 18 pancakes?
3. Find the batting average for 168 hits in 575 times at bat.

1. 300 km   2. $2\frac{2}{5}$ cups   3. 0.292

Problem Solving—Careers (Page 325)

Teaching Hints Real-life problems encountered by machinists, involving multiplication and division by fractions, are presented in this lesson.

Simplifying Expressions (Pages 326–327)

Teaching Hints Examples 1 and 2 require the use of the distributive property of multiplication. It might be helpful to have it displayed for the students in both forms:

$$(a + b)c = ac + bc \qquad ab + ac = a(b + c)$$

In Example 3, point out that the associative property of multiplication is used. In Examples 4 and 5, you might have to remind students that adding the opposite of an integer is the same as subtracting that integer.

Cumulative Review

1. Add. $-12 + (-15)$ 2. Add. $-19 + 17$
3. Subtract. $-14 - (-14)$

1. -27 2. -2 3. 0

Solving Equations (Pages 328–329)

Teaching Hints Stress that equations involving integers are solved in the same way as equations involving whole numbers. Undo multiplication by division and undo division by multiplication. Insist that students check their solutions in the original equations.

Cumulative Review

Solve and check.

1. $n + (-3) = -7$ 2. $3 + x = -6$
3. $n - (-5) = 2$

1. -4 2. -9 3. -3

Non-Routine Problems (Page 330)

Since there is no specific strategy for solving these problems, they are non-routine. Students should be reminded to read them several times, thoughtfully, before they attempt to solve these problems. Common sense and ingenuity should be applied. See TE Answer Section for suggested solutions.

Computer Activities (Page 331)

Whenever writing a program to evaluate a fractional expression it is a good technique to represent the numerator and denominator separately. This enables the computer to examine the denominator before trying to evaluate the fraction. Point out the error caused by placing the instruction on line 90 before line 80. You do not want the computer to try to find Q before it has determined that Q will be undefined due to division by 0.

Chapter 14

Equations and Inequalities

Chapter Objectives

To solve equations and inequalities
To evaluate and simplify algebraic expressions containing exponents
To classify and simplify polynomials
To multiply a monomial by a polynomial
To factor polynomials

Chapter Overview

Students learn to use more than one property to solve equations. The properties of inequalities are presented, and students solve simple inequalities requiring the use of one property. They then evaluate and simplify algebraic expressions with exponents and simplify and factor polynomials.

Chapter Opener (Page 334)

Problem Formulation

The picture shows Kitt Peak National Observatory, located southwest of Tucson, Arizona. The observatory is located on Kitt Peak, which is 6,875 ft or 2,096 m above sea level. It has three telescopes, two optical and one radio. The solar telescope, known as the Robert R. McMath Telescope, is 500 ft long and is the largest telescope of its kind in the world. The three mirrors it utilizes produce an image of the sun 86 cm in diameter.

The optical telescope, known as the N.U. Mayall Telescope, was completed in 1973, and its main mirror is 401 cm in diameter and is 61 cm thick. The mirror weighs about 14 metric tons.

Have students use the preceding data to formulate and solve at least two problems.

Estimation

The solar telescope used at the Kitt Peak National Laboratory produces the sun's image of 86 cm in diameter. Estimate, to the nearest centimeter, the area of the sun's image.

Application

Have students, working in groups, undertake a project of deciding what telescope they should use to study faint galaxies. Have them discuss the advantages and disadvantages of each telescope. They should decide on the data to be collected to make the best decision.

Solving Equations (Pages 335–336)

Teaching Hints Through Example 1, students discover that when the variable appears on both sides of the equation they can use the addition or the subtraction property to isolate the variable on either the left or the right side. There is a natural tendency to isolate the variable on the left side. However, students might want to think in terms of isolating the variable on the side in which the coefficient of the variable would be positive rather than negative.

Enrichment Have students solve equations such as the following: $7x + 2 = 2x + x - 14$. Point out that the terms containing a variable should be combined on the right side of the equation before the subtraction is applied. ($7x + 2 = 3x - 14$)

Cumulative Review

1. It takes 4 hours to cut the grass. What part can be cut in 3 hours?
2. Find the batting average for 175 hits in 490 times at bat.
3. Al drove 450 km in 6 hours. How far can he drive in 8 hours?

1. $\frac{3}{4}$ 2. 0.357   3. 600 km

Equations with Parentheses (Pages 337–338)

Teaching Hints Before students study Example 3 you might want to review the property, $-a = -1(a)$. They will apply it in rewriting $-(-6x - 3x)$ as $-1(-6x - 3x)$. Then they can apply the distributive property.

Enrichment For those students who seem ready, you might suggest the following alternative method for solving the equation in Example 1: $5(x - 2) = 20$. First divide each side by 5 to get $x - 2 = 4$.

Cumulative Review

Solve each proportion. Check.

1. $\frac{3}{5} = \frac{n}{15}$ 2. $\frac{x}{8} = \frac{3}{2}$ 3. $n:9 = 5:3$

1. 9   2. 12   3. 15

Number Problems (Pages 339–341)

Teaching Hints In Examples 1–3, students have the opportunity to practice translating word problems into equations without being concerned about solving them. If, however, some students wish to solve the equation and find the number, they should be encouraged to do so.

Preventing Common Errors Point out that when checking the solution to a word problem, the tentative solution should be replaced in the original word problem, not in the equation. If the equation is translated incorrectly from the word problem, the solution would check in the equation but not in the original problem.

Cumulative Review

1. Add. $-14 + (-15)$
2. Subtract. $-18 - 19$
3. Subtract. $-34 - (-14)$

1. -29 2. -37 3. -20

Problem Solving—Applications (Pages 342–343)

Teaching Hints Real-life problems using baby-sitting as the theme are presented in this problem-solving applications lesson. Review the four-step problem-solving strategy as it applies to the algebraic solution to word problems. Stress the **PLAN** step in which the variable is assigned to the unknown and the equation is written.

Properties for Inequalities (Pages 344–346)

Teaching Hints The three properties for inequalities are summarized.

Enrichment Some students might want to examine the cases involving 0. When 0 is added to each side of a true inequality, there is no change in the inequality.

Cumulative Review

1. Multiply. $-41(-14)$
2. Perform the operation as indicated. $3.17 + (-25) \cdot 2$
3. Divide. $-45 \div (-5)$

1. 574   2. -46.83 3. 9

Solving Inequalities (Pages 347–349)

Teaching Hints Each inequality, whether it has the symbol $>$ or $<$ in it, has an infinite number of integral solutions. Therefore, it is impossible to check all the solutions. But notice that in checking some numbers it is advisable to choose one number that is the boundary case to be sure that the final simple inequality we arrive at is the correct one. For example, in the case of $x > -5$, we should check -4 for x, since -4 is the first integer greater than -5.

Preventing Common Errors You might find it necessary to convince some students that -4 is indeed greater than -5 because the point on a number line corresponding to -4 is to the right of the point corresponding to -5.

Cumulative Review

1. Simplify. $-4a - (-9a)$
2. Solve and check. $-6x = 36$
3. Solve and check. $-4y = -44$

1. $5a$ 2. -6 3. 11

Problem Solving—Careers (Page 350)

Teaching Hints Real-life problems involving railroad brakers are dealt with on this page.

Exponents (Pages 351–352)

Teaching Hints For Example 2, you might want to have students examine three different cases, $-4y^2$, $-(4y)^2$, and $(-4y)^2$ when y is replaced by -5. Examples such as these illustrate the importance of parentheses.

Enrichment Some students might be interested in observing the pattern of alternating signs with odd and even exponents when the variable is replaced by a negative number such as -2. Have students fill in this chart.

x^2	x^3	x^4	x^5	x^6
$(-2)^2$	$(-2)^3$	$(-2)^4$	$(-2)^5$	$(-2)^6$
4	-8	16	-32	64
pos.	neg.	pos.	neg.	pos.

Cumulative Review

1. Factor completely. Write using exponents. 18
2. Evaluate. $10x^5$ if $x = 3$
3. Multiply. $395,105 \times 41$

1. $2 \cdot 3^2$ 2. 2,430   3. 16,199,305

Properties of Exponents (Pages 353–355)

Teaching Hints Examples 1, 4, and 6 can be used to develop the three properties of exponents. Examples 2, 3, 5, and 7 are applications of the properties.

Preventing Common Errors Students tend to make this error: $a^2 \cdot b^5 = (ab)^7$. When working with Examples 1–3, be sure to stress that the product of powers property only pertains to cases where the base is the same. For example,

$$a^2 \cdot a^5 = a^{2+5} = a^7$$

But for a case such as $a^2 \cdot b^5$, we may *not* apply the property since a and b may represent different numbers.

Cumulative Review

1. What % of 60 is 15?
2. 5 is what % of 15?
3. 9 is 75% of what number?

1. 25%   2. $33\frac{1}{3}$%   3. 12

Polynomials (Pages 356–357)

Teaching Hints You might want to remind students that it is the distributive property that enables us to combine like terms. For example, we combine $-2x$ and $5x$ as follows:

$$-2x + 5x = (-2 + 5)x = 3x$$

Preventing Common Errors Many students erroneously combine terms that are not like terms. When working with simplifying polynomials, you might want to give students some examples that cannot be simplified, such as:

$$3y^2 - 5y + 4y^3 - 1 \text{ and } 6x + 3x^3 + 4 - 7x^2$$

Cumulative Review

1. Find the range. Find the mean. 90, 75, 80, 95, 70
2. Find the median. 82, 94, 86, 74, 88, 92
3. Find the mode(s). 8, 9, 8, 9, 8, 7, 9, 6, 5, 9, 10, 8

1. 25, 82   2. 87   3. 8, 9

Simplifying Polynomials (Pages 358–359)

Teaching Hints You might want to suggest using a vertical form for combining like terms once parentheses have been removed. For Example 3, students might prefer to write the terms as follows:

$$
\begin{array}{r}
a^2 + 5a - 4 \\
3a^2 - 3a + 6 \\
\hline
4a^2 + 2a + 2
\end{array}
$$

Preventing Common Errors For cases like those shown in Examples 4 and 5, students often forget to apply the property $-a = -1 \cdot a$. Point out that they must distribute the -1 before combining like terms.

Cumulative Review

Change as indicated.

1. 5.3 km to m   2. 2 dm to m
3. 12 cm to m

1. 5,300 m   2. 0.2 m   3. 0.12 m

Common Factors (Pages 360–362)

Teaching Hints You may find that some students need additional review in finding the greatest common factor (GCF) of two numbers. If so, refer them to pages 90 and 91.

Preventing Common Errors Some students may factor a polynomial like $6x^3 - 8x^2 + 2x$ as $x(6x^2 - 8x + 2)$. Since the term *factor* means to use the GCF, $2x(3x^2 - 4x + 1)$ is the correct form.

Cumulative Review

Solve each proportion.

1. $\dfrac{x}{2} = \dfrac{9}{5}$ 2. $\dfrac{6}{7} = \dfrac{3}{n}$ 3. $x{:}7 = 3{:}9$

1. 3.6   2. 3.5   3. $2\frac{1}{3}$

Mathematics Aptitude Test (Page 363)

The problems on this page are structured to resemble those found on standardized tests. The format is multiple choice and can provide students with practice in transferring their answers to a master sheet. Have them use the sheet provided in the blackline master package.

Computer Activities (Page 364)

Students learn to program a computer to solve equations. To do so, students must generalize specific equations. Thus, the computer is used to enhance the understanding of mathematics, paving the way for the teaching of literal equations in Algebra 1.

Point out that the constants in the general equation $Ax + B = Cx + D$ can be any integer. Have them identify the constants in the equation $-2x + 10 = 4x - 8$ ($A = -2$, $B = 10$, $C = 4$, $D = -8$). Note that A and D are negative integers.

Project Ask the class to write a program to solve any equation of the types $3x - 4 = 11$ and $7x + 7 = 37$. Then have them use the computer to solve these two equations.

```
10 INPUT "TYPE IN VALUE FOR A "; A
20 INPUT "TYPE IN VALUE FOR B "; B
30 INPUT "TYPE IN VALUE FOR C "; C
40 LET X = (C − B)/A
50 PRINT "SOLUTION IS X = "X
60 END
```

Chapter 15

Some Ideas from Geometry

Chapter Objectives

To describe points and lines
To measure and classify angles
To tell whether two angles are complementary or supplementary

To identify and determine the measure of vertical, alternate interior, and corresponding angles
To construct a perpendicular line from a point not on the given line
To classify polygons and triangles
To identify perfect squares
To use a table to approximate a square root
To use the Pythagorean theorem

Chapter Overview

Students are introduced to the basic figures in geometry. They learn to construct a perpendicular line through a point not on the given line and use the Pythagorean theorem to solve right triangles.

Chapter Opener (Page 367)

Problem Formulation

The picture shows Mount Rushmore National Memorial located in the Black Hills of South Dakota. This memorial stands in the mountains 25 m from Rapid City. The entire memorial occupies 1,278.45 acres and is the second largest of the 23 memorials in the United States.

The figures in the memorial are the largest of any statue in the world. The head of George Washington is equal to the height of a five-story building, or about 60 ft. In that scale, a person would be 465 ft tall.

The work on the memorial began in 1927 and continued for 14 years. The actual work took $6\frac{1}{2}$ years, but due to lapses in the work, it was not completed until 1941. The memorial is 5,725 ft above sea level and about 500 ft above the valley.

Have students use the preceding data to formulate and solve at least two problems.

Estimation

Estimate the number of times the total human body is longer than its head.

Application

To create the Mount Rushmore Memorial, the figures were cut from the granite cliff. Have students, working in groups, gather data and formulate problems that might have had to been solved to have accomplished this task.

Points, Lines, and Planes (Pages 368–369)

Teaching Hints Mention to students that when naming a line or a segment, the order in which the points are given does not matter. Thus, $\overleftrightarrow{AB}$ and $\overleftrightarrow{BA}$ name the same line.

Enrichment Some students might enjoy constructing a regular tetrahedron from a pattern of four equilateral triangles.

pattern →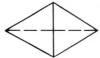

Cumulative Review

1. Add. $4\frac{3}{5}$
$+ 3\frac{1}{3}$

2. Subtract. $5\frac{2}{3} - 2\frac{1}{2}$

3. Solve. Simplify if possible. $x + \frac{6}{7} = \frac{1}{8}$

1. $7\frac{14}{15}$ 2. $3\frac{1}{6}$ 3. $-\frac{41}{56}$

Angles (Pages 370–371)

Teaching Hints You might want to make a sheet showing an acute, a right, and an obtuse angle with the sides extended so that students can measure them easily with a protractor.

Preventing Common Errors In Example 3, emphasize that after solving the equation, students must find the measures of *both* angles.

Cumulative Review

1. Evaluate $9t$ if $t = 7$.
2. Factor 18 into primes.
3. Evaluate x^4 if $x = 3$.
1. 63   2. $2 \cdot 3 \cdot 3 = 2 \cdot 3^2$ 3. 81

Problem Solving—Applications (Page 372)

Teaching Hints Real-life problems including pricing, earnings, profit, and inventory are presented.

Parallel Lines and Angles (Pages 373–374)

Teaching Hints Show non-parallel lines *and* parallel lines, each cut by a transversal. Ask students which lines have pairs of congruent alternate interior and corresponding angles. parallel lines

Enrichment Ask students to find all the angle measures for the figure below.

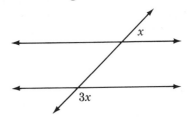

(*Hint:* $x + 3x = 180$)
Angles measure 45° and 135°.

Cumulative Review

1. Add. $14 + (-15)$ **2.** Subtract. $18 - 19$
3. Subtract. $34 - (-14)$

1. -1 2. -1 3. 48

Perpendicular Lines (Pages 375–376)

Teaching Hints Point out that students can draw perpendicular lines with the aid of a T-square or a protractor. But for a geometric *construction,* they can only use a compass and a straightedge.

Cumulative Review

Compare. Use > or <.

1. -5 __?__ -3 **2.** -4 __?__ 1 **3.** 7 __?__ -10

1. $<$ 2. $<$ 3. $>$

Triangles (Pages 377–378)

Teaching Hints Have students cut out a large triangle, tear off the three angles, and fit them next to each other. They should discover that the three angles form a straight line. This means that the sum of the measures of the three angles in a triangle is 180°.

Cumulative Review

1. Find the GCF. 54; 63
2. Find the LCM. 3; 5
3. Compute. $3 \cdot 2 + 4 \cdot 3 + 5 \cdot 4$

1. 9   2. 15   3. 38

Polygons (Pages 379–380)

Teaching Hints The Venn diagram below may help students in classifying quadrilaterals.

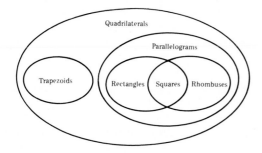

Enrichment Ask students to construct a square and a regular hexagon. Some may want to try the construction of a regular pentagon.

Cumulative Review

1. 18 is 40% of what number?
2. 39.9 is what percent of 70?
3. Jack earns $250.00 a week plus 12% commission on his sales. His sales last week were $975.00. Find his total earnings.

1. 45   2. 57%   3. $367.00

Problem Solving—Careers (Page 381)

Teaching Hints The real-life problems encountered by tile setters, involving area measure, are presented in this problem-solving lesson.

Square Roots (Pages 382–384)

Teaching Hints Before showing your students the table of square roots on page 475 have them approximate square roots. For example, approximate $\sqrt{72}$ to the nearest tenth. $8 \times 8 = 64$ and $9 \times 9 = 81$. 72 is about halfway between 64 and 81, so $\sqrt{72}$ is about halfway between 8 and 9, which is 8.5.

Cumulative Review

Solve.
1. $x + (-3) > -9$ 2. $x - (-4) < 15$

3. $x + 7 > -8$

1. $x > -6$ 2. $x < 11$ 3. $x > -15$

The Pythagorean Theorem (Pages 385–387)

Teaching Hints When developing the Pythagorean theorem, you might want to have students examine this model of the 3-4-5 right triangle.

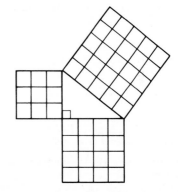

By counting the small squares, students can verify that $9 + 16 = 25$, or $3^2 + 4^2 = 5^2$.

Enrichment At this point students might want to generate Pythagorean triples. Help them to discover this pattern:

$$n, \frac{n^2 - 1}{2}, \frac{n^2 + 1}{2}$$

The pattern gives three numbers that satisfy the Pythagorean theorem when n is replaced by an odd number.

Cumulative Review

Divide, if possible.

1. $-45 \div 3$ **2.** $-35 \div (-7)$ **3.** $-95 \div 0$

1. -15 2. 5   3. not possible

Non-Routine Problems (Page 388)

Since there is no specific strategy for solving these problems, they are non-routine. Students should be reminded to read them several times, thoughtfully, before they attempt to solve these problems. Common sense and ingenuity should be applied. See the TE Answer Section for suggested solutions.

Computer Activities (Page 389)

The **BASIC** symbol for square root is **SQR( )**. That is, the square root of 16 is written in **BASIC** as **SQR(16)**. Note that there is no space between **SQR** and **(**. This will be used in finding the hypotenuse of a right triangle given the lengths of two legs. Review the Pythagorean theorem, and guide students to express the formula for finding the hypotenuse, $c = \sqrt{a^2 + b^2}$.

Chapter

Coordinate Geometry

Chapter Objectives

To graph integers, equations, and inequalities on a number line

To identify and graph ordered pairs for a point on a coordinate plane

To determine whether a line is horizontal or vertical, given the coordinates of two points on the line

To graph equations and inequalities on a coordinate plane

To find the slope of a line, given two points on the line

To find the slope and y-intercept of a line, given its equation

To solve a system of equations by graphing

Chapter Overview

In this chapter students graph on the number line and on the coordinate plane. They learn to find the slope and y-intercept from the equation and to solve systems of equations by graphing.

Chapter Opener (Page 392)

Problem Formulation

The picture shows one of the dams built by the Tennessee Valley Authority (TVA). The TVA was created in 1933 to build dams and reservoirs that would improve navigation and control floods on the Tennessee River. The nine dams in the following chart are on the Tennessee River.

Main River Projects	Dam Height (feet)	Dam Length (feet)	Lake Length (miles)	Lake Area (acres)
Kentucky	206	8,422	185	160,300
Pickwick Landing	113	7,715	53	43,100
Wilson	137	4,541	16	15,500
Wheeler	72	6,342	74	67,100
Guntersville	94	3,979	76	67,900
Nickajack	81	3,767	46	10,370
Chickamauga	129	5,800	59	36,400
Watts Bar	112	2,960	96	39,000
Fort Loudoun	122	4,190	61	14,600

Have students use the preceding data to formulate and solve at least two problems.

Estimation

Have students estimate the total area of the nine lakes. Have them round the areas to the nearest 10,000 acres.

Application

Have students formulate and solve problems related to the construction of dams.

Graphing on a Number Line (Pages 393–395)

Teaching Hints When graphing integers greater than -2, we use an arrow over the last integer to indicate that the points go on forever to the right.

Remind students that beginning with Example 4, we are dealing not only with the integers but all the numbers between each consecutive pair of integers. All of these numbers make up what are called the real numbers.

Preventing Common Errors When graphing the integers between two given integers, point out that the two given integers are not included.

Cumulative Review

1. Multiply. $3\frac{1}{5} \cdot \frac{10}{3}$ 2. Divide. $9\frac{1}{3} \div \frac{10}{9}$

3. Solve. $\frac{5}{6}x = \frac{7}{12}$

1. $10\frac{2}{3}$ 2. $8\frac{2}{5}$ 3. $\frac{7}{10}$

More on Graphing Inequalities (Pages 396–398)

Teaching Hints As students work through the four examples in this lesson, they should realize that the process of solving an inequality is identical to that of solving an equation, with the exception of the rule for multiplying or dividing each side by a negative number.

Note that it is always possible to avoid multiplying or dividing each side of an inequality by a negative number. Example 2 could be solved by first adding $3a$ to each side.

As students work the Calculator activity, they should should see that the square of the sum of two whole numbers is greater than the sum of the squares of the same two numbers.

Enrichment Give some examples where students can apply the multiplication property for inequalities. Graph:

1. $\frac{1}{2}x + 3 < -5$ 2. $-\frac{2}{3}y - 7 > 3$

Cumulative Review

1. Add. $84.7 + 101.35 + 0.007$

2. Subtract. 193.07
    $- 49.78$

3. Multiply. $(86.3)(1.005)$

1. 186.057 2. 143.29 3. 86.7315

Problem Solving—Applications (Pages 399–400)

Teaching Hints Real-life problems encountered at a gas station are presented in this lesson. Review the four-step problem-solving strategy as it applies to solving multiple-step problems. Emphasize the importance of reading the problem carefully and planning the solution that may require several steps.

Graphing Points (Pages 401–403)

Teaching Hints As a practical application of the coordinate plane, you might want to illustrate the system of numbered avenues and streets used in many cities.

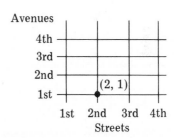

Corner of 2nd Street and 1st Avenue

Preventing Common Errors Students often consider ordered pairs such as (4, 3) and (3, 4) to be the same. To help students grasp the importance of the order in an ordered pair, you might graph several pairs of points such as (4, 3) and (3, 4) and have students give the ordered pairs. Stress that the x-coordinate is always first and the y-coordinate second.

Cumulative Review

1. Find the percent 65 is of 147 to the nearest tenth.
2. 18 is what percent of 90?
3. 20 is 40% of what number?

1. 44.2%   2. 20%   3. 50%

Horizontal and Vertical Lines (Pages 404–405)

Teaching Hints You might want to begin the lesson by reviewing the following concepts:

Horizontal Line	Vertical Line
parallel to x-axis	parallel to y-axis
perpendicular to y-axis	perpendicular to x-axis
same y-coordinate	same x-coordinate

Cumulative Review

1. Multiply. $(-19)(-30)$
2. Divide. $-4\frac{1}{3} \div \left(-\frac{5}{6}\right)$
3. Solve and check. $-9n = -63$

1. 570   2. $5\frac{1}{5}$ 3. 7

Graphing Equations and Inequalities (Pages 406–408)

Teaching Hints When graphing an equation such as $y = 3x - 5$ or an inequality such as $y \leq 3x - 5$, students are to draw a solid line through the points that satisfy the equation. When graphing an inequality such as $y < 3x - 5$, they are to draw a dotted line for $y = 3x - 5$ to show that it is not part of the graph of the inequality. For all inequalities the appropriate area is shaded after the line is drawn.

Cumulative Review

1. Write in mathematical terms. 10 less than x
2. Write an equation. A number plus 14 is 35.
3. Three times a number, increased by 8, is equal to 20. Find the number.

1. $x - 10$ 2. $n + 14 = 35$ 3. 4

Problem Solving—Careers (Page 409)

Teaching Hints Real-life problems that may be encountered by hospital technicians are presented in this lesson. Concepts taught thus far are reviewed.

Slope and Y-Intercept (Pages 410–412)

Teaching Hints As you go over Examples 1 and 2, guide students through the two interpretations of slope: (1) rise over run, and (2) the difference of the y-coordinates over the difference of the x-coordinates.

Preventing Common Errors As students find slopes, caution them to beware of positive and negative signs. Whichever way students choose to subtract the y-coordinates, they must subtract the x-coordinates in the same order. For two points (x_1, y_1) and (x_2, y_2) the slope is

$$\frac{y_2 - y_1}{x_2 - x_1} \quad \text{or} \quad \frac{y_1 - y_2}{x_1 - x_2}$$

Cumulative Review

1. Solve. $x - 8 = -4$ 2. Solve. $-6x = 42$
3. Solve. $-2x - 4 = 10$

1. 4   2. -7 3. -7

Solving Systems of Equations (Pages 413–414)

Teaching Hints Emphasize that the solution of a system of two equations in two variables is an *ordered pair of numbers,* not a point. The ordered pair is the coordinates of the point where the graphs of the two equations intersect.

Enrichment Ask students to solve the following system:

$$y = 3x + 1$$
$$y = 3x - 2$$

When the two graphs are drawn, students will find that the two lines are parallel. Since parallel lines do not intersect, there is no solution for the system.

Cumulative Review

1. Add. $-16 + (-19)$
2. Subtract. $-25 - (-14)$
3. Solve and check. $x + (-15) = -32$

1. -35 2. -11 3. -17

Computer Activities (Page 415)

The use of expressions like X1 may be strange to students. Point out that X2 must not be confused with X^2. The 2 in X2 is called a subscript; it appears in mathematics textbooks as x_2. However, since there is no way to lower the 2 on a computer keyboard we are forced to write the X and 2 alongside each other. Emphasize the importance of writing the **IF X2 − X1 = 0 THEN** . . . statement before line 40. Another alternative is to **LET N = Y2 − Y1** and **D = X2 − X1**. Then use the **IF D = 0 THEN** . . . statement. In this case the slope will be indicated by **LET S = N/D**.

INPUT (Pages 420–422)

Provide some motivation for using the word **INPUT** in programming. Ask students to look up the meaning of the word in the dictionary. The **INPUT** command distinguishes the capabilities of a computer from the calculator. The computer uses the general formula $S = 7 + 2 *Y$ to find the value of S for any value of Y. Simply type in the value of Y. You do not have to type in the instructions for how to find S for each new value of Y.

Some students may become confused when using two **INPUT** statements as in Example 2 on page 421. When the first **?** appears on the screen, the student may not be sure what to type in, **L** or **W**. Assure students that this problem will be solved in the lesson "More on **PRINT and INPUT**" on page 426. In this first lesson we chose to soften the introduction by keeping the typing down to a minimum. Students are easily frustrated in the beginning by their typing errors and lack of familiarity with the keyboard.

Loops (Pages 423–425)

Emphasize the following concepts when teaching loops;

1. There are *two* lines in using the loop:
   FOR I = 1 TO some number
   NEXT I
2. If **I** is used then any other variables in the program must differ from **I**.

Each time the computer reaches the **NEXT I** line, it goes back to the **FOR I = 1 TO some number** line until **I** equals that number plus one. Then it drops to the next command in the program. In the Examples in this lesson the command is **END**. The commands between and including the **FOR-NEXT** commands make up the **LOOP** in the program. Point out that other variables besides **I** can be used for the **FOR-NEXT** commands. However, once a variable has been chosen for the **FOR** command that same variable must be used in the **NEXT** command.

More on PRINT and INPUT (Pages 426–428)

Students tend to become confused by commands such as **PRINT "P = "P**. Emphasize that the computer will print exactly what is in quotes. For example, if P equals 5, the computer will print **P = 5**. The computer prints **P =** , as it appears inside the quotes. Then it prints the value of **P** because the final **P** appears outside the quotes in the command statement.

By now students should appreciate the value of the **enhanced INPUT**. The **enhanced INPUT** makes it clear just what is called for. When faced with **?** on the screen, the student might not know what to type in. But with the command **INPUT "WHAT IS P? "; P**, a specific question, **WHAT IS P?**, will appear on the screen when the program is **RUN**.

Emphasize the importance of leaving a space before the second " in an **enhanced INPUT** statement. If the space is not provided for, there will be no space between the printed **P** and the value of **P** on the screen. The printout will appear as **WHAT IS P?5.**

The IF . . . THEN Statement (Pages 429–431)

The **IF . . . THEN** and **GOTO** statements are two commands used for branching in computer programs. Branching allows the computer to perform numerous repetitive calculations. Conditional branching allows the computer to perform a variety of different tasks in different situations. The **IF . . . THEN** statement directs the computer to go to another line in the program, or to perform a given command, only when a specified condition is true. The **GOTO** statement is an unconditional command. It directs the computer to perform the instructions contained in another line in the program.

Sequential Input (Pages 432–433)

Emphasize that in sequential looping the **FOR** statement plays two roles:

1. determines the number of reentries into the loop;
2. provides input of data.

Point out that the variable in the **FOR** statement is the same variable used in the formula in line 30. Students will be surprised by the speed at which the output is produced when the program is **RUN.** They will complain that the program **RUNS** too rapidly; they cannot read the data. Show how this **RUN** can be slowed down by the use of a clock. The following illustrates the use of a **nested** loop: a loop **nested** within another loop. Ask students to list their programs. Then type the following two lines:

45 FOR J = 1 to 1,000
47 NEXT J

This will cause the computer to count from 1 to 1,000 before going to the next value of **A** in line 50. This slows the computer down. Have students list the program. Note that the two new lines will appear in proper sequence in the program. Have students experiment with the following changes in line 45:

FOR J = 1 TO 2,000 or **FOR J = 1 TO 500**

They will discover that increasing the number after **TO** slows the program, whereas decreasing the value speeds up the program.

INT (Page 434)

One of the main applications of the **INT** function is to provide the necessary tools for development of the number function, **RND,** in the next and last lesson. The **INT** function is used in Chapter 4, page 106, to write a program to simplify fractions. Emphasize that the **INT** function does not round decimals to the nearest whole number. The **INT** and **RND** of the next lesson are used together in Chapter 10, page 256, to write a program to simulate the tossing of a die.

RND (Page 435)

An excellent application of **RND** to probability is provided in Chapter 10, page 256. Note that the RND functions are defined differently on the TRS-80. The TRS-80 uses **RND(0)** for random numbers that are less than 1. On the TRS-80 random numbers from 1 to 6 can be generated by **RND(6)**; in this case the **INT** function is not necessary. The Apple will reject **RND(A)** if **A** is any number other than 1.

Answer Section

Page 38 ORAL EXERCISES

1. Divide each side by 6. **2.** Multiply each side by 12. **3.** Divide each side by 7.
4. Multiply each side by 4. **5.** Multiply each side by 4. **6.** Divide each side by 19.
7. Divide each side by 16. **8.** Multiply each side by 3.

Page 62 ORAL EXERCISES

1. First, subtract 4 from each side. Then divide each side by 2. **2.** First, add 6 to each side. Then divide each side by 3. **3.** First, subtract 4 from each side. Then divide each side by 8. **4.** First, add 4 to each side. Then divide each side by 2. **5.** First, subtract 5 from each side. Then multiply each side by 7.
6. First, add 4 to each side. Then multiply each side by 2.

Page 94 CHALLENGE

Separate the nine quarters into groups of three. There will be three groups. Place one group of three quarters on one side of the scale. Place the second group of three quarters on the other side of the scale. There will be one group of three quarters left over. Place this group of three quarters on the side. Look at the scale. If the scale balances, the counterfeit quarter is among the three quarters left on the side. If the scale does not balance, determine which is the lighter side. The group of quarters on that side of the scale contains the counterfeit quarter.

There are now two groups of quarters which you know do not contain the counterfeit coin and one group of three quarters which does. The problem has now been reduced to finding the counterfeit quarter among three quarters. Separate the group of quarters containing the counterfeit coin as follows. Place one coin on one side of the scale. Place a second coin on the other side of the scale. Place the third quarter on the side. Look at the scale. If the scale balances, then the counterfeit quarter is the one left on the side. If the scale does not balance, determine which is the lighter side. The quarter on that side must be counterfeit.

Page 106 EXERCISES

6.
```
100 HOME
110 INPUT "TYPE IN 1ST NUMERATOR ";X
120 INPUT "TYPE IN 1ST DENOMINATOR ";Y
130 INPUT "TYPE IN 2ND NUMERATOR ";Z
140 INPUT "TYPE IN 2ND DENOMINATOR ";W
150 LET A = X * Z
160 LET B = Y * W
170 IF A / B < = 1 THEN LET C = A: GOTO 190
180 LET C = B
190 FOR I = C TO 1 STEP - 1
200 LET N = A / I
210 LET D = B / I
220 IF N = INT (N) AND D = INT (D) THEN 250
230 NEXT I
240 PRINT : PRINT
250 PRINT A"/"B" = "N"/"D
260 IF D = 1 THEN PRINT "OR "N
270 END

JRUN
TYPE IN 1ST NUMERATOR 2
TYPE IN 1ST DENOMINATOR 5
TYPE IN 2ND NUMERATOR 4
TYPE IN 2ND DENOMINATOR 8
8/40 = 1/5
```

Page 124 EXERCISES

1. $10\frac{2}{5}$ 2. $14\frac{2}{3}$ 3. $13\frac{1}{2}$ 4. $5\frac{1}{3}$ 5. $1\frac{1}{2}$

6. $8\frac{4}{5}$ 7. $12\frac{19}{20}$ 8. $4\frac{14}{15}$ 9. $5\frac{5}{8}$ 10. $10\frac{13}{18}$

11. $5\frac{1}{4}$ 12. $3\frac{1}{2}$ 13. $7\frac{1}{8}$ 14. $6\frac{3}{10}$ 15. $1\frac{7}{12}$

16. $12\frac{1}{2}$ 17. $5\frac{1}{4}$ 18. $10\frac{5}{8}$ 19. $15\frac{1}{2}$ 20. $5\frac{3}{4}$

21. $8\frac{1}{2}$ 22. $12\frac{1}{3}$ 23. $16\frac{11}{12}$ 24. $10\frac{2}{9}$

25. $15\frac{3}{8}$ 26. $16\frac{1}{3}$ 27. $18\frac{1}{5}$ 28. 16

29. $11\frac{2}{5}$ 30. $14\frac{3}{8}$ 31. 14 32. $10\frac{11}{12}$

33. $17\frac{13}{24}$

Page 136 EXERCISES 5-8

```
10  HOME
20  INPUT "WRITE 1ST NUMERATOR ";X
30  INPUT "WRITE 1ST DENOMINATOR ";Y
40  INPUT "WRITE 2ND NUMERATOR ";Z
50  INPUT "WRITE 2ND DENOMINATOR ";W
60  LET A = X * W + Y * Z
70  LET B = Y * W
80  IF A / B < = 1 THEN  LET C = A: GOTO 100
90  LET C = B
100 FOR I = C TO 1 STEP  - 1
110 LET N = A / I
120 LET D = B / I
130 IF N =  INT (N) AND D =  INT (D) THEN 160
140 NEXT I
150 PRINT : PRINT
160 PRINT A"/"B" = "N"/"D
170 IF D = 1 THEN  PRINT "OR "N
180 END
```

Page 161 EXERCISES

2.
```
10  PRINT "LENGTH","WIDTH","PERIMETER"
20  FOR K = 1 TO 3
30  READ L,W
40  LET P = L * W
50  DATA  5,8,19,23,42,17
60  PRINT L,W,P
70  NEXT K
80  END
```

```
]RUN
```

LENGTH	WIDTH	PERIMETER
5	8	40
19	23	437
42	17	714

Page 167 ORAL EXERCISES

1. two and four tenths 2. thirty-six and twenty-one hundredths 3. forty-two and seven hundredths 4. five hundredths
5. two hundred seventy-one and six hundred thirty-two thousandths 6. three hundred six and one hundred forty-five ten-thousandths
7. seventy-nine and six thousandths 8. nine and twenty-four thousand eight hundred sixty-three hundred-thousandths 9. one thousand three hundred fifty-two and sixty-thousandths 10. four million six hundred twenty-eight thousand and nine tenths
11. five hundred ninety-two millionths
12. one hundred sixty and eighty-three hundred-thousandths

Page 174 EXERCISES

1. 0.0324 2. 231.68 3. 0.3392 4. 0.0104
5. 0.0418 6. 0.9072 7. 0.9331 8. 1.7408
9. 7.31 10. 3.38 11. 1.645 12. 0.1886
13. 362.1 14. 11.454 15. 36.33
16. 4.3734 17. 4.944 18. 2.75847
19. 0.14859 20. 0.02295 21. 0.32092
22. 0.0295386 23. 35.213

Page 174 CHALLENGE

First move	Move s to pole 2.
Second move	Move m to pole 3.
Third move	Move s on top of m.
Fourth move	Move L to pole 2.
Fifth move	Move s back to pole 1.
Sixth move	Move m on top of L at pole 2.
Seventh move	Move s on top of m on pole 2.

Page 183 ORAL EXERCISES

1. Subtract 1.8 from each side. 2. Divide each side by 0.7. 3. Subtract 12.6 from each side. 4. Add 4.2 to each side. 5. Multiply each side by 7.9.

6. Divide each side by 12.03. **7.** Add 3.8 to each side. **8.** Multiply each side by 1.07. **9.** Divide each side by 0.3. **10.** Multiply each side by 8.1. **11.** Add 9.2 to each side. **12.** Subtract 0.09 from each side.

Page 186 EXERCISES

2.
```
10  FOR K = 1 TO 3
20  INPUT "TYPE NUMBER SOLD ":N
30  IF N > = 8 THEN 60
40  LET T = 185
50  GOTO 70
60  LET T = 185 + 10 * (N - 7)
70  PRINT "TOTAL PAY IS "T
80  NEXT K
90  END
```

3.
```
10   FOR K = 1 TO 3
20   INPUT "TYPE NUMBER SOLD ":N
30   IF N > = 10 GOTO 90
40   IF N > 5 GOTO 70
50   LET T = 220 + 6 * N
60   GOTO 100
70   LET T = 220 + 8 * (N - 5) + 30
80   GOTO 100
90   LET T = 220 + 9 * (N - 9) + 62
100  PRINT "THE TOTAL PAY IS "T
110  NEXT K
120  END
```

Page 201 EXERCISES

1. 300 **2.** 80,000 **3.** 60,000 **4.** 7 **5.** 6
6. 4 **7.** 61,000 **8.** 420 **9.** 200 **10.** 0.76
11. 9.3 **12.** 390,000 **13.** 2,600 **14.** 76.4
15. 830,000 **16.** 79,000 **17.** 0.45 **18.** 3.8
19. 4.2 **20.** 340 **21.** 6,700 **22.** 3.75
23. 4,200 **24.** 0.71

Page 211 EXERCISES

1.
```
10  FOR K = 1 TO 6
20  READ C
30  LET F = 9 / 5 * C + 32
40  PRINT "FAHRENHEIT IS "F
50  NEXT K
60  DATA  10,8,26,45,67,90
70  END

]RUN
FAHRENHEIT IS 50
FAHRENHEIT IS 46.4
FAHRENHEIT IS 78.8
FAHRENHEIT IS 113
FAHRENHEIT IS 152.6
FAHRENHEIT IS 194
```

Page 242 EXERCISES

1. range: 9; mean: 89 **2.** range: 8; mean: 12
3. range: 8; mean: 18 **4.** range: 13; mean: 96
5. range: 5; mean: 13 **6.** range: 6; mean: 19
7. range: 32; mean: 38 **8.** range: 19; mean:
81 **9.** range: 9; mean: 72 **28.** range: 26;
mean: 38; median: 36; mode: 36 **29.** range: 2;
mean: 8.43; median: 9; mode: 9 **30.** range:
11; mean: 52.17; median: 52; mode: 52
31. range: 8; mean: $92.3\overline{3}$; median: 92.5;
mode: 91, 94 **32.** range: 7; mean: 89.6;
median: 90; mode: 87, 90 **33.** range: 3; mean:
8; median: 8; mode: 8 **34.** range: 2; mean:
9.14; median: 9; mode: 10 **35.** range: 4;
mean: 28.2; median: 28; mode: 30 **36.** range:
14; mean: $96.8\overline{3}$; median: 98.5; mode: 98, 100

Page 256 EXERCISES

4.
```
5   FOR X = 1 TO 5
10    LET H = 0
20    FOR K = 1 TO 200
30    LET R =  INT (2 * RND (1) + 1)
40    IF R = 1 THEN 60
50    PRINT "TOSS "K."TAILS COMES UP."
55    GOTO 80
60    LET H = H + 1
70    PRINT "TOSS "K."HEADS COMES UP."
80    NEXT K
90    PRINT "HEADS OCCURS "H" TIMES"
100   NEXT X
110   END
```

Page 272 CHALLENGE

1. The sentence is true because $\frac{1}{5} = \frac{1}{5}, \frac{1}{4} = \frac{1}{4}$, $\frac{1}{3} = \frac{1}{3}$, and $\frac{1}{6} < \frac{1}{2}$.   2. The sentence is true because $\frac{1}{9} > \frac{1}{10}, \frac{1}{7} > \frac{1}{8}, \frac{1}{5} > \frac{1}{6}$, and $\frac{1}{3} > \frac{1}{4}$.

3. The sentence is true because $\frac{1}{10} = 0.10$, $\frac{1}{20} = 0.05, \frac{1}{5} = 0.20$, and $\frac{1}{4} = 0.25$.

Page 293 ORAL EXERCISES

1. positive 3 add positive 9   2. positive 6 add positive 8   3. positive 9 add positive 4
4. negative 6 add negative 2   5. negative 7 add negative 5   6. positive 3 add negative 7
7. positive 1 add negative 9   8. negative 2 add positive 8   9. negative 7 add positive 3
10. positive 8 add negative 3   11. negative 2 add positive 2   12. positive 7 add negative 7
13. positive 9 add 0   14. 0 add negative 2
15. negative 9 add 0

Page 297 ORAL EXERCISES

1. positive 4 subtract positive 2
2. positive 7 subtract positive 5
3. positive 9 subtract positive 4
4. positive 5 subtract positive 2
5. positive 3 subtract positive 5
6. positive 2 subtract positive 8
7. positive 4 subtract positive 6
8. positive 5 subtract positive 7
9. negative 2 subtract positive 3
10. negative 1 subtract positive 4
11. negative 3 subtract positive 7
12. negative 4 subtract positive 6
13. negative 4 subtract negative 2
14. negative 1 subtract negative 1
15. negative 2 subtract negative 3
16. negative 7 subtract negative 2
17. positive 5 subtract negative 1
18. positive 3 subtract negative 3
19. positive 7 subtract negative 2
20. positive 4 subtract negative 6
21. 0 subtract negative 2
22. 0 subtract positive 5
23. 0 subtract negative 4
24. 0 subtract positive 1

Page 327

1. $-4x$ 2. $-13y$ 3. $13z$ 4. $-6a$
5. $-12c$ 6. $-4m$ 7. $-10n$ 8. $2c$ 9. $-3t$
10. $-12p$ 11. x 12. $-7x$ 13. $8u$
14. $-12v$ 15. 0 16. $14y$ 17. $18h$
18. $19k$ 19. $2r$ 20. $-10d$ 21. $21m$
22. $45d$ 23. $-14b$ 24. $-72n$ 25. $36t$
26. $6x + 9$ 27. $10y - 2$ 28. $-8x - 2$
29. $-12z + 8$ 30. $12h + 6$ 31. $4n - 8$
32. $3p + 21$ 33. $-5x + 15$ 34. $-2y - 4$
35. $7a - 7$ 36. $-r + 3$ 37. $-8 + z$
38. $-4 + 2b$ 39. $-4 - 3w$ 40. $-5y + 3$
41. $-16m + 8p + 5$ 42. $-6m - 6n - 10$

Page 329 CHALLENGE

1. Let x = the number of chickens. Let y = the number of goats. $x + y = 43$ $2x + 4y = 108$ $2(43 - y) + 4y = 108$ $86 - 2y + 4y = 108$ $86 + 2y = 108$ $2y = 108 - 86$ $2y = 22$ $y = 11$ $x = 43 - 11 = 32$ There are 11 goats and 32 chickens.   2. Fill up the 4-liter container and then pour it into the 3-liter container. The 1 liter left in the 4-liter container should be poured into the 7-liter container. Now fill up the 4-liter container and then pour it into the 7-liter container, too. Now there are 5 liters in the 7-liter container.

Page 349 EXERCISES

1. $x > 2$ 2. $y \le 1$ 3. $z \le 4$ 4. $a > 3$
5. $c \ge -4$ 6. $w < -2$ 7. $d \ge -5$ 8. $x < -1$ 9. $m < -7$ 10. $e \le -5$ 11. $y > -8$
12. $n \le -6$ 13. $r > 8$ 14. $k \ge 5$
15. $z \ge 13$ 16. $b < 8$ 17. $s < 4$ 18. $n < 3$
19. $p \ge 3$ 20. $a \ge 1$ 21. $v \ge -9$
22. $x \le -2$ 23. $t < -2$ 24. $z > -4$
25. $n \ge -3$ 26. $x < 2$ 27. $y \le -3$
28. $a > -6$ 29. $k < -2$ 30. $n \ge 1$
31. $d \le 3$ 32. $z > -2$ 33. $m \ge -5$
34. $x < 1$ 35. $c > -4$ 36. $s \ge -1$
37. $t < 2$ 38. $s \le 4$ 39. $v > -2$ 40. $n \ge 1$
41. $x > -2$ 42. $y > 6$ 43. $a \ge -10$
44. $m < -28$ 45. $a \le -12$ 46. $c \ge 10$
47. $d < -30$ 48. $w > 12$ 49. $c > 12$
50. $s \ge -5$ 51. $y \le 6$ 52. $z > -40$

Page 357 EXERCISES

1. binomial 2. trinomial 3. monomial
4. binomial 5. trinomial 6. monomial
7. monomial 8. trinomial 9. binomial
10. trinomial 11. binomial 12. monomial
13. $3x^2 - 7x + 3$ 14. $7y^2 - y + 4$
15. $a^2 + 3a - 5$ 16. $5n^2 + 3n - 1$
17. $-3y^2 + 7$ 18. $2x + 7$ 19. $-4y^2 + 11y - 2$ 20. $-4a^2 - 4a - 1$ 21. $3x^2 - 5x - 1$ 22. $3c^2 - c - 4$ 23. $5z^2 - 2z + 4$
24. $2x^2 - 9$ 25. $-6y^2 - y + 3$ 26. $-9n + 5$ 27. $4y^3 - 2y^2 + 5y - 3$ 28. $-2r^4 + 7r^2 - 3r - 3$ 29. $5x^3 - 6x^2 + 2x + 2$
30. $3x^4 - 5x + 4$ 31. $7c^3 - 2c - 5$
32. $-2a^4 + 8a^2 + 5a - 9$ 33. $-7z^3 - 2z^2 + 6z$ 34. $5a^3 + 2a - 1$ 35. $4x^4 - 8x^3 + 2x^2 + 2x - 3$ 36. $3a^3 - 9a^2 + 2a + 12$

Page 359 EXERCISES

1. $3x^3 + 2x$ 2. $4y^2 - y$ 3. $-5a^3 + a^2$
4. $2x^3 + 2x$ 5. $6c^2 - 6c$ 6. $-8y^3 + 12y^2$
7. $15c^3 + 6c^2 - 9c$ 8. $-4x^3 - 24x^2 - 8x$
9. $5a^3 + 25a^2 - 10a$ 10. $2z^4 - 8z^3 + 4z^2$
11. $3y^4 + 4y^3 - 5y^2$ 12. $3b^4 - 6b^3 + 3b^2$
13. $-x^2 - 3x$ 14. $-r^2 + 6r - 2$
15. $-3c^2 - 2c - 5$ 16. $2a^2 - 4a + 3$
17. $-6x^2 + 5x + 4$ 18. $4y^2 + y - 1$
19. $5x^2 - 10x - 13$ 20. $4y^2 - 9y + 1$

21. $4z^2 - 16z + 5$ 22. $22c + 20$
23. $-3x^2 - 27x + 8$ 24. $y^2 - 4y - 17$
25. $2a^2 + 6a - 3$ 26. $-c^2 - 10c + 10$
27. $-3z^2 - z + 3$ 28. $-5y^2 + 1$
29. $-5x^2 - 2x + 2$ 30. $2z^2 - 10z + 1$
31. $5x^4 - 16x^3 + 10x^2$
32. $13y^4 - 13y^3 + 9y^2$ 33. $10x^4 - 23x^3 - 20x^2 + 5x$ 34. $29a^4 + 38a^3 - 4a^2$

Page 362 EXERCISES

1. $4(n + 2)$ 2. $3(2x - 3)$ 3. $3(c + 4)$
4. $2(4y - 1)$ 5. $2(d - 5)$ 6. $5(3y + 1)$
7. $7(x^2 - 2)$ 8. $8(2c^2 + 1)$ 9. $2(x^2 + 2x - 4)$
10. $3(y^2 - 3y + 2)$ 11. $5(x^2 - 2x + 4)$
12. $4(a^2 - 3a - 5)$ 13. $3(2z^2 + 4z + 1)$
14. $2(4x^2 - 3x + 1)$ 15. $5(2n^2 + n - 3)$
16. $4(3y^2 - 2y - 1)$ 17. $y(y + 2)$
18. $x(x - 3)$ 19. $c^2(c + 6)$ 20. $n(n^2 - 5)$
21. $z(3z^2 - 2z + 4)$ 22. $y(7y^2 + 2y - 3)$
23. $n(6n^2 - 2n + 1)$ 24. $x(5x^2 - 4x - 1)$
25. $2x(x + 2)$ 26. $3y(y + 2)$ 27. $2a(4a - 1)$
28. $10d(d - 2)$ 29. $7n(2n + 1)$
30. $4y^2(2y + 1)$ 31. $6n^2(n - 2)$
32. $5b^2(3b - 1)$ 33. $2x(2x^2 + x - 3)$
34. $3c(c^2 - 2c + 4)$ 35. $5r(r^2 + 3r - 2)$
36. $4y(y^2 - 2y - 3)$ 37. $2c(5c^2 - 2c + 1)$
38. $10n(2n^2 - 3n + 1)$ 39. $6a(a^2 + 3a - 1)$
40. $2x(2x^3 - 4x^2 + 3x - 1)$ 41. $3y(y^3 - 3y^2 + 2y + 1)$ 42. $5c(2c^3 - 4c^2 + 6c - 1)$
43. $6n(2n^3 - n^2 - 3n + 1)$

Page 374 PRACTICE

vertical angles: $\angle 2 \cong \angle 3$, $\angle 1 \cong \angle 4$, $\angle 6 \cong \angle 7$, $\angle 5 \cong \angle 8$; alternate interior angles: $\angle 4 \cong \angle 5$, $\angle 3 \cong \angle 6$; corresponding angles: $\angle 2 \cong \angle 6$, $\angle 1 \cong \angle 5$, $\angle 4 \cong \angle 8$, $\angle 3 \cong \angle 7$

Page 389 EXERCISES

4.
```
10  INPUT "TYPE IN LENGTH OF LEG ";A
20  INPUT "TYPE IN LENGTH OF HYPOTENUSE ";C
30  LET B = SQR (C ^ 2 - A ^ 2)
40  PRINT "THE LENGTH OF THE OTHER LEG IS ";B
50  END


]RUN
TYPE IN LENGTH OF LEG 10
TYPE IN LENGTH OF HYPOTENUSE 26
THE LENGTH OF THE OTHER LEG IS 24
```

2. All four points are collinear; three points are collinear; any two points determine exactly one line.   **5. a.** Answers may vary. Some samples are given. Vertical angles: $\angle 2 \cong \angle 4$, $\angle 6 \cong \angle 8$; alternate interior angles: $\angle 3 \cong \angle 5$, $\angle 4 \cong \angle 6$; corresponding angles: $\angle 2 \cong \angle 6$, $\angle 1 \cong \angle 5$ **7.** rectangle: four right angles; rhombus: four congruent sides; square: four congruent sides and four right angles   **8. a.** right: one right angle; scalene: no two sides congruent   **b.** obtuse: one obtuse angle; isosceles: two congruent sides   **c.** acute: all angles acute; equilateral: all sides congruent

Page 391   CHAPTER TEST

5. a. Answers may vary. Some samples are given. vertical angles: $\angle 1 \cong \angle 4$, $\angle 6 \cong \angle 7$; alternate interior angles: $\angle 3 \cong \angle 6$, $\angle 4 \cong \angle 5$; corresponding angles: $\angle 2 \cong \angle 6$, $\angle 4 \cong \angle 8$ **8. a.** obtuse: one obtuse angle; isosceles: two congruent sides   **b.** right: one right angle; scalene: no two sides congruent   **c.** acute: all angles acute; equilateral: all sides congruent

Page 398

1. all reals > 5 **2.** all reals < -3 **3.** all reals ≤ 1 **4.** all reals > -2 **5.** all reals < 3 **6.** all reals ≥ -3 **7.** all reals ≥ 0 **8.** all reals > 1 **9.** all reals ≥ 2 **10.** all reals < -7 **11.** all reals > 3 **12.** all reals ≥ 4 **13.** all reals ≥ 3 **14.** all reals > 3 **15.** all reals < 5 **16.** all reals > -1 **17.** all reals ≤ 4 **18.** all reals ≥ 2 **19.** all reals > 3 **20.** all reals < -4 **21.** all reals ≤ -4 **22.** all reals ≥ 0 **23.** all reals ≤ 5 **24.** all reals > -3 **25.** all reals < 3 **26.** all reals ≥ 9 **27.** all reals > -2 **28.** all reals < 4 **29.** all reals ≤ 6 **30.** all reals ≥ 4

31. all reals > -6 **32.** all reals < -7 **33.** all reals ≥ 5 **34.** all reals ≥ 1 **35.** all reals > 1 **36.** all reals < 5 **37.** all reals ≤ -5 **38.** all reals < -2 **39.** all reals < -1 **40.** all reals ≤ 1 **41.** all reals < 4 **42.** all reals ≥ 5

COMPUTER SECTION
Page 422   EXERCISES

12.
```
10  INPUT S
20  LET A = S ^ 2
30  PRINT A
40  END

]RUN
?6
36
```

13.
```
10  INPUT S
20  LET B = .006 * S ^ 2
30  PRINT B
40  END

]RUN
?45
12.15
```

Page 425   EXERCISES

6.
```
10  FOR K = 1 TO 2
20  INPUT P
30  INPUT R
40  INPUT T
50  LET I = P * R * T
60  PRINT I
70  NEXT K
80  END
```

7.
```
10  FOR K = 1 TO 5
20  INPUT X
30  LET Y = 4 * X ^ 2 + 3 * X
40  PRINT Y
50  NEXT K
60  END
```

Page 428   EXERCISES

15.
```
10  PRINT "PROGRAM FINDS QUOTIENTS OF ANY FOUR
        PAIRS OF NONZERO NUMBERS"
20  FOR I = 1 TO 4
30  INPUT "TYPE IN DIVIDEND ":N
40  INPUT "TYPE IN DIVISOR ":D
50  LET Q = N / D
60  PRINT N" DIVIDED BY "D" IS "Q
70  NEXT I
80  END
```

Page 430   EXERCISES

4.
```
10   FOR K = 1 TO 3
20   INPUT "TYPE IN NUMBER OF PENS ":N
30   IF N > 6 THEN 60
40   LET T = 1.50 * N
50   GOTO 70
60   LET T = 1.25 * N
70   PRINT "TOTAL COST IS "T
80   PRINT : PRINT
90   NEXT K
100  END
```

```
]RUN
TYPE IN NUMBER OF PENS 2
TOTAL COST IS 3


TYPE IN NUMBER OF PENS 8
TOTAL COST IS 10


TYPE IN NUMBER OF PENS 6
TOTAL COST IS 9
```

Page 431   EXERCISES

12.
```
10   FOR I = 1 TO 2
20   INPUT "TYPE IN THE FIRST GRADE ":F
30   INPUT "TYPE IN THE SECOND GRADE ":S
40   INPUT "TYPE IN THE THIRD GRADE ":T
50   INPUT "TYPE IN THE FOURTH GRADE ":L
60   LET A = (F + S + T + L) / 4
70   PRINT "THE AVERAGE IS "A
80   IF A < 70 THEN 90
85   GOTO 100
90   PRINT "SEND A WARNING NOTICE"
100  NEXT I
110  END
```

```
]RUN
TYPE IN THE FIRST GRADE 72
TYPE IN THE SECOND GRADE 60
TYPE IN THE THIRD GRADE 80
TYPE IN THE FOURTH GRADE 60
THE AVERAGE IS 68
SEND A WARNING NOTICE
TYPE IN THE FIRST GRADE 80
TYPE IN THE SECOND GRADE 90
TYPE IN THE THIRD GRADE 90
TYPE IN THE FOURTH GRADE 80
THE AVERAGE IS 85
```

Page 433   EXERCISES

1.
```
10  PRINT "NUMBER","SQUARE OF NUMBER"
20  FOR X = 4 TO 16 STEP 3
30  LET S = X ^ 2
40  PRINT X,S
50  NEXT X
60  END
```

```
]RUN
NUMBER          SQUARE OF NUMBER
4               16
7               49
10              100
13              169
16              256
```

2.
```
10  PRINT "NUMBER"."CUBE OF NUMBER"
20  FOR X = 1 TO 9
30  LET C = X ^ 3
40  PRINT X.C
50  NEXT X
60  END
```

```
]RUN
NUMBER           CUBE OF NUMBER
1                1
2                8
3                27
4                64
5                125
6                216
7                343
8                512
9                729
```

4.
```
10  PRINT "VALUE OF X"."VALUE OF X^2 + 3"
20  FOR X = 1 TO 37 STEP 6
30  LET Y = X ^ 2 + 3
40  PRINT X.Y
50  NEXT X
60  END
```

```
]RUN
VALUE OF X       VALUE OF X^2 + 3
1                4
7                52
13               172
19               364
25               628
31               964
37               1372
```

3.
```
10  PRINT "VALUE OF B"."VALUE OF 4B + 3"
20  FOR B = 6 TO 16 STEP 2
30  LET Y = 4 * B + 3
40  PRINT B.Y
50  NEXT B
60  END
```

```
]RUN
VALUE OF B       VALUE OF 4B + 3
6                27
8                35
10               43
12               51
14               59
16               67
```

5.
```
10  PRINT "NUMBER"."SQUARE OF NUMBER"
20  FOR X = 14 TO 2 STEP  - 2
30  LET Y = X ^ 2
40  PRINT X.Y
50  NEXT X
60  END
```

```
]RUN
NUMBER           SQUARE OF NUMBER
14               196
12               144
10               100
8                64
6                36
4                16
2                4
```

Solutions to Non-Routine Problems
Chapter 1 PAGE 21

1. Make a table. Point out to students that the maximum number of $50 bills is 10 (10 × $50 = $500).

Number of $50 bills	Number of $10 bills	Total number of bills	Total value
10	10	20	$600
9	11	20	$560
8	12	20	$520

There are eight $50 bills and twelve $10 bills in the drawer.

2. Make a table.

Week	Amount saved	Week	Amount saved
1	$0.01	9	$2.56
2	$0.02	10	$5.12
3	$0.04	11	$10.24
4	$0.08	12	$20.48
5	$0.16	13	$40.96
6	$0.32	14	$81.92
7	$0.64	15	$163.84
8	$1.28	16	$327.68

The amount saved by the above method is $655.35. If you save $60 per week for 16 weeks, the amount is $960.00. So, the friend is wrong.

3. This is a trick question. Note that:

1 costs 10¢	54 cost 20¢
700 cost 30¢	5,000 cost 40¢

Observe that each digit costs 10¢. Thus, the friend might be buying digits, perhaps to put them on a house as an address.

4. Have students choose a convenient number to work with, for example, $100.

	Henry	Peter
Start:	$100	$100
Last year:	$100 + $\left(\frac{1}{10} \cdot \$100\right)$	$100 − $\left(\frac{1}{10} \cdot \$100\right)$
	$100 + 10 = $110	$100 − 10 = $90
This year:	$110 − $\left(\frac{1}{10} \cdot \$110\right)$	$90 + $\left(\frac{1}{10} \cdot \$90\right)$
	$110 − 11 = $99	$90 + 9 = $99

Henry and Peter now earn the same amount.

5. Have students use the trial-and-error method for solving this problem. Have a student guess the cost of the math book, for example, $5.00. Then the science book would cost $6.50 and the social studies book would cost $7.50. So, $5.00 + 6.50 + 7.50 = $19.00. The guess was too low. Have a student take another guess for the cost of the math book, for example, $8.00. Then the science book would cost $9.50 and the social studies book would cost $10.50. So, $8.00 + 9.50 + 10.50 = $28.00. The guess is still too low. However, if you add $1.00 to the cost of each book, the total will be $31.00. So, the math book costs $9.00, the science books costs $10.50, and the social studies book costs $11.50.

Chapter 3 PAGE 74

1. Make a table.

Age of first child	Age of second child	Sum	Product
5	10	15	50
4	11	15	44
3	12	15	36

So, the children's ages are 3 years and 12 years.

2. Follow these steps.

   (1) Double the number of pennies in each of the two piles.
                 Pile 1: $10 \cdot 2 = 20$ Pile 2: $18 \cdot 2 = 36$
   (2) Add these two amounts. $20 + 36 = 56$
   (3) Subtract this amount from 70. $70 - 56 = 14$

So, double the number of pennies in the third pile is 14. Therefore, there are 7 pennies in the third pile.

3. Make a table. Point out to students that since there are 36 heads in all, there are 36 animals in all. Therefore, if there are 3 cows, there must be 33 chickens, and so on.

Cows	Legs	Chickens	Legs	Total Heads	Total Legs
3	12	33	66	36	78
4	16	32	64	36	80
5	20	31	62	36	82
6	24	30	60	36	84

So, there are 6 cows and 30 chickens.

4. Make a table.

Math Boxes	Math Books	Science Boxes	Science Books	Total Books
7	84	2	32	116
4	48	2	32	80
5	60	2	32	92
5	60	3	48	108
3	36	6	96	132
2	24	6	96	120
1	12	6	96	108

So, there are two possible solutions: 5 boxes are math and 3 are science, or
1 box is math and 6 are science.

5. Make a table.

Numbers	Quotient	Difference
10 and 5	2	5
8 and 4	2	4
6 and 3	2	3
4 and 2	2	2

So, the numbers are 4 and 2.

6. Make a table of the amounts in the savings accounts.

Week	Lester	Rhonda
1	$23 + 1 = $24	$50 + 1 = $51
2	$24 + 1 = $25	$51 + 1 = $52
3	$25 + 1 = $26	$52 + 1 = $53
4	$26 + 1 = $27	$53 + 1 = $54

So, after 4 weeks, the amount in Rhonda's account will be twice the amount
in Lester's account.

7. Make a table.

Silver buttons	Cost of silver buttons	Gold buttons	Cost of gold buttons	Total cost
10	$0.80	10	$1.00	$1.80
15	$1.20	15	$1.50	$2.70
20	$1.60	15	$1.50	$3.10

So, there are 20 silver buttons and 15 gold buttons in the package.

Note that for problems 1–3, students' solutions may be different from the sample solutions.

1. $3 = \dfrac{4 + 4 + 4}{4}$

$5 = \sqrt{4} + \sqrt{4} + \dfrac{4}{4}$

$7 = 4 + 4 - \dfrac{4}{4}$

$9 = 4 + 4 + \dfrac{4}{4}$

$4 = \sqrt{4} + \sqrt{4} + 4 - 4$

$6 = 4 + 4 - 4 + \sqrt{4}$

$8 = \sqrt{4} + \sqrt{4} + \sqrt{4} + \sqrt{4}$

$10 = 4 + 4 + 4 - \sqrt{4}$

2. a. $6 = 9 - \dfrac{9}{\sqrt{9}}$

c. $81 = 9 \cdot \sqrt{9} \cdot \sqrt{9}$

e. $4 = \dfrac{\sqrt{9} + 9}{\sqrt{9}}$

g. $12 = \sqrt{9} \cdot \sqrt{9} + \sqrt{9}$

b. $18 = \sqrt{9} \cdot \sqrt{9} + 9$

d. $27 = \sqrt{9} \cdot \sqrt{9} \cdot \sqrt{9}$

f. $24 = 9 \cdot \sqrt{9} - \sqrt{9}$

h. $10 = 9 + \dfrac{9}{9}$

3. a. $0 = \dfrac{6 - 6}{6}$

c. $60 = 66 - 6$

e. $11 = \dfrac{66}{6}$

g. $7 = 6\dfrac{6}{6}$

b. $6 = 6 + 6 - 6$

d. $18 = 6 + 6 + 6$

f. $30 = 6 \cdot 6 - 6$

h. $5 = 6 - \dfrac{6}{6}$

4.

Transaction	Status
buy for $4.00	$-\$4$
sell for $5.00	$+\$1 \ (-4 + 5)$
buy for $6.00	$-\$5 \ (+1 - 6)$
sell for $7.00	$+\$2 \ (-5 + 7)$

Therefore, $2.00 was made on the entire transaction.

5. Make a table.

Age of first child	Age of second child	Product	Sum
2	16	32	18
4	8	32	12

So, their ages are 4 years and 8 years.

6. Use the trial-and-error method of solution. Make a guess, for example, 10.

Check: $\dfrac{10 \times 6}{9} + 7 = \dfrac{60}{9} + 7$ This is not a whole number and is less than 14.

Try a greater number, for example, 12.

Check: $\dfrac{12 \times 6}{9} + 7 = 15$ So, the number is 12.

7. Make a table of amounts in the savings accounts.

Week	Sandy	Maria
1	$8 + 1 = $9	$12 + 3 = $15
2	$9 + 1 = $10	$15 + 3 = $18
3	$10 + 1 = $11	$18 + 3 = $21
4	$11 + 1 = $12	$21 + 3 = $24

So, after 4 weeks, the amount in Maria's account is twice the amount in Sandy's account.

Chapter 7   PAGE 185

1. There are 500 1's in the difference, so the difference is 500.

2. Have students suppose the three parts are 1, 3, and 8. Follow these steps.
   (1) Find the sum. $1 + 3 + 8 = 12$
   (2) Divide. $96 \div 12 = 8$
   (3) Multiply each guess by 8. $1 \cdot 8 = 8; 3 \cdot 8 = 24; 8 \cdot 8 = 64$
The parts are 8, 24, and 64. These numbers are in the ratio 1 to 3 to 8 and the sum is 96.

3. Make a table.

Guess	Conditions	Sum
16 and 24	$\frac{1}{2} \cdot 16 = 8; \frac{1}{3} \cdot 24 = 8$	40
32 and 48	$\frac{1}{2} \cdot 32 = 16; \frac{1}{3} \cdot 48 = 16$	80
40 and 60	$\frac{1}{2} \cdot 40 = 20; \frac{1}{3} \cdot 60 = 20$	100
42 and 63	$\frac{1}{2} \cdot 42 = 21; \frac{1}{3} \cdot 63 = 21$	105

So, the numbers are 42 and 63.

4. Have the students guess an amount, for example, $7.

1st son: $\frac{1}{2} \cdot \$7 + \$0.50 = \$4.00$ (left $3.00)

2nd son: $\frac{1}{2} \cdot \$3 + \$0.50 = \$2.00$ (left $1.00)

3rd son: $\frac{1}{2} \cdot \$1 + \$0.50 = \$1.00$ (left 0)

So, the father gave away $7.

5. If the number triples each minute, then it was one-third full one minute before the 15 minutes. It means that the bottle was one-third full after 14 minutes.

6. Try three numbers, for example, 7, 8, and 9. Since 8 has a factor of 2 and 9 has a factor of 3, the product has a factor of 6 and therefore is divisible by 6. Try another three numbers, for example, 19, 20, and 21. The same argument applies here. This argument applies to each triplet of consecutive numbers with the exception of 0, 1, and 2.

Chapter 9   PAGE 235

1. Ms. Sobel traveled 100 mi during the first 2 h. So, she also traveled 100 mi on the return trip, but at 40 mi/h.
Compute the time for the return trip, using the distance formula.

$t = \dfrac{100}{40}$ or $2\dfrac{1}{2}$ h   The total trip was 200 mi and took $4\dfrac{1}{2}$ h.

Now compute the average speed for the entire trip, using the distance formula.

$\dfrac{200}{4\frac{1}{2}} = \dfrac{200}{\frac{9}{2}} = \dfrac{400}{9}$ or $44\dfrac{4}{9}$ mi/h

2. To cover 2 mi at the speed of 40 mi/h would take $\dfrac{2}{40}$ or $\dfrac{1}{20}$ h, which is 3 min. Traveling 1 mi at 20 mi/h takes $\dfrac{1}{20}$ h or 3 min. Since the entire 3 minutes are used up on the first mile, it is impossible to have the speed of 40 mph for the 2 miles.

3. The Osbornes start traveling when the Dearings have covered 30 miles. Since the Osbornes are traveling 10 mi/h faster, they close the 30-mile gap by 10 miles each hour. It will take $\dfrac{30}{10}$ or 3 h to close the gap. So, the Osbornes will catch up with the Dearings 3 h after the Osbornes began their trip, or 4 h after the Dearings began theirs.

4. Both cars together cover 80 mi/h. They have to travel 400 mi, so it will take them $\dfrac{400}{80}$ or 5 h.

5. To travel 400 mi with the speed of 50 mi/h, the train needs $\dfrac{400}{50}$ or 8 h. Since the train used 5 h on the first 200 mi, it has 8 − 5, or 3 h left for the remaining 200 mi. It must travel with the speed of $\dfrac{200}{3}$ or $66\dfrac{2}{3}$ mi/h.

6. Since the first train covered 90 miles during 2 hours, it traveled with the speed of $\dfrac{90}{2}$ or 45 mi/h. Since the second train's speed is 10 mi/h faster, it is traveling at 55 mi/h. So, the second train covered 110 miles in 2 hours. Thus, the trains were 90 + 110, or 200 miles apart when they started.

7. Your friend travels at 12 mi/h, so it will take your friend $\frac{3}{12}$ or $\frac{1}{4}$ h to cover the 3 mi. This is 15 min. You will take $\frac{3}{15}$ or $\frac{1}{5}$ h, that is, 12 min to cover the distance. Since it takes you $15 - 12$, or 3 min less time, you should start out 3 min later in order for both of you to arrive at the same time.

Chapter 11 PAGE 283

1. List all possible combinations.
(1) 1¢ + 5¢ = 6¢   (6) 5¢ + 25¢ = 30¢
(2) 1¢ + 10¢ = 11¢   (7) 5¢ + 50¢ = 55¢
(3) 1¢ + 25¢ = 26¢   (8) 10¢ + 25¢ = 35¢
(4) 1¢ + 50¢ = 51¢   (9) 10¢ + 50¢ = 60¢
(5) 5¢ + 10¢ = 15¢   (10) 25¢ + 50¢ = 75¢
So, there are 10 different amounts of money you can obtain by forming 2-coin combinations.

2. There are 2 possible routes from Niceville to Warsaw, and for each of these there are 3 different routes from Warsaw to Farmville. Thus, there are altogether $2 \cdot 3$, or 6 different routes from Niceville to Farmville.

3. a. To form the smallest possible number, use the smaller digits for the larger place values. 12,359
b. To form the largest possible number, use the larger digits for the larger place values. 95,321
c. There are 5 choices for the first place, 4 choices for the second place, 3 choices for the third, 2 choices for the fourth, and 1 choice for the fifth place. Thus, there are $5 \cdot 4 \cdot 3 \cdot 2 \cdot 1$, or 120 choices, and 120 different numbers can be formed.
d. There are 5 choices for the first place, 4 choices for the second, 3 choices for the third, and 2 choices for the fourth. Therefore, there are $5 \cdot 4 \cdot 3 \cdot 2$, or 120 choices, and 120 different numbers can be formed.
e. There are 5 choices for the first place, 4 choices for the second place, and 3 choices for the third place. Therefore, there are $5 \cdot 4 \cdot 3$, or 60 choices, and 60 different numbers can be formed.
f. There are 5 choices for the first place and 4 choices for the second place. Therefore, there are $5 \cdot 4$, or 20 choices, and 20 different numbers can be formed.

4. There are 5 colors (choices) available for the top portion. For each of the colors, there are 4 colors (choices) available for the bottom portion. So, altogether there can be formed $5 \cdot 4$, or 20 two-color flags.

5. There are 26 choices for the letter spot. For each of the 26 letters, there are 10 choices of digits. So, there are 26 · 10, or 260 different one letter-one digit combinations. For each of these plates, there are 10 choices of digits available for the second digit. So, there are 260 · 10, or 2,600 different one letter-two digit plates.

Chapter 13   *PAGE 330*

Let *a*, *b*, and *c* be the variables for numbers arranged as follows:

1. The pattern is $(a \div b) + c$.
   So, the missing number is $(6 \div 3) + 4$, or 6.
2. The pattern is $(a + b) \cdot c$.
   So, the missing number is $(2 + 3) \cdot 4$, or 20.
3. The pattern is $(a - b) \div c$.
   So, the missing number is $(18 - 2) \div 2$, or 8.
4. The pattern is $(a \cdot b) + c$.
   So, the missing number is $(3 \cdot 2) + 0$, or 6.
5. The pattern is $(a \cdot b) - c$.
   So, the missing number is $(6 \cdot 1) - 2$, or 4.
6. The pattern is $a \div (b + c)$.
   So, the missing number is $16 \div (2 + 2)$, or 4.

Assignments for Reinforcement and Maintenance

The following is a suggested schedule of assignments for 170 school days, allowing some days for testing. It is designed to provide a program of reinforcement and maintenance of the major concepts and skills.

Assignments are arranged on two levels:
Level 1 for the basic program;
Level 2 for an enriched program.

Each day's assignment consists of three parts:
(1) Selected practice for the current lesson;

(2) Selected maintenance for the previous lesson;

(3) Selected maintenance for the lesson taught two days prior to the current lesson.

It is intended that the teacher exercise his or her judgment in using this reinforcement and maintenance program to suit particular needs.

In addition to this program, optional materials are provided. These materials can be used for further enrichment in different ways:

1. additional examples for teaching the lesson;

2. Practice Exercises for seat work or chalkboard work;

3. items for quizzes;

4. Practice Exercises for extra help outside of class time;

5. extra testing;

6. Enrichment Exercises.

	Optional Materials	
Practice	**Testing**	**Enrichment**
Cumulative Reviews in Pupils' Edition and Teacher's Edition Extra Practice Exercises Algebra Maintenance	Math Aptitude Tests Competency Tests Extra Chapter Tests Cumulative Review Tests (Essay & Multiple-Choice) Final Tests	Calculator Challenge Computer Activities Activities, Projects, and Enrichment in Teacher's Edition Non-Routine Problems

Day	Level 1	Level 2
1	2 Ex. Odd 1–25	2 Ex. Even 2–24
2	4 Ex. Even 2–26 2 Ex. 2, 18, 22	4 Ex. Odd 1–27 2 Ex. 5, 17, 23
3	7 Ex. Odd 1–29 4 Ex. 1, 3, 13, 19 2 Ex. 10, 20, 24	7 Ex. Even 2–30, 31, 32 4 Ex. 2, 4, 16, 22 2 Ex. 7, 19, 25
4	10 Ex. Odd 1–23 7 Ex. 2, 10, 20 4 Ex. 5, 23, 25	10 Ex. Even 2–24, 25 7 Ex. 1, 9, 19 4 Ex. 6, 24, 26
5	12 Ex. Odd 1–29 10 Ex. 2, 6, 22 7 Ex. 4, 12, 22	12 Ex. Even 2–30 10 Ex. 1, 7, 23 7 Ex. 3, 11, 21
6	15 Ex. 1–10 12 Ex. 14, 22 10 Ex. 4	15 Ex. 1–10, 15 12 Ex. 7, 17, 31 10 Ex. 3
7	17 Ex. 1–8 15 Ex. 11, 12 12 Ex. 16, 24	17 Ex. 1–8 15 Ex. 11, 12, 15 12 Ex. 9, 19, 32
8	19 Ex. Odd 1–7 17 Ex. 9 15 Ex. 13, 14	19 Ex. Odd 1–7 17 Ex. 9 15 Ex. 13, 14, 16
9	23 Chapter Review 19 Ex. 2, 4, 6 17 Ex. 10	23 Chapter Review 19 Ex. 2, 4, 6 17 Ex. 10
10	24 Chapter Test	24 Chapter Test
11	27 Odd 1–29	27 Even 2–32
12	29 Ex. 1–4, 7–19 27 Ex. 6, 26	29 Ex. 1–4, 7–19, 22 27 Ex. 9, 25, 31
13	32 Ex. Odd 1–33 29 Ex. 5, 20 27 Ex. 12, 28	32 Ex. Even 2–34 29 Ex. 5, 20 27 Ex. 11, 29
14	34 Ex. Odd 1–39 32 Ex. 4, 28 29 Ex. 6, 21	34 Ex. Even 2–44 32 Ex. 13, 27, 35 29 Ex. 6, 21
15	38 Ex. 1–20 34 Ex. 16, 34 32 Ex. 14, 32	38 Ex. 1–20, 25 34 Ex. 23, 41, 43 32 Ex. 15, 29

Day	Level 1	Level 2
16	41 Ex. 1–14 38 Ex. 21, 22 34 Ex. 20, 38	41 Ex. 1–14, 19 38 Ex. 21, 22, 26 34 Ex. 39, 45
17	43 Ex. 1–6 41 Ex. 15–16 38 Ex. 23, 24	43 Ex. 1–6 41 Ex. 15–16, 20 38 Ex. 23, 24
18	44 Ex. 1–15 43 Ex. 7 41 Ex. 17, 18	44 Ex. 1–15, 19 43 Ex. 7 41 Ex. 17, 18
19	45 Ex. 1–10 44 Ex. 16, 17 43 Ex. 8	45 Ex. 1–10 44 Ex. 16, 20 43 Ex. 8
20	47 Ex. Odd 1–13 45 Ex. 11, 12 44 Ex. 18	47 Ex. Even 2–16, 15 45 Ex. 11, 12 44 Ex. 17, 18
21	50 Chapter Review 47 Ex. 8, 10, 14 45 Ex. 13, 14	50 Chapter Review 47 Ex. 9, 11, 13 45 Ex. 13, 14
22	51 Chapter Test	51 Chapter Test
23	55 Ex. Odd 1–21	55 Ex. Even 2–22
24	57 Ex. 1–12 55 Ex. 2, 10, 14	57 Ex. 1–12, 16 55 Ex. 3, 9, 19, 23
25	59 Ex. Odd 1–33 57 Ex. 13, 14 55 Ex. 6, 12, 20	59 Ex. Even 2–32 57 Ex. 13, 14, 17 55 Ex. 5, 11, 21
26	60 Ex. 1–6 59 Ex. 14, 30 57 Ex. 15	60 Ex. 1–6 59 Ex. 15, 31 57 Ex. 15, 18
27	62 Ex. Odd 1–27 60 Ex. 7 59 Ex. 16, 32	62 Ex. Even 2–30 60 Ex. 7 59 Ex. 17, 33
28	64 Ex. 1–17 62 Ex. 6, 26 60 Ex. 8	64 Ex. 1–17, 21 62 Ex. 7, 17, 29 60 Ex. 8
29	67 Ex. 1–8 64 Ex. 18, 19 62 Ex. 8, 28	67 Ex. 1–8, 13 64 Ex. 18, 22 62 Ex. 9, 27

Day	Level 1	Level 2
30	71 Ex. Odd 1–51 67 Ex. 9, 10 64 Ex. 20	71 Ex. Even 2–58 67 Ex. 9, 10, 14 64 Ex. 19, 20
31	73 Ex. Odd 1–45 71 Ex. 8, 28, 46 67 Ex. 11, 12	73 Ex. Even 2–48 71 Ex. 49, 53, 55 67 Ex. 11, 12
32	76 Chapter Review 73 Ex. 8, 24, 42 71 Ex. 10, 30, 50	76 Chapter Review 73 Ex. 39, 47, 49 71 Ex. 51, 57, 59
33	77 Chapter Test	77 Chapter Test
34	81 Ex. Odd 1–33	81 Ex. Even 2–36
35	84 Ex. Odd 1–43, 47 81 Ex. 2, 10, 30	84 Ex. Even 2–48 81 Ex. 5, 9, 35
36	87 Ex. 1–12, 21 84 Ex. 14, 32, 48 81 Ex. 6, 12, 32	87 Ex. 1–13, 17, 21 84 Ex. 17, 45, 47 81 Ex. 11, 31, 37
37	89 Ex. Odd 1–35 87 Ex. 13, 14, 22 84 Ex. 22, 36, 44	89 Ex. Even 2–40 87 Ex. 14, 18, 22 84 Ex. 19, 35, 43
38	91 Ex. Odd 1–27 89 Ex. 8, 16, 32 87 Ex. 15, 16	91 Ex. Even 2–30 89 Ex. 9, 15, 37 87 Ex. 15, 16, 19
39	94 Ex. Odd 1–27, 31 91 Ex. 12, 26 89 Ex. 10, 18, 34	94 Ex. Even 2–32 91 Ex. 11, 29 89 Ex. 11, 17, 39
40	97 Ex. Odd 1–25, Odd 29–35 94 Ex. 24, 32 91 Ex. 14, 28	97 Ex. Even 2–34 94 Ex. 29, 31 91 Ex. 13, 27
41	98 Ex. 1–7 97 Ex. 22, 30, 34 94 Ex. 8, 22	98 Ex. 1–7 97 Ex. 27, 31, 35 94 Ex. 11, 21
42	100 Ex. Odd 1–19, 24 98 Ex. 8 97 Ex. 24, 32	100 Ex. Even 2–24 98 Ex. 8 97 Ex. 25, 33

Day	Level 1	Level 2
43	101 Ex. 1–7 100 Ex. 12, 25 98 Ex. 9	101 Ex. 1–7, 10 100 Ex. 13, 23, 25 98 Ex. 9
44	103 Ex. Odd 1–25 101 Ex. 8 100 Ex. 14, 20	103 Ex. Even 2–30 101 Ex. 8, 11 100 Ex. 15, 21
45	105 Ex. 1–8, 10 103 Ex. 14, 22 101 Ex. 9	105 Ex. 1–8, 10, 12 103 Ex. 15, 27 101 Ex. 9
46	107 Chapter Review 105 Ex. 9, 11 103 Ex. 16, 24	107 Chapter Review 105 Ex. 9, 11, 13 103 Ex. 17, 29
47	108 Chapter Test	108 Chapter Test
48	113 Ex. Odd 1–31, 36	113 Ex. Even 2–36
49	115 Ex. Odd 1–23 113 Ex. 18, 30, 37	115 Ex. Even 2–26 113 Ex. 17, 33, 37
50	118 Ex. Odd 1–29, 33 115 Ex. 2, 20 113 Ex. 20, 32	118 Ex. Even 2–34 115 Ex. 13, 26 113 Ex. 19, 31, 35
51	121 Ex. 1–16, 21 118 Ex. 8, 28, 34 115 Ex. 4, 22	121 Ex. 1–16, 21 118 Ex. 13, 31, 33 115 Ex. 15, 21
52	124 Ex. Odd 1–37 121 Ex. 17, 18, 22 118 Ex. 10, 24, 30	124 Ex. Even 2–36 121 Ex. 17, 18, 22 118 Ex. 9, 15, 25
53	125 Ex. 1–5 124 Ex. 12, 26, 34 121 Ex. 19, 20	125 Ex. 1–5 124 Ex. 13, 31, 35 121 Ex. 19, 20
54	128 Ex. 1–16, 22, 23 125 Ex. 6 124 Ex. 14, 28, 36	128 Ex. 1–16, 22, 23 125 Ex. 6 124 Ex. 15, 33, 37
55	130 Ex. 1–9 128 Ex. 17, 18, 24 125 Ex. 7	130 Ex. 1–9, 13 128 Ex. 17, 18, 24 125 Ex. 7

Day	Level 1	Level 2
56	134 Ex. 1–15 130 Ex. 10, 11 128 Ex. 19, 20, 25	134 Ex. 1–16, 19, 20 130 Ex. 10, 14 128 Ex. 19, 20, 25
57	137 Chapter Review 134 Ex. 16–18 130 Ex. 12	137 Chapter Review 134 Ex. 17, 18, 21 130 Ex. 11, 12
58	138 Chapter Test	138 Chapter Test
59	142 Ex. 1–4	142 Ex. 1–4
60	144 Ex. 1–5, 10	144 Ex. 1–5, 11
61	146 Ex. 1–8 144 Ex. 6–9	146 Ex. 1–8 144 Ex. 6–9
62	149 Ex. 1–6, 11 146 Ex. 9, 10	149 Ex. 1–6, 11 146 Ex. 9, 10 144 Ex. 10
63	151–152 Ex. 1–4, 14–17 149 Ex. 7–10 146 Ex. 11	151–152 Ex. 1–4, 14–17 149 Ex. 7–10 146 Ex. 11
64	156 Ex. 1–4, 8–11 152 Ex. 5–13 149 Ex. 12	156 Ex. 1–4, 8–11 152 Ex. 5–13 149 Ex. 12
65	158–159 Ex. 1–10 156 Ex. 5–7 152 Ex. 18–21	158–159 Ex. 1–16 156 Ex. 5–7 152 Ex. 18–21
66	162 Chapter Review 159 Ex. 11, 12	162 Chapter Review 159 Ex. 17, 18
67	163 Chapter Test	163 Chapter Test
68	167 Ex. Odd 1–47	167 Ex. Even 2–50
69	170 Ex. Odd 1–9, Odd 13–33, 37 167 Ex. 22, 34, 46	170 Ex. Even 2–38 167 Ex. 21, 33, 49
70	172 Ex. Odd 1–15 170 Ex. 8, 22, 38 167 Ex. 24, 36, 48	172 Ex. Even 2–16, 17 170 Ex. 11, 21, 37 167 Ex. 23, 35, 47
71	174 Ex. Odd 1–19, Odd 25–29 172 Ex. 4, 10, 14 170 Ex. 10, 24, 34	174 Ex. Even 2–28 172 Ex. 3, 9, 13 170 Ex. 9, 23, 33
72	176 Ex. 1–11 174 Ex. 18, 24, 28 172 Ex. 6, 12, 16	176 Ex. 1–11 174 Ex. 21, 25, 29 172 Ex. 5, 11, 15

Day	Level 1	Level 2
73	179 Ex. Odd 1–25, 28 176 Ex. 12, 13 174 Ex. 8, 20, 26	179 Ex. Even 2–28 176 Ex. 12, 13 174 Ex. 23, 27
74	181 Ex. Odd 1–27 179 Ex. 18, 22, 29 176 Ex. 14, 15	181 Ex. Even 2–28 179 Ex. 19, 27, 29 176 Ex. 14, 15
75	183 Ex. Odd 1–29 181 Ex. 10, 22, 26 179 Ex. 20, 24	183 Ex. Even 2–30 181 Ex. 9, 21, 25 179 Ex. 21, 25
76	187 Chapter Review 183 Ex. 4, 22, 26 181 Ex. 12, 24, 28	187 Chapter Review 183 Ex. 11, 21, 29 181 Ex. 11, 23, 27
77	188 Chapter Test	188 Chapter Test
78	191 Ex. Odd 1–15	191 Ex. Even 2–16
79	194 Ex. Odd 1–23 191 Ex. 2, 6, 10	194 Ex. Even 2–24 191 Ex. 3, 7, 11
80	197 Ex. Odd 1–25 194 Ex. 4, 16, 20 191 Ex. 8, 12, 16	197 Ex. Even 2–26 194 Ex. 3, 15, 21 191 Ex. 5, 9, 13
81	199 Ex. 1–12 197 Ex. 18, 22 194 Ex. 6, 18, 24	199 Ex. 1–12 197 Ex. 19, 23, 27 194 Ex. 5, 17, 23
82	201 Ex. Odd 1–17, Odd 25–33 199 Ex. 13, 14 197 Ex. 20, 24	201 Ex. Even 2–36 199 Ex. 13, 14 197 Ex. 21, 25
83	204 Ex. 1–14, 21 201 Ex. 16, 26, 28 199 Ex. 15, 16	204 Ex. 1–18, 21 201 Ex. 19, 21, 33 199 Ex. 15, 16
84	207 Ex. Odd 1–23, 28, 29 204 Ex. 15, 22 201 Ex. 18, 30, 32	207 Ex. Even 2–30 204 Ex. 19, 22 201 Ex. 23, 31, 35
85	210 Ex. Odd 1–23 207 Ex. 8, 22, 30 204 Ex. 16, 17	210 Ex. Even 2–28 207 Ex. 9, 25, 29 204 Ex. 20
86	212 Chapter Review 210 Ex. 12, 24 207 Ex. 10, 24, 31	212 Chapter Review 210 Ex. 11, 25, 27 207 Ex. 11, 27, 31
87	213 Chapter Test	213 Chapter Test

Day	Level 1	Level 2
88	218 Ex. Odd 1–49	218 Ex. Even 2–52
89	221 Ex. Odd 1–15, 19, 21 218 Ex. 16, 46, 50	221 Ex. Even 2–22 218 Ex. 15, 45, 49
90	223 Ex. Odd 1–13, Odd 17–23 221 Ex. 2, 16, 20 218 Ex. 18, 48	223 Ex. Even 2–24 221 Ex. 13, 17, 19 218 Ex. 17, 47, 51
91	225 Ex. Odd 1–11, 15, 17 223 Ex. 10, 18, 20 221 Ex. 8, 14, 22	225 Ex. Even 2–18 223 Ex. 15, 17, 19 221 Ex. 11, 15, 21
92	228 Ex. 1–10 225 Ex. 4, 10, 16 223 Ex. 14, 22, 24	228 Ex. 1–10, 12 225 Ex. 3, 13, 15 223 Ex. 13, 21, 23
93	231 Ex. Odd 1–9, Odd 13–17 228 Ex. 11, 13 225 Ex. 6, 12, 18	231 Ex. Even 2–16 228 Ex. 11, 13 225 Ex. 5, 11, 17
94	233 Ex. 1–13 231 Ex. 4, 12, 14 228 Ex. 14	233 Ex. 1–13 231 Ex. 9, 11, 13 228 Ex. 14
95	237 Chapter Review 233 Ex. 14–15 231 Ex. 6, 8, 16	237 Chapter Review 233 Ex. 14–15 231 Ex. 7, 15, 17
96	238 Chapter Test	238 Chapter Test
97	242 Ex. Odd 1–35	242 Ex. Even 2–38
98	245 Ex. 1–3 242 Ex. 34	245 Ex. 1–3 242 Ex. 33, 37
99	246 Ex. 1 245 Ex. 4 242 Ex. 36	246 Ex. 1 245 Ex. 4 242 Ex. 35
100	247 Ex. 1–7 246 Ex. 2 245 Ex. 5	247 Ex. 1–7 246 Ex. 2 245 Ex. 5
101	251 Ex. 1–8 247 Ex. 8–9 246 Ex. 3	251 Ex. 1–8, 13 247 Ex. 8–9 246 Ex. 3
102	253 Ex. 1–4 251 Ex. 9, 10 247 Ex. 10	253 Ex. 1–7 251 Ex. 9, 10, 14 247 Ex. 10

Day	Level 1	Level 2
103	257 Chapter Review 253 Ex. 5 251 Ex. 11, 12	257 Chapter Review 253 Ex. 8 251 Ex. 11, 12, 15
104	258 Chapter Test	258 Chapter Test
105	262 Ex. 1–16	262 Ex. 1–16, 21
106	265 Ex. Odd 1–33 262 Ex. 17–18	265 Ex. Even 2–32 262 Ex. 17–18, 22
107	268 Ex. 1–12 265 Ex. 2, 12, 30 262 Ex. 19–20	268 Ex. 1–12, 17 265 Ex. 3, 13, 29 262 Ex. 19–20
108	269 Ex. 1–6 268 Ex. 13–14 265 Ex. 4, 14, 32	269 Ex. 1–6 268 Ex. 13–14, 18 265 Ex. 5, 15, 31
109	272 Ex. 1–2 269 Ex. 7 268 Ex. 15–16	272 Ex. 1–2 269 Ex. 7 268 Ex. 15–16
110	275–276 Ex. 1–16 272 Ex. 3 269 Ex. 8	275–276 Ex. 1–16, 30 272 Ex. 3 269 Ex. 8
111	279 Ex. 1–5, 7 276 Ex. 17–22 272 Ex. 4	279 Ex. 1–5, 7 276 Ex. 17–22, 31 272 Ex. 4
112	282 Ex. Odd 1–33 279 Ex. 6 276 Ex. 23–29	282 Ex. Even 2–36 279 Ex. 6, 8 276 Ex. 23–29, 32
113	285 Chapter Review 282 Ex. 6, 18, 30 279 Ex. 8	285 Chapter Review 282 Ex. 23, 33, 35 279 Ex. 9
114	286 Chapter Test	286 Chapter Test
115	290 Ex. Odd 1–45	290 Ex. Even 2–48
116	293 Ex. Odd 1–43 290 Ex. 6, 14, 22	293 Ex. Even 2–44 290 Ex. 7, 15, 47
117	294 Ex. 1–7 293 Ex. 6, 26, 38 290 Ex. 30, 34, 40	294 Ex. 1–7 293 Ex. 9, 27, 41 290 Ex. 27, 35, 49
118	297 Ex. Odd 1–43 294 Ex. 8–9 293 Ex. 8, 28, 40	297 Ex. Even 2–44 294 Ex. 8–9 293 Ex. 11, 29, 43

Day	Level 1	Level 2
119	300 Ex. 1–10, 13–16, 19–28 297 Ex. 6, 30, 42 294 Ex. 10–11	300 Ex. 1–10, 13–16, 19–28 297 Ex. 9, 29, 41 294 Ex. 10–11
120	302 Ex. Odd 1–35 300 Ex. 11, 17, 29 297 Ex. 8, 32, 44	302 Ex. Even 2–36, 37 300 Ex. 11, 17, 29 297 Ex. 11, 31, 43
121	304 Ex. Odd 1–29 302 Ex. 10, 18, 30 300 Ex. 12, 18, 30	304 Ex. Even 2–34 302 Ex. 9, 23, 33 300 Ex. 12, 18, 30
122	306 Ex. Odd 1–27 304 Ex. 12, 22, 24 302 Ex. 12, 20, 32	306 Ex. Even 2–32 304 Ex. 19, 31, 33 302 Ex. 11, 25, 35
123	308 Chapter Review 306 Ex. 22, 24, 26 304 Ex. 18, 28, 30	308 Chapter Review 306 Ex. 27, 29, 31, 33 304 Ex. 17, 27, 29
124	309 Chapter Test	309 Chapter Test
125	313 Ex. 1–12, 17	313 Ex. 1–12, 19–20
126	315 Ex. 1–14 313 Ex. 13, 14, 18	315 Ex. 1–14, 19 313 Ex. 13, 14, 17
127	318 Ex. Odd 1–33 315 Ex. 15–16 313 Ex. 15, 16, 19	318 Ex. Even 2–34 315 Ex. 15, 16, 20 313 Ex. 15, 16, 18
128	319 Ex. 1–6 318 Ex. 8, 16, 24, 32 315 Ex. 17–18	319 Ex. 1–6 318 Ex. 7, 15, 23, 31 315 Ex. 17, 18, 21
129	322 Ex. Odd 1–57 319 Ex. 7 318 Ex. 10, 18, 26, 34	322 Ex. Even 2–60 319 Ex. 7 318 Ex. 9, 17, 25, 33
130	324 Ex. 1–13 322 Ex. 18, 34, 54 319 Ex. 8	324 Ex. 1–13 322 Ex. 21, 47, 59 319 Ex. 8
131	327 Ex. Odd 1–39 324 Ex. 14, 15 322 Ex. 20, 36, 56	327 Ex. Even 2–42 324 Ex. 14, 15 322 Ex. 31, 41, 57
132	329 Ex. 1–18 327 Ex. 6, 22, 38 324 Ex. 16, 17	329 Ex. 1–18 327 Ex. 11, 23, 41 324 Ex. 16, 17

Day	Level 1	Level 2
133	332 Chapter Review 329 Ex. 19, 20 327 Ex. 8, 24, 40	332 Chapter Review 329 Ex. 19, 20 327 Ex. 13, 25, 39
134	333 Chapter Test	333 Chapter Test
135	336 Ex. 1–15	336 Ex. 1–15, 19–20
136	338 Ex. Odd 1–25 336 Ex. 16–17	338 Ex. Even 2–34 336 Ex. 16–17, 21–22
137	341 Ex. Odd 1–25 338 Ex. 12, 18 336 Ex. 18	341 Ex. Even 2–28 338 Ex. 17, 27, 29 336 Ex. 18, 23–24
138	343 Ex. Odd 1–15 341 Ex. 12, 20, 24 338 Ex. 14, 26	343 Ex. Even 2–16 341 Ex. 9, 19, 27 338 Ex. 21, 31, 33
139	346 Ex. Odd 1–27 343 Ex. 6, 14 341 Ex. 10, 22, 26	346 Ex. Even 2–30 343 Ex. 5, 13 341 Ex. 11, 21, 25
140	349 Ex. Odd 1–51 346 Ex. 8, 14, 24 343 Ex. 8, 16	349 Ex. Even 2–52 346 Ex. 11, 17, 29 343 Ex. 7, 15
141	352 Ex. Odd 1–33 349 Ex. 26, 46 346 Ex. 10, 16, 26	352 Ex. Even 2–46 349 Ex. 25, 45 346 Ex. 11, 17, 29
142	355 Ex. Odd 1–43 352 Ex. 14, 26 349 Ex. 28, 48	355 Ex. Even 2–48 352 Ex. Odd 35–41 349 Ex. 27, 47
143	357 Ex. Odd 1–33 355 Ex. 16, 34 352 Ex. 16, 28	357 Ex. Even 2–36 355 Ex. 11, 45 352 Ex. Odd 43–47
144	359 Ex. Odd 1–29 357 Ex. 10, 26 355 Ex. 18, 36	359 Ex. Even 2–34 357 Ex. 11, 35 355 Ex. 31, 47
145	362 Ex. Odd 1–39 359 Ex. 8, 24 357 Ex. 12, 32	362 Ex. Even 2–40, 41 359 Ex. 11, 31 357 Ex. 27, 31
146	365 Chapter Review 362 Ex. 8, 38 359 Ex. 12, 28	365 Chapter Review 362 Ex. 27, 43 359 Ex. 27, 33

Day	Level 1	Level 2
147	366 Chapter Test	366 Chapter Test
148	369 Ex. 1–3	369 Ex. 1–4
149	371 Ex. 1–9	371 Ex. 1–9
150	372 Ex. 1–8	372 Ex. 1–8
151	374 Ex. 1–2 372 Ex. 9	374 Ex. 1–3 372 Ex. 9
152	376 Ex. 1 372 Ex. 10	376 Ex. 1–2 372 Ex. 10
153	378 Ex. 1–5, 7, 13–14, 16–17	378 Ex. 1–9, 13–20 376 Ex. 3
154	380 Ex. 1–4, 8–9 378 Ex. 6, 8 376 Ex. 4	380 Ex. 1–4, 8–9 378 Ex. 10, 11 376 Ex. 4
155	381 Ex. 1–3 380 Ex. 5, 7 378 Ex. 15, 18	381 Ex. 1–3 380 Ex. 5, 7 378 Ex. 12, 21
156	384 Ex. Odd 1–35 381 Ex. 4 380 Ex. 6	384 Ex. Even 2–38 381 Ex. 4 380 Ex. 6
157	387 Ex. Odd 1–25 384 Ex. 8, 24, 32 381 Ex. 5	387 Ex. Even 2–30 384 Ex. 9, 23, 37 381 Ex. 5
158	390 Chapter Review 387 Ex. 20, 26 384 Ex. 14, 30, 36	390 Chapter Review 387 Ex. 21, 27, 29 384 Ex. 15, 29, 33
159	391 Chapter Test	391 Chapter Test
160	395 Ex. Odd 1–7, Odd 11–27, Odd 31–45	395 Ex. Even 2–50
161	398 Ex. Odd 1–35 395 Ex. 4, 24, 42	398 Ex. Even 2–42 395 Ex. 9, 23, 29
162	400 Ex. 1–6 398 Ex. 14, 30 395 Ex. 6, 26, 44	400 Ex. 1–6 398 Ex. 19, 37 395 Ex. 43, 47, 49
163	403 Ex. Odd 1–15, Odd 19–33 400 Ex. 7 398 Ex. 16, 32	403 Ex. Even 2–38 400 Ex. 7 398 Ex. 39, 41

Day	Level 1	Level 2
164	405 Ex. 1–8 403 Ex. 12, 28 400 Ex. 8	405 Ex. 1–8, 11, 12 403 Ex. 17, 35 400 Ex. 8
165	408 Ex. Odd 1–27 405 Ex. 9 403 Ex. 14, 30	408 Ex. Even 2–40 405 Ex. 9, 13 403 Ex. 29, 37
166	409 Ex. 1–5 408 Ex. 14, 24 405 Ex. 10	409 Ex. 1–5 408 Ex. 29, 31, 33 405 Ex. 10, 14
167	412 Ex. 1–15 409 Ex. 6 408 Ex. 16, 26	412 Ex. 1–15, 20–23 409 Ex. 6 408 Ex. 35, 37, 39
168	414 Ex. Odd 1–11 412 Ex. 16–17 409 Ex. 7	414 Ex. Even 2–12 412 Ex. 16–17, 24 409 Ex. 7
169	416 Chapter Review 414 Ex. 6, 12 412 Ex. 18–19	416 Chapter Review 414 Ex. 5, 11 412 Ex. 18–19, 25
170	417 Chapter Test	417 Chapter Test

HOLT
PRE-ALGEBRA

Authors

Eugene D. Nichols

Robert O. Lawton Distinguished
Professor of Mathematics Education
Florida State University
Tallahassee, Florida

Mervine L. Edwards

Chairman of Mathemathics Department
Shore Regional High School
West Long Branch, New Jersey

Sylvia A. Hoffman

Resource Consultant in Mathematics
Illinois State Board of Education
State of Illinois

Albert Mamary

Superintendent of Schools for Instruction
Johnson City Central School District
Johnson City, New York

HOLT
PRE-ALGEBRA

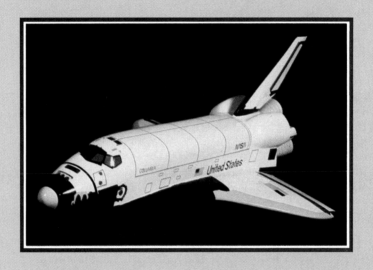

Eugene D. Nichols • Mervine L. Edwards
Sylvia A. Hoffman • Albert Mamary

HOLT, RINEHART AND WINSTON, PUBLISHERS
New York • Toronto • Mexico City • London • Sydney • Tokyo

Photo Credits

ISBN: 0-03-001858-7

901 040 987

Contents

4 Multiplying and Dividing Fractions

5 Adding and Subtracting Fractions

6 Organizing Data

7 Decimals

8 Measurement

9 Percents

Equations – Adding and Subtracting

Up to 62,000 people can be seated in the Houston Astrodome, the first domed stadium.

Introduction to Variables

To identify the variables,
   constants, and terms in an
   expression
To evaluate algebraic expressions
   for given values of variables

Compute $18 - 5 + 6$.

$$18 - 5 + 6$$
$$13 + 6$$
$$19$$

Think of a number. Add 2 to the number.

$$\text{Number} \;+\; 2$$
$$\downarrow \downarrow$$
$$x + 2$$
$$\uparrow$$
$$\text{Variable}$$

Use x to represent the number.

A *variable* is a letter that may be replaced by different values.

Example 1

To evaluate means to find the value.

Substitute 8 for y.

Evaluate $y + 7$ if $y = 8$.
$$y + 7$$
$$\downarrow$$
$$8 + 7$$
$$15$$
So, the value of $y + 7$ is 15 if $y = 8$.

practice ▷ Evaluate if $x = 2$, $m = 3$, $y = 5$, and $a = 2$.

1. $x + 5$ 7   2. $7 + m$ 10   3. $6 - y$ 1   4. $a + 9$ 11

Example 2

$$\begin{array}{r} 90 \\ -23 \\ \hline \end{array} \qquad \begin{array}{r} \overset{8\;\;10}{\cancel{9}\;\cancel{0}} \\ -2\;3 \\ \hline 7 \end{array} \qquad \begin{array}{r} \overset{8\;\;10}{\cancel{9}\;\cancel{0}} \\ -2\;3 \\ \hline 6\;7 \end{array}$$

Evaluate $c - d$ if $c = 90$ and $d = 23$.
$$c - d$$
$$90 - 23$$
$$67$$
So, the value of $c - d$ is 67 if $c = 90$ and $d = 23$.

practice ▷ Evaluate if $x = 40$, $y = 26$, $a = 56$, $b = 39$, and $d = 13$.

5. $x - y$ 14   6. $a + b$ 95   7. $a - b$ 17   8. $b + d$ 52

Example 3

Evaluate $a + b - c + d$ if $a = 17$, $b = 48$, $c = 19$, and $d = 8$.

$$
\begin{array}{r}
\overset{1}{17} \\
+48 \\
\hline
65
\end{array}
$$

$$
\begin{array}{r}
65 \\
-19 \\
\hline
46
\end{array}
\qquad
\begin{array}{r}
{}^{5\ \ 15} \\
\cancel{6}\,\cancel{5} \\
-1\,9 \\
\hline
4\,6
\end{array}
$$

$$
\begin{array}{ccccc}
a & + & b & - & c & + & d \\
17 & + & 48 & - & 19 & + & 8
\end{array}
$$

$$\underbrace{65 \ - \ 19} + 8$$

$$\underbrace{46 \quad + 8}$$

$$54$$

So, the value of $a + b - c + d$ is 54.

---

practice ▷ Evaluate if $x = 13$, $y = 59$, $z = 32$, $w = 6$, $a = 48$, $b = 44$, $c = 17$, and $d = 5$.

9. $x + y - z + w$ 46   **10.** $a + b - c + d$ 80   **11.** $y - a + z - d$ 38

The expression $x + y - 6$ has three *terms:* x, y, and 6.
Terms are added or subtracted.
The *variables* are x and y; 6 is a *constant*.

Example 4

Name the terms, variables, and constants of $x - 7 + y$.
The terms are x, 7, and y.
The variables are x and y. The constant is 7.

---

practice ▷ Name the terms, variables, and constants of each expression.

12. $11 - p$ **13.** $t + 4 - u$ **14.** $x + 8 - y$ **15.** $x - 9 + y - 4$
      11, p; p; 11   t, 4, u; t, u; 4   x, 8, y; x, y; 8   x, 9, y, 4; x, y; 9, 4

◇ EXERCISES ◇

Evaluate the following expressions if $r = 7$, $s = 9$, $t = 4$, and $w = 5$.

1. $r + 5$ 12   **2.** $9 + t$ 13   **3.** $11 - w$ 6   **4.** $s + 6$ 15   **5.** $17 - t$ 13
6. $r + t$ 11   **7.** $s - w$ 4   **8.** $w + r$ 12   **9.** $s + 22$ 31   **10.** $19 - w$ 14

Evaluate the following expressions if $a = 70$, $b = 42$, $c = 73$, and $d = 49$.

11. $b + 19$ 61   **12.** $c - 19$ 54   **13.** $a + b$ 112   **14.** $c - d$ 24
15. $a + b - 13$ 99   **16.** $a - b + d$ 77   **17.** $d - b + a$ 77   **18.** $39 + a - d$ 60
19. $a + c - b + d$ 150 **20.** $c - d + a - b$ 52 **21.** $d - b + c - a$ 10

Name the terms, variables, and constants of each expression.

22. $a - 4$ **23.** $p - 12 + t$ **24.** $5 + a + 7 - n$
     a, 4; a; 4   p, 12, t; p, t; 12   5, a, 7, n; a, n; 5, 7

Properties of Addition

To recognize the commutative, associative, and additive identity properties

To use the commutative and associative properties to make computation easier

$8 + 3 = 11$ and $3 + 8 = 11$

So, $8 + 3 = 3 + 8$.

When adding two numbers, you can change the order.

The recall suggests this.

commutative property of addition
For all numbers a and b, $a + b = b + a$.

Example 1

Add inside parentheses first.

Answers are the same.

Show that $(8 + 2) + 5 = 8 + (2 + 5)$.

$(8 + 2) + 5$	$=$	$8 + (2 + 5)$
$10 + 5$		$8 + 7$
15	$=$	15

So, $(8 + 2) + 5 = 8 + (2 + 5)$

When adding, you can change the way numbers are grouped.

associative property of addition
For all numbers a, b, and c,
$(a + b) + c = a + (b + c)$.

Example 2

The result of addition is called the sum.

Add.

$6 + 0$	$0 + 4$	$5 + 0$	$0 + 5$
6	4	5	5

So, the sum of any number and 0 is that number.

property of additive identity
For any number a, $a + 0 = a$ and $0 + a = a$.

Example 3

Which property is illustrated?

$8 + 9 = 9 + 8$ commutative property of addition

$(6 + 4) + 2 = 6 + (4 + 2)$ associative property of addition

$12 + 0 = 12$ property of additive identity

Which property of addition is illustrated?

1. $(7 + 6) + 9 = 7 + (6 + 9)$ **2.** $6 + 8 = 8 + 6$ **3.** $7 + 0 = 7$
associative   commutative   identity

Example 4

Group 297 and 3 together since their sum is easier to find.

Rewrite the sum $297 + 18 + 3$ to make the computation easier. Then compute.

$$297 + 18 + 3 = 297 + 3 + 18 \quad \text{commutative property}$$
$$= (297 + 3) + 18 \quad \text{associative property}$$
$$= 300 + 18$$
$$= 318$$

If 18 and 3 are grouped together, the answer is the same, but the computation is more difficult.

$$297 + 18 + 3 = 297 + (18 + 3)$$
$$= 297 + 21$$
$$= 318$$

---

practice ▷ **Rewrite to make the computation easier. Then compute.**

4. $49 + 28 + 1$
$49 + 1 + 28 = 78$

5. $93 + 62 + 7$
$93 + 7 + 62 = 162$

6. $2 + 19 + 198$
$198 + 2 + 19 = 219$

◇ EXERCISES ◇

Which property of addition is illustrated?

2. associative   4. associative

1. $16 + 9 = 9 + 16$ commutative

2. $(14 + 7) + 26 = 14 + (7 + 26)$

3. $38 + 0 = 38$ identity

4. $19 + (5 + 14) = (19 + 5) + 14$

5. $0 + 49 = 49$ identity

6. $72 + 28 = 28 + 72$ commutative

Rewrite to make the computation easier. Then compute.   Check students' answers.

7. $99 + 66 + 1$ 166   **8.** $3 + 79 + 97$ 179   **9.** $47 + 50 + 3$ 100

10. $8 + 37 + 2$ 47   **11.** $199 + 39 + 1$ 239   **12.** $68 + 40 + 2$ 110

13. $25 + 49 + 75$ 149   **14.** $72 + 20 + 8$ 100   **15.** $87 + 36 + 3$ 126

16. $195 + 49 + 5$ 249   **17.** $149 + 29 + 1$ 179   **18.** $139 + 47 + 1$ 187

19. $7 + 62 + 93$ 162   **20.** $398 + 76 + 2$ 476   **21.** $55 + 89 + 45$ 189

22. $175 + 89 + 25$ 289   **23.** $17 + 38 + 3 + 2$ 60   **24.** $18 + 49 + 2 + 1$ 70

25. $78 + 19 + 2 + 1$ 100   **26.** $1 + 98 + 79 + 2 + 399$ 579

27. Does the associative property hold for subtraction?
Hint: Is $(13 - 5) - 4 = 13 - (5 - 4)$ true?   no

Open Sentences

To recognize open sentences
To recognize the solution(s) of an
  open sentence

Evaluate $x + 5$ if $x = 3$.
$x + 5$
$3 + 5$ Substitute 3 for x.
8

A sentence like "She was prime minister of India." is
neither true nor false. It is called an *open sentence*.

Example 1

In the open sentence "She was prime minister of
India." replace *She* with the name of a person to
make a true sentence. Then replace *She* to make a
false sentence.

true   *Indira Ghandi* was prime minister of India.
false   *Margaret Thatcher* was prime minister of India.

It is true or false depending upon
the value of x.

A sentence like $x + 4 = 10$ is an open sentence. It
contains a variable x.

Example 2

Identify the sentence as open or not open. For each
sentence that is not open, indicate whether it is true
or false.

$7 + 2 = 9$	$x - 4 = 11$	$8 - 5 = 4$
not open; true	open	not open; false

practice ▷   **Identify the sentence as open or not open. For each sentence
that is not open, indicate whether the sentence is true or false.**

1. $7 - 3 = 2$
   not open; false

2. $x + 9 = 14$ open

3. $5 + 11 = 16$
   not open; true

Example 3

Substitute the indicated values for x in $x + 5 = 12$.
Are the resulting sentences true or false?

Substitute 3 for x.

$x + 5 = 12$	
$3 + 5$	12
8	12
8 $= 12$ false	

Substitute 7 for x.

$x + 5 = 12$	
$7 + 5$	12
12	
$12 = 12$ true	

Substitute 3 for x.

A value of a variable that makes an open sentence true is called a *solution* of the open sentence.

Example 4

Which of the values 4, 6, or 8 is a solution of $7 + x = 13$?

Substitute 4 for x.	Substitute 6 for x.	Substitute 8 for x.
$7 + x = 13$	$7 + x = 13$	$7 + x = 13$
$7 + 4$ ∣ 13	$7 + 6$ ∣ 13	$7 + 8$ ∣ 13
11	13	15
$11 = 13$	$13 = 13$	$15 = 13$
false	true	false

So, 6 is a solution of $7 + x = 13$.

practice ▷ **Which of the given values of the variables is a solution of the open sentence?**

4. $5 + x = 14$; 3, ⑨ 11

5. $x - 12 = 9$; 14, 20, ㉑

Sometimes an open sentence contains an inequality symbol, $<$ or $>$.

$$4 < 8$$
4 *is less than* 8.

$$7 > 1$$
7 *is greater than* 1.

Example 5

Which of the values 0, 1, or 3 are solutions of $x + 6 < 8$?

Substitute 0 for x.	Substitute 1 for x.	Substitute 3 for x.
$x + 6 < 8$	$x + 6 < 8$	$x + 6 < 8$
$0 + 6$ ∣ 8	$1 + 6$ ∣ 8	$3 + 6$ ∣ 8
6	7	9
$6 < 8$	$7 < 8$	$9 < 8$
true	true	false

So, 0 and 1 are solutions of $x + 6 < 8$.

practice ▷ **Which of the given values of the variable is a solution of the open sentence?**

6. $x + 9 < 11$; ⓪ ① 2

7. $7 > m + 4$; ⓪ ① ② 5

◇ *ORAL EXERCISES* ◇

Which are open sentences? For each sentence that is not open, indicate whether it is true or false.

not open; false
1. $5 + 4 = 3$
2. $x - 7 = 10$ open
not open; true
3. $7 + 2 = 9$
open
4. $3 + x > 8$
5. $14 = x + 8$ open
6. $4 + 3 > 1$
not open; true
7. $x < 12$ open
8. $7 = 4 + 3$
not open; true

◇ *EXERCISES* ◇

Identify the sentence as open or not open. For each sentence that is not open, indicate whether the sentence is true or false.

not open; false
1. $7 - 3 = 2$
2. $x + 9 = 14$ open
not open; true
3. $8 < 10$
open
4. $x + 5 > 3$
5. $9 + 1 > 2$
not open; true
6. $x + 2 < 6$ open
7. $4 + 5 < 2$
not open; false
8. $10 = x + 5$
open

Which of the given values of the variable, if any, is a solution of the open sentence?

9. $x + 8 = 17;$ 6, ⑨, 10
10. $11 = x - 8;$ 9, 12, ⑲
11. $x + 5 = 14;$ 6, 8, ⑨
12. $8 = y - 9;$ 4, 11, ⑰
13. $7 + p = 25;$ 5, 12, ⑱
14. $a + 9 = 34;$ 16, 27, 29 none
15. $y + 5 = 16;$ 10, ⑪, 12
16. $14 = t + 7;$ 6, ⑦, 9
17. $p - 7 = 15;$ 18, 20, ㉒
18. $8 + b = 39;$ 30, ㉛, 32
19. $t + 5 < 14;$ ⑥, ⑧, 10
20. $y + 9 < 12;$ ⓪, ①, ②, 4
21. $t + 4 > 7;$ 2, ④, ⑤
22. $m + 2 > 15;$ 6, 7, ⑭
23. $a - 5 > 6;$ 8, 9, 10 none
24. $x + 5 < 8;$ 4, 5, 6 none
25. $x + 5 = 7 + 9;$ 9, 10, ⑪
26. $19 - 5 = x + 6;$ 5, ⑧, 9
27. $14 - 9 = 2 + g;$ 1, ③, 7
28. $t + 5 > 6 + 4;$ 5, ⑧, ⑨
29. $7 + g > 3 + 6;$ 2, ④, ⑥
30. $y - 6 < 12 + 5;$ ⑦, ⑨, ⑳

★ 31. Are there any whole numbers for which $x + 5 < 5$ is true? Why or why not? No, $0 + 5 = 5$; x would have to be negative.

Challenge

Add the numbers in the horizontal row. Add the numbers in the diagonal. Supply the numbers in the empty boxes so that the sum of the numbers in any row, column, or diagonal is the same.

16	10	28	6
30	4	18	8
2	24	14	20
12	22	0	26

OPEN SENTENCES

Solving Equations

—◇ **OBJECTIVES** ◇—

To solve equations by using the addition property of equations

To solve equations by using the subtraction property of equations

—◇ **RECALL** ◇—

Evaluate $x + 9$ if $x = 4$.

$$4 + 9$$
$$13$$

Mavis weighs 120 lb. She gains or adds 5 lb. She can *undo* the addition of 5 lb by losing or subtracting 5 lb.

Think: $120 + 5 - 5$ takes her back to her original weight of 120 lb.

Addition and subtraction are *inverse* operations. Inverse operations undo each other.

Example 1

Subtracting 6 from 7, then adding 6 to the result gets back to 7.

Simplify each expression.

$$7 - 6 + 6 \qquad \qquad x + 5 - 5$$
$$\downarrow \qquad \qquad \qquad \downarrow$$
$$7 \qquad \qquad \qquad x$$

---

practice ▷ **Simplify each expression.**

1. $9 + 2 - 2$ 9
2. $8 - 4 + 4$ 8
3. $x + 13 - 13$ x

The sentence $9 + 6 = 15$ is an *equation*. A sentence containing $=$ is an equation.

Example 2

The equation $9 + 6 = 15$ is true. Show that each of the following operations on the equation results in a true equation.

Add 4 to each side.

$$9 + 6 + 4 = 15 + 4$$
$$19 \quad = \quad 19$$
$$\text{true}$$

Subtract 2 from each side.

$$9 + 6 - 2 = 15 - 2$$
$$13 \quad = \quad 13$$
$$\text{true}$$

So, adding 4 to each side of $9 + 6 = 15$ or subtracting 2 from each side results in a true equation.

CHAPTER ONE

You can add or subtract the same number from each side of an equation.

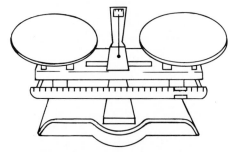

addition/subtraction properties of equations
For all numbers a, b, and c,

$$\text{if } a = b$$
$$\text{then } a + c = b + c$$
$$\text{and } a - c = b - c.$$

In the equation $x - 5 = 14$, x is not alone on the left side of the $=$ symbol. 5 is subtracted from x. To solve, *undo* the subtraction by adding 5 to each side, as shown below.

Example 3

Solve $x - 5 = 14$. Check the solution.

5 is subtracted from x.
Undo the subtraction by adding 5 to each side.

$$x - 5 = 14$$
$$x - 5 + 5 = 14 + 5$$
$$x = 19$$

To check, substitute 19 for x.

Check.
$$\begin{array}{c|c} x - 5 = 14 \\ \hline 19 - 5 & 14 \\ 14 \\ 14 & = 14 \quad \text{true} \end{array}$$

So, 19 is the solution of $x - 5 = 14$.

---

practice ▷   **Solve and check.**

4. $x - 3 = 5$ 8   **5.** $t - 18 = 3$ 21   **6.** $g - 9 = 1$ 10

Example 4

Solve $a + 6 = 27$. Check the solution.

Undo the addition of 6 to a. Subtract 6 from each side.

$$a + 6 = 27$$
$$a + 6 - 6 = 27 - 6$$
$$a = 21$$

Check.
$$\begin{array}{c|c} a + 6 = 27 \\ \hline 21 + 6 & 27 \\ 27 \\ 27 & = 27 \quad \text{true} \end{array}$$

To check, substitute 21 for a.

So, 21 is the solution of $a + 6 = 27$.

---

practice ▷   **Solve and check.**

7. $m + 7 = 13$ 6   **8.** $x + 5 = 25$ 20   **9.** $b + 3 = 18$ 15

SOLVING EQUATIONS

9

Example 5

Solve $29 = y + 8$. Check the solution.

Undo the addition of 8.
Subtract 8 from each side.

$$29 = y + 8$$
$$29 - 8 = y + 8 - 8$$
$$21 = y$$
$$\text{or} \quad y = 21$$

Check.
$$\begin{array}{c|c} 29 = y + 8 \\ \hline 29 & 21 + 8 \\ & 29 \\ 29 = & 29 \qquad \text{true} \end{array}$$

So, 21 is the solution of $29 = y + 8$.

Example 6

Solve $28 = g - 6$. Check the solution.

Undo the subtraction of 6.
Add 6 to each side.

$$28 = g - 6$$
$$28 + 6 = g - 6 + 6$$
$$34 = g$$
$$\text{or} \quad g = 34$$

Check.
$$\begin{array}{c|c} 28 = g - 6 \\ \hline 28 & 34 - 6 \\ & 28 \\ 28 = & 28 \qquad \text{true} \end{array}$$

So, 34 is the solution of $28 = g - 6$.

practice ▷ **Solve and check.**

10. $x - 8 = 11$ 19   **11.** $y + 4 = 14$ 10   **12.** $27 = y + 8$ 19

◇ ORAL EXERCISES ◇

To solve each equation, what must be done to each side?

subtract 7	add 14	subtract 9	add 15
1. $x + 7 = 19$	**2.** $a - 14 = 8$	**3.** $m + 9 = 13$	**4.** $y - 15 = 14$
5. $9 = x + 4$	**6.** $14 = t + 13$	**7.** $25 = b - 8$	**8.** $29 = h + 14$
subtract 4	subtract 13	add 8	subtract 14

◇ EXERCISES ◇

Simplify each expression.

1. $8 + 12 - 12$ 8   **2.** $17 - 4 + 4$ 17   **3.** $27 - 19 + 19$ 27   **4.** $46 + 13 - 13$ 46

Solve each equation. Check the solution.

5. $x + 7 = 19$ 12   **6.** $y - 14 = 7$ 21   **7.** $z - 18 = 2$ 20   **8.** $w + 11 = 19$ 8

9. $9 = a + 7$ 2   **10.** $11 = b - 17$ 28   **11.** $c - 12 = 14$ 26   **12.** $y + 3 = 17$ 14

13. $x + 14 = 16$ 2   **14.** $17 = b + 13$ 4   **15.** $19 = p - 16$ 35   **16.** $t - 7 = 21$ 28

17. $x + 6 = 38$ 32   **18.** $32 = f - 9$ 41   **19.** $g + 8 = 23$ 15   **20.** $43 = u - 7$ 50

21. $x - 27 = 18$ 45   **22.** $y + 13 = 45$ 32   **23.** $33 = a + 13$ 20   **24.** $41 = y + 19$ 22

★ **25.** $7 + x + 5 = 49 + 17$ 54   ★ **26.** $18 + 16 = 12 - 4 + g + 1$ 25

Algebraic Phrases

To write English phrases in
  mathematical form

English Phrase	Mathematical Symbol
increase	+
decrease	−

Example 1

Write in mathematical form.

7 increased by 6	x increased by 9
7 + 6	x + 9

5 decreased by 2	y decreased by 4
5 − 2	y − 4

---

practice ▷ **Write in mathematical form.**

1. 6 increased by 7
      $6 + 7$
2. 5 decreased by 3
      $5 - 3$
3. x increased by 11
      $x + 11$

Example 2

Write in mathematical form.
the sum of x and 12
  $x + 12$

Example 3

8 less than 14 does not mean 8 − 14.
  It means 14 − 8.

8 less than 14   7 less than x
14 ← − → 8   x ← − → 7

9 more than 3 means 3 made
  greater by 9.

9 more than 3   2 more than y
3 ← + → 9   y ← + → 2

---

practice ▷ **Write in mathematical form.**

4. the sum of x and 4 $x + 4$
6. 5 less than b $b - 5$
5. the sum of 6 and y $6 + y$
7. 8 more than t $t + 8$

Example 4

Write in mathematical form.
a number decreased by 2 | 6 less than a number

Let x represent the number.

x   decreased by 2 | 6 less than   x

x   −   2 | x ← − → 6

practice ▷ **Write in mathematical form.**

8. a number decreased by 13 $x - 13$ **9.** 5 less than a number $x - 5$

Summary

English Phrase	Mathematical Form
x increased by 4	$x + 4$
the sum of y and 3	$y + 3$
1 more than w	$w + 1$
x decreased by 4	$x - 4$
4 less than x	$x - 4$

Reading in Math

Match each English phrase with its mathematical form.
(A letter may be used more than once.)

1. x increased by 10   b
2. 10 less than x d
3. the sum of x and 15   c
4. 10 more than x b
5. x decreased by 15   a

a. $x - 15$
b. $x + 10$
c. $x + 15$
d. $x - 10$
e. $15 - x$
f. $10 - x$

◇ EXERCISES ◇

Write in mathematical form.

1. 5 increased by 4 $5 + 4$
2. 7 increased by 2 $7 + 2$
3. 15 increased by 7 $15 + 7$
4. 8 increased by 7 $8 + 7$
5. x decreased by 6 $x - 6$
6. 16 increased by g $16 + g$
7. the sum of x and 2 $x + 2$
8. 4 less than t $t - 4$
9. 5 more than w $w + 5$
10. 8 decreased by m $8 - m$
11. the sum of y and 12 $y + 12$
12. 3 more than b $b + 3$
13. 15 less than w $w - 15$
14. u increased by 14 $u + 14$
15. x decreased by 20 $x - 20$
16. the sum of 6 and 12 $6 + 12$
17. 19 less than w $w - 19$
18. y increased by 13 $y + 13$
19. a number increased by 4 $x + 4$
20. 7 more than a number $x + 7$
21. 29 less than a number $x - 29$
22. 17 increased by a number $17 + x$
23. the sum of a number and 14 $x + 14$
24. a number decreased by 23 $x - 23$
25. 15 less than a number $x - 15$
26. 18 more than a number $x + 18$
27. 45 increased by a number $45 + x$
28. 42 more than a number $x + 42$
★ **29.** 7 increased by 3 more than x
★ **30.** 2 increased by 6 less than a number
★ **31.** 15 added to 3 more than a number $x + 3 + 15$
★ **32.** 14 increased by the sum of a number and 4 $14 + (x + 4)$

29. $7 + x + 3$
30. $2 + x - 6$

Solving Word Problems

---◇ **OBJECTIVE** ◇---   ---------◇ **RECALL** ◇---------

To solve word problems by using a four-step problem-solving method

$$\underbrace{x \text{ decreased by } 4}$$
$$x \quad - \quad 4$$

10 less than s
$$s \leftarrow - \rightarrow 10$$

Four steps for solving a word problem are *Read, Plan, Solve,* and *Interpret.*

Example I

If the cost of a shirt is decreased by $4, the sale price is $13. Find the cost of the shirt.

Read

Identify the given.   Given: The sale price is $13.
                   The cost is decreased by $4.

Identify the unknown.   Find: the cost.

Plan

Choose a variable for the unknown.   Let c = the cost.

The cost decreased by $4 is $13.

Write an equation. $c \quad - \quad 4 = 13$

Solve

Solve the equation.

$$c - 4 = 13$$
$$c - 4 + 4 = 13 + 4 \quad \text{Undo the subtraction.}$$
$$c = 17 \quad \text{Add 4 to each side.}$$

Interpret

Check your solution.

Reread the word problem. Is the answer reasonable?

Check: cost decreased by 4 is 13.

$$\begin{array}{c|c} 17 - 4 & 13 \\ 13 & \\ 13 & = 13 \end{array}$$ Substitute 17 for c in the equation.

true

So, the cost of the shirt is $17.

---

practice ▷

1. Hank's weight decreased by 4 lb is 148 lb. Find his weight.   152 lb

2. Tanya's salary increased by $40 is $115. Find her salary. $75

Example 2

10 less than Leroy's age is 14. How old is Leroy?

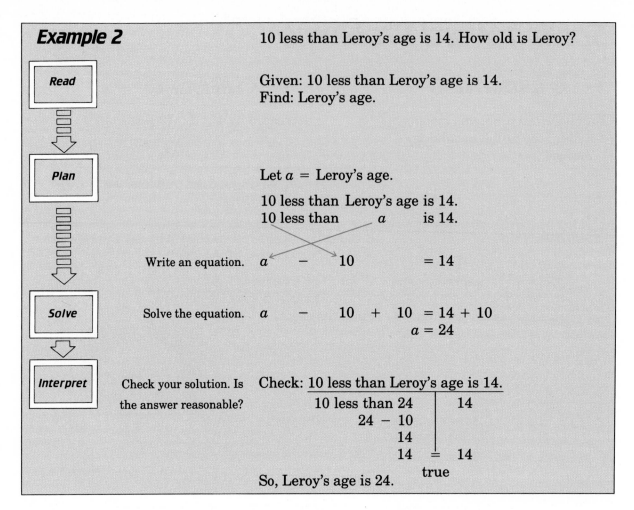

Read

Given: 10 less than Leroy's age is 14.
Find: Leroy's age.

Plan

Let a = Leroy's age.

10 less than Leroy's age is 14.
10 less than a is 14.

Write an equation. $a \quad - \quad 10 \quad = 14$

Solve

Solve the equation. $a \quad - \quad 10 + 10 = 14 + 10$
$$a = 24$$

Interpret

Check your solution. Is the answer reasonable?

Check: 10 less than Leroy's age is 14.

10 less than 24	14
$24 - 10$	
14	
14 $=$	14

true

So, Leroy's age is 24.

practice ▷

3. 5 more than Susan's age is 20. How old is Susan?   15 yr old

4. 6 less than Bernard's age is 8. How old is Bernard?   14 yr old

Example 3

The sum of Fay's salary and a $15 commission is $175. What is Fay's salary?

Given: commission is $15; total pay is $175.
Find: the salary.

Let s = the salary.

Sum of salary and $15 commission is $175.

$$s \quad + \quad 15 \qquad\qquad = 175$$
$$s \quad + \quad 15 - 15 \qquad = 175 - 15$$
$$s \qquad\qquad\qquad = 160$$

Check the answer.
Is the answer reasonable?

◇ EXERCISES ◇

Solve these problems.

1. A number increased by 14 is 39. What is the number?   25

2. A number decreased by 17 is 43. What is the number?   60

3. 30 less than Maria's age is 25. How old is Maria?   55 yr old

4. The cost of a shirt increased by $5 is $19. Find the cost. $14

5. 40 more than Herbie's bowling score is 310. What is his score?   270

6. 10 less than Matt's age is 24. How old is Matt?   34 yr old

7. The sum of a number and 14 is 38. What is the number?   24

8. 32 is the same as 18 less than some number. Find the number.   50

9. The sum of a $3 sales tax and a restaurant bill is $60. What is the bill? $57

10. The sum of Rob's salary and a $25 commission is $175. What is Rob's salary? $150

11. The $360 selling price of a stereo is $50 more than the cost. What is the cost of the stereo? $310

12. The $65 selling price of a camera is the cost increased by a profit of $15. What is the cost? $50

13. 12 lb less than Joe's weight is 139 lb. How much does he weigh?   151 lb

14. 5°F less than the temperature is 45°F. What is the temperature?   50°F

★ 15. If 6 is increased by 5 more than a number, the result is 38. What is the number?   27

★ 16. The sum of 7 more than a number, and 19, is the same as 16 increased by 38. What is the number?   28

Calculator

When using a calculator to compute, mistakes are made by hitting a wrong key. A good practice is to estimate the answer before using the calculator. Then see if your calculator result is close to the estimate.

For example: Estimate 79 + 31 + 199 as

$$\downarrow \quad \downarrow \quad \downarrow$$
$$80 + 30 + 200$$
$$\downarrow \quad \quad \downarrow$$
$$110 \quad + 200 \quad \text{or} \quad 310.$$

Estimate each sum. Use a calculator to check your estimate.

1. 399 + 19 + 188   606
   400 + 20 + 190 = 610

2. 3,999 + 2,107 + 33,022   39,128
   4,000 + 2,100 + 33,000 = 39,100

3. 39 + 11 + 58 + 89 + 798   995
   40 + 10 + 60 + 90 + 800 = 1,000

SOLVING WORD PROBLEMS   15

Problem Solving – Applications
Jobs for Teenagers

In solving word problems, key words or phrases can help you
understand what to do in each problem-solving step. For example, the
following key words and phrases tell you when to add or subtract.

Phrases that tell you to *add:*
  Find the *total*
  How many *combined*
  How many *in all*
  How many *altogether*

Phrases that tell you to *subtract:*
  Find the *change*
  How much *greater than* ⎫
  How much *less than* ⎬ Phrases that compare
  Find the *difference* ⎭

Example 1

A customer bought a $1.95 hamburger, juice for
$0.95, and salad for $1.05. The check was paid with
a $20 bill. How much change should the customer
receive?

Given: The cost of three items; check was paid with
a $20 bill.
Find: The change returned to the customer.

Look for key words to decide what operations to use.
Find the total: *Add* the cost of each item.
Find the change: *Subtract* the total from $20.00.

First: Find the total of the purchases.
$$\begin{array}{r}\$\ 1.95 \\ 0.95 \\ \text{Add} \quad +\ 1.05 \\ \hline \$\ 3.95 \quad \text{total}\end{array}$$

Second: Find the change from $20.00.
$$\begin{array}{r}\$20.00 \\ \text{Subtract} \quad -3.95 \\ \hline \$16.05 \quad \text{change}\end{array}$$

Was the total bill correct?
Check the addition.
Was the change correct?
Check the subtraction by adding.

$$\begin{array}{r}\$16.05 \quad \text{change} \\ +\ 3.95 \quad \text{total bill} \\ \hline \$20.00 \quad \text{paid for with \$20 bill}\end{array}$$

So, the change returned should be $16.05.

Solve these problems.

1. Earl works part time at the luncheonette. He earned $15.35 on Monday, $14.35 on Friday, and $21.55 on Saturday. How much did he earn altogether? $51.25

2. Last week, the cooks prepared 257 breakfasts, 197 lunches, and 244 dinners. How many meals did they prepare in all? 698

3. Mary's tips totaled $14.74 last weekend. This weekend her tips totaled $23.49. How much more did she receive in tips this weekend? $8.75 more

4. How much change from a twenty-dollar bill must the cashier give to a customer who ordered a meal for $4.95 and a cup of coffee for $0.75? $14.30 change

5. One restaurant can seat 235 people. Another restaurant can seat 179 people inside and an additional 72 people in its sidewalk cafe. What is the difference between the number of people the two restaurants can seat? 16

6. Wanda is working part time at the luncheonette. She saved $375 in the fall, $410 in the spring, and $975.59 in the summer. How much did she save in all? $1,760.59

7. Abe treated his two friends to lunches, which cost $4.75 and $4.79. His lunch cost $4.35. How much was the total bill? $13.89

8. The Monday luncheon special costs $4.95. This same lunch on other days costs $6.50. How much less does the lunch cost on Monday? $1.55 less

9. Juice costs $0.50, a hamburger $1.65, and salad $1.35. How much greater than the $3.50 daily special is the total cost of these three items? They both cost the same.

10. The local bakery delivered 78 loaves of bread to the luncheonette during one week, 102 during the next, and 97 during the third. The bakery billed the luncheonette owner for 267 loaves. Was this correct? No; there were 277 loaves.

Formulas

To evaluate formulas
To write formulas describing
  mathematical relationships

Evaluate $x + 7$ if $x = 13$.
$13 + 7$
20

See Geometry Supplement on page 420 for more information.

Example 1

A formula for the perimeter of a triangle is
$P = a + b + c$. Find the perimeter of triangle STW
if $a = 8$ in., $b = 3$ in., and $c = 9$ in.

*Perimeter P of a triangle is the distance
around the triangle; a, b, and c
represent the lengths of the sides
of the triangle.*

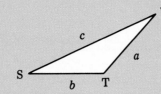

$P = a + b + c$
$P = 8 + 3 + 9$
$P = 20$

So, the perimeter of triangle STW is 20 in.

Example 2

A formula for the perimeter of a rectangle is
$P = l + w + l + w$. Find the perimeter of a rectangle
if the length (l) is 14 cm and the width (w) is 5 cm.

To find the perimeter, replace l by 14
and w by 5 in the formula and add.

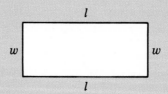

$P = l + w + l + w$
$P = 14 + 5 + 14 + 5$
$P = 38$

So, the perimeter of the rectangle is 38 cm.

---

practice ▷   Find P.

1. $P = a + b + c$

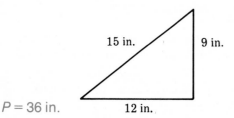

15 in.   9 in.

$P = 36$ in.   12 in.

2. $P = l + w + l + w$

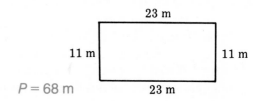

23 m

11 m   11 m

$P = 68$ m   23 m

Example 3

An important formula in business is selling price = cost plus profit.

$$S = C + P$$

Find S if C is \$14.00 and P is \$6.00.

$S = C + P$
$S = 14 + 6$
$S = 20$

So, the selling price, S, is \$20.00.

practice ▷ **3.** Find S in $S = C + P$ if C is \$47 and P is \$13. $S = \$60$

4. Find C in $C = S - P$ if S is \$73 and P is \$19. $C = \$54$

Example 4

Write a formula.

Customer charge is selling price plus tax.

$$C = S + T$$

practice ▷ **Write a formula.**

5. The winning record is the total games minus the losses.
$W = T - L$

6. The class income is the dues added to the profit from the dances. $I = d + p$

◇ EXERCISES ◇

Find P.

1. $P = a + b + c$

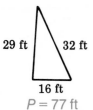
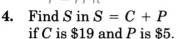

29 ft   32 ft
16 ft
$P = 77$ ft

2. $P = l + w + l + w$

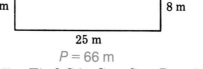

25 m
8 m   8 m
25 m
$P = 66$ m

3. $P = a + b + c + d$

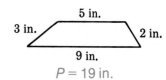

5 in.
3 in.   2 in.
9 in.
$P = 19$ in.

4. Find S in $S = C + P$ if C is \$19 and P is \$5.
$S = \$24$

5. Find C in $C = S - P$ if S is \$43 and P is \$7.
$C = \$36$

6. Find T in $T = A + B - C$ if A is 8, B is 7, and C is 3.
$T = 12$

Write a formula.

7. Money left in a checking account is the balance minus the withdrawal. $M = b - w$

Problem Solving – Careers
Hotel Management

1. The manager of the Lowatt Regency Hotel advertises the following weekend special in the newspaper: "For the month of May stay in a deluxe room for two nights, Saturday and Sunday, for only $120 total." If the regular room rate is $85.00 per night, how much less will the hotel make on each weekend special? $50 less

2. The manager of the restaurant of a hotel must check on the performance of a new waitress. The waitress wrote a check for $34.65, including tax. The customer gave her $40. She gave him back $5.45. Is the arithmetic correct?   No; it should be $5.35.

3. Ms. Joyce Hammer is a hotel chain manager. She attends the Hotel Administrators Convention for two days. She is allowed an expense account for the meeting. Joyce submits the following expense voucher: room $135, food $44, and transportation $23. Find the total of her expenses. $202

4. Frank Childs is a hotel clerk at the Happyway Inn. The manager at the Harriet Motel has called to find out if rooms are available for an over-booking of guests. Frank knows that 57 of the 85 rooms at Happyway are occupied. How many rooms does he have available?   28 rooms

5. The administrator of a large hotel wants to compare this summer's total guest registration with that of last summer. This summer there were 755 guests in June; 1,055 in July; and 1,145 in August. Last summer there was a total of 3,415 guests. How many fewer guests were there this summer than last?   There were 460 fewer guests this summer.

CHAPTER ONE

Non-Routine Problems

There is no simple method for solving these problems. You may need to try different methods until you find one that works.
See TE Answer Section for sample solutions to these problems.

1. The teller's drawer has some $10 bills and some $50 bills, 20 bills altogether. These 20 bills are worth $520. How many $10 bills and how many $50 bills are there?   12, $10 bills and 8, $50 bills

2. You decided to save money the painless way: 1¢ the first week, 2¢ the second week, 4¢ the third week, and so on, for 16 weeks. Your friend says that you will save more this way than if you save $60 every week for 16 weeks. Is your friend right? First guess the answer, then compute to see if your friend was right.   No; $655.35 vs. $960.

3. Your friend was buying something. The clerk told your friend that one of these would cost 10¢, fifty-four would cost 20¢, seven hundred would cost 30¢, and five thousand would cost 40¢. What was your friend buying?   Digits; each digit costs 1 dime.

4. Henry and Peter started working for two different companies at the same salary. Last year Henry had a raise of one-tenth of his salary. Peter's company had hard financial times, and Peter had to have his salary reduced by one-tenth. This year Henry's company had to reduce his salary by one-tenth. Peter was lucky, and his salary was raised by one-tenth. Who is making more this year?
Neither; they are again earning the same amount.

5. You bought one math book, one science book, and one social studies book. You paid $31 for the three books. The science book costs $1.50 more than the math book, and the social studies book cost $1.00 more than the science book. How much did each of the three books cost?   math book, $9.00; science book, $10.50; social studies book, $11.50

COMPUTER ACTIVITIES

A computer program is a set of written instructions for a computer. If you have access to a computer that uses the language of **BASIC**, you may use the computer to type in the program in the Example below.

Example Write a program to find S for S = 219 − 43 + 124
Then type the program into a computer and **RUN** it. (At the end of each line, press the **RETURN** key to get to the next line.)

Write a formula telling the computer how to find S. This uses the LET command.	10 LET S = 219 - 43 + 124
The PRINT command tells the computer to print S.	20 PRINT S
END tells the computer that the program is finished.	30 END
The computer will do nothing until you tell it to **RUN** the program. Type **RUN**	]RUN
The computer now displays the value of S.	300

Notice that the lines of the program are numbered by 10's. This allows you to go back and insert more lines in between any two lines if necessary. To clear the present program from the computer's memory:

Press **RETURN**

Type **NEW**

Press **RETURN**

Exercises

Find the value of each by first doing your own computation. Then write and **RUN** a program to check that the computer's results are the same as your own.

1. Find A. A = 76 + 49 − 18   107
2. Find B. B = 82 − 54 + 16   44
3. Find A. A = 69 + 49   118
4. Find Y. Y = 114 − 39 + 73   14
5. Find P. P = 89 − 62 + 43 − 26   44
6. Find B. B = 95 + 46 − 7   134

Chapter Review

Name the terms, variables, and constants of each expression. [1]

1. $m + 19$ *m*, 19; *m*; 19
2. $23 + y - 14$
   23, *y*, 14; *y*; 23, 14
3. $x + 17 - w + 39$
   x, 17, *w*, 39; *x*, *w*; 17, 39

Evaluate if $a = 19$, $b = 25$, $c = 17$, and $d = 32$. [1]

4. $a + 29$ 48
5. $42 - c$ 25
6. $a + b + d - c$
   59

Which property of addition is illustrated? [3]

7. $43 + 0 = 43$
   identity
8. $(8 + 6) + 4 = 8 + (6 + 4)$
   associative
9. $142 + 63 = 63 + 142$
   commutative

Identify whether the sentence is open or not. For each sentence that is not open, indicate whether the sentence is true or false. [5]

10. $x + 8 = 23$ open
11. $12 - 9 = 1$ not open; false
12. $25 = 18 + 7$
    not open; true

Which of the given values of the variable is a solution of the open sentence? [5]

13. $x + 9 = 16$; 5, ⑦, 9
14. $t + 8 < 11$; ⓪, ①, ②, 3
15. $9 - 5 = x + 3$; ①, 4, 5

Solve each equation. Check the solution. [8]

16. $x - 6 = 17$ 23
17. $y + 8 = 29$ 21
18. $39 = b + 7$ 32
19. $33 = k - 7$
    40

Solve these problems. [13]

20. 7 more than Tina's age is 29. How old is Tina?   22 yr old

21. 49 is the same as a number decreased by 12. What is the number?   61

Find P. [18]

18 in.   17 in. $P = 55$ in.

20 in.

22. $P = a + b + c$

23. Find y in $y = a + b - c$ if $a = 27$, $b = 63$, and $c = 19$. $y = 71$

Write a formula. [18]

24. The amount paid by a customer is the selling price minus the discount. $A = s - d$

Name the terms, variables, and constants of each expression.

1. $t + 14$ *t, 14; t; 14*

2. $x - 19 + y$ *x, 19, y; x, y; 19*

3. $c - d + 4 + f$ *c, d, 4, f; c, d, f; 4*

Evaluate if $a = 15$, $b = 29$, and $c = 19$.

4. $a + 17$ *32*

5. $51 - b$ *22*

6. $a + b - c$ *25*

Which property of addition is illustrated?

7. $(7 + 5) + 28 = 7 + (5 + 28)$
   associative

8. $19 + 129 = 129 + 19$
   commutative

9. $0 + 32 = 32$
   identity

Identify whether the sentence is open or not. For each sentence that is not open, indicate whether the sentence is true or false.

10. $8 + 4 = 12$ *not open; true*

11. $t - 3 = 11$ *open*

12. $7 + 3 < 4$
    not open; false

Which of the given values is a solution of the open sentence?

13. $a + 12 = 32$; 18, ⟨20⟩, 24

14. $m + 5 < 10$; ⟨2⟩, ⟨3⟩, 5, 6

Solve each equation. Check the solution.

15. $x + 7 = 12$ *5*

16. $19 = a + 14$ *5*

17. $y - 9 = 23$ *32*

18. $14 = m - 6$ *20*

Solve these problems.

19. $20 less than John's salary is $130. Find his salary. *$150*

20. The sum of a number and 19 is 32. What is the number? *13*

Find P.

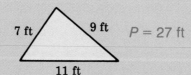

$P = 27$ ft

21. $P = a + b + c$

22. Find g in $g = x - y + z$, if $x = 32$, $y = 19$, and $z = 4$. *g = 17*

Write a formula.

23. The score in a game is the earned points minus the penalty points. *S = e − p*

Equations–Multiplying and Dividing

Diners atop Seattle's Space Needle can view the city from a height of 607 ft.

Multiplication and Division

◇ OBJECTIVE ◇

To evaluate expressions like $6x$, $\frac{x}{3}$, $\frac{18}{y}$, and $\frac{5a}{2}$ for given value(s) of the variable(s)

◇ RECALL ◇

$$4 \times 7 = 28 \qquad 36 \div 4 \text{ means } 4\overline{)36}^{\,9}$$

$$\uparrow \qquad\qquad\qquad\qquad \uparrow$$

multiply   divide

Let n represent any number.
You can represent 5 times any number as

$$\downarrow$$
$$5 \cdot n$$

or, you can omit the centered dot.

$$5n$$

Example 1

Evaluate.

$4a$ if $a = 3$	$16b$ if $b = 24$

$4a$ means $4 \cdot a$. $4 \cdot a$ $16 \cdot b$

Substitute 3 for a. $4 \cdot 3$ $16 \cdot 24$

$\qquad\qquad\qquad\qquad 12 \qquad\qquad\quad 384$

$$\begin{array}{r} 24 \\ \times 16 \\ \hline 144 \\ 24 \\ \hline 384 \end{array}$$

---

practice ▷   **Evaluate if $x = 5$, $b = 12$, $m = 32$, and $p = 23$.**

1. $7x$ 35   **2.** $9b$ 108   **3.** $13m$ 416   **4.** $14p$ 322

Example 2

Evaluate.

$\frac{x}{3}$ if $x = 24$	$\frac{438}{m}$ if $m = 6$

Substitute 24 for x. $\frac{24}{3}$ $\frac{438}{6}$

$\frac{24}{3}$ means $24 \div 3$.   8   73

$$\begin{array}{r} 73 \\ 6\overline{)438} \\ 42 \\ \hline 18 \\ 18 \\ \hline \end{array}$$

---

practice ▷ **Evaluate if $a = 8$, $b = 5$, $c = 6$, and $d = 28$.**

5. $\dfrac{a}{4}$ 2   **6.** $\dfrac{20}{b}$ 4   **7.** $\dfrac{204}{c}$ 34   **8.** $\dfrac{d}{7}$ 4

An expression like $\dfrac{5a}{2}$ involves both multiplication $(5 \cdot a)$ and division $(5 \cdot a$ divided by 2).

Example 3

Evaluate $\dfrac{5a}{2}$ if $a = 4$.

Substitute 4 for a. $\dfrac{5 \cdot a}{2} = \dfrac{5 \cdot 4}{2}$

Multiply. $5 \cdot 4 = 20$ $= \dfrac{20}{2}$

Divide. $20 \div 2 = 10$ $= 10$

---

practice ▷ **Evaluate if $x = 4$, $y = 9$, $a = 8$, and $b = 5$.**

9. $\dfrac{3x}{2}$ 6   **10.** $\dfrac{4y}{3}$ 12   **11.** $\dfrac{5a}{4}$ 10   **12.** $\dfrac{6b}{10}$ 3

◇ EXERCISES ◇

Evaluate if $a = 6$, $b = 12$, $c = 26$, $d = 36$, $x = 5$, and $y = 7$.

1. $9a$ 54   **2.** $3b$ 36   **3.** $7y$ 49   **4.** $5a$ 30

5. $12c$ 312   **6.** $13d$ 468   **7.** $23c$ 598   **8.** $12d$ 432

9. $14x$ 70   **10.** $32c$ 832   **11.** $15d$ 540   **12.** $18b$ 216

13. $\dfrac{a}{2}$ 3   **14.** $\dfrac{35}{x}$ 7   **15.** $\dfrac{b}{6}$ 2   **16.** $\dfrac{138}{a}$ 23

17. $\dfrac{252}{a}$ 42   **18.** $\dfrac{315}{x}$ 63   **19.** $\dfrac{d}{9}$ 4   **20.** $\dfrac{405}{x}$ 81

21. $\dfrac{3b}{4}$ 9   **22.** $\dfrac{13a}{2}$ 39   **23.** $\dfrac{7b}{4}$ 21   **24.** $\dfrac{3d}{2}$ 54

25. $\dfrac{3a}{6}$ 3   **26.** $\dfrac{5b}{4}$ 15   **27.** $\dfrac{15b}{2}$ 90   **28.** $\dfrac{3c}{2}$ 39

29. $\dfrac{25a}{3}$ 50   **30.** $\dfrac{17b}{3}$ 68   ★ **31.** $\dfrac{5d}{3}$ 10   ★ **32.** $\dfrac{14ad}{3y}$ 144

MULTIPLICATION AND DIVISION

Properties of Multiplication

To recognize the commutative, associative, and identity properties of multiplication

To use the commutative and associative properties to make computation easier

$4 \cdot 5 = 20$ and $5 \cdot 4 = 20$

So, $4 \cdot 5 = 5 \cdot 4$

When multiplying two numbers, you can change the order.

commutative property of multiplication
For all numbers a and b, $a \cdot b = b \cdot a$.

Example 1

Multiply inside parentheses first.
Both answers are the same.

Show that $(4 \cdot 2) \cdot 3 = 4 \cdot (2 \cdot 3)$.

$(4 \cdot 2) \cdot 3$	$4 \cdot (2 \cdot 3)$
$8 \cdot 3$	$4 \cdot 6$
24	24

So, $(4 \cdot 2) \cdot 3 = 4 \cdot (2 \cdot 3)$.
When multiplying, you can change the way numbers are grouped.

associative property of multiplication
For all numbers a, b, and c, $(a \cdot b) \cdot c = a \cdot (b \cdot c)$.

Example 2

Multiply.

$6 \cdot 1$	$1 \cdot 6$	$7 \cdot 1$	$1 \cdot 7$
6	6	7	7

Product means multiply.   So, the product of any number and 1 is that number.

property of multiplicative identity
For any number a, $a \cdot 1 = a$ and $1 \cdot a = a$.

Example 3

Which property is illustrated?

$6 \cdot 8 = 8 \cdot 6$	commutative property of multiplication
$(7 \cdot 5) \cdot 3 = 7 \cdot (5 \cdot 3)$	associative property of multiplication
$1 \cdot a = a$	property of multiplicative identity

Which property of multiplication is illustrated?

    1. $(8 \cdot 7) \cdot 9 = 8 \cdot (7 \cdot 9)$ **2.** $16 \cdot 43 = 43 \cdot 16$ **3.** $39 \cdot 1 = 39$
       associative   commutative   identity

Example 4

Rewrite the product $25 \cdot 6 \cdot 4$ to make the computation easier. Then compute.

Group 25 and 4 together, since their product is easier to find.

$$\begin{aligned} 25 \cdot 6 \cdot 4 &= 25 \cdot 4 \cdot 6 \quad \text{commutative property} \\ &= (25 \cdot 4) \cdot 6 \quad \text{associative property} \\ &= 100 \cdot 6 \\ &= 600 \end{aligned}$$

If 6 and 4 are grouped together, the answer is the same, but the computation is more difficult.

$$\begin{aligned} 25 \cdot 6 \cdot 4 &= 25 \cdot (6 \cdot 4) \\ &= 25 \cdot 24 \\ &= 600 \end{aligned}$$

---

practice ▷   **Rewrite to make the computation easier. Then compute.**

    4. $25 \cdot 17 \cdot 4$ **5.** $2 \cdot 19 \cdot 50$ **6.** $5 \cdot 33 \cdot 20$
        $25 \cdot 4 \cdot 17 = 1{,}700$ $2 \cdot 50 \cdot 19 = 1{,}900$ $5 \cdot 20 \cdot 33 = 3{,}300$

◇ ORAL EXERCISES ◇

Which two factors would you combine to make the computation easier?

    1. $50 \cdot 63 \cdot 2$ **2.** $4 \cdot 39 \cdot 25$ **3.** $2 \cdot 86 \cdot 50$ **4.** $25 \cdot 165 \cdot 4$
       50 and 2   4 and 25   2 and 50   25 and 4

◇ EXERCISES ◇

Which property of multiplication is illustrated?
                                           associative

    1. $38 \cdot 1 = 38$ identity   **2.** $(25 \cdot 4) \cdot 7 = 25 \cdot (4 \cdot 7)$ **3.** $1 \cdot 49 = 49$ identity
    4. $(35 \cdot 6) \cdot 5 = 35 \cdot (6 \cdot 5)$ **5.** $249 \cdot 1 = 249$ identity   **6.** $76 \cdot 63 = 63 \cdot 76$
      associative   commutative

Rewrite to make the computation easier. Then compute.

    7. $4 \cdot 23 \cdot 5$ 460   **8.** $50 \cdot 14 \cdot 2$ 1,400   **9.** $2 \cdot 47 \cdot 50$ 4,700   **10.** $8 \cdot 3 \cdot 50$ 1,200
    11. $5 \cdot 26 \cdot 2$ 260   **12.** $4 \cdot 17 \cdot 25$ 1,700   **13.** $50 \cdot 43 \cdot 2$ 4,300   **14.** $20 \cdot 62 \cdot 5$ 6,200
    15. $25 \cdot 28 \cdot 4$ 2,800   **16.** $150 \cdot 7 \cdot 2$ 2,100   **17.** $4 \cdot 7 \cdot 50$ 1,400   **18.** $125 \cdot 3 \cdot 4$ 1,500
    19. $2 \cdot 165 \cdot 25 \cdot 2$ 16,500   **20.** $500 \cdot 78 \cdot 2 \cdot 2$ 156,000   **21.** $5 \cdot 39 \cdot 2 \cdot 5 \cdot 2$ 3,900

 ★ **22.** Does the associative property hold for division?
        (*Hint:* Does $(12 \div 6) \div 2 = 12 \div (6 \div 2)$?)   no

---

Order of Operations

◇ **OBJECTIVES** ◇

To compute numerical
expressions using the rules for
order of operations
To compute numerical
expressions containing
parentheses

◇ **RECALL** ◇

$8 \cdot 4$ means $8 \times 4 = 32$.
$12 \div 3$ means $3\overline{)12} = 4$.

Example 1

What is the value of $4 \cdot 6 + 3$?

There are two possibilities.

Multiply first.	Add first.
Then add.	Then multiply.
$4 \cdot 6 + 3$	$4 \cdot 6 + 3$
$24 + 3$	$4 \cdot \quad 9$
27	36

Which is correct, 27 or 36?

We make this agreement in order to
avoid confusion.

order of operations
When several operations occur,
1. Compute all multiplications and divisions first in
   order from left to right.
2. Then compute all additions and subtractions in
   order from left to right.

Thus, in Example 1 above, $\underbrace{4 \cdot 6} + 3$ Multiply first.

$\qquad\qquad\qquad\qquad 24 + 3 \qquad$ Then add.

$\qquad\qquad\qquad\qquad\qquad 27$

Example 2

Compute.

Multiply first.	$9 + \underbrace{4 \cdot 2}$	$14 - \underbrace{6 \div 2}$	Divide first.
Then add.	$9 + \quad 8$	$14 - \quad 3$	Then subtract.
	17	11	

---

practice ▷ **Compute.**

1. $4 \cdot 7 + 2$ 30   **2.** $20 \div 4 - 3$ 2   **3.** $7 + 8 \cdot 5$ 47   **4.** $2 + 12 \div 6$ 4

Sometimes all four basic operations are involved in the same expression.

Example 3

Compute $4 \cdot 3 + 9 - 12 \div 4 + 2 \cdot 5$.

Do all multiplications and divisions first.	$4 \cdot 3 + 9 - 12 \div 4 + 2 \cdot 5$

$$4 \cdot 3 + 9 - 12 \div 4 + 2 \cdot 5$$
$$12 + 9 - 3 + 10$$

Then do all additions and subtractions.

$$21 - 3 + 10$$
$$18 + 10$$
$$28$$

---

practice ▷ **Compute.**

5. $5 \cdot 2 + 6 - 10 \div 2 + 4 \cdot 6$ 35   **6.** $12 + 6 \cdot 3 - 8 \div 4 - 1 \cdot 3$ 25

7. $8 \cdot 3 - 21 \div 7 + 5 \cdot 7$ 56

If operations occur within parentheses, they are to be done first. Then the rules for order of operations are followed.

Example 4

Compute $5 + (6 + 3) \cdot 2$.

Compute within parentheses first. $5 + (6 + 3) \cdot 2$

Multiply. $5 + 9 \cdot 2$

Then add. $5 + 18$
$$23$$

When parentheses are used as a grouping symbol, the multiplication symbol may be omitted.

Example 5

Compute $8(5 - 1) + 18 \div 6$.

Compute within parentheses first. $8(5 - 1) + 18 \div 6$ $8(5 - 1)$ means $8 \cdot (5 - 1)$.

Do multiplications and divisions next. $8(4) + 18 \div 6$

Then add. $32 + 3$
$$35$$

---

practice ▷ **Compute.**

    8. $8 + (5 + 2)3$ 29 **9.** $7(3 - 1) + 4$ 18 **10.** $6(2 + 3) - 16 \div 4$ 26

◇ ORAL EXERCISES ◇

Tell the order of operations you must perform to compute each.

1. $5 + \underbrace{7 \cdot 3}_{1^{st}}$ **2.** $\underbrace{4 \cdot 7}_{1^{st}} - 1$ **3.** $6 + \underbrace{9 \div 3}_{1^{st}}$ **4.** $\underbrace{3 \cdot 6}_{1^{st}} + 7$ **5.** $9 - \underbrace{8 \div 4}_{1^{st}}$

6. $7 - \underbrace{15 \div 5}_{1^{st}}$ **7.** $\underbrace{8 \cdot 2}_{1^{st}} + 5$ **8.** $45 - \underbrace{4 \cdot 6}_{1^{st}}$ **9.** $10 + \underbrace{14 \div 7}_{1^{st}}$ **10.** $11 + \underbrace{9 \cdot 3}_{1^{st}}$

◇ EXERCISES ◇

Compute.

1. $5 \cdot 2 + 7$ 17   **2.** $6 + 5 \cdot 3$ 21   **3.** $10 - 2 \cdot 3$ 4

4. $16 - 8 \div 4$ 14   **5.** $7 \cdot 6 - 5$ 37   **6.** $9 \cdot 4 + 5$ 41

7. $2 \cdot 9 - 8$ 10   **8.** $8 - 12 \div 6$ 6   **9.** $9 - 14 \div 2$ 2

10. $8 - 20 \div 5$ 4   **11.** $16 - 15 \div 3$ 11   **12.** $13 - 4 \cdot 3$ 1

13. $9 \div 3 + 5$ 8   **14.** $7 + 4 \cdot 6$ 31   **15.** $9 \cdot 7 - 3$ 60

16. $4 \cdot 2 + 8 - 15 \div 5 + 8$ 21   **17.** $8 - 12 \div 2 + 7 \cdot 6$ 44   **18.** $30 \div 5 - 1 \cdot 3 + 9$ 12

19. $13 - 16 \div 2 + 7 \cdot 4$ 33   **20.** $6 \cdot 8 - 21 \div 3 + 4 \cdot 9$ 77   **21.** $3 + 7 \cdot 5 - 25 \div 5$ 33

22. $4 \cdot 12 - 28 \div 14 + 7 \cdot 3$ 67   **23.** $4 \cdot 9 - 35 \div 7 + 7 \cdot 7$ 80   **24.** $9 \cdot 5 + 27 \div 3 - 1$ 53

25. $7 + (6 + 4)2$ 27   **26.** $6(11 - 5) + 3$ 39   **27.** $7(5 - 2) - 45 \div 9$ 16

28. $(9 - 6)2 + 7$ 13   **29.** $7(8 - 6) + 3$ 17   **30.** $11 - 2(18 - 14)$ 3

31. $4(9 - 7) + 49 \div 7$ 15   **32.** $3(6 + 5) - 18 \div 2$ 24   **33.** $35 \div 7 + (7 + 2)2$ 23

Compute. (*Hint:* Start within the parentheses, then work within the brackets.)

★ **34.** $24 \div [14 - 2(3 + 1)] + 17$ 21   ★ **35.** $48 - 6[20 - 3(54 \div 9)] - 4 \cdot 9$ 0

Calculator

Copy and complete.
$$1 \cdot 9 + 2 = \underline{\quad 11 \quad}$$
$$12 \cdot 9 + 3 = \underline{\quad 111 \quad}$$
$$123 \cdot 9 + 4 = \underline{\quad 1{,}111 \quad}$$
$$1{,}234 \cdot 9 + 5 = \underline{\quad 11{,}111 \quad}$$

If the same pattern is continued, what is the next line?
Continue the pattern as far as you can.
$12{,}345 \cdot 9 + 6 = 111{,}111$
$123{,}456 \cdot 9 + 7 = 1{,}111{,}111$
$1{,}234{,}567 \cdot 9 + 8 = 11{,}111{,}111$
and so on.

Order of Operations: Variables

─── ◇ *OBJECTIVE* ◇ ───   ─────── ◇ *RECALL* ◇ ───────

To use the rules for order of
  operations to evaluate
  algebraic expressions for
  given value(s) of the
  variable(s)

Compute $7 \cdot 4 + 8 - 3 \cdot 2$

$$28 + 8 - 6 \leftarrow \text{Multiply first.}$$
$$36 - 6 \leftarrow \text{Then add and subtract.}$$
$$30$$

Example 1

Evaluate each for the given value of the variable.

$6x + 4$ if $x = 2$	$\frac{x}{3} + 2$ if $x = 15$
$6 \cdot x + 4$	

Substitute 2 for x. $6 \cdot 2 + 4$ $\frac{15}{3} + 2$

Multiply first. $12 + 4$ $5 + 2$

Then add. 16 7

---

practice ▷   **Evaluate if $x = 3$, $y = 2$, and $a = 4$.**

   1. $7x + 3$ 24   **2.** $6 + 2y$ 10   **3.** $\frac{a}{2} + 3$ 5

Example 2

Evaluate $9a - 4 + 3b$ if $a = 6$ and $b = 2$.
$$9a - 4 + 3b$$
$$9 \cdot a - 4 + 3 \cdot b$$

Substitute 6 for a and 2 for b. $9 \cdot 6 - 4 + 3 \cdot 2$

$$54 - 4 + 6$$
$$50 + 6$$
$$56$$

---

practice ▷   **Evaluate if $x = 5$ and $y = 2$.**

   4. $9x - 2 + 13y$ 69   **5.** $7 + 5x + 2y$ 36   **6.** $5 + 4x + \frac{1}{2}y$ 26

Example 3

Evaluate $4xy - 10$ if $x = 2$ and $y = 5$.
$$4xy - 10$$

$4xy$ means $4 \cdot x \cdot y$. $4 \cdot x \cdot y - 10$

Substitute 2 for x and 5 for y. $4 \cdot 2 \cdot 5 - 10$

$4 \cdot 2 \cdot 5 = 8 \cdot 5 = 40$. $40 - 10$

$$30$$

---

practice ▷ Evaluate if $a = 3$, $b = 2$, and $c = 5$.

7. $3ab - 5$ 13   **8.** $6 + 5bc$ 56   **9.** $2ac + 7$ 37

◇ ORAL EXERCISES ◇

Evaluate each for the indicated value of x.

1. $2x + 1$ if $x = 3$ 7   **2.** $1 + 3x$ if $x = 3$ 10   **3.** $2x - 1$ if $x = 4$ 7

4. $\frac{x}{4} + 5$ if $x = 12$ 8   **5.** $\frac{x}{6} - 1$ if $x = 24$ 3   **6.** $7 + \frac{x}{2}$ if $x = 6$ 10

◇ EXERCISES ◇

Evaluate if $x = 2$, $y = 4$, $z = 1$, $a = 3$, $b = 9$, and $c = 5$.

1. $3x + 7$ 13   **2.** $8 + 5x$ 18   **3.** $\frac{b}{3} + 6$ 9

4. $5y - 1$ 19   **5.** $4 + 3y$ 16   **6.** $8y - 1$ 31

7. $7z - 2$ 5   **8.** $\frac{y}{2} + 5$ 7   **9.** $3b - 8$ 19

10. $6 + \frac{c}{5}$ 7   **11.** $3 + 8b$ 75   **12.** $4b + 12$ 48

13. $2a + 3b$ 33   **14.** $5x + 3z$ 13   **15.** $4b + 3c$ 51

16. $2x + 4y$ 20   **17.** $7a + \frac{x}{2}$ 22   **18.** $5a + 3c$ 30

19. $2x + 4c$ 24   **20.** $7x + \frac{b}{3}$ 17   **21.** $\frac{y}{2} + 6z$ 8

22. $3a - 5 + 7c$ 39   **23.** $5y + 2z - 4$ 18   **24.** $3 + 8x + 2y$ 27

25. $7z + 8 + 3y$ 27   **26.** $3 + 5a + 6b$ 72   **27.** $9x + 3b - 8$ 37

28. $5bz - 4$ 41   **29.** $7bc - 2$ 313   **30.** $11 + 2yz$ 19

31. $5xy - 3$ 37   **32.** $7 + 9bc$ 412   **33.** $7xy - 9$ 47

★ **34.** $ab(3x + 2y)$ 378   ★ **35.** $xyz(5ab + 4c)$ 1,240   ★ **36.** $3bc(4xyz - 2a)$ 3,5

★ **37.** $7bc(5xyz - 3a)$ 9,765   ★ **38.** $2ac(3ab - 2xy)$ 1,950   ★ **39.** $(8ab - 3y)2az$ 1,22

Answers may vary. Some possible answers are given.

Step 1 Choose four digits. Write them in descending order.   8,53

Step 2 Reverse the order.   1,358

Step 3 Subtract. $8{,}531 - 1{,}358 = 7{,}173$

Step 4 Reverse the order again.   3,717

Step 5 Add. What is the result? $7{,}173 + 3{,}717 = 10{,}890$

Repeat the five steps above with four different digits. Guess what the answer will be.   The final answer should always be 10,890.

Problem Solving – Careers
Computer Technicians

Most microcomputers store information on *floppy* disks. Floppy disks are made from a flexible plastic material called *mylar*. One letter of the alphabet takes one *byte* of memory on a disk. Thus, the address 137 Abbott Avenue uses up 17 bytes of memory. (Allow one byte of memory for each space between words.) The number of bytes available on a disk depends upon how the disk is manufactured. For example, a $5\frac{1}{4}$ inch single-sided, single-density disk will contain 125,000 bytes. A double-sided, double-density disk can hold 4 times that amount.

1. The words *computer program* require how many bytes of disk memory?   16

Use the following facts to answer Exercises 2–5 below.
A particular kind of disk is made up of 40 *tracks*. Three of these tracks are used for information required to operate the computer. This leaves 37 tracks available for use by the programmer. Each *track* contains 16 *sectors*. Each sector can store 256 bytes of information.

2. A computer technician needs 5 sectors for a program. How many bytes of memory will be used?   1,280

3. How many sectors will be required for 768 bytes of memory?   3

4. Find the total number of bytes of usable memory available on the disk.   151,552

5. Use the result of Exercise 4 above to answer the following: José writes programs for games. His new Space Wars game requires 3,265 bytes of memory. How many bytes are left for another program on the same disk?   148,287

6. Matt has stored a program that requires 9,285 bytes of memory. How many programs of the same size could he store on a $5\frac{1}{4}$ inch single-sided, single-density disk?   13

7. The letter *K* is used to represent approximately 1,000 bytes of memory. A computer having 48*K* memory size can hold approximately 48,000 bytes. Represent the number of bytes that can be stored on a $5\frac{1}{4}$ inch double-sided, double-density disk in terms of *K*.   500K

Solving Equations

◇ OBJECTIVES ◇

To solve equations by using the division property of equations

To solve equations by using the multiplication property of equations

◇ RECALL ◇

Inverse operations are operations that undo each other.

Addition and subtraction are inverse operations.

Multiplication and division are inverse operations.
Inverse operations undo each other.
Division will undo multiplication.

$$\frac{4 \cdot 8}{4} = \frac{32}{4} = 8 \quad \text{8 remains unchanged.}$$

Similarly, multiplication will undo division.

$$2 \cdot \frac{6}{2} = 2 \cdot 3 = 6$$

Example 1

Simplify each expression.

$$\frac{7 \cdot 3}{7} \quad\Big|\quad \frac{6x}{6} \quad\Big|\quad 8 \cdot \frac{16}{8} \quad\Big|\quad 5 \cdot \frac{x}{5}$$

Multiplying 3 by 7, then dividing the result by 7 leaves 3 unchanged.

$$\frac{7 \cdot 3}{7} \quad\Big|\quad \frac{6 \cdot x}{6} \quad\Big|\quad 8 \cdot \frac{16}{8} \quad\Big|\quad 5 \cdot \frac{x}{5}$$

$$\downarrow \qquad\quad \downarrow \qquad\qquad \downarrow \qquad\qquad \downarrow$$

$$3 \qquad\quad x \qquad\qquad 16 \qquad\qquad x$$

---

practice ▷ **Simplify each expression.**

1. $\dfrac{6 \cdot 8}{6}$ 8

2. $\dfrac{5k}{5}$ k

3. $9 \cdot \dfrac{25}{9}$ 25

4. $8 \cdot \dfrac{x}{8}$ x

Example 2

The equations $5 \cdot 2 = 10$ and $\frac{12}{3} = 4$ are true.

Show that each of the following operations on each equation results in a true equation.

$$5 \cdot 2 = 10 \qquad\qquad\qquad \frac{12}{3} = 4$$

Dividing or multiplying each side of a true equation by the same number results in a true equation.

Divide each side by 5.

$$\frac{5 \cdot 2}{5} = \frac{10}{5}$$

$$2 = 2$$

true

Multiply each side by 3.

$$3 \cdot \frac{12}{3} = 3 \cdot 4$$

$$12 = 12$$

true

You can multiply or divide each side of of an equation by the same number.

multiplication property of equations:

For all numbers a, b, and c, if $a = b$, then $c \cdot a = c \cdot b$.

division property of equations:

For all numbers a, b, and c ($c \neq 0$), if $a = b$, then $\frac{a}{c} = \frac{b}{c}$.

Example 3

Solve $4x = 20$. Check the solution.

Undo multiplication by 4.　　$4x = 20$

Divide each side by 4.　　$\dfrac{4 \cdot x}{4} = \dfrac{20}{4}$

$x = 5$

Check:

$$
\begin{array}{c|c}
4x & = 20 \\
\hline
4 \cdot 5 & 20 \\
20 & \\
20 & = 20 \quad \text{true}
\end{array}
$$

Substitute 5 for x.

So, the solution is 5.

---

practice ▷　**Solve each equation. Check the solution.**

5. $3t = 15$　5　　**6.** $9x = 27$　3　　**7.** $7p = 14$　2　　**8.** $4m = 32$　8

Example 4

Solve $\frac{a}{5} = 6$. Check the solution.

Undo division by 5.　　$\dfrac{a}{5} = 6$

Multiply each side by 5.　　$5 \cdot \dfrac{a}{5} = 5 \cdot 6$

$a = 30$

Check:

$$
\begin{array}{c|c}
\frac{a}{5} & = 6 \\
\hline
\frac{30}{5} & 6 \\
6 & \\
6 & = 6 \quad \text{true}
\end{array}
$$

Substitute 30 for a.

So, the solution is 30.

---

practice ▷　**Solve each equation. Check the solution.**

9. $\frac{x}{8} = 7$　56　　**10.** $\frac{b}{4} = 4$　16　　**11.** $\frac{m}{5} = 9$　45　　**12.** $\frac{w}{6} = 3$　18

SOLVING EQUATIONS

Example 5

Solve each equation.

Undo multiplication by 7.	$28 = 7p$	$5 = \dfrac{b}{3}$	Undo division by 3.
Divide each side by 7.	$\dfrac{28}{7} = \dfrac{7 \cdot p}{7}$	$3 \cdot 5 = 3 \cdot \dfrac{b}{3}$	Multiply each side by 3.
	$4 = p$	$15 = b$	
Check the solution.	or $p = 4$	or $b = 15$	

practice ▷ **Solve each equation. Check the solution.**

13. $6g = 30$ 5   **14.** $\dfrac{a}{4} = 10$ 40   **15.** $45 = 9k$ 5   **16.** $7 = \dfrac{y}{9}$ 63

To solve an equation that shows multiplication, divide each side by the same number.

To solve an equation that shows division, multiply each side by the same number.

◇ ORAL EXERCISES ◇

To solve each equation, what must be done to each side?   See TE Answer Section.

1. $6x = 60$ **2.** $\dfrac{m}{12} = 2$ **3.** $7z = 63$ **4.** $\dfrac{a}{4} = 15$

5. $35 = \dfrac{b}{4}$ **6.** $19c = 38$ **7.** $48 = 16y$ **8.** $17 = \dfrac{w}{3}$

◇ EXERCISES ◇

Simplify each expression.

1. $\dfrac{7 \cdot 9}{7}$ 9   **2.** $\dfrac{14t}{14}$ t **3.** $7 \cdot \dfrac{12}{7}$ 12   **4.** $13 \cdot \dfrac{m}{13}$ m

Solve each equation. Check the solution.

5. $11t = 22$ 2   **6.** $7m = 56$ 8   **7.** $44 = 11b$ 4   **8.** $27 = 9m$ 3

9. $24 = 12f$ 2   **10.** $5b = 45$ 9   **11.** $72 = 9j$ 8   **12.** $3p = 33$ 11

13. $\dfrac{x}{8} = 4$ 32   **14.** $\dfrac{m}{6} = 3$ 18   **15.** $8 = \dfrac{y}{3}$ 24   **16.** $5 = \dfrac{w}{12}$ 60

17. $3 = \dfrac{s}{12}$ 36   **18.** $\dfrac{r}{13} = 2$ 26   **19.** $24 = \dfrac{t}{2}$ 48   **20.** $6 = \dfrac{u}{13}$ 78

21. $16k = 320$ 20   **22.** $13 = \dfrac{b}{14}$ 182   **23.** $245 = 7u$ 35   **24.** $23 = \dfrac{a}{11}$ 253

★ **25.** $8 - 3 \cdot 2 + 24 = 13x$ 2   ★ **26.** $3(4 + 2) = \dfrac{x}{5}$ 90

Solving Word Problems

To write English phrases in
   mathematical form
To solve word problems
   involving multiplication or
   division

Solve $5n = 35$

$$\frac{5 \cdot n}{5} = \frac{35}{5}$$

$$n = 7$$

Solve $\frac{x}{3} = 8$

$$3 \cdot \frac{x}{3} = 3 \cdot 8$$

$$x = 24$$

Example 1

Write in mathematical form.

8 times n	twice y	x divided by 3
$8 \cdot n$	2 times y	$\frac{x}{3}$
or $8n$	$2y$	

---

practice ▷   **Write in mathematical form.**

1. twice 9 $2 \cdot 9$ **2.** p divided by 9 $\frac{p}{9}$ **3.** 13 times w $13w$

In Chapter 1, you learned four basic steps for solving
word problems. These steps are used below in
solving word problems involving multiplication or
division.

Example 2

5 times a number is 35. What is the number?

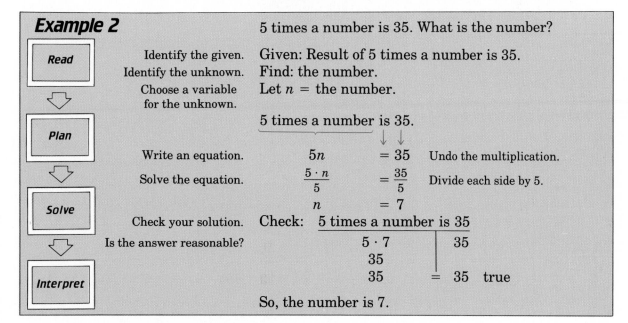

Read

  Identify the given.   Given: Result of 5 times a number is 35.
  Identify the unknown.   Find: the number.
  Choose a variable   Let n = the number.
  for the unknown.

5 times a number is 35.

Plan

  Write an equation. $5n = 35$ Undo the multiplication.

  Solve the equation. $\frac{5 \cdot n}{5} = \frac{35}{5}$ Divide each side by 5.

$$n = 7$$

Solve

  Check your solution.   Check: 5 times a number is 35

  Is the answer reasonable? $5 \cdot 7$ | 35
          35
          35   =   35   true

Interpret

So, the number is 7.

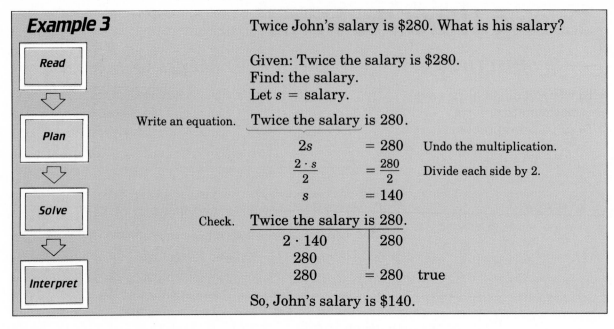

Example 3

Twice John's salary is $280. What is his salary?

Read

Given: Twice the salary is $280.
Find: the salary.
Let s = salary.

Plan

Write an equation. Twice the salary is 280.

$$2s = 280$$ Undo the multiplication.

$$\frac{2 \cdot s}{2} = \frac{280}{2}$$ Divide each side by 2.

$$s = 140$$

Solve

Check. Twice the salary is 280.

$2 \cdot 140$	280
280	
280	= 280 true

Interpret

So, John's salary is $140.

practice ▷ Solve.

4. 7 times a number is 21. Find the number.   3

5. Twice Mary's age is 40. How old is Mary?   20

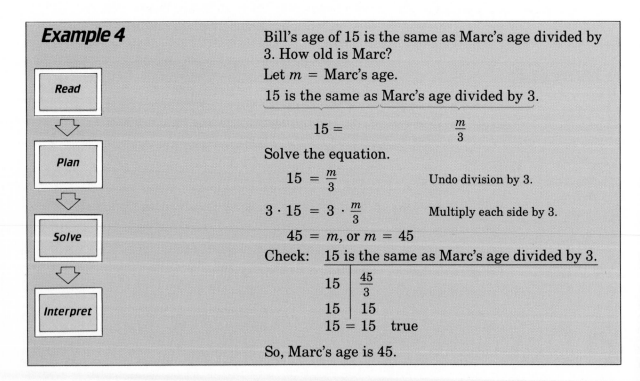

Example 4

Bill's age of 15 is the same as Marc's age divided by 3. How old is Marc?

Let m = Marc's age.

15 is the same as Marc's age divided by 3.

Read

$$15 = \frac{m}{3}$$

Solve the equation.

Plan

$$15 = \frac{m}{3}$$ Undo division by 3.

$$3 \cdot 15 = 3 \cdot \frac{m}{3}$$ Multiply each side by 3.

$$45 = m, \text{ or } m = 45$$

Solve

Check: 15 is the same as Marc's age divided by 3.

15	$\frac{45}{3}$
15	15
15 = 15	true

Interpret

So, Marc's age is 45.

practice ▷ **6.** 8 is the same as a number divided by 5. Find the number. 40

7. Bill's age divided by 2 is 14. How old is Bill? 28

◇ ORAL EXERCISES ◇

Give an equation for each sentence.

1. 6 times n is the same as 18. $6n = 18$
2. 40 is twice b. $40 = 2b$
3. x divided by 4 is the same as 3. $\frac{x}{4} = 3$
4. 25 is the same as y divided by 6. $25 = \frac{y}{6}$
5. 27 is the same as 9 times t. $27 = 9t$
6. r divided by 6 is the same as 12. $\frac{r}{6} = 12$
7. 9 times w is 45. $9w = 45$
8. 32 is the same as u divided by 4. $32 = \frac{u}{4}$
9. q divided by 6 is 14. $\frac{q}{6} = 14$
10. 39 is the same as 13 times k. $39 = 13k$

◇ EXERCISES ◇

Write in mathematical form.

1. 12 times t $12t$
2. n divided by 14 $\frac{n}{14}$
3. twice r $2r$
4. twice u $2u$
5. 24 times g $24g$
6. p divided by 8 $\frac{p}{8}$

Solve these problems.

7. 4 times a number is 44. Find the number. 10

8. Twice a number is 18. Find the number. 9

9. 5 times a number is 35. Find the number. 7

10. 42 is twice a number. Find the number. 21

11. Ruiz's 12 home runs this year is 4 times his record of last season. Find the number of home runs last season. 3

12. Henry's savings of $120 is the same as 4 times his sister's savings. How much did his sister save? $30

13. 12 is the same as a number divided by 4. Find the number. 48

14. A number divided by 7 is 8. Find the number. 56

15. Mona's age of 13 is the same as Joan's age divided by 3. How old is Joan? 39 yr old

16. Pearl's age of 12 is the same as her grandfather's age divided by 6. How old is her grandfather? 72 yr old

17. Car insurance of $800 is twice what it was a decade ago. Find the cost of insurance 10 yr ago. $400

18. The cost of a pizza divided among 4 people is $3.00 each person. Find the cost of the pizza. $12.00

★ **19.** 7 times a number is the same as 24 decreased by 3. Find the number. 3

★ **20.** 8 more than twice 3 is the same as 7 times a number. Find the number. 2

SOLVING WORD PROBLEMS   **41**

Problem Solving – Applications
Jobs for Teenagers

Certain key phrases or situations tell you whether
to multiply or divide.

Situations calling for multiplication:	Situations calling for division:
Given the cost of one item, find the cost of *several*.	Given the cost of several items, find the cost of *each* or *one*.
Given the number of items in one carton, find the number in *several* cartons.	Given a total length to be cut into a number of equal parts, find the length of *each* part.

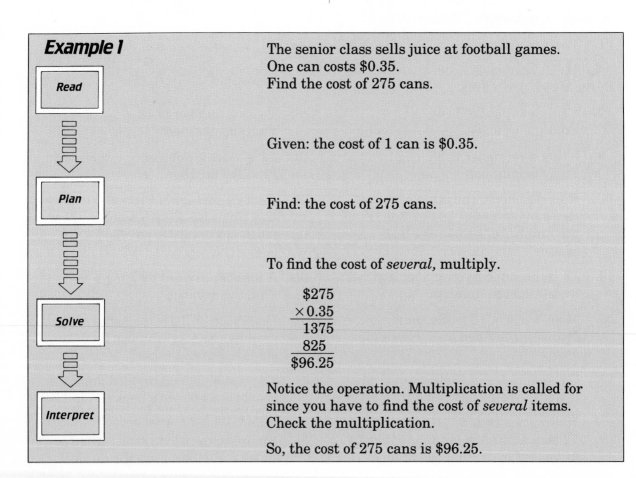

Example 1

Read

The senior class sells juice at football games.
One can costs $0.35.
Find the cost of 275 cans.

Given: the cost of 1 can is $0.35.

Plan

Find: the cost of 275 cans.

To find the cost of *several*, multiply.

$$\begin{array}{r} \$275 \\ \times\,0.35 \\ \hline 1375 \\ 825 \\ \hline \$96.25 \end{array}$$

Solve

Interpret

Notice the operation. Multiplication is called for
since you have to find the cost of *several* items.
Check the multiplication.

So, the cost of 275 cans is $96.25.

Example 2

The senior class has decided to sell greeting cards. A committee of 15 volunteers is to sell 630 boxes. How many boxes must each student sell?

Given: 630 boxes to be sold by 15 students.
Find: the number of boxes to be sold by each student.

To find the number to be sold by *each,* use division.

$$\begin{array}{r} 42 \\ 15\overline{)630} \\ \underline{60} \\ 30 \\ \underline{30} \end{array}$$

Check: 15 students to sell 42 boxes each.

$$\begin{array}{r} 42 \\ \times\,15 \\ \hline 210 \\ 42 \\ \hline \end{array}$$

Total to be sold is 630. The answer checks. 630

So, each student must sell 42 boxes.

Solve these problems.

1. Tickets are ordered for a senior activity. A box contains 125 tickets. How many tickets are there in all if 25 boxes are ordered? 3,125 tickets

2. 800 yearbooks are delivered in 50 cartons. How many yearbooks should be in each carton? 16 books

3. Hamburgers come 36 to a box. The fund-raising committee ordered 15 boxes. How many hamburgers were ordered in all? 540 hamburgers

4. A decorating committee is building shelves for a booth. A 60-in. board is to be cut into 4 equal pieces. How long is each piece? 15 in. each

5. Sandwiches were sold at a game for $1.35 each. The senior class sold 315. How much did the class collect in all for the day? $425.25

6. The total cost of the senior class prom dinner is $2,400. There are 120 seniors. How much should each senior be charged? $20 each

7. The class sells pretzels for $0.35 each. If 145 are sold at a football game, how much money is collected in all? $50.75

8. In a fund-raising raffle, $125 in prize money is to be shared equally by 5 winners. How much money must be given to each winner? $25 each

PROBLEM SOLVING

Mixed Types of Equations

⎯⎯ ◇ OBJECTIVE ◇ ⎯⎯   ⎯⎯⎯⎯ ◇ RECALL ◇ ⎯⎯⎯⎯

To solve equations when different types are mixed

You have already learned to solve equations like

$$14 = c + 9 \qquad 3d = 39$$
$$x - 7 = 2 \qquad 6 = \frac{m}{2}$$

Recall that in order to solve equations like those above, you must use one of the following strategies:

1. Undo addition   Subtract the same number from each side.
2. Undo subtraction   Add the same number to each side.
3. Undo multiplication   Divide each side by the same number.
4. Undo division   Multiply each side by the same number.

Example 1

Solve each equation.

Undo subtraction.
$$x - 7 = 2$$
$$x - 7 + 7 = 2 + 7$$
$$x = 9$$

$$14 = x + 9$$
$$14 - 9 = x + 9 - 9$$
$$5 = x \text{ or } x = 5$$
Undo addition.

Example 2

Solve each equation.

Undo multiplication.
$$3x = 39$$

Divide each side by 3.
$$\frac{3 \cdot x}{3} = \frac{39}{3}$$

$$x = 13$$

$$6 = \frac{x}{7}$$

$$7 \cdot 6 = 7 \cdot \frac{x}{7}$$

$$42 = x \text{ or } x = 42$$

Undo division.

Multiply each side by 7.

◇ EXERCISES ◇

Solve.

1. $5x = 25$ 5
2. $x + 9 = 13$ 4
3. $\frac{a}{4} = 10$ 40
4. $11 = y - 6$ 17

5. $14 = p + 3$ 11
6. $x - 14 = 16$ 30
7. $6 = \frac{b}{9}$ 54
8. $17 = k - 3$ 20

9. $48 = 4g$ 12
10. $24 = m + 9$ 15
11. $\frac{t}{7} = 8$ 56
12. $17 = y - 9$ 26

13. $84 = 4w$ 21
14. $w + 12 = 25$ 13
15. $72 = 6n$ 12
16. $x - 13 = 46$ 59

17. $38 = g - 4$ 42
18. $16 = \frac{y}{3}$ 48
★ 19. $2(3 + 7) = 4t$ 5
★ 20. $x - 7 = 6 \cdot 3 + 4$ 29

Problem Solving: Mixed Types

—◇ **OBJECTIVE** ◇—

To solve mixed types of word problems

Review the four-step strategy for solving word problems on page 39. Also review the key phrases. This should help you to solve the word problems that follow.

Key Phrases

ENGLISH PHRASE	MATHEMATICAL FORM
x increased by 4	$x + 4$
x decreased by 4	$x - 4$
4 more than x	$x + 4$
4 less than x	$x - 4$
the sum of x and 4	$x + 4$
twice x	$2x$
4 times x	$4x$
x divided by 4	$\frac{x}{4}$

Solve these problems.

1. A number decreased by 7 is 18. Find the number.   25

2. 7 more than a number is 12. Find the number.   5

3. 45 is 5 times some number. Find the number.   9

4. A number divided by 8 is the same as 9. What is the number?   72

5. The sum of a number and 12 is 18. What is the number?   6

6. 42 is the same as 7 times a number. What is the number?   6

7. Hazel's weight decreased by 30 lb is 100 lb. Find her weight.   130 lb

8. Twice Rod's age is 46. How old is Rod?   23

9. Henry's score of 12 baskets is 6 baskets less than Harry's. What is Harry's score?   18

10. The per person cost for a dinner is $8.00. Find the total charge for 7 people at a dinner party. $56

11. Tina's age of 55 is 11 times her nephew's age. How old is her nephew?   5

12. The sum of a restaurant bill of $35 and its sales tax is $38. What is the sales tax? $3

13. Sonya's salary of $215 is $35 more than her husband's salary. Find her husband's salary. $180

14. If Bill can increase his bowling score by 40 pins his score will be 225. Find his score.   185

Evaluating Geometric Formulas

To evaluate geometric formulas
involving multiplication and
division

Evaluate xy if $x = 5$ and $y = 6$.

$$5 \cdot 6$$
$$30$$

See Geometry Supplement on page 420 for more information.

Example 1

A formula for the area of a rectangle is $A = lw$.
Find the area of rectangle $ABCD$ if $l = 7$ in. and
$w = 5$ in.

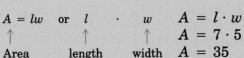

$A = lw$ or $l \quad \cdot \quad w$
$\qquad\uparrow \qquad\qquad \uparrow \qquad\quad \uparrow$
Area   length   width

$A = l \cdot w$
$A = 7 \cdot 5$
$A = 35$

in.^{2} means *square inches.*   So, the area of the rectangle is 35 in.^{2}

D — $l = 7$ in. — C
$w = 5$ in.
A — B

Example 2

A formula for the area of a triangle is $A = \dfrac{bh}{2}$. Find
the area of a triangle if $b = 14$ cm and $h = 7$ cm.

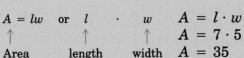

base   height
$A \overset{}{=} \dfrac{bh}{2}$
$\uparrow$
Area

$A = \dfrac{b \cdot h}{2}$

$A = \dfrac{14 \cdot 7}{2}$

$A = \dfrac{98}{2}$

$A = 49$

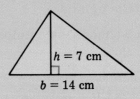

$h = 7$ cm
$b = 14$ cm

So, the area is 49 cm^2.

practice ▷   **Find the area.**

1. $A = lw$

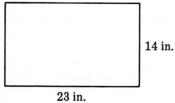

$A = 322$ in.^{2}

14 in.

23 in.

2. $A = \dfrac{bh}{2}$

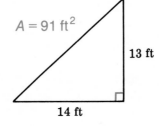

$A = 91$ ft^2

13 ft

14 ft

3. $A = \dfrac{bh}{2}$

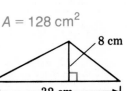

$A = 128$ cm^2

8 cm

— 32 cm —

Example 3

A formula for the perimeter of a rectangle is $P = 2l + 2w$. Find the perimeter of a rectangle if $l = 12$ in. and $w = 4$ in.

$$P = 2l + 2w$$

Substitute 12 for l, 4 for w. $\quad P = \quad 2 \cdot l + 2 \cdot w$

$l = 12$ in.

$w = 4$ in.

Do the multiplications first. $\quad P = 2 \cdot 12 + 2 \cdot 4$

Then add. $\quad P = \quad 24 \quad + \quad 8$

$$P = \quad 32$$

So, the perimeter of the rectangle is 32 in.

practice ▷ $P = 2l + 2w$. **Find P for the given values of l and w.**

   4. $l = 9$ cm, $w = 3$ cm   24 cm   **5.** $l = 13$ ft, $w = 19$ ft   64 ft

---

◇ EXERCISES ◇

$A = lw$. **Find A for the given values of l and w.**

1. l is 15 ft, w is 8 ft   120 ft^2 **2.** l is 20 cm, w is 14 cm   **3.** l is 13 in., w is 17 in.

280 cm^2

221 in.^{2}

$A = \dfrac{bh}{2}$. **Find A for the given values of b and h.**

140 in.^{2}

4. b is 14 in., h is 20 in.   **5.** b is 21 m, h is 8 m   84 m^2 **6.** b is 6 ft, h is 24 ft

72 ft^2

$P = 2l + 2w$. **Find P for the given values of l and w.**

7. l is 18 yd, w is 14 yd 64 yd **8.** l is 24 m, w is 13 m   74 m **9.** l is 40 in., w is 24 in.

128 in.

The formula for the volume of a rectangular solid is $V = lwh$ or $V = l \cdot w \cdot h$. Find V for the given values of l, w, and h.

10. l is 4 in., w is 7 in., h is 8 in.   224 in.^{3} **11.** l is 9 cm, w is 12 cm, h is 4 cm

432 cm^3

Solve these problems.

12. The circumference of a circle is the distance around it. d is the diameter. Evaluate $C = 3.14d$ if d is 8.2   25.748

13. The circumference of a circle is $C = 3.14d$ where d is the diameter. Find C if d is 5.6.   17.584

★ **14.** The area of a rectangle is 54 in.^{2} The length is 9 in. Find the width.   6 in.

★ **15.** The area of a right triangle is 24 ft^2. The height is 8 ft. Find the base.   6 ft

EVALUATING GEOMETRIC FORMULAS

Mathematics Aptitude Test

**This test will prepare you for taking standardized tests.
Choose the best answer for each question.**

1. How many numbers between 1 and 100 end in 8?
   a. 9   **b.** 10   **c.** 19   **d.** 20

2. Six people meet in a room and each shakes hands with everyone. How many handshakes are there?
   a. 6   **b.** 21   **c.** 30   **d.** 720

3. $6 + a + b = 18$, where a and b are 1-digit whole numbers. What is the lowest value of a?
   a. 0   **b.** 1   **c.** 2   **d.** 3

4. The first number is 12. The second number is less than 8. Which of the following is true?
   a. The first number is greater than the second.
   b. The second number is greater than the first.
   c. The sum of the numbers is 20.
   d. Cannot be determined which number is greatest.

5. Snow is falling at the rate of 15 cm/h. How many centimeters of snow will fall in 40 minutes?
   a. 10   **b.** 45   **c.** 60   **d.** 80

6. Which number is 10 more than 999?
   a. 1,000   **b.** 989   **c.** 1,009   **d.** 1,109

7. Find the missing number in the following sequence.
   7, 13, 25, 49, ___?___
   a. 67   **b.** 97   **c.** 73   **d.** 81

8. The outer diameter of a plastic pipe is 3.05 inches. The inner diameter is 2.45 inches. What is the thickness of the plastic?
   a. 1.20 in.   **b.** 0.60 in.   **c.** 5.50 in.   **d.** 0.30 in.

9. How many whole numbers between 10 and 100 have two identical digits?
   a. 8   **b.** 9   **c.** 10   **d.** 20

You have seen that the **BASIC** symbols for addition and subtraction are the same as in arithmetic: + and −. The **BASIC** symbols for × and ÷ are:

Multiply by: ∗
Divide by: /

The order of operations rule you have used in this chapter also holds for expressions written in **BASIC**.

Example   A = 12 + 8 / (4 − 2) − 3 ∗ 5. Find A first by computation. Then write a program to find A. Type and **RUN** it.

First find A by computation.

A = 12 + 8 / (4 − 2) − 3 ∗ 5   Compute within parentheses first.

A = 12 + 8 / 2   − 3 ∗ 5   Do all multiplications and divisions next.

A = 12 +   4   − 15   Do all additions and subtractions last.

A =   16   − 15

A =   1

Type in this program.

```
10   LET A = 12 + 8 / (4 - 2) - 3 * 5
20   PRINT A
30   END
```

Now **RUN** the program. No line number is needed. The screen displays the answer.

```
]RUN
1
```

Exercises

Compute.

1. 6 ∗ 4 _24_   **2.** 18 / 9 _2_   **3.** 8 ∗ 3 + 2 _26_   **4.** 13 + 4 / 2 _15_   **5.** 10 + 8 / 4 − 2 _10_
6. 5 + 4 ∗ 2 − 12 / 3 _9_   **7.** (10 + 8) / (4 − 2) _9_   **8.** 10 + 8 / (4 − 2) _14_

Write a program to find the value of each variable. Then type and **RUN** the program. Remember to type **NEW** when you have finished **RUN**ning each program and are ready to begin the next.

9. A = 4 + 6 ∗ 7 − 24 / 6 _42_   **10.** B = 6 ∗ 7 − 8 + 15 / 5 _37_   **11.** K = 45 / 9 + 8 − 2 ∗ 5 _3_
12. T = (40 + 8) / (3 + 1) _12_   **13.** K = 40 + 8 / (3 + 1) _42_   **14.** W = (40 + 8) / 4 + 1 _13_

Chapter Review

Evaluate if $x = 9$, $a = 2$, $k = 6$, and $t = 10$. [26]

1. $7x$ 63

2. $\frac{32}{a}$ 16

3. $\frac{258}{k}$ 43

4. $\frac{3t}{2}$ 15

What property of multiplication is illustrated? [28]

5. $(9 \cdot 3) \cdot 4 = 9 \cdot (3 \cdot 4)$
associative

6. $1 \cdot 7 = 7$
identity

7. $16 \cdot 8 = 8 \cdot 16$
commutative

8. $24 = 24 \cdot 1$
identity

Compute. [30]

9. $8 + 4 \cdot 3$ 20

10. $18 - 14 \div 7$ 16

11. $19 - 6 \div 2$ 16

12. $9 + 3(4 + 7)$ 42

13. $32 \div 8 + 4(2 + 5)$ 32

14. $2(7 + 4) - 35 \div 7$
17

Evaluate if $x = 3$, $y = 2$, $a = 6$, and $b = 4$. [33]

15. $7x + 3$ 24

16. $\frac{a}{3} + 7$ 9

17. $6a - 2 + 5b$ 54

18. $3 + 2xy$ 15

19. $8ab + 14$ 206

★ 20. $(6xy - 2a)4ay$
1,152

Solve each equation. Check the solution. [36]

21. $9x = 45$ 5

22. $7 = \frac{a}{10}$ 70

23. $68 = 34b$ 2

24. $\frac{y}{6} = 14$ 84

25. $408 = 8m$ 51

★ 26. $8(2 + 7) = 18p$
4

Write in mathematical form. [39]

27. 12 times x 12x

28. twice q 2q

29. m divided by 18
$\frac{m}{18}$

Solve these problems. [39]

30. 8 times a number is 56. Find the number.   7

31. The $40 selling price of a radio is twice the cost. Find the cost. $20

32. Mr. Ruiz's savings of $4,000 is the same as his father's divided by 4. Find his father's savings. $16,000

33. This year's basketball team has 30 wins. This is twice last year's. How many games were won last year?   15

34. Maria's age of 3 is the same as her cousin's age divided by 6. How old is her cousin?   18

★ 35. 7 times a number is the same as 20 increased by 8. Find the number.   4

36. $A = lw$. Find A if l is 22 in. and w is 14 in. [46]   308 in.²

37. $A = \frac{bh}{2}$. Find A if b is 4 m and h is 18 m. [46]   36 m²

38. $P = 2l + 2w$. Find P if l is 32 ft and w is 14 ft. [46]   92 ft

39. $P = 2l + 2w$. Find P if l is 28 cm and w is 42 cm. [46]   140 cm

Evaluate if $x = 8$, $a = 3$, and $t = 12$.

1. $4x$ 32

2. $\dfrac{168}{a}$ 56

3. $\dfrac{5t}{2}$ 30

What property of multiplication is illustrated?

4. $1 \cdot 9 = 9$ identity

5. $17 \cdot 23 = 23 \cdot 17$
   commutative

6. $(8 \cdot 4) \cdot 7 = 8 \cdot (4 \cdot 7)$
   associative

Compute.

7. $11 + 6 \cdot 3$ 29

8. $44 - 33 \div 11$ 41

9. $6 + 4(7 + 2)$ 42

Evaluate if $a = 8$ and $b = 2$.

10. $7a + 9$ 65

11. $3a - 4 + 2b$ 24

12. $5 + 4ab$ 69

Solve each equation. Check the solution.

13. $7m = 42$ 6

14. $48 = 4t$ 12

15. $\dfrac{a}{5} = 12$ 60

Write in mathematical form.

16. 15 times g 15g

17. p divided by 29 $\dfrac{p}{29}$

18. twice r 2r

Solve these problems.

19. Three times Maria's age is 45. How old is Maria?   15

20. Jared's 42 baskets for the season is the same as O'Ruark's divided by 2. Find O'Ruark's baskets for the season.   84 baskets

21. $A = lw$. Find A if l is 13 in. and w is 9 in.   117 in.²

22. $A = \dfrac{bh}{2}$. Find A if b is 16 m and h is 4 m.   32 m²

23. $P = 2l + 2w$. Find P if l is 43 ft and w is 19 ft.   124 ft

Expressions and Formulas

New England's covered bridges are supported by triangular trusses.

The Distributive Property

To rewrite expressions using the distributive property

$5(4 + 2)$ means $5 \cdot (4 + 2)$.

Example 1

Compute $5(2 + 4)$ and $5 \cdot 2 + 5 \cdot 4$.
Answers are the same.

Show that $5(2 + 4)$ is the same as $5 \cdot 2 + 5 \cdot 4$.

$5(2 + 4)$	$5 \cdot 2 + 5 \cdot 4$
$5 \cdot 6$	$10 + 20$
30	30

So, $5(2 + 4) = 5 \cdot 2 + 5 \cdot 4$.

Example 2

Answers are the same.

Show that $(5 - 2)7 = 5 \cdot 7 - 2 \cdot 7$.

$(5 - 2)7$	$5 \cdot 7 - 2 \cdot 7$
$3 \cdot 7$	$35 - 14$
21	21

So, $(5 - 2)7 = 5 \cdot 7 - 2 \cdot 7$.

distributive property
For all numbers a, b, and c,
$a(b + c) = a \cdot b + a \cdot c$ and
$\qquad (b + c)a = b \cdot a + c \cdot a$;
$a(b - c) = a \cdot b - a \cdot c$ and
$\qquad (b - c)a = b \cdot a - c \cdot a$.

Example 3

Distribute the 6 over 7 and 1.

Rewrite $6(7 + 1)$ using the distributive property.
Then compute both expressions.

$6(7 + 1) = 6 \cdot 7 + 6 \cdot 1$

Use the rules for order of operations.

$6(7 + 1)$	$6 \cdot 7 + 6 \cdot 1$
$6 \cdot 8$	$42 + 6$
48	48

So, the results are the same, 48.

Example 4

Distribute 8 over 9 and 6.

Rewrite $(9 - 6)8$ using the distributive property.

$(9 - 6)8 = 9 \cdot 8 - 6 \cdot 8$

THE DISTRIBUTIVE PROPERTY

Rewrite using the distributive property. Then compute both expressions.

1. 9(7 + 5)   108
     9 · 7 + 9 · 5 = 108

2. (5 − 3)4   8
     5 · 4 − 3 · 4 = 8

3. 6(7 + 9)   96
     6 · 7 + 6 · 9 = 9(

You can use the distributive property in reverse to rewrite 5 · 6 + 5 · 8.

5 · 6 + 5 · 8

The 5 is distributed as a multiplier over 6 and 8.

So, **5 · 6 + 5 · 8** is the same as

5(6 + 8).

Example 5

Rewrite using the distributive property.

The 8 is distributed over 4 and 2.   **8 · 4 + 8 · 2**

8(4 + 2)

The 5 is distributed over 7 and 1.   **7 · 5 − 1 · 5**

(7 − 1)**5**

practice ▷ **Rewrite using the distributive property.**

4. 8 · 6 + 8 · 11
     8(6 + 11)

5. 13 · 3 − 10 · 3
     (13 − 10)3

6. 14 · 12 + 14 · 9
     14(12 + 9)

You can use the distributive property to simplify expressions that contain variables.

Example 6

Simplify.

Distribute the 4.   4(3a + 5)   7(2b − 3)   Distribute the 7.

3a means 3 · a.   4 · 3a + 4 · 5   7 · 2b − 7 · 3

4 · 3a = 12 · a = 12a   12a + 20   14b − 21

practice ▷ **Simplify.**

                                                    20x − 30
7. 8(2x + 3)   16x + 24   **8.** 7(2b − 5)   14b − 35   **9.** 5(4x − 6)

What number must be distributed as a multiplier?

1. $6(3 + 12)$ 6
2. $(8 + 9)4$ 4
3. $5(14 - 3)$ 5

What number has been distributed as a multiplier?

4. $8 \cdot 7 + 8 \cdot 12$ 8
5. $9 \cdot 6 - 7 \cdot 6$ 6
6. $4 \cdot 8 + 4 \cdot 13$ 4

◇ *EXERCISES* ◇

Rewrite using the distributive property. Then compute both expressions.

1. $4(7 + 1)$ $4 \cdot 7 + 4 \cdot 1 = 32$
2. $6(8 - 3)$ $6 \cdot 8 - 6 \cdot 3 = 30$
3. $3(9 + 4)$ $3 \cdot 9 + 3 \cdot 4 = 39$
4. $(9 + 6)2$ $9 \cdot 2 + 6 \cdot 2 = 30$
5. $(5 - 4)7$ $5 \cdot 7 - 4 \cdot 7 = 7$
6. $(9 + 3)8$ $9 \cdot 8 + 3 \cdot 8 = 96$

Rewrite using the distributive property.

7. $5(4 + 6)$ $5 \cdot 4 + 5 \cdot 6$
8. $6(4 + 9)$ $6 \cdot 4 + 6 \cdot 9$
9. $3(6 - 2)$ $3 \cdot 6 - 3 \cdot 2$
10. $(3 + 8)6$ $3 \cdot 6 + 8 \cdot 6$
11. $4 \cdot 6 + 4 \cdot 2$ $4(6 + 2)$
12. $9 \cdot 6 - 9 \cdot 4$ $9(6 - 4)$

Simplify.

13. $3(2a + 5)$ $6a + 15$
14. $4(6m - 7)$ $24m - 28$
15. $7(4a + 3)$ $28a + 21$
16. $8(6b - 3)$ $48b - 24$
17. $4(3x - 5)$ $12x - 20$
18. $6(2m + 10)$ $12m + 60$
19. $12(2a - 4)$ $24a - 48$
20. $11(2m - 3)$ $22m - 33$
21. $14(2a + 3)$ $28a + 42$

★ 22. Is division distributive over addition? no

Does $36 \div (6 + 3) = (36 \div 6) + (36 \div 3)$? $36 \div 9 = 4 \neq 6 + 12 = 18$

★ 23. Is addition distributive over multiplication? no

Does $8 + (6 \cdot 3) = (8 + 6) \cdot (8 + 3)$? $8 + 18 = 26 \neq 14 \cdot 11 = 154$

Algebra Maintenance

Evaluate if $x = 4$, $m = 7$, $k = 9$, and $d = 10$.

1. $k - x$ 5
2. $4d + m$ 47
3. $\dfrac{121}{x + m}$ 11

Solve each equation.

4. $7a = 56$ 8
5. $42 = x + 9$ 33
6. $34 = 2x$ 17
7. $\dfrac{m}{5} = 12$ 60
8. $45 = c - 12$ 57
9. $5 = \dfrac{r}{3}$ 15

Solve these problems.

10. The $32 selling price of a set of computer disks is twice the cost. What is the cost? $16

11. Tom's age of 7 is the same as his sister's age divided by 3. What is his sister's age? 21

Combining Like Terms

To simplify algebraic
  expressions by combining like
  terms
To simplify algebraic
  expressions using the property
  of multiplicative identity

Rewrite using the distributive property.
$5 \cdot 6 + 2 \cdot 6$ ← 6 is distributed over 5 and 2.

$(5 + 2)6$

The expression $3x + 4y$ has two *terms*, $3x$ and $4y$.
In the term $3x$, 3 is the *coefficient* of x.
A *coefficient* is a multiplier of a variable.

$5a + 5a$ $7m - 5m$ $3x + 4y$

like terms   like terms   *un*like terms

Like terms are terms that are exactly alike or differ
only by their *coefficients*.

Example 1

Rewrite $5a + 4a$ using the distributive property.
Simplify the result.

distributive property $\quad$ $5a + 4a = 5 \cdot a + 4 \cdot a$ $\qquad$ a is distributed over 5 and 4.
$= (5 + 4) \cdot a$
$= 9 \cdot a$
$9a$

You can use a shortcut to combine like terms. Like
terms can be combined by adding or subtracting
their coefficients. Thus, $5a + 4a = 9a$ and
$7x - 5x = 2x$.

Example 2

Simplify if possible.

Rearrange to group like terms.
$3a + 4 + 7a$
$3a + 7a + 4$
$10a \quad + 4$

$5x + 2y + 3$
The terms $5x$, $2y$, and 3 are
unlike terms.
They cannot be combined.

So, $3a + 4 + 7a = 10a + 4$; $5x + 2y + 3$ cannot be
simplified.

Example 3

Simplify $7 + 3x + 5y + 4 - 2y + 6x$.

$7 + 3x + 5y + 4 - 2y + 6x$

Rearrange to group like terms. $3x + 6x + 5y - 2y + 4 + 7$

$9x \quad + \quad 3y \quad + \quad 11$

practice ▷ **Simplify if possible.**

1. $5t + 4t$
$9t$

2. $8x + 3y + 5$
not possible

3. $6 + 2m + 9n + 1 + 11m - 5n$
$13m + 4n + 7$

Recall the property of multiplicative identity. Multiplication of any number by 1 gives that number.

Example 4

Simplify $8y + 3 + y$.

The coefficient of $8y$ is 8. $8y + 3 + y$

Rewrite y as $1 \cdot y$. $8y + 3 + 1 \cdot y \quad 1 \cdot y = y$

$8y + 1y + 3$

Add the coefficients of the *like* terms. $9y \quad + 3$

practice ▷ **Simplify.**

$4a + 9$

4. $6a + 4 + a$ $7a + 4$ **5.** $p + 8 + 7p$ $8p + 8$ **6.** $2 + 3a + 7 + a$

◇ ORAL EXERCISES ◇

For each, name the like terms. Name the coefficient of each of these terms.

1. $7x + 5y + 3x$ 7x, 3x, 7, 3

2. $3a + 4b + 7a + 6b$ 3a, 7a, 3, 7; 4b, 6b, 4, 6

3. $12y + 4t + 2y$ 12y, 2y, 12, 2

4. $9g + 3r + 7g + 13r$ 9g, 7g, 9, 7; 3r, 13r, 3, 13

5. $x + 5c + 3x$ x, 3x, 1, 3

6. $2t + 3g + t + g$ 2t, t, 2, 1; 3g, g, 3, 1

◇ EXERCISES ◇

Simplify if possible.

1. $3x + 5x$ 8x

2. $8y + 7y$ 15y

3. $6m - 4m + 3$ 2m + 3

4. $9a + 3b - 5$ not possible

5. $7 + 4x + 3y$ not possible

6. $8 + 11x + 3 - 4x$ 11 + 7x

7. $3 + 5x + 4 + 2x$ 7 + 7x

8. $2a + 3x + 4a + 8x$

9. $6k + 9y - 2k + 4y$ 4k + 5y

10. $3x + x + 4$ 4x + 4

11. $5 + y + 13y$ 5 + 14y

12. $h + 21 + 6h - 4$ 7h + 17

13. $3 + 8g + 1 + g$ 4 + 9g

14. $y + 14 + 3y + 2$ 4y + 16

15. $x + 6y + 7x + y$ 8x + 7y

★ **16.** $1 + 19x + x + 4 \cdot 5$
21 + 20x

★ **17.** $(2 + 3 + 4)y + 23y$
32y

★ **18.** $x + 3(3 + 2) + 5x + 8$
6x + 23

8. $6a + 11x$

COMBINING LIKE TERMS

57

Simplifying and Evaluating Expressions

To simplify and then evaluate
algebraic expressions for
given values of the variables

Evaluate $4x + 6$ if $x = 2$.

$4 \cdot 2 + 6$ ⟵ Substitute 2 for x.

$8 + 6$

14

Example 1

Simplify $3a + 4 + 5a$. Then evaluate if $a = 2$.

$$3a + 4 + 5a$$

Rearrange to group like terms. $\quad 3a + 5a + 4$

Combine like terms. $\qquad 8a + 4$

Substitute 2 for a. $\qquad 8 \cdot 2 + 4$

Multiply first. $\qquad 16 + 4$

Then add. $\qquad 20$

Example 2

Simplify $7 + x - 6 + 4x$. Then evaluate if $x = 6$.

First rewrite x as $1x$. $\qquad 7 + 1x - 6 + 4x$

Rearrange to group like terms. $\quad 1x + 4x + 7 - 6$

Combine like terms. $\qquad 5x + 1$

Substitute 6 for x. $\qquad 5 \cdot 6 + 1$

Use the rules for order of operations. $\quad 30 + 1$

31

---

practice ▷ **Simplify. Then evaluate for $a = 2$, $b = 3$, and $c = 4$.**

   1. $4a + 5 + 3a$ 2. $2b + 5 + 6b$ 3. $6c + 8 + c + 9$
      $7a + 5$; 19 $8b + 5$; 29 $7c + 17$; 45

Example 3

Simplify $3a + b + a + b$. Then evaluate if $a = 3$ and $b = 7$.

$$3a + b + a + b$$

Coefficient of 1 is understood. $\quad 3a + 1b + 1a + 1b$

$$3a + 1a + 1b + 1b$$

$ 4a + 2b$

Substitute 3 for a, 7 for b. $\quad 4 \cdot 3 + 2 \cdot 7$

$ 12 + 14$

$ 26$

Example 4

Simplify $4 + 3a + a + 6b + 5 + b$. Then evaluate if $a = 6$ and $b = 3$.

$$4 + 3a + a + 6b + 5 + b$$

$$4 + 3a + 1a + 6b + 5 + 1b$$

$$\underline{4 + 5} + \underline{3a + 1a} + \underline{6b + 1b}$$

$$9 \quad + \quad 4a \quad + \quad 7b$$

Substitute 6 for a, 3 for b.
$$9 \quad + \quad 4 \cdot 6 \quad + \quad 7 \cdot 3$$
$$9 \quad + \quad 24 \quad + \quad 21$$
$$54$$

practice ▷ **Simplify. Then evaluate if $x = 5$ and $y = 6$.**

4. $x + y + x + y$ $2x + 2y$; 22
5. $3x + 5y + 4x + 2y$ $7x + 7y$; 77
6. $x + y + 5x + y$ $6x + 2y$; 42
7. $7 + 2y + y + 3x + 4 + x$ $4x + 3y + 11$; 49
8. $x + 2x + 5 + 9y + 4 - 2y$ $3x + 7y + 9$; 66

◇ EXERCISES ◇

Simplify. Then evaluate for these values of the variables:
$a = 4, b = 5, c = 2, x = 7, y = 9, z = 6.$

1. $2x + 4 + 7x$ $9x + 4$; 67
2. $8a + 3 - 5a$ $3a + 3$; 15
3. $6c + 9 + 3c$ $9c + 9$; 27

4. $8a + 1 + 3a$ $11a + 1$; 45
5. $7b - 2 + 5b$ $12b - 2$; 58
6. $8z - 2z + 5$ $6z + 5$; 41

7. $9 + 2y + 3y$ $5y + 9$; 54
8. $9c + 4 + 2c$ $11c + 4$; 26
9. $8 + 4a + 4a$ $8a + 8$; 40

10. $8 + y - 4 + 3y$ $4y + 4$; 40
11. $2b + 5 + b + 4$ $3b + 9$; 24
12. $7 + z + 3 + z$ $2z + 10$; 22

13. $x + 6 + x + 9$ $2x + 15$; 29
14. $3 + b + 9 + 2b$ $3b + 12$; 27
15. $8z + 1 + z + 6$ $9z + 7$; 61

16. $8 + 11x + 3 - 4x$ $7x + 11$; 60
17. $8x + 5 + x + 1$ $9x + 6$; 69
18. $4 + 3c + 7 + c$ $4c + 11$; 19

19. $b + z + b + 2z$ $2b + 3z$; 28
20. $4x + y + 3x + y$ $7x + 2y$; 67
21. $z + 3a + a + 5z$ $4a + 6z$; 52

22. $c + 9b + c + b$ $10b + 2c$; 54
23. $7y + a + a + y$ $8y + 2a$; 80
24. $a + 3a + 4b + b$ $4a + 5b$; 41

25. $9a + c + c + a$ $10a + 2c$; 44
26. $5a + a + y + 4y$ $6a + 5y$; 69
27. $9z + 3y + y + 4z$ $13z + 4y$; 114

28. $7x + 3y + x + 4y + y$ $8x + 8y$; 128
29. $a + 3b + 4a + b + 5a$ $10a + 4b$; 60

30. $z + 3y + z + 4y + 5z + y$ $8y + 7z$; 114
31. $4 + 3x + 5 - 2x + y + y$ $x + 2y + 9$; 34

32. $6a + a + 5 + 3b + 2 + 2b$ $7a + 5b + 7$; 60
33. $5a + b + 2 + 3a + 4b + 6$ $8a + 5b + 8$; 65

Problem Solving – Applications
Jobs for Teenagers

1. A musical group started playing at 8:15 P.M. They finished at 1:15 A.M. How many hours did the group play? 5

2. One evening the Keynotes played for three hours. Each song lasted an average of 4 minutes. How many songs did they play? 45

3. A band earns an average of $450 a weekend. They wanted to buy new equipment that costs $2,700. How many weekends must they work in order to buy the new equipment? 6

4. A three-piece band played from 9:00 P.M. to midnight on Friday, from 8:00 P.M. to 2:00 A.M. on Saturday, and from 9:00 P.M. to 11:45 P.M. on Sunday. How many hours did they play altogether on that weekend? $11\frac{3}{4}$

5. A rock group charges a flat fee of $620. They played for 5 hours. How much did they make per hour? $124

6. The sophomore class hires a band that charges $350. Students are charged $3.00 a ticket. 275 students attend the dance. Find the class profit. $475

7. A piano player earned $20 per hour for the first three hours of playing and $24 per hour for each hour over 3 hours. How much did she earn for a 4-hour engagement? $84

8. A drummer earns $19.50 per hour for playing until midnight. She is then paid $22.00 per hour for playing after midnight. She played from 8:00 P.M. to 2:00 A.M. How much did she earn? $122

Solving Equations: Using Two Properties

◇ OBJECTIVE ◇

To solve equations using more than one property of equations

◇ RECALL ◇

Solve $x + 3 = 8$.   Undo the addition.
$x + 3 - 3 = 8 - 3$ Subtract 3 from
$x = 5$ each side.

Solve $3y = 15$.   Undo the multiplication.
$\dfrac{3y}{3} = \dfrac{15}{3}$ Divide each side by 3.
$y = 5$

So far you have solved equations using only one operation. In the equation $2x + 3 = 11$, x is multiplied by 2. Then 3 is added to the result. This is an equation with two operations. Solving this equation requires undoing the addition and then undoing the multiplication.

Example 1

Solve $2x + 3 = 11$. Check the solution.

First undo addition of 3. $2x + 3 = 11$
Subtract 3 from each side. $2x + 3 - 3 = 11 - 3$
Undo multiplication of x by 2. $2x = 8$
Divide each side by 2. $\dfrac{2x}{2} = \dfrac{8}{2}$
$x = 4$

Check:
$$2x + 3 = 11$$
$$2 \cdot 4 + 3 \mid 11$$
$$8 + 3$$
$$11 = 11 \quad \text{true}$$

So, 4 is the solution of $2x + 3 = 11$.

Example 2

Solve $13 = 5k - 2$.   Check the solution.

Undo subtraction of 2. $13 = 5k - 2$ Check: $13 = 5k - 2$
Add 2 to each side. $13 + 2 = 5k - 2 + 2$ $13 \mid 5 \cdot 3 - 2$
Undo multiplication of k by 5. $15 = 5k$ $15 - 2$
Divide each side by 5. $\dfrac{15}{5} = \dfrac{5k}{5}$ $13 = 13 \quad \text{true}$
$3 = k$

So, 3 is the solution of $13 = 5k - 2$.

practice ▷ **Solve each equation. Check the solution.**

1. $3y + 4 = 13$ 3

2. $2t + 5 = 15$ 5

3. $13 = 5g - 2$ 3

Example 3

Solve $\frac{x}{2} + 3 = 6$. Check the solution.

Undo addition of 3. Subtract 3 from each side.	$\frac{x}{2} + 3 - 3 = 6 - 3$
Undo division by 2.	$\frac{x}{2} = 3$
Multiply each side by 2.	$2 \cdot \frac{x}{2} = 3 \cdot 2$
	$x = 6$

Check: $\frac{x}{2} + 3 = 6$

$$\frac{6}{2} + 3 \;\Big|\; 6$$
$$3 + 3$$
$$6 \;=\; 6 \quad \text{true}$$

So, 6 is the solution of $\frac{x}{2} + 3 = 6$.

---

practice ▷ **Solve each equation. Check the solution.**

4. $\frac{x}{3} + 2 = 7$ 15

5. $\frac{x}{4} - 1 = 11$ 48

6. $9 = \frac{x}{6} + 4$ 30

◇ ORAL EXERCISES ◇

Tell what must be done to solve each equation.   See TE Answer Section.

1. $2z + 4 = 14$

2. $12 = 3d - 6$

3. $8a + 4 = 28$

4. $6 = 2b - 4$

5. $\frac{x}{7} + 5 = 8$

6. $\frac{x}{2} - 4 = 8$

◇ EXERCISES ◇

Solve each equation. Check the solution.

1. $2x + 1 = 5$ 2

2. $3y + 1 = 10$ 3

3. $14 = 4u + 6$ 2

4. $2m - 2 = 8$ 5

5. $6k - 4 = 8$ 2

6. $12 = 5y - 3$ 3

7. $\frac{a}{8} + 2 = 6$ 32

8. $\frac{b}{6} - 2 = 1$ 18

9. $9 = \frac{x}{3} + 2$ 21

Solve each equation.

10. $8x - 3 = 29$ 4

11. $24 = 5y - 1$ 5

12. $14 = 8c - 2$ 2

13. $3y - 9 = 12$ 7

14. $15 = 5m - 10$ 5

15. $12 = 4z + 4$ 2

16. $2t - 8 = 24$ 16

17. $18 = 7g - 3$ 3

18. $11u - 8 = 14$ 2

19. $3k + 8 = 23$ 5

20. $28 = 9y + 1$ 3

21. $44 = 6p + 2$ 7

22. $\frac{g}{5} + 7 = 17$ 50

23. $8 = \frac{k}{4} + 2$ 24

24. $\frac{b}{7} - 2 = 3$ 35

25. $4 = \frac{i}{6} - 3$ 42

26. $\frac{w}{5} + 6 = 15$ 45

27. $\frac{d}{12} - 5 = 2$ 84

★ **28.** $4x - 1 = 5 \cdot 4 + 3$ 6

★ **29.** $2(5 + 3) = 4x - 12$ 7

★ **30.** $\frac{y}{7} + 2 = 2(6 - 3)$ 28

English Phrases to Algebra

── ◇ *OBJECTIVE* ◇ ──   ── ◇ *RECALL* ◇ ──────

To rewrite in mathematical form
  English phrases involving two
  operations

English Phrases	Mathematical Form
n increased by 6	$n + 6$
6 more than n	$n + 6$
n decreased by 6	$n - 6$
6 less than n	$n - 6$
6 times n	$6n$
n divided by 6	$\dfrac{n}{6}$

You can easily extend the ideas of the recall to
English phrases involving two operations.

Example 1

Write in mathematical form.

$\underbrace{\text{3 times } n,}\ \underbrace{\text{increased by 5.}}$

$\quad 3 \cdot n \qquad\qquad + 5$

or $3n + 5$

Example 2

Write in mathematical form.
Twice a number, decreased by 4.

Let n represent the number. $\underbrace{\text{twice } n}\quad \underbrace{\text{decreased by 4}}$

$\qquad 2 \cdot n \qquad\qquad - 4$

or $2n - 4$

practice ▷ **Write in mathematical form.**

1. 4 times x, increased by 3
   $4x + 3$

2. twice a number, decreased by 9
   $2x - 9$

Example 3

Write in mathematical form.
8 less than a number divided by 4.

Let x represent the number.   8 less than x divided by 4.

8 less than $\dfrac{x}{4}$.

$\dfrac{x}{4} \quad - \quad 8$

ENGLISH PHRASES TO ALGEBRA   **63**

practice ▷ **Write each in mathematical form.**

3. 7 less than a number
   divided by 6 $\frac{n}{6} - 7$

4. 6 more than a number
   divided by 7 $\frac{x}{7} + 6$

◇ ORAL EXERCISES ◇

Express in mathematical form.

1. $5x$ increased by 3 $5x + 3$

2. 9 less than $2b$ $2b - 9$

3. $12k$ decreased by 5 $12k - 5$

4. 12 more than twice y $2y + 12$

5. 15 less than $\frac{y}{7}$ $\frac{y}{7} - 15$

6. $\frac{t}{2}$ increased by 14 $\frac{t}{2} + 14$

7. 18 more than $\frac{b}{3}$ $\frac{b}{3} + 18$

8. 1 less than $\frac{y}{6}$ $\frac{y}{6} - 1$

◇ EXERCISES ◇

Write in mathematical form.

1. 5 times y, increased by 4 $5y + 4$
2. twice x, increased by 7 $2x + 7$
3. 9 less than 8 times b $8b - 9$
4. 4 more than 6 times p $6p + 4$
5. twice g, increased by 12 $2g + 12$
6. 16 times y, decreased by 13 $16y - 13$
7. 18 less than 5 times k $5k - 18$
8. 11 times w, decreased by 1 $11w - 1$
9. 12 times a number, increased by 6
10. 15 more than 3 times a number $3k + 15$
11. 4 times a number, decreased by 2 $4x - 2$
12. twice a number, increased by 19 $2a + 19$
13. 2 less than a number divided by 5
14. 17 more than a number divided by 3 $\frac{y}{3} +$
15. 6 more than a number divided by 13
16. 14 less than a number divided by 2 $\frac{m}{2} -$
17. n increased by 7 times n $n + 7n$
18. 4 times x, decreased by twice x $4x - 2x$
19. 8 times a number, increased by that
   number $8r + r$
20. 6 times a number, decreased by that
   number $6n - n$
★ **21.** 4 more than twice a number,
   increased by 8 $2x + 4 + 8$
★ **22.** 9 increased by 5 less than 3 times a
   number $9 + 3n - 5$

9. $12n + 6$ **13.** $\frac{b}{5} - 2$ **15.** $\frac{x}{13} + 6$

Calculator

Simplify. Then evaluate for these values of the variables: $a = 349$, $b = 75$, $c = 1{,}234$, $d = 992$.

1. $13a + 29b + 16b + 112a$ $54{,}200$
2. $195a + 315c + 58a + 77c$ $572{,}025$
3. $249d + 154c + 8c + 77d$ $523{,}300$
4. $1{,}345d + 77c + 879d + 145c$ $2{,}480{,}156$
5. $456b + 762a + 448b + 1{,}123a$ $725{,}665$
6. $4{,}675a + 158 + 59a + 789$ $1{,}653{,}113$

Word Problems: Two Operations

—◇ **OBJECTIVE** ◇—   ———◇ **RECALL** ◇ ———

To solve word problems
involving two operations

There are four basic steps for solving all
word problems.

READ: { Identify what is given.
         { Identify what is to be found.

PLAN: { Analyze the information.
         { Represent the data.
         { Write an equation.

SOLVE: Solve the equation.

INTERPRET: { Check the solution.
           { Decide if the answer is reasonable.

The four basic steps above can now be applied to
solving word problems involving *two* operations.

Example 1

8 less than twice a number is 4.
What is the number?

Given: 8 less than twice a number is 4.
Find: the number.

Represent the number. Let n = the number.
8 less than twice a number is 4.

Write an equation.   8 less than $2n$ is 4

                      $2n$ — → 8   = 4

Undo subtraction of 8. $2n - 8 = 4$
Add 8 to each side. $2n - 8 + 8 = 4 + 8$
Undo multiplication. $2n = 12$
Divide each side by 2. $\dfrac{2 \cdot n}{2} = \dfrac{12}{2}$
                               $n = 6$

Check: 8 less than twice a number is 4

         8 less than twice 6 | 4
              12 − 8 |
                4 | = 4   true

So, the number is 6.

Example 2

Rosa's age divided by 4, increased by 5, is 10. How old is Rosa?

Represent her age with a variable.

Let a = Rosa's age.

age divided by 4, increased by 5 is 10

Write an equation.

$$\frac{a}{4} + 5 = 10$$

Undo addition of 5.
Subtract 5 from each side.

$$\frac{a}{4} + 5 = 10$$

$$\frac{a}{4} + 5 - 5 = 10 - 5$$

Undo division.

$$\frac{a}{4} = 5$$

Multiply each side by 4.

$$4 \cdot \frac{a}{4} = 5 \cdot 4$$

$$a = 20$$

Check: age divided by 4, increased by 5, is 10

$\frac{a}{4} + 5$	10
$\frac{20}{4} + 5$	
10	= 10 true

So, Rosa's age is 20.

practice ▷

1. 6 less than twice a number is 14. Find the number.   10

2. If Tina's age divided by 3 is decreased by 2, the result is 10. How old is Tina?   32

Example 3

The $300 selling price of a stereo is the same as the cost increased by 3 times the cost. What is the cost?

Let c = the cost.
300 is the same as cost increased by 3 times the cost.

Write an equation.

$$300 = c + 3c$$

$c = 1c$.

$$300 = 1c + 3c$$

Combine like terms.

$$300 = 4c$$

Undo multiplication by 4.

$$\frac{300}{4} = \frac{4c}{4}$$

Divide each side by 4.

$$75 = c$$

Check on your own.

So, the cost of the stereo is $75.

practice ▷ 3. The $90 selling price of a camera is the same as the cost increased by twice the cost. What is the cost? $30

4. Dan's score increased by 4 times his score is 120. What is his score? 24

◇ ORAL EXERCISES ◇

Give an equation for each sentence.

1. 7 more than 3 times x is 28. $3x + 7 = 28$

2. 4 times n, decreased by 8 is 12. $4n - 8 = 12$

3. 24 is 6 less than 3 times y. $24 = 3y - 6$

4. t, increased by twice t, is 27. $t + 2t = 27$

5. 48 is 8 more than y divided by 10.
$48 = \dfrac{y}{10} + 8$

6. 15 is 10 less than p divided by 3.
$15 = \dfrac{p}{3} - 10$

◇ EXERCISES ◇

Solve these problems.

1. 10 less than twice a number is 14. Find the number. 12

2. 16 is 4 more than twice a number. Find the number. 6

3. 2 more than Lisa's age divided by 5 is 8. How old is Lisa? 30

4. A number divided by 3, increased by 4, is 7. Find the number. 9

5. 26 is 4 less than 5 times a number. Find the number. 6

6. 50 more than 3 times Jo's bowling score is 500. Find her score. 150

7. If the cost of a shirt is increased by twice the cost, the result is $27. Find the cost. $9

8. The $80 selling price of a cassette radio is the cost increased by 3 times the cost. Find the cost. $20

9. 40 lb more than twice Royal's weight is 280 lb. Find his weight. 120 lb

10. If 5 times Renee's age is decreased by 20, the result is 30. Find her age. 10

11. If Lou's savings were increased by 5 times his savings, the result would be $30,000. How much has he saved? $5,000

12. 8 times the temperature decreased by the temperature was 14°. Find the temperature. 2°

★ 13. If 3 times Pete's age is increased by 4 more than twice his age, the result is 34. How old is Pete? 6

★ 14. 5 times a number, increased by 2 less than twice the number, is 40. Find the number. 6

WORD PROBLEMS: TWO OPERATIONS

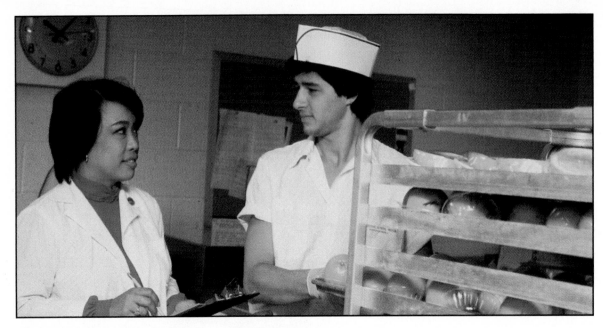

Problem Solving – Careers
Dietitians

Breakfast	Calories	Lunch	Calories	Dinner	Calories
orange juice	54	1 hamburger	425	roast beef	250
2 scrambled eggs	212	French fries	225	mashed potatoes	
				with butter	180
1 slice of buttered toast		milk	166	peas	35
with jam	168	apple	66	salad with dressing	80
		vegetable soup	74	2 rolls with butter	163
				apple pie	331
				skim milk	87

1. How many calories are there in one slice of apple pie?   331 calories

2. Find the total number of calories in the dinner.   1,126 calories

3. Find the total number of calories in all 3 meals together.   2,516 calories

4. An average teenage boy needs about 3,000 calories each day. How many more calories is this than the total number of calories for the three meals?   484 calories

5. In order to lose 2 lb of fat in one week, your diet should be about 7,000 calories a week below your body needs. How many fewer calories is this each day?   1,000 calories

6. The starting salary of a hospital dietitian is about $13,000 a year. How much is this per week? $250 per week

Factors and Exponents

To write expressions like
2 · 2 · 4 · 4 · 4 using
exponents
To evaluate expressions like x^4
and $6x^2$ if $x = 2$

Compute 7×6 and $6 \cdot 4$.
$$7 \times 6 = 42; 6 \cdot 4 = 24$$
Multiplication can be shown using "$\times$"
or a raised dot "·."

In the product $7 \cdot 3$, 7 and 3 are *factors*. *Factors* are
numbers that are *multiplied*. The product
$2 \cdot 2 \cdot 2 \cdot 2 \cdot 2$ consists of the factor 2 used as a
multiplier 5 times.

The exponent indicates the number
of times the base is used as a factor.

A convenient way of writing $2 \cdot 2 \cdot 2 \cdot 2 \cdot 2$ is

base $\rightarrow 2^5$. $\leftarrow$ exponent

Example 1

3^4 can be read as the 4th power of 3, or
3 to the 4th power.

Find the value of 3^4.
$$\begin{aligned} 3^4 &= 3 \cdot 3 \cdot 3 \cdot 3 \\ &= 9 \cdot 9 \\ &= 81 \end{aligned}$$

So, $3^4 = 81$.

---

practice ▷ **Find the value.**

1. 2^4 16 **2.** 3^2 9 **3.** 5^3 125 **4.** 4^3 64 **5.** 8^1 8 **6.** 1^4 1 **7.** 6^3
216

Example 2

Write $6 \cdot 6 \cdot 6 \cdot 6$ using an exponent.
6 is used 4 times as a factor.

base $\rightarrow 6^4$ $\leftarrow$ exponent

So, $6 \cdot 6 \cdot 6 \cdot 6 = 6^4$.

---

practice ▷ **Write using an exponent.**

8. $5 \cdot 5 \cdot 5$ **9.** $8 \cdot 8 \cdot 8 \cdot 8$ **10.** $9 \cdot 9$ **11.** $7 \cdot 7 \cdot 7$ **12.** $6 \cdot 6 \cdot 6 \cdot 6$
5^3 8^4 9^2 7^3 6^4

Example 3

Write $4 \cdot 4 \cdot 4 \cdot 5 \cdot 5$ using exponents.

$$4 \cdot 4 \cdot 4 \cdot 5 \cdot 5$$

4 is used 3 times   5 is used 2 times

$$4^3 \cdot 5^2$$

So, $4 \cdot 4 \cdot 4 \cdot 5 \cdot 5 = 4^3 \cdot 5^2$

practice ▷ **Write using exponents.**

13. $5 \cdot 5 \cdot 5 \cdot 5$ **14.** $7 \cdot 7 \cdot 7 \cdot 8 \cdot 8$ **15.** $a \cdot a \cdot a \cdot a \cdot a \cdot b \cdot b$
5^4 $7^3 \cdot 8^2$ $a^5 \cdot b^2$

Example 4

Evaluate a^3 if $a = 4$.

a^3

Use a as a factor 3 times. $a \cdot a \cdot a$

Substitute 4 for a. $4 \cdot 4 \cdot 4$

$16 \cdot 4$

64

So, $a^3 = 64$ if $a = 4$.

practice ▷ **Evaluate.**

16. n^3 if $n = 2$ 8   **17.** a^2 if $a = 8$ 64   **18.** b^5 if $b = 3$ 243

Example 5

Evaluate $5x^2$ if $x = 3$.

$5x^2$

Substitute 3 for x. $5 \cdot 3^2$

Use 3 as a factor twice. $5 \cdot 3 \cdot 3$

$5 \cdot 9$

45

So, $5x^3 = 45$ if $x = 3$.

practice ▷ **Evaluate.**

19. $4x^5$ if $x = 2$ 128   **20.** $7n^4$ if $n = 1$ 7   **21.** $5x^2$ if $x = 9$ 405

State as a product of factors.

1. x^4
$x \cdot x \cdot x \cdot x$

2. b^3 $b \cdot b \cdot b$

3. y^5
$y \cdot y \cdot y \cdot y \cdot y$

4. t^8
$t \cdot t \cdot t \cdot t \cdot t \cdot t \cdot t \cdot t$

5. g^1 g

6. w^2 $w \cdot w$

7. m^6
$m \cdot m \cdot m \cdot m \cdot m \cdot m$

State using exponents.

8. $5 \cdot 5 \cdot 5 \cdot 3 \cdot 3$ $5^3 \cdot 3^2$

9. $3 \cdot 3 \cdot 3 \cdot 3 \cdot 2 \cdot 2$ $3^4 \cdot 2^2$

10. $a \cdot a \cdot a \cdot a \cdot b \cdot b$ $a^4 \cdot b^2$

11. $3 \cdot 3 \cdot 3 \cdot 4 \cdot 4 \cdot 4$
$3^3 \cdot 4^3$

12. $a \cdot a \cdot b \cdot b$ $a^2 \cdot b^2$

13. $t \cdot t \cdot t \cdot y \cdot y \cdot y \cdot y$ $t^3 \cdot y^4$

◇ *EXERCISES* ◇

Find the value.

1. 1^5 1
2. 3^3 27
3. 2^5 32
4. 7^1 7
5. 6^2 36
6. 8^2 64
7. 9^2 81
8. 8^3 512
9. 10^2 100
10. 11^2 121
11. 7^3 343
12. 10^3 1,000
13. 4^5 1,024
14. 9^3 729
15. 20^2 400

Write using exponents.

16. $9 \cdot 9 \cdot 9 \cdot 9$ 9^4
17. $12 \cdot 12 \cdot 12$ 12^3
18. $b \cdot b \cdot b \cdot b \cdot b \cdot b$ b^6
19. $7 \cdot 7 \cdot 7 \cdot 7$ 7^4
20. $11 \cdot 11 \cdot 11$ 11^3
21. $10 \cdot 10 \cdot 10 \cdot 10 \cdot 10 \cdot 10 \cdot 10$ 10^7
22. $k \cdot k$ k^2
23. $m \cdot m \cdot m \cdot m$ m^4
24. $15 \cdot 15 \cdot 15$ 15^3
25. $2 \cdot 2 \cdot 4 \cdot 4 \cdot 4$
26. $5 \cdot 5 \cdot 5 \cdot 5 \cdot 8 \cdot 8 \cdot 8$
27. $t \cdot t \cdot w \cdot w \cdot w \cdot w$ $t^2 \cdot w^4$
28. $7 \cdot 7 \cdot 6 \cdot 6 \cdot 6 \cdot 6$
29. $g \cdot g \cdot h$ $g^2 \cdot h$
30. $r \cdot r \cdot r \cdot k \cdot k$ $r^3 \cdot k^2$
31. $x \cdot x \cdot y \cdot y \cdot y$ $x^2 \cdot y^3$
32. $12 \cdot 12 \cdot 14$ $12^2 \cdot 14$
33. $d \cdot d \cdot d \cdot d \cdot d \cdot y$ $d^5 \cdot y$

25. $2^2 \cdot 4^3$
26. $5^4 \cdot 8^3$
28. $7^2 \cdot 6^4$

Evaluate.

34. y^3 if $y = 3$ 27
35. x^4 if $x = 3$ 81
36. t^2 if $t = 5$ 25
37. w^5 if $w = 2$ 32
38. r^4 if $r = 10$ 10,000
39. t^6 if $t = 2$ 64
40. y^3 if $y = 7$ 343
41. m^4 if $m = 5$ 625
42. a^8 if $a = 2$ 256
43. $5x^2$ if $x = 4$ 80
44. $3x^3$ if $x = 2$ 24
45. $4b^2$ if $b = 5$ 100
46. $6a^5$ if $a = 2$ 192
47. $8n^3$ if $n = 3$ 216
48. $9n^2$ if $n = 3$ 81
49. $10y^2$ if $y = 5$ 250
50. $9a^6$ if $a = 2$ 576
51. $15b^3$ if $b = 1$ 15

Evaluate.

★ 52. $4a^2b^3$ if $a = 3$ and $b = 2$ 288
★ 53. $6x^4y$ if $x = 3$ and $y = 4$ 1,944
★ 54. $5ab^4$ if $a = 7$ and $b = 2$ 560
★ 55. $8t^4u^3$ if $t = 2$ and $u = 3$ 3,456
★ 56. $7x^2y^3$ if $x = 3$ and $y = 2$ 504
★ 57. $100x^5y^3$ if $x = 2$ and $y = 5$ 400,000
★ 58. $11pt^5$ if $p = 9$ and $t = 1$ 99
★ 59. $7a^4b^4$ if $a = 2$ and $b = 2$ 1,792

FACTORS AND EXPONENTS

Exponents in Formulas

To evaluate formulas containing
exponents

Evaluate x^3 if $x = 2$.

$$2^3$$
$$2 \cdot 2 \cdot 2$$
$$8$$

See Geometry Supplement on page 420 for more information.

Example 1

The formula for the area of a square is $A = s^2$.
Find A if $s = 7$.

$A = s^2$

Substitute 7 for s. $A = 7^2 = 7 \cdot 7$ or 49

So, A is 49 if $s = 7$.

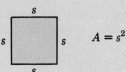

practice ▷ $A = s^2$. **Find A for the given value of s.**

1. 6   36
2. 8   64
3. 13   169
4. 19   361

Example 2

The formula for the area of a circle is $A = \pi r^2$.
Find A in terms of π if $r = 9$.

$A = \pi r^2$

Substitute 9 for r. $A = \pi \cdot 9^2$
$9^2 = 9 \cdot 9 = 81$ $A = \pi \cdot 81$, or 81π

So, $A = 81\pi$ if $r = 9$.

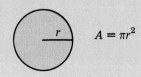

practice ▷ $A = \pi r^2$. **Find A in terms of π for the given values of r.**

5. 4 16π
6. 3 9π
7. 10 100π
8. 11 121π

Example 3

The formula for the volume of a cylinder is
$V = \pi r^2 h$. Find V in terms of π if $r = 4$ and $h = 10$.

$V = \pi r^2 h$

Substitute 4 for r and 10 for h. $V = \pi \cdot 4^2 \cdot 10$
$V = \pi \cdot 16 \cdot 10$
$160 \cdot \pi = 160\pi$ $V = \pi \cdot 160$ or 160π

So, $V = 160\pi$ if $r = 4$ and $h = 10$.

practice ▷ $V = \pi r^2 h$. Find V in terms of π for the given values of r and h.

9. $r = 5, h = 8$ 200π

10. $r = 3, h = 10$ 90π

11. $r = 10, h = 4$ 400π

12. $r = 7, h = 3$ 147π

Example 4

The formula for the surface area of a cube is $6e^2$ where e is the length of an edge. Find S if $e = 5$.

$$S = 6e^2$$

Substitute 5 for e. $\quad S = 6 \cdot 5^2$

$\quad\quad\quad 25 \quad\quad\quad S = 6 \cdot 25$ or 150

$\quad\quad\quad \underline{\times\ 6}$

$\quad\quad\quad 150 \quad\quad$ So, $S = 150$ if $e = 5$.

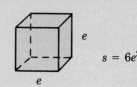

$s = 6e^2$

---

practice ▷ $S = 6e^2$. Find S for the given value of e.

13. 3 $\quad$ 54 $\quad\quad$ **14.** 6 $\quad$ 216 $\quad\quad$ **15.** 30 $\quad$ 5,400 $\quad\quad$ **16.** 10 $\quad$ 600

◇ EXERCISES ◇

$A = s^2$. Find A for the given value of s.

1. 2 $\quad$ 4 $\quad\quad$ **2.** 5 $\quad$ 25 $\quad\quad$ **3.** 3 $\quad$ 9 $\quad\quad$ **4.** 4 $\quad$ 16 $\quad\quad$ **5.** 18 $\quad$ 324

6. 10 $\quad$ 100 $\quad\quad$ **7.** 16 $\quad$ 256 $\quad\quad$ **8.** 21 $\quad$ 441 $\quad\quad$ **9.** 40 $\quad$ 1,600 $\quad\quad$ **10.** 15 $\quad$ 225

$A = \pi r^2$. Find A in terms of π for the given value of r.

11. 5 $\quad$ 25π $\quad\quad$ **12.** 6 $\quad$ 36π $\quad\quad$ **13.** 8 $\quad$ 64π $\quad\quad$ **14.** 12 $\quad$ 144π $\quad\quad$ **15.** 2 $\quad$ 4π

16. 14 $\quad$ 196π $\quad\quad$ **17.** 22 $\quad$ 484π $\quad\quad$ **18.** 20 $\quad$ 400π $\quad\quad$ **19.** 17 $\quad$ 289π $\quad\quad$ **20.** 30 $\quad$ 900π

$V = \pi r^2 h$. Find V in terms of π for the given values of r and h.

21. $r = 3, h = 3$ $\quad$ 27π $\quad\quad$ **22.** $r = 6, h = 5$ $\quad$ 180π $\quad\quad$ **23.** $r = 8, h = 7$ $\quad$ 448π

24. $r = 6, h = 8$ $\quad$ 288π $\quad\quad$ **25.** $r = 5, h = 7$ $\quad$ 175π $\quad\quad$ **26.** $r = 20, h = 11$ $\quad$ $4,400\pi$

$S = 6e^2$. Find S for the given value of e.

27. 4 $\quad$ 96 $\quad\quad$ **28.** 7 $\quad$ 294 $\quad\quad$ **29.** 1 $\quad$ 6 $\quad\quad$ **30.** 9 $\quad$ 486 $\quad\quad$ **31.** 8 $\quad$ 384

32. 12 $\quad$ 864 $\quad\quad$ **33.** 14 $\quad$ 1,176 $\quad\quad$ **34.** 19 $\quad$ 2,166 $\quad\quad$ **35.** 21 $\quad$ 2,646 $\quad\quad$ **36.** 42 $\quad$ 10,584

The formula for the volume of a cube is $V = e^3$ where e is the length of a side. Find V for the given value of e.

37. 6 $\quad$ 216 $\quad\quad$ **38.** 2 $\quad$ 8 $\quad\quad$ **39.** 5 $\quad$ 125 $\quad\quad$ **40.** 7 $\quad$ 343 $\quad\quad$ **41.** 10 $\quad$ 1,000

42. 15 $\quad$ 3,375 $\quad\quad$ **43.** 11 $\quad$ 1,331 $\quad\quad$ **44.** 18 $\quad$ 5,832 $\quad\quad$ **45.** 25 $\quad$ 15,625 $\quad\quad$ **46.** 13 $\quad$ 2,197

The formula for the surface area of a cylinder is $S = 2\pi r^2 + 2\pi rh$. Find S in terms of π.

★ **47.** $r = 4, h = 10$ $\quad$ 112π $\quad\quad$ ★ **48.** $r = 8, h = 6$ $\quad$ 224π $\quad\quad$ ★ **49.** $r = 10, h = 3$ $\quad$ 260π

EXPONENTS IN FORMULAS

Non-Routine Problems

There is no simple method used to solve these problems. Use whatever methods you can. Do not expect to solve these easily.

See TE Answer Section for sample solutions to these problems.

1. The Gordons have two children. The sum of their ages is 15 and the product is 36. What are their ages?   3 yr and 12 yr

2. You have three piles of pennies. There are 10 pennies in the first pile, 18 pennies in the second pile, and a certain number in the third pile. If you double the number of pennies in each pile, you will have 70 pennies. How many are there in the third pile?   7 pennies

3. A farmer has a herd of cows and some chickens. Altogether there are 36 heads and 84 legs. How many cows and how many chickens are there?
   6 cows and 30 chickens

4. There are books packed in a certain number of boxes. The math books are packed 12 to a box and science books 16 to a box. There are altogether 108 books. How many boxes contain math books, and how many boxes contain science books?

5. A third grader was given a problem of dividing one number by another. Instead of dividing, the student subtracted one number from the other. The teacher said that the answer was right. What numbers was the student asked to divide?   4 and 2

6. Lester has $23 in his savings account. Rhonda has $50 in hers. Beginning now, each of them will add $1 each week to their accounts. After how many weeks will Rhonda have twice as much as Lester?   4

7. A package of buttons sells for $3.10. There are silver and gold buttons in the package. Each silver button costs $0.08. Each gold button costs $0.10. How many silver and how many gold buttons are there in the package?
   20 silver buttons and 15 gold buttons

4. 5 boxes contain math books and 3 boxes contain science books or 1 box contains math books and 6 boxes contain science books.

You have used the **BASIC** operations of $+$, $-$, $*$, and $/$.
Many algebraic expressions also contain exponents.
But there is no way for a computer keyboard to "raise" the 3 in an expression
like 4^3. Another device is used. 4^3 is written in **BASIC** as $4 \wedge 3$, where 3 is
the exponent and the symbol $\wedge$ is used to show exponentiation.
The order of operations rule can now be extended as follows:

     1st: raise to a power; $\wedge$
     2nd: do multiplication or division; $*$ or $/$
     3rd: do addition or subtraction; $+$ or $-$

Example $A = 2 * 4 \wedge 3 - 6 \wedge 2$. Find the value of A by computation. Write
        a program to find A. Then type and **RUN** it.
            By computation: $A = 2 * 4 \wedge 3 - 6 \wedge 2$
        Do exponentiation first: $A = 2 * 64 - 36$
           $4 \wedge 3 = 4^3 = 64$ $A = 128 - 36$
           $6 \wedge 2 = 6^2 = 36$ $A = 92$

Type in this program.

```
10   LET A = 2 * 4 ^ 3 - 6 ^ 2
20   PRINT A
30   END
```

          RUN the program.]RUN
  The screen displays the answer. 92

Exercises

Compute.

1. $7 \wedge 2$ **2.** $4 * 3 \wedge 2 - 4$ **3.** $19 + 4 \wedge 3 - 6 * 7$ **4.** $49 - 7 \wedge 2 + 8 * 6 - 5$
   49   32   41   43

Find the value of each variable by computation. Then write a program to find
the value. Type and **RUN** it.

5. $A = 8 * 4 \wedge 2$ **6.** $B = 6 * 3 \wedge 2 + 4 * 2 \wedge 5$ **7.** $T = 3 * 2 \wedge 4 + 8 \wedge 2$
   128   182   112

You are now ready to learn more about the techniques of programming. The
Computer Section beginning on page 420 will provide you with an introduction
to the concepts of programming.

Rewrite using the distributive property. [53]

1. $8(6 + 1)$ $8 \cdot 6 + 8 \cdot 1$
2. $(5 + 2)6$ $5 \cdot 6 + 2 \cdot 6$
3. $7(5 - 4)$ $7 \cdot 5 - 7 \cdot 4$

Rewrite using the distributive property. [53]

4. $8 \cdot 7 + 8 \cdot 13$ $8(7 + 13)$
5. $3 \cdot 6 - 3 \cdot 2$ $3(6 - 2)$
6. $17 \cdot 2 + 15 \cdot 2$ $(17 + 15)2$

Simplify. [53, 56]

7. $6(5x + 2)$ $30x + 12$
8. $3(5m - 1)$ $15m - 3$
9. $8k + 5k$ $13k$
10. $6w + 5y + 2 + 3y + 1 + 5w$ $11w + 8y + 3$
11. $5k + 2 + 4m$ not possible
12. $7a + 2 - 2a$ $5a + 2$

Simplify. Then evaluate for these values of the variables:
$a = 5, b = 2, c = 6.$ [58]

13. $3a + 5 + 8a$ 60
14. $4b + 2 + 3b$ 16
15. $9 + 2c + c$ 27
16. $3a + b + 4a + b$ 39
17. $c + 5a + c + 4 - 3a$ 26
18. $8 + 2b + 3 + b + a + c + 3a + 1$ 44

Solve each equation. Check the solution. [61]

19. $2h + 8 = 14$ 3
20. $14 = 5m - 1$ 3
21. $9y - 2 = 16$ 2
22. $\frac{x}{3} + 1 = 10$ 27

Write in mathematical form. [63]

23. 8 times k, decreased by 5 $8k - 5$
24. 9 less than a number divided by 2 $\frac{n}{2} - 9$

Solve these problems. [65]

25. 1 less than 7 times a number is 13. Find the number.   2
26. 36 is the same as a number increased by twice the number. Find the number.   12

Find the value. [69]

27. 2^3 8
28. 1^9 1
29. 6^3 216

Evaluate. [69]

30. x^4 if $x = 7$ 2,401
31. $7a^3$ if $a = 3$ 189
32. $20y^2$ if $y = 4$ 320

Find A in terms of π if $r = 6$. [72]

33. $A = \pi r^2$ $A = 36\pi$

Find V in terms of π if $r = 3$ and $h = 4$.

34. $V = \pi r^2 h$ $V = 36\pi$

Rewrite using the distributive property.

1. $7(9 + 3)$
   $7 \cdot 9 + 7 \cdot 3$

2. $(8 - 4)6$
   $8 \cdot 6 - 4 \cdot 6$

3. $7 \cdot 9 - 7 \cdot 3$
   $7(9 - 3)$

4. $13 \cdot 5 + 11 \cdot 5$
   $(13 + 11)5$

Simplify.

5. $7(2k + 4)$ $14k + 28$

6. $8(5m - 2)$ $40m - 16$

7. $9t + 4t$ $13t$

8. $c + 5 + 9d + 3 - 4d + 3c$
   $4c + 5d + 8$

Simplify. Then evaluate for these values of the variables: $a = 5, b = 4.$

9. $7a + 2 + 3a$ $10a + 2, 52$

10. $2a + 3b + 4$
    not possible, 26

11. $a + 3b + 6 + b + 4a + 1$
    $5a + 4b + 7, 48$

Solve each equation. Check the solution.

12. $2x + 10 = 14$ 2

13. $14 = 3k - 7$ 7

14. $8t - 3 = 21$ 3

15. $\frac{x}{4} - 5 = 6$ 44

Write in mathematical form.

16. a number divided by 5 is increased by 9 $\frac{w}{5} + 9$

17. 13 less than twice a number $2n - 13$

Solve these problems.

18. 5 less than 4 times a number is 15. Find the number. 5

19. 40 is the same as a number increased by 3 times the number. Find the number. 10

Find the value.

20. 3^4 81

21. 8^4 4,096

Evaluate.

22. x^4 if $x = 5$ 625

23. $6a^2$ if $a = 4$ 96

Find A in terms of π, if $r = 9$.

24. $A = \pi r^2$ 81π

Find V in terms of π if $r = 4$ and $h = 9$.

25. $V = \pi r^2 h$ 144π

Multiplying and Dividing Fractions

4

Bryce Canyon National Park in Utah covers 36,010 acres.

Introduction to Fractions

— ◇ **OBJECTIVES** ◇ —

To find numbers corresponding
to points on a number line

To find products like $6 \cdot \frac{1}{7}$

— ◇ **RECALL** ◇ —

Points on a number line correspond to numbers.

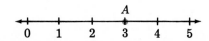

The number 3 corresponds to point A.

Example 1

What number corresponds to point A?

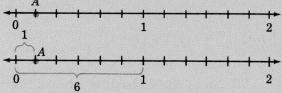

There are 6 equal subdivisions
from 0 to 1.

1 of 6 equal parts A corresponds to 1 of 6 equal parts.

$\frac{1}{6}$ So, $\frac{1}{6}$ corresponds to A.

---

practice ▷   **What number corresponds to point A?**

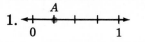

1. $\frac{1}{4}$ 2. $\frac{1}{9}$

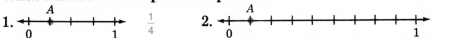

Example 2

What number corresponds to point A?

Count the number of equal
subdivisions. There are 9 equal
subdivisions from 0 to 1.

4 of 9 equal parts

$\frac{4}{9}$

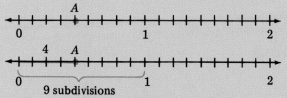

A corresponds to 4 of 9 equal parts.

So, $\frac{4}{9}$ corresponds to A.

---

practice ▷   **What number corresponds to point A?**

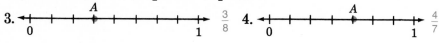

3. $\frac{3}{8}$ 4. $\frac{4}{7}$

Recall that numbers such as $\frac{4}{9}$ or $\frac{1}{6}$ are fractions.

4 is the numerator.———→$\frac{4}{}$
9 is the denominator.———→$\frac{}{9}$

A fraction like $\frac{6}{0}$ is undefined.

Example 3

Think: Check by multiplication.

Show that $\frac{6}{0}$ is undefined.

$\frac{6}{0}$ means $6 \div 0$.

Let $6 \div 0 = a$.
Then $0 \cdot a = 6$.
But, $0 \cdot a = 0$, **not** 6.

So, $\frac{6}{0}$ is undefined.

The denominator of a fraction *cannot* be 0. Division by 0 is undefined.

practice ▷ **Which of the following are undefined?**

5. $\frac{7}{9}$ no

6. $\frac{0}{2}$ no

7. $\frac{5}{0}$ yes

Example 4

Divide the number line into 6 equal parts from 0 to 1. Count 5 subdivisions from 0 to $\frac{5}{6}$.

Use a number line to show that $5 \cdot \frac{1}{6} = \frac{5}{6}$.

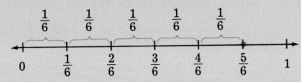

So, $5 \cdot \frac{1}{6} = \frac{5}{6}$.

Example 5

Multiply. $4 \cdot \frac{1}{7}$

$4 \cdot \frac{1}{7}$

$\frac{4 \cdot 1}{7} = \frac{4}{7}$

Multiply. $\frac{1}{5} \cdot 5$

$\frac{1}{5} \cdot 5$ means $5 \cdot \frac{1}{5}$.

$\frac{5 \cdot 1}{5} = \frac{5}{5}$, or 1

So, $4 \cdot \frac{1}{7} = \frac{4}{7}$ and $\frac{1}{5} \cdot 5 = 1$.

practice ▷ **Multiply.**

8. $3 \cdot \frac{1}{5}$ $\frac{3}{5}$
9. $4 \cdot \frac{1}{9}$ $\frac{4}{9}$
10. $\frac{1}{6} \cdot 5$ $\frac{5}{6}$
11. $\frac{1}{8} \cdot 8$ 1
12. $\frac{1}{15} \cdot 15$ 1

CHAPTER FOUR

Multiply.

1. $3 \cdot \dfrac{1}{4}$ $\dfrac{3}{4}$ **2.** $\dfrac{1}{3} \cdot 2$ $\dfrac{2}{3}$ **3.** $8 \cdot \dfrac{1}{8}$ 1   **4.** $7 \cdot \dfrac{1}{11}$ $\dfrac{7}{11}$ **5.** $\dfrac{11}{2} \cdot 2$ 11

Which of the following are undefined?

6. $\dfrac{7}{8}$ no   **7.** $\dfrac{0}{4}$ no   **8.** $\dfrac{9}{0}$ yes   **9.** $\dfrac{71}{10}$ no   **10.** $\dfrac{14}{0}$ yes

◇ EXERCISES ◇

What number corresponds to point A?

1. 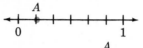 $\dfrac{1}{7}$ **2.** 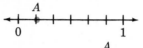 $\dfrac{1}{6}$ **3.**  $\dfrac{1}{8}$

4. 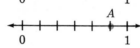 $\dfrac{5}{7}$ **5.** 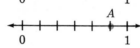 $\dfrac{5}{6}$ **6.** 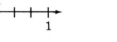 $\dfrac{2}{5}$

Multiply.

7. $3 \cdot \dfrac{1}{7}$ $\dfrac{3}{7}$ **8.** $\dfrac{1}{5} \cdot 2$ $\dfrac{2}{5}$ **9.** $7 \cdot \dfrac{1}{9}$ $\dfrac{7}{9}$ **10.** $\dfrac{1}{4} \cdot 3$ $\dfrac{3}{4}$ **11.** $6 \cdot \dfrac{1}{6}$ 1

12. $\dfrac{1}{15} \cdot 4$ $\dfrac{4}{15}$ **13.** $7 \cdot \dfrac{1}{8}$ $\dfrac{7}{8}$ **14.** $\dfrac{1}{9} \cdot 5$ $\dfrac{5}{9}$ **15.** $\dfrac{1}{8} \cdot 7$ $\dfrac{7}{8}$ **16.** $7 \cdot \dfrac{1}{11}$ $\dfrac{7}{11}$

17. $\dfrac{1}{4} \cdot 4$ 1   **18.** $\dfrac{1}{3} \cdot 2$ $\dfrac{2}{3}$ **19.** $\dfrac{1}{13} \cdot 5$ $\dfrac{5}{13}$ **20.** $3 \cdot \dfrac{1}{3}$ 1   **21.** $\dfrac{1}{7} \cdot 5$ $\dfrac{5}{7}$

22. $\dfrac{1}{12} \cdot 7$ $\dfrac{7}{12}$ **23.** $4 \cdot \dfrac{1}{15}$ $\dfrac{4}{15}$ **24.** $\dfrac{1}{12} \cdot 12$ 1   **25.** $5 \cdot \dfrac{1}{8}$ $\dfrac{5}{8}$ **26.** $\dfrac{1}{10} \cdot 3$ $\dfrac{3}{10}$

27. $\dfrac{1}{7} \cdot 7$ 1   **28.** $\dfrac{1}{12} \cdot 5$ $\dfrac{5}{12}$ ★ **29.** $3 \cdot 4 \cdot \dfrac{1}{17}$ $\dfrac{12}{17}$ ★ **30.** $7 \cdot \dfrac{1}{15} \cdot 2$ $\dfrac{14}{15}$ ★ **31.** $4 \cdot \dfrac{1}{9} \cdot 2$ $\dfrac{8}{9}$

Is it possible to make five squares of the same size by moving only three sticks? How? Move sticks labeled A, B, and C to their new position in the figure below.

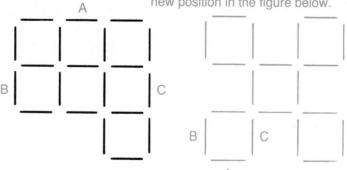

Multiplying by a Whole Number

──◇ **OBJECTIVES** ◇──

To multiply expressions like

$12 \cdot \frac{2}{3}$ and $1\frac{1}{2} \cdot 3$

To change a mixed number to a
fraction

────◇ **RECALL** ◇────

$2 \cdot \frac{1}{3} = \frac{2}{3}$

$\frac{3}{8} = 3 \cdot \frac{1}{8}$

Example 1

Multiply. $8 \cdot \frac{3}{4}$

$$8 \cdot \frac{3}{4}$$

$\frac{3}{4} = 3 \cdot \frac{1}{4}$

$$8 \cdot 3 \cdot \frac{1}{4}$$

$$24 \cdot \frac{1}{4}$$

$$\frac{24 \cdot 1}{4} = \frac{24}{4}, \text{ or } 6$$

So, $8 \cdot \frac{3}{4} = \frac{24}{4}$, or 6.

Example 2

Multiply. $7 \cdot \frac{2}{5}$

$$7 \cdot \frac{2}{5} = \frac{7 \cdot 2}{5}$$

In $\frac{14}{5}$, the numerator is larger than
the denominator.
Write the remainder over the
divisor as a fraction.

$\frac{14}{5}$ ← improper fraction

$$\frac{14}{5} = 5\overline{)14}^{\,2}, \text{ or } 2\frac{4}{5} \leftarrow \text{mixed number}$$

$$\underline{10}$$

divisor   4 ← remainder

So, $7 \cdot \frac{2}{5} = \frac{14}{5}$, or $2\frac{4}{5}$.

---

practice ▷ **Multiply.**

1. $6 \cdot \frac{2}{3}$ 4   **2.** $10 \cdot \frac{2}{5}$ 4   **3.** $5 \cdot \frac{5}{6}$ $4\frac{1}{6}$ **4.** $8 \cdot \frac{2}{3}$ $5\frac{1}{3}$ **5.** $4 \cdot \frac{3}{7}$ $1\frac{5}{7}$

In Example 2, division was used to change $\frac{14}{5}$ to a mixed number. In the next example, the mixed number $4\frac{3}{5}$ is changed to a fraction.

Example 3

Change $4\frac{3}{5}$ to a fraction.

Multiply 5 by 4. Then add 3.

$$4\frac{3}{5} \qquad 4 \underset{\cdot}{\longrightarrow} \overset{3}{\underset{5}{}} +$$

$$\frac{(5 \cdot 4) + 3}{5}$$

$$\frac{20 + 3}{5} = \frac{23}{5}$$

So, $4\frac{3}{5} = \frac{23}{5}$.

practice ▷ **Change to a fraction.**

6. $3\frac{4}{5}$ $\frac{19}{5}$ 7. $5\frac{2}{3}$ $\frac{17}{3}$ 8. $4\frac{2}{5}$ $\frac{22}{5}$ 9. $6\frac{1}{3}$ $\frac{19}{3}$ 10. $4\frac{5}{6}$ $\frac{29}{6}$ 11. $9\frac{3}{4}$ $\frac{39}{4}$

Example 4

Multiply. $3\frac{2}{3} \cdot 2$

Change $3\frac{2}{3}$ to a fraction.

$$3\frac{2}{3} = \frac{(3 \cdot 3) + 2}{3} = \frac{9 + 2}{3} = \frac{11}{3}$$

Multiplication can be done in any order.

$$\begin{array}{r} 7 \\ 3\overline{)22} \\ \underline{21} \\ 1 \end{array} = 7\frac{1}{3}$$

$$3\frac{2}{3} \cdot 2$$

$$\frac{11}{3} \cdot 2$$

$$2 \cdot \frac{11}{3}$$

$$\frac{22}{3}$$

$$7\frac{1}{3}$$

So, $3\frac{2}{3} \cdot 2 = 7\frac{1}{3}$.

practice ▷ **Multiply.**

12. $3\frac{1}{3} \cdot 2$ $6\frac{2}{3}$ 13. $2\frac{2}{3} \cdot 2$ $5\frac{1}{3}$ 14. $1\frac{1}{5} \cdot 2$ $2\frac{2}{5}$ 15. $5 \cdot 1\frac{1}{2}$ $7\frac{1}{2}$ 16. $3 \cdot 1\frac{1}{4}$ $3\frac{3}{4}$

◇ ORAL EXERCISES ◇

Multiply.

1. $3 \cdot \frac{2}{7}$ $\frac{6}{7}$ **2.** $4 \cdot \frac{3}{13}$ $\frac{12}{13}$ **3.** $2 \cdot \frac{4}{9}$ $\frac{8}{9}$ **4.** $3 \cdot \frac{2}{11}$ $\frac{6}{11}$ **5.** $4 \cdot \frac{2}{13}$ $\frac{8}{13}$ **6.** $2 \cdot \frac{4}{17}$ $\frac{8}{17}$

7. $2 \cdot \frac{5}{13}$ $\frac{10}{13}$ **8.** $5 \cdot \frac{3}{19}$ $\frac{15}{19}$ **9.** $2 \cdot \frac{6}{13}$ $\frac{12}{13}$ **10.** $\frac{5}{19} \cdot 2$ $\frac{10}{19}$ **11.** $\frac{2}{9} \cdot 4$ $\frac{8}{9}$ **12.** $\frac{4}{13} \cdot 3$ $\frac{12}{13}$

Change to a fraction.

13. $3\frac{1}{2}$ $\frac{7}{2}$ **14.** $2\frac{1}{3}$ $\frac{7}{3}$ **15.** $4\frac{1}{2}$ $\frac{9}{2}$ **16.** $3\frac{1}{3}$ $\frac{10}{3}$ **17.** $5\frac{1}{4}$ $\frac{21}{4}$ **18.** $2\frac{1}{8}$ $\frac{17}{8}$

19. $4\frac{1}{5}$ $\frac{21}{5}$ **20.** $2\frac{1}{6}$ $\frac{13}{6}$ **21.** $3\frac{1}{4}$ $\frac{13}{4}$ **22.** $6\frac{1}{2}$ $\frac{13}{2}$ **23.** $7\frac{1}{2}$ $\frac{15}{2}$ **24.** $4\frac{1}{7}$ $\frac{29}{7}$

◇ EXERCISES ◇

Multiply.

1. $10 \cdot \frac{3}{5}$ 6 **2.** $4 \cdot \frac{3}{4}$ 3 **3.** $3 \cdot \frac{2}{3}$ 2 **4.** $9 \cdot \frac{2}{3}$ 6 **5.** $14 \cdot \frac{2}{7}$ 4 **6.** $15 \cdot \frac{2}{5}$ 6

7. $12 \cdot \frac{3}{4}$ 9 **8.** $10 \cdot \frac{2}{5}$ 4 **9.** $14 \cdot \frac{1}{7}$ 2 **10.** $15 \cdot \frac{2}{3}$ 10 **11.** $9 \cdot \frac{2}{3}$ 6 **12.** $18 \cdot \frac{2}{3}$ 12

13. $2 \cdot \frac{3}{5}$ $1\frac{1}{5}$ **14.** $4 \cdot \frac{3}{5}$ $2\frac{2}{5}$ **15.** $2 \cdot \frac{2}{3}$ $1\frac{1}{3}$ **16.** $5 \cdot \frac{3}{7}$ $2\frac{1}{7}$ **17.** $3 \cdot \frac{4}{5}$ $2\frac{2}{5}$ **18.** $2 \cdot \frac{4}{7}$ $1\frac{1}{7}$

19. $7 \cdot \frac{2}{3}$ $4\frac{2}{3}$ **20.** $5 \cdot \frac{4}{7}$ $2\frac{6}{7}$ **21.** $8 \cdot \frac{2}{3}$ $5\frac{1}{3}$ **22.** $7 \cdot \frac{2}{5}$ $2\frac{4}{5}$ **23.** $6 \cdot \frac{3}{5}$ $3\frac{3}{5}$ **24.** $3 \cdot \frac{5}{7}$ $2\frac{1}{7}$

Change to a fraction.

25. $6\frac{2}{3}$ $\frac{20}{3}$ **26.** $4\frac{2}{3}$ $\frac{14}{3}$ **27.** $9\frac{3}{4}$ $\frac{39}{4}$ **28.** $6\frac{3}{5}$ $\frac{33}{5}$ **29.** $4\frac{3}{8}$ $\frac{35}{8}$ **30.** $7\frac{7}{8}$ $\frac{63}{8}$

31. $5\frac{5}{9}$ $\frac{50}{9}$ **32.** $4\frac{2}{7}$ $\frac{30}{7}$ **33.** $6\frac{3}{7}$ $\frac{45}{7}$ **34.** $5\frac{2}{9}$ $\frac{47}{9}$ **35.** $8\frac{5}{6}$ $\frac{53}{6}$ **36.** $6\frac{6}{7}$ $\frac{48}{7}$

Multiply.

37. $3\frac{1}{3} \cdot 2$ $6\frac{2}{3}$ **38.** $1\frac{1}{3} \cdot 5$ $6\frac{2}{3}$ **39.** $1\frac{1}{2} \cdot 7$ $10\frac{1}{2}$ **40.** $3\frac{2}{3} \cdot 2$ $7\frac{1}{3}$ **41.** $1\frac{1}{4} \cdot 3$ $3\frac{3}{4}$

42. $2 \cdot 2\frac{1}{3}$ $4\frac{2}{3}$ **43.** $5 \cdot 1\frac{1}{2}$ $7\frac{1}{2}$ **44.** $2 \cdot 1\frac{1}{3}$ $2\frac{2}{3}$ ★**45.** $8 \cdot 2\frac{1}{2} \cdot 3$ 60 ★**46.** $6 \cdot 3\frac{1}{3} \cdot 15$ 30

Solve these problems.

47. Jason works on the yearbook $\frac{3}{4}$ hour each day. How much time does he put in during a 5-day week? $3\frac{3}{4}$ h

48. Janet is a clerk at Walling Drugs. She works $2\frac{1}{2}$ h each night. How many hours does she work during a 5-day week? $12\frac{1}{2}$ h

Multiplying by a Fraction

───◇ *OBJECTIVE* ◇───   ───◇ *RECALL* ◇───

To multiply expressions like
$\frac{2}{5} \cdot \frac{3}{4}$ and $3\frac{1}{2} \cdot \frac{1}{8}$

$$8 \cdot \frac{1}{8} = \frac{8 \cdot 1}{8} = \frac{8}{8} = 1$$

Since $8 \cdot \frac{1}{8} = 1$, 8 and $\frac{1}{8}$ are reciprocals of each other. Two numbers are *reciprocals* if their product is 1.

Example 1

Two numbers are reciprocals if their product is 1.

Find the reciprocal of 3.
Think: 3 times what number is 1?

$$3 \cdot \frac{1}{3} = 1$$

So, $\frac{1}{3}$ is the reciprocal of 3.

Example 2

Show that $\frac{1}{3} \cdot \frac{1}{2} = \frac{1}{6}$.

First show that the reciprocal of 6 is $\frac{1}{3} \cdot \frac{1}{2}$.

$$6 \quad \cdot \frac{1}{3} \cdot \frac{1}{2}$$

$3 \cdot 2 = 6$ $3 \cdot 2 \cdot \frac{1}{3} \cdot \frac{1}{2}$

Regroup. $\left(3 \cdot \frac{1}{3}\right)\left(2 \cdot \frac{1}{2}\right)$

$3 \cdot \frac{1}{3} = 1$ $\vdots$ $2 \cdot \frac{1}{2} = 1$ $\underbrace{}_{(1)} \quad \underbrace{}_{(1)}$

$$1$$

$$6 \cdot \frac{1}{3} \cdot \frac{1}{2} = 1$$

The reciprocal of 6 is $\frac{1}{3} \cdot \frac{1}{2}$.

Since $6 \cdot \frac{1}{3} \cdot \frac{1}{2} = 1$ and $6 \cdot \frac{1}{6} = 1$,

The reciprocal of 6 is also $\frac{1}{6}$.

$$\frac{1}{3} \cdot \frac{1}{2} = \frac{1}{6}.$$

A number has exactly one reciprocal.

So, $\frac{1}{3} \cdot \frac{1}{2} = \frac{1}{6}$.

Example 3

Multiply. $\frac{3}{7} \cdot \frac{4}{5}$

$$\frac{3}{7} \cdot \frac{4}{5}$$

$\frac{3}{7} = 3 \cdot \frac{1}{7}$ and $\frac{4}{5} = 4 \cdot \frac{1}{5}$

$3 \cdot \frac{1}{7} \cdot 4 \cdot \frac{1}{5}$

Reorder.

$3 \cdot 4 \cdot \frac{1}{7} \cdot \frac{1}{5}$

$\frac{1}{7} \cdot \frac{1}{5} = \frac{1}{35}$

$12 \cdot \frac{1}{35} = \frac{12}{35}$

So, $\frac{3}{7} \cdot \frac{4}{5} = \frac{3 \cdot 4}{7 \cdot 5}$, or $\frac{12}{35}$.

In Example 3, we found that $\frac{3}{7} \cdot \frac{4}{5} = \frac{3 \cdot 4}{7 \cdot 5}$.

Example 4

Multiply. $\frac{3}{5} \cdot \frac{7}{2}$

Multiply the numerators.
Multiply the denominators.

$\frac{3}{5} \cdot \frac{7}{2} = \frac{3 \cdot 7}{5 \cdot 2} = \frac{21}{10}$

$\begin{array}{r} 2 \\ 10\overline{)21} \\ 20 \\ \hline 1 \end{array} = 2\frac{1}{10}$ So, $\frac{3}{5} \cdot \frac{7}{2} = 2\frac{1}{10}$.

practice ▷ **Multiply.**

1. $\frac{2}{3} \cdot \frac{5}{3}$ $1\frac{1}{9}$ **2.** $\frac{5}{3} \cdot \frac{4}{3}$ $2\frac{2}{9}$ **3.** $\frac{3}{4} \cdot \frac{5}{7}$ $\frac{15}{28}$ **4.** $\frac{1}{3} \cdot \frac{7}{2}$ $1\frac{1}{6}$ **5.** $\frac{3}{4} \cdot \frac{5}{2}$ $1\frac{7}{8}$

Example 5

Multiply. $3\frac{1}{2} \cdot \frac{1}{8}$

$3\frac{1}{2} = \frac{(2 \cdot 3) + 1}{2} = \frac{6+1}{2} = \frac{7}{2}$

$3\frac{1}{2} \cdot \frac{1}{8} = \frac{7}{2} \cdot \frac{1}{8} = \frac{7 \cdot 1}{2 \cdot 8} = \frac{7}{16}$

So, $3\frac{1}{2} \cdot \frac{1}{8} = \frac{7}{16}$.

practice ▷ **Multiply.**

6. $2\frac{1}{2} \cdot \frac{1}{3}$ $\frac{5}{6}$ **7.** $3\frac{1}{3} \cdot \frac{1}{9}$ $\frac{10}{27}$ **8.** $2\frac{1}{4} \cdot \frac{1}{7}$ $\frac{9}{28}$ **9.** $\frac{2}{5} \cdot 1\frac{1}{5}$ $\frac{12}{25}$ **10.** $2\frac{2}{3} \cdot \frac{1}{3}$ $\frac{8}{9}$

Multiply.

1. $\frac{3}{5} \cdot \frac{2}{5}$ $\frac{6}{25}$
2. $\frac{1}{7} \cdot \frac{2}{3}$ $\frac{2}{21}$
3. $\frac{3}{4} \cdot \frac{1}{2}$ $\frac{3}{8}$
4. $\frac{5}{8} \cdot \frac{1}{3}$ $\frac{5}{24}$
5. $\frac{1}{2} \cdot \frac{7}{9}$ $\frac{7}{18}$

6. $\frac{4}{7} \cdot \frac{2}{3}$ $\frac{8}{21}$
7. $\frac{3}{5} \cdot \frac{1}{5}$ $\frac{3}{25}$
8. $\frac{4}{9} \cdot \frac{2}{3}$ $\frac{8}{27}$
9. $\frac{2}{5} \cdot \frac{3}{7}$ $\frac{6}{35}$
10. $\frac{2}{9} \cdot \frac{1}{3}$ $\frac{2}{27}$

◇ *EXERCISES* ◇

Multiply.

1. $\frac{3}{5} \cdot \frac{1}{4}$ $\frac{3}{20}$
2. $\frac{1}{3} \cdot \frac{7}{2}$ $1\frac{1}{6}$
3. $\frac{7}{2} \cdot \frac{1}{2}$ $1\frac{3}{4}$
4. $\frac{3}{5} \cdot \frac{7}{2}$ $2\frac{1}{10}$
5. $\frac{5}{2} \cdot \frac{5}{3}$ $4\frac{1}{6}$

6. $\frac{3}{2} \cdot \frac{5}{8}$ $\frac{15}{16}$
7. $\frac{3}{4} \cdot \frac{3}{5}$ $\frac{9}{20}$
8. $\frac{2}{7} \cdot \frac{1}{3}$ $\frac{2}{21}$
9. $\frac{3}{2} \cdot \frac{3}{7}$ $\frac{9}{14}$
10. $\frac{4}{9} \cdot \frac{1}{5}$ $\frac{4}{45}$

11. $\frac{3}{2} \cdot \frac{7}{2}$ $5\frac{1}{4}$
12. $\frac{2}{5} \cdot \frac{8}{3}$ $1\frac{1}{15}$
13. $\frac{4}{3} \cdot \frac{5}{3}$ $2\frac{2}{9}$
14. $4\frac{1}{2} \cdot \frac{1}{7}$ $\frac{9}{14}$
15. $1\frac{2}{3} \cdot \frac{5}{2}$ $4\frac{1}{6}$

16. $\frac{2}{3} \cdot 2\frac{2}{3}$ $1\frac{7}{9}$
★ 17. $\frac{2}{3} \cdot \frac{7}{3} \cdot \frac{5}{3}$ $2\frac{16}{27}$
★ 18. $\frac{4}{5} \cdot \frac{8}{3} \cdot \frac{8}{5}$ $3\frac{31}{75}$
★ 19. $3\frac{1}{2} \cdot 1\frac{1}{4} \cdot \frac{3}{4}$ $3\frac{9}{32}$

Solve these problems.

20. John worked $\frac{1}{2}$ hour. Lucy worked $1\frac{1}{2}$ times as many hours. How many hours did Lucy work? $\frac{3}{4}$ h

21. Maria walked $\frac{3}{4}$ mile. Joe walked $\frac{1}{2}$ as far. How far did Joe walk? $\frac{3}{8}$ mi

Algebra Maintenance

Simplify.

$15j - 15$

1. $8 \cdot 5k$ $40k$
2. $7(3x + 4)$ $21x + 28$
3. $5(3j - 3)$

4. $4x + 3 + 7x + 4$ $11x + 7$
5. $2a + b + 3b + c + 4a + 4b$ $6a + 8b + c$

Solve each equation. Check.

6. $3x + 4 = 22$ 6
7. $15 = 7y - 6$ 3
8. $12 = 4z - 8$ 5

Write in mathematical form.

9. 7 times m, decreased by 4 $7m - 4$
10. 4 less than three times a number $3x - 4$

MULTIPLYING BY A FRACTION

Prime Factorization

◇ OBJECTIVES ◇

To factor a number into primes
To determine whether a number
  is prime, composite, or neither

◇ RECALL ◇

$$2 \cdot 2 \cdot 2 \cdot 2 \quad = 2^4$$
$$3 \cdot 3 \qquad\quad = 3^2 \qquad\quad \text{exponents}$$
$$2 \cdot 2 \cdot 3 \cdot 3 \cdot 3 \; = 2^2 \cdot 3^3$$
$$3 \cdot 4 \qquad\quad = \quad 12$$

$$\underset{\text{factors}}{\uparrow\;\uparrow} \qquad\qquad \underset{\text{product}}{\uparrow}$$

A *prime number* is a whole number greater than 1 whose only factors are itself and 1.

A *composite number* is a whole number greater than 1 that has factors other than itself and 1.

Thus, the numbers 0 and 1 are neither prime nor composite.

Example 1

5 and 3 are numbers whose product is 15; each is prime.

Factor 15 into primes.
$$15 = 5 \cdot 3 \text{ or } 3 \cdot 5$$
A number is factored into primes when each of its factors is prime. The order in which the factors is written does not matter.

---

practice ▷ **Factor into primes.**

1. 6 $3 \cdot 2$ **2.** 10 $2 \cdot 5$ **3.** 21 $3 \cdot 7$ **4.** 35 $5 \cdot 7$

A number may have more than one pair of factors. To factor the number into primes, you can begin by using any pair of its factors.

Example 2

6 is a composite number. Factor 6 into $3 \cdot 2$.

Factor 12 into primes.
You can begin in more than one way.
$$12 = 6 \cdot 2 \qquad\qquad 12 = 3 \cdot 4$$

$$= 3 \cdot 2 \cdot 2 \qquad\qquad = 3 \cdot 2 \cdot 2$$

Each factorization gives the same prime factors.
In each case, 3 is used as a factor once. 2 is used as a factor two times.
So, $12 = 2 \cdot 2 \cdot 3$ or $2^2 \cdot 3$.

Example 3

Factor 7 into primes, if possible.
The only factors of 7 are 7 and 1.
7 is prime. It can be factored no further.

practice ▷ **Factor into primes, if possible.**

    5. 20 $2^2 \cdot 5$ **6.** 13 Prime   **7.** 18 $2 \cdot 3^2$ **8.** 8 2^3

Example 4

Factor 36 into primes.
Write using exponents.

First find two factors. $36 = 4 \cdot 9$ $36 = 3 \cdot 12$ $36 = 6 \cdot 6$

Then continue to factor any composite numbers. $= 2 \cdot 2 \cdot 3 \cdot 3$ $= 3 \cdot 4 \cdot 3$ $3 \cdot 2 \cdot 3 \cdot 2$

    $= 3 \cdot 2 \cdot 2 \cdot 3$

The order of the factors does not matter.   The three methods give the same result.
So, $36 = 3 \cdot 3 \cdot 2 \cdot 2$ or $3^2 \cdot 2^2$.

practice ▷ **Factor into primes. Write using exponents.**

    9. 24 $2^3 \cdot 3$ **10.** 40 $2^3 \cdot 5$ **11.** 32 2^5 **12.** 54 $2 \cdot 3^3$

◇ ORAL EXERCISES ◇

Tell whether each number is prime, composite, or neither.

  1. 11 P   **2.** 22 C   **3.** 1 N   **4.** 80 C   **5.** 13 P

◇ EXERCISES ◇

Prime, composite, or neither?

1. 25 C	**2.** 36 C	**3.** 31 P	**4.** 75 C	**5.** 120 C	**6.** 15 C
7. 1 N	**8.** 48 C	**9.** 5 P	**10.** 32 C	**11.** 17 P	**12.** 26 C

Factor into primes, if possible.

13. 4 2^2 **14.** 22 $2 \cdot 11$ **15.** 19 Prime **16.** 9 3^2 **17.** 49 7^2 **18.** 39 $3 \cdot 13$

Factor into primes, if possible. Write using exponents.

19. 24 $2^3 \cdot 3$ **20.** 28 $2^2 \cdot 7$ **21.** 27 3^3 **22.** 45 $3^2 \cdot 5$ **23.** 30 $2 \cdot 3 \cdot 5$ **24.** 5 Prime

25. 44 $2^2 \cdot 11$ **26.** 50 $2 \cdot 5^2$ **27.** 48 $2^4 \cdot 3$ **28.** 75 $3 \cdot 5^2$ **29.** 99 $3^2 \cdot 11$ **30.** 16 2^4

31. 60 $2^2 \cdot 3 \cdot 5$ **32.** 72 $2^3 \cdot 3^2$ **33.** 81 3^4 **34.** 13 Prime **35.** 100 $2^2 \cdot 5^2$ **36.** 64 2^6

★ **37.** 400 $2^4 \cdot 5^2$ ★ **38.** 120 $2^3 \cdot 3 \cdot 5$ ★ **39.** 140 $2^2 \cdot 5 \cdot 7$ ★ **40.** 225 $3^2 \cdot 5^2$

Greatest Common Factor (GCF)

To find the greatest common
factor (GCF) of two numbers

Factor 20 into primes.
$$20 = 5 \cdot 4 \quad \text{or} \quad 20 = 10 \cdot 2$$
$$5 \cdot 2 \cdot 2 \qquad\qquad = 5 \cdot 2 \cdot 2$$

Example 1

Find the greatest common factor (GCF) of 12
and 30.
Factor each number into primes.
$$12 = 6 \cdot 2 \qquad\qquad 30 = 10 \cdot 3$$
$$= 2 \cdot 3 \cdot 2 \qquad\qquad = 2 \cdot 5 \cdot 3$$
There are at *most* one 2 and one 3 common to 12
and 30.

So, the greatest common factor is $2 \cdot 3$, or 6.

Example 2

Find the GCF of 10 and 21.
Factor each number into primes.

$$10 = 5 \cdot 2 \qquad 21 = 7 \cdot 3$$

Note: 1 is a factor of every number.   There are no common *prime* factors.

So, the GCF of 10 and 21 is 1.

---

practice ▷   **Find the greatest common factor (GCF).**

1. 4 and 16   4   **2.** 9 and 21   3   **3.** 6 and 25   1   **4.** 6 and 14   2

Example 3

Find the GCF of 40 and 24.
First factor each number into primes.
$$40 = 10 \cdot 4 \qquad\qquad 24 = 12 \cdot 2$$
$$= 5 \cdot 2 \cdot 4 \qquad\qquad = 4 \cdot 3 \cdot 2$$
$$= 5 \cdot 2 \cdot 2 \cdot 2 \qquad\qquad = 2 \cdot 2 \cdot 3 \cdot 2$$
There are at *most* three 2's common to 40 and 24.

So, the GCF of 40 and 24 is $2 \cdot 2 \cdot 2$, 2^3, or 8.

Find the greatest common factor (GCF).

5.   20 and 30   10 **6.**   18 and 45   9 **7.**   28 and 32   4 **8.**   12 and 18   6

◇ ORAL EXERCISES ◇

Find the greatest common factor (GCF), given the prime factorization.

1. $2 \cdot 3 \cdot 2 \cdot 5$ and $2 \cdot 7 \cdot 3$ 6
2. $3 \cdot 3 \cdot 2 \cdot 5$ and $3 \cdot 5 \cdot 5 \cdot 3$ 45
3. $2 \cdot 7 \cdot 7 \cdot 5 \cdot 5$ and $7 \cdot 7 \cdot 3$ 49
4. $5 \cdot 5 \cdot 5 \cdot 3$ and $5 \cdot 5 \cdot 3 \cdot 2 \cdot 3$ 75
5. $11 \cdot 7 \cdot 2 \cdot 3 \cdot 2$ and $11 \cdot 7 \cdot 7 \cdot 3$ 231
6. $3 \cdot 5 \cdot 7 \cdot 2$ and $11 \cdot 13$ 1

◇ EXERCISES ◇

Find the greatest common factor (GCF).

1. 6; 15   3
2. 12; 42   6
3. 10; 15   5
4. 35; 45   5
5. 28; 42   14
6. 30; 48   6
7. 18; 55   1
8. 27; 81   27
9. 14; 25   1
10. 44; 60   4
11. 25; 40   5
12. 16; 28   4
13. 35; 54   1
14. 15; 40   5
15. 14; 30   2
16. 15; 33   3
17. 24; 28   4
18. 32; 20   4
19. 16; 20   4
20. 63; 72   9
21. 44; 66   22
22. 32; 26   2
23. 27; 18   9
24. 26; 65   13
25. 36; 42   6
26. 32; 80   16
27. 64; 24   8
28. 28; 35   7
★ **29.** 12; 18; 24   6
★ **30.** 33; 55; 77   11

Calculator

The prime factors of a number can be found by a series of divisions by primes until the answer is 1. Below is a way to factor 84 into primes.

$84 \div 2 = 42$
$42 \div 2 = 21$
$21 \div 3 = 7$
$7 \div 7 = 1$

Divide by the least prime contained in the number.

The prime factors of 84 are 2, 2, 3, and 7.
So, $84 = 2 \cdot 2 \cdot 3 \cdot 7$ or $2^2 \cdot 3 \cdot 7$.

Factor into primes.

1. 90
$3^2 \cdot 2 \cdot 5$

2. 120
$2^3 \cdot 3 \cdot 5$

3. 220
$2^2 \cdot 5 \cdot 11$

4. 1,680
$2^4 \cdot 3 \cdot 5 \cdot 7$

Simplest Form

◇ **OBJECTIVE** ◇

To simplify fractions

◇ **RECALL** ◇

$$\frac{4}{5} \cdot \frac{2}{3} = \frac{4 \cdot 2}{5 \cdot 3} = \frac{8}{15} \quad \bigg| \quad \frac{7}{7} = 1$$

Example 1

Write $\frac{10}{14}$ in simplest form.

One way	More compact form
$\frac{10}{14}$	$\frac{10}{14}$

Factor 10 into primes.
Factor 14 into primes.

$\dfrac{5 \cdot 2}{2 \cdot 7}$ | $\dfrac{5 \cdot 2}{2 \cdot 7}$

Rearrange the like factors, the 2's, so that they are over each other.

$\dfrac{2 \cdot 5}{2 \cdot 7}$ | $\dfrac{5 \cdot \overset{1}{\cancel{2}}}{\underset{1}{\cancel{2}} \cdot 7}$

$\dfrac{2}{2} \cdot \dfrac{5}{7}$ | $\dfrac{5 \cdot 1}{1 \cdot 7}$

$\dfrac{2}{2} = 1;\ 1 \cdot \dfrac{5}{7} = \dfrac{5}{7}$ $1 \cdot \dfrac{5}{7}$, or $\dfrac{5}{7}$ | $\dfrac{5}{7}$

So, $\dfrac{10}{14} = \dfrac{5}{7}$.

Example 2

Write in simplest form. $\dfrac{6}{15}; \dfrac{10}{21}$

$\dfrac{6}{15}$	$\dfrac{10}{21}$

Factor 6 into primes.
Factor 15 into primes.

$\dfrac{3 \cdot 2}{5 \cdot 3}$ | $\dfrac{5 \cdot 2}{7 \cdot 3}$ Factor 10 into primes.
Factor 21 into primes.

Divide out like factors.

$\dfrac{\overset{1}{\cancel{3}} \cdot 2}{5 \cdot \underset{1}{\cancel{3}}}$ | There are no like factors, except 1.

$\dfrac{10}{21}$ is in simplest form.

$\dfrac{1 \cdot 2}{5 \cdot 1}$, or $\dfrac{2}{5}$

So, $\dfrac{6}{15} = \dfrac{2}{5}; \dfrac{10}{21}$ is in simplest form.

---

practice ▷ **Write in simplest form.**

1. $\dfrac{6}{10}$ $\dfrac{3}{5}$ **2.** $\dfrac{9}{14}$ $\dfrac{9}{14}$ **3.** $\dfrac{10}{15}$ $\dfrac{2}{3}$ **4.** $\dfrac{14}{25}$ $\dfrac{14}{25}$ **5.** $\dfrac{6}{9}$ $\dfrac{2}{3}$ **6.** $\dfrac{10}{25}$ $\dfrac{2}{5}$

Example 3

Simplify $\frac{20}{6}$.

$$\frac{20}{6}$$

$20 = 2 \cdot 10$ $6 = 3 \cdot 2$ $\dfrac{2 \cdot 2 \cdot 5}{3 \cdot 2}$

$2 \cdot 2 \cdot 5$

Simplify means write in simplest form.

Divide out like factors. $\dfrac{\overset{1}{\cancel{2}} \cdot 2 \cdot 5}{3 \cdot \underset{1}{\cancel{2}}}$

$$\frac{1 \cdot 2 \cdot 5}{3 \cdot 1}$$

$\begin{array}{r} 3 \\ 3\overline{)10} \\ \underline{9} \\ 1 \end{array}$, or $3\frac{1}{3}$ $\dfrac{10}{3}$, or $3\frac{1}{3}$

So, $\dfrac{20}{6} = 3\frac{1}{3}$.

practice ▷ **Simplify.**

7. $\dfrac{12}{9}$ $1\frac{1}{3}$ **8.** $\dfrac{18}{10}$ $1\frac{4}{5}$ **9.** $\dfrac{15}{8}$ $1\frac{7}{8}$ **10.** $\dfrac{30}{20}$ $1\frac{1}{2}$ **11.** $\dfrac{21}{12}$ $1\frac{3}{4}$ **12.** $\dfrac{12}{10}$ $1\frac{1}{5}$

Example 4

Simplify $\frac{18}{30}$.

Two ways

$\dfrac{18}{30}$ $\dfrac{18}{30}$

$18 = 6 \cdot 3$ $30 = 15 \cdot 2$

$3 \cdot 2 \cdot 3$ $5 \cdot 3 \cdot 2$

$\dfrac{6 \cdot 3}{15 \cdot 2} = \dfrac{3 \cdot 2 \cdot 3}{5 \cdot 3 \cdot 2}$ $\dfrac{9 \cdot 2}{10 \cdot 3} = \dfrac{3 \cdot 3 \cdot 2}{5 \cdot 2 \cdot 3}$

Divide out like factors. $\dfrac{3 \cdot \overset{1}{\cancel{2}} \cdot \overset{1}{\cancel{3}}}{5 \cdot \underset{1}{\cancel{3}} \cdot \underset{1}{\cancel{2}}}$ $\dfrac{\overset{1}{\cancel{3}} \cdot 3 \cdot \overset{1}{\cancel{2}}}{5 \cdot \underset{1}{\cancel{2}} \cdot \underset{1}{\cancel{3}}}$

$\dfrac{3 \cdot 1 \cdot 1}{5 \cdot 1 \cdot 1}$ $\dfrac{1 \cdot 3 \cdot 1}{5 \cdot 1 \cdot 1}$

$\dfrac{3}{5}$ ←——— same ———→ $\dfrac{3}{5}$

So, $\dfrac{18}{30} = \dfrac{3}{5}$.

practice ▷ **Simplify.**

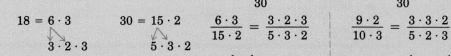

13. $\dfrac{12}{20}$ $\frac{3}{5}$ **14.** $\dfrac{12}{18}$ $\frac{2}{3}$ **15.** $\dfrac{15}{20}$ $\frac{3}{4}$ **16.** $\dfrac{8}{30}$ $\frac{4}{15}$ **17.** $\dfrac{16}{27}$ $\frac{16}{27}$

SIMPLEST FORM

Simplify.

1. $\dfrac{2 \cdot 3}{5 \cdot 3}$ $\dfrac{2}{5}$
2. $\dfrac{7 \cdot 5}{5 \cdot 11}$ $\dfrac{7}{11}$
3. $\dfrac{3 \cdot 5}{13 \cdot 3}$ $\dfrac{5}{13}$
4. $\dfrac{5 \cdot 11}{11 \cdot 7}$ $\dfrac{5}{7}$
5. $\dfrac{9 \cdot 2}{7 \cdot 9}$ $\dfrac{2}{7}$

6. $\dfrac{7 \cdot 2}{2 \cdot 11}$ $\dfrac{7}{11}$
7. $\dfrac{3 \cdot 5}{5 \cdot 5}$ $\dfrac{3}{5}$
8. $\dfrac{3 \cdot 11}{13 \cdot 3}$ $\dfrac{11}{13}$
9. $\dfrac{2 \cdot 2}{5 \cdot 2}$ $\dfrac{2}{5}$
10. $\dfrac{3 \cdot 7}{7 \cdot 17}$ $\dfrac{3}{17}$

11. $\dfrac{7 \cdot 2 \cdot 5}{2 \cdot 5 \cdot 11}$ $\dfrac{7}{11}$
12. $\dfrac{3 \cdot 2 \cdot 5}{2 \cdot 5 \cdot 7}$ $\dfrac{3}{7}$
13. $\dfrac{3 \cdot 7 \cdot 11}{11 \cdot 3 \cdot 13}$ $\dfrac{7}{13}$
14. $\dfrac{5 \cdot 2 \cdot 2}{2 \cdot 3 \cdot 5}$ $\dfrac{2}{3}$
15. $\dfrac{4 \cdot 5 \cdot 3}{5 \cdot 7 \cdot 4}$ $\dfrac{3}{7}$

Simplify.

1. $\dfrac{4}{10}$ $\dfrac{2}{5}$
2. $\dfrac{2}{6}$ $\dfrac{1}{3}$
3. $\dfrac{4}{15}$ $\dfrac{4}{15}$
4. $\dfrac{6}{9}$ $\dfrac{2}{3}$
5. $\dfrac{5}{10}$ $\dfrac{1}{2}$
6. $\dfrac{6}{14}$ $\dfrac{3}{7}$

7. $\dfrac{14}{4}$ $3\dfrac{1}{2}$
8. $\dfrac{8}{6}$ $1\dfrac{1}{3}$
9. $\dfrac{12}{10}$ $1\dfrac{1}{5}$
10. $\dfrac{18}{10}$ $1\dfrac{4}{5}$
11. $\dfrac{20}{15}$ $1\dfrac{1}{3}$
12. $\dfrac{27}{6}$ $4\dfrac{1}{2}$

13. $\dfrac{18}{20}$ $\dfrac{9}{10}$
14. $\dfrac{8}{30}$ $\dfrac{4}{15}$
15. $\dfrac{8}{20}$ $\dfrac{2}{5}$
16. $\dfrac{18}{27}$ $\dfrac{2}{3}$
17. $\dfrac{12}{25}$ $\dfrac{12}{25}$
18. $\dfrac{12}{20}$ $\dfrac{3}{5}$

19. $\dfrac{10}{12}$ $\dfrac{5}{6}$
20. $\dfrac{14}{6}$ $2\dfrac{1}{3}$
21. $\dfrac{8}{15}$ $\dfrac{8}{15}$
22. $\dfrac{10}{4}$ $2\dfrac{1}{2}$
23. $\dfrac{14}{18}$ $\dfrac{7}{9}$
24. $\dfrac{20}{8}$ $2\dfrac{1}{2}$

25. $\dfrac{15}{25}$ $\dfrac{3}{5}$
26. $\dfrac{12}{28}$ $\dfrac{3}{7}$
27. $\dfrac{28}{6}$ $4\dfrac{2}{3}$
★ 28. $\dfrac{16}{50}$ $\dfrac{8}{25}$
★ 29. $\dfrac{36}{24}$ $1\dfrac{1}{2}$
★ 30. $\dfrac{32}{48}$ $\dfrac{2}{3}$

Solve. Write answers in simplest form.

31. A basketball team won 14 out of 20 games. What fractional part of the games did they win? $\dfrac{7}{10}$

32. A pizza was cut into 8 slices. The Yings ate 6 slices. What fractional part of the pizza did they eat? $\dfrac{3}{4}$

Bill has 9 quarters. They all appear to be genuine, but one is counterfeit. It weighs less than the others. Use a balance scale to find the counterfeit quarter in exactly two weighings. See TE Answer Section.

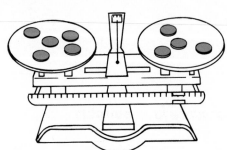

Simplifying Products

To simplify products

2 cases of 6 bottles each

$2 \cdot 6 = 12$ ←*of* means *times.*

Example 1

Simplify. $\dfrac{3}{5} \cdot \dfrac{10}{9}$

$$\dfrac{3}{5} \cdot \dfrac{10}{9} = \dfrac{3 \cdot 10}{5 \cdot 9}$$

Factor 10: $10 = 5 \cdot 2$
Factor 9: $9 = 3 \cdot 3$

$$\dfrac{3 \cdot 5 \cdot 2}{5 \cdot 3 \cdot 3}$$

Divide out like factors.

$$\dfrac{\overset{1}{\cancel{3}} \cdot \overset{1}{\cancel{5}} \cdot 2}{\underset{1}{\cancel{5}} \cdot \underset{1}{\cancel{3}} \cdot 3}$$

$$\dfrac{2}{3}$$

So, $\dfrac{3}{5} \cdot \dfrac{10}{9} = \dfrac{2}{3}$.

Example 2

Simplify. $2\dfrac{2}{5} \cdot \dfrac{1}{18}$

$2\dfrac{2}{5} = \dfrac{(5 \cdot 2) + 2}{5} = \dfrac{10 + 2}{5} = \dfrac{12}{5}$ $2\dfrac{2}{5} \cdot \dfrac{1}{18} = \dfrac{12}{5} \cdot \dfrac{1}{18} = \dfrac{12 \cdot 1}{5 \cdot 18}$

$12 = 2 \cdot 6$ $18 = 9 \cdot 2$ $\dfrac{2 \cdot 2 \cdot 3 \cdot 1}{5 \cdot 3 \cdot 3 \cdot 2}$
$\quad 2 \cdot 2 \cdot 3$ $\quad 3 \cdot 3 \cdot 2$

$$\dfrac{\overset{1}{\cancel{2}} \cdot 2 \cdot \overset{1}{\cancel{3}} \cdot 1}{5 \cdot \underset{1}{\cancel{3}} \cdot 3 \cdot \underset{1}{\cancel{2}}}$$

$1 \cdot 2 \cdot 1 \cdot 1 = 2$ $\dfrac{2}{15}$
$5 \cdot 1 \cdot 3 \cdot 1 = 15$

So, $2\dfrac{2}{5} \cdot \dfrac{1}{18} = \dfrac{2}{15}$.

---

practice ▷   **Simplify.**

1. $\dfrac{3}{5} \cdot \dfrac{5}{9}$ $\quad \dfrac{1}{3}$ **2.** $\dfrac{1}{3} \cdot \dfrac{6}{5}$ $\quad \dfrac{2}{5}$ **3.** $1\dfrac{1}{3} \cdot \dfrac{8}{12}$ $\quad \dfrac{8}{9}$ **4.** $3\dfrac{1}{2} \cdot \dfrac{1}{14}$ $\quad \dfrac{1}{4}$

Example 3

Simplify. $2\frac{1}{7} \cdot 1\frac{1}{6}$

$2\frac{1}{7} = \frac{(7 \cdot 2) + 1}{7} = \frac{14 + 1}{7} = \frac{15}{7}$

$1\frac{1}{6} = \frac{(6 \cdot 1) + 1}{6} = \frac{6 + 1}{6} = \frac{7}{6}$

$2\frac{1}{7} \cdot 1\frac{1}{6} = \frac{15}{7} \cdot \frac{7}{6} = \frac{15 \cdot 7}{7 \cdot 6}$

$\frac{5 \cdot 3 \cdot 7}{7 \cdot 3 \cdot 2}$

$15 = 5 \cdot 3$ Divide out
$6 = 3 \cdot 2$ like factors.

$\dfrac{5 \cdot \overset{1}{\cancel{3}} \cdot \overset{1}{\cancel{7}}}{\underset{1}{\cancel{7}} \cdot \underset{1}{\cancel{3}} \cdot 2}$

$\frac{5}{2}$, or $2\frac{1}{2}$

So, $2\frac{1}{7} \cdot 1\frac{1}{6} = 2\frac{1}{2}$.

practice ▷ **Simplify.**

5. $2\frac{1}{4} \cdot 1\frac{1}{3}$ 3   **6.** $1\frac{3}{5} \cdot 1\frac{1}{4}$ 2   **7.** $3\frac{1}{3} \cdot 2\frac{1}{4}$ $7\frac{1}{2}$ **8.** $4\frac{1}{2} \cdot \frac{2}{3}$ 3

Example 4

Simplify. $\frac{3}{4}$ of 18

of means *times.* $\frac{3}{4}$ of 18 means $\frac{3}{4} \cdot 18$.

$18 = \frac{18}{1}$ $\frac{3}{4} \cdot \frac{18}{1} = \frac{3 \cdot 18}{4 \cdot 1}$

$18 = 3 \cdot 3 \cdot 2$ $\frac{3 \cdot 3 \cdot 3 \cdot 2}{2 \cdot 2 \cdot 1}$
$4 = 2 \cdot 2$

Divide out like factors. $\dfrac{3 \cdot 3 \cdot 3 \cdot \overset{1}{\cancel{2}}}{2 \cdot \underset{1}{\cancel{2}} \cdot 1}$

$\frac{27}{2}$, or $13\frac{1}{2}$

So, $\frac{3}{4}$ of 18 $= 13\frac{1}{2}$.

practice ▷ **Simplify.**

9. $\frac{3}{4}$ of 14 $10\frac{1}{2}$ **10.** $\frac{5}{6}$ of 8 $6\frac{2}{3}$ **11.** $\frac{3}{5}$ of 20   12   **12.** $\frac{2}{3}$ of 20 $13\frac{1}{3}$

Simplify.

1. $\frac{3}{2}$ $1\frac{1}{2}$ **2.** $\frac{7}{3}$ $2\frac{1}{3}$ **3.** $\frac{5}{2}$ $2\frac{1}{2}$ **4.** $\frac{7}{2}$ $3\frac{1}{2}$ **5.** $\frac{5}{3}$ $1\frac{2}{3}$ **6.** $\frac{7}{5}$ $1\frac{2}{5}$ **7.** $\frac{11}{3}$ $3\frac{2}{3}$ **8.** $\frac{8}{5}$ $1\frac{3}{5}$

9. $\frac{5}{4}$ $1\frac{1}{4}$ **10.** $\frac{9}{5}$ $1\frac{4}{5}$ **11.** $\frac{9}{2}$ $4\frac{1}{2}$ **12.** $\frac{8}{3}$ $2\frac{2}{3}$ **13.** $\frac{9}{7}$ $1\frac{2}{7}$ **14.** $\frac{13}{5}$ $2\frac{3}{5}$ **15.** $\frac{7}{4}$ $1\frac{3}{4}$ **16.** $\frac{8}{7}$ $1\frac{1}{7}$

◇ *EXERCISES* ◇

Simplify.

1. $\frac{3}{5} \cdot \frac{10}{9}$ $\frac{2}{3}$ **2.** $\frac{2}{3} \cdot \frac{6}{5}$ $\frac{4}{5}$ **3.** $\frac{5}{6} \cdot \frac{3}{10}$ $\frac{1}{4}$ **4.** $\frac{3}{7} \cdot \frac{4}{15}$ $\frac{4}{35}$ **5.** $\frac{2}{9} \cdot \frac{3}{4}$ $\frac{1}{6}$

6. $\frac{2}{7} \cdot \frac{14}{6}$ $\frac{2}{3}$ **7.** $\frac{4}{21} \cdot \frac{7}{6}$ $\frac{2}{9}$ **8.** $\frac{3}{14} \cdot \frac{10}{15}$ $\frac{1}{7}$ **9.** $\frac{5}{8} \cdot \frac{6}{10}$ $\frac{3}{8}$ **10.** $\frac{12}{15} \cdot \frac{3}{4}$ $\frac{3}{5}$

11. $1\frac{1}{5} \cdot \frac{5}{4}$ $1\frac{1}{2}$ **12.** $3\frac{1}{3} \cdot \frac{1}{2}$ $1\frac{2}{3}$ **13.** $\frac{7}{10} \cdot 2\frac{1}{2}$ $1\frac{3}{4}$ **14.** $1\frac{3}{4} \cdot \frac{1}{14}$ $\frac{1}{8}$ **15.** $3\frac{1}{2} \cdot 2$ 7

16. $1\frac{1}{3} \cdot \frac{9}{12}$ 1 **17.** $\frac{8}{15} \cdot 2\frac{1}{2}$ $1\frac{1}{3}$ **18.** $1\frac{1}{6} \cdot \frac{12}{14}$ 1 **19.** $\frac{2}{5} \cdot 3\frac{1}{3}$ $1\frac{1}{3}$ **20.** $2\frac{6}{7} \cdot \frac{14}{18}$ $2\frac{2}{9}$

21. $2\frac{1}{2} \cdot 1\frac{2}{5}$ $3\frac{1}{2}$ **22.** $2\frac{1}{3} \cdot 1\frac{4}{7}$ $3\frac{2}{3}$ **23.** $1\frac{1}{3} \cdot 1\frac{1}{5}$ $1\frac{3}{5}$ **24.** $2\frac{1}{4} \cdot 1\frac{2}{3}$ $3\frac{3}{4}$ **25.** $2\frac{2}{3} \cdot 7\frac{1}{2}$ 20

★ **26.** $2\frac{1}{2} \cdot \frac{7}{15} \cdot \frac{3}{28}$ $\frac{1}{8}$ ★ **27.** $1\frac{3}{5} \cdot 3\frac{1}{3} \cdot \frac{5}{12}$ $2\frac{2}{9}$ ★ **28.** $6\frac{2}{5} \cdot \frac{15}{24} \cdot 1\frac{1}{3}$ $5\frac{1}{3}$

29. $\frac{5}{6}$ of 4 $3\frac{1}{3}$ **30.** $\frac{3}{8}$ of 14 $5\frac{1}{4}$ **31.** $\frac{4}{5}$ of 20 16 **32.** $\frac{2}{3}$ of 4 $2\frac{2}{3}$ **33.** $\frac{3}{7}$ of 14 6

Solve these problems.

34. Mary bought $12\frac{1}{4}$ pounds of meat. Her family ate $\frac{1}{3}$ of it. How much meat did they eat? $4\frac{1}{12}$ lb

35. Bill rode his bike for $1\frac{1}{4}$ hours. He averaged $12\frac{4}{5}$ miles per hour. How far did he ride?   16 mi

Calculator

Most calculators cannot display fractions. We can find products like $9 \cdot \frac{26}{6}$ on a calculator as shown below.

$$9 \cdot 26 \div 6 \qquad \text{Press } 9 \otimes 26 \ominus 6$$
$$234 \div 6 \qquad \text{Display } 39$$

Find the product.

1. $9 \cdot \frac{40}{6}$ 60   **2.** $12 \cdot \frac{28}{8}$ 42   **3.** $\frac{15}{6} \cdot 14$ 35

Problem Solving – Applications
Jobs for Teenagers

Use the chart to find the charge in Exercises 1–4.

HAPPY TIME CAMPGROUND	
Item	Rates per Hour
Bike riding	$ 1.50
Canoeing	$ 3.00
Archery	$ 2.00
Horseback riding	$15.00

1. $1\frac{1}{2}$ hours of bike riding $2.25

2. $1\frac{1}{4}$ hours of archery $2.50

3. $2\frac{1}{2}$ hours of canoeing $7.50

4. $3\frac{1}{2}$ hours of horseback riding $52.50

5. Bill rents umbrellas at the beach at $1.80 an hour. Find the charge for $\frac{3}{4}$ hour. Find the change from a $5 bill.
$1.35; $3.65

6. Carmine operates a horseback-riding concession. The rate is $16.00 an hour. Find the charge for $\frac{1}{2}$ hour. Find the change from a $10 bill. $8.00; $2.00

7. José runs a canoe-renting concession. The rate is $2.80 an hour. Find the charge for $\frac{1}{2}$ hour. Find the change from a $20 bill. $1.40; $18.60

8. Jim works in the office at a campground. The rate for a tent site is $4.50 per day. Find the charge for 3 day. Find the change from $20. $13.50; $6.50

9. Gray Cloud works at a parking lot at an entertainment center. The rate is $1.40 an hour. Find the charge for $4\frac{1}{2}$ hours. Find the change from a $10 bill.
$6.30; $3.70

Dividing Fractions

──◇ **OBJECTIVE** ◇── ──────◇ **RECALL** ◇──────

To divide fractions

$$\frac{4 \cdot 5}{5 \cdot 4} = \frac{\overset{1}{\cancel{4}} \cdot \overset{1}{\cancel{5}}}{\cancel{5} \cdot \cancel{4}} = 1 \qquad 8 \div 2 \text{ means } \frac{8}{2}.$$

Example 1

Show that $\frac{3}{5} \div \frac{2}{7}$ means $\frac{3}{5} \cdot \frac{7}{2}$.

$$\frac{3}{5} \div \frac{2}{7} \text{ means } \frac{\dfrac{3}{5}}{\dfrac{2}{7}}.$$

We can multiply the numerator and the denominator by the same number, $\frac{7}{2}$.

$$\frac{2}{7} \cdot \frac{7}{2} = \frac{\overset{1}{\cancel{2}} \cdot \overset{1}{\cancel{7}}}{\cancel{7} \cdot \cancel{2}} = 1$$

$$\frac{\dfrac{3}{5} \cdot \dfrac{7}{2}}{\dfrac{2}{7} \cdot \dfrac{7}{2}}$$

$$\frac{\dfrac{3}{5} \cdot \dfrac{7}{2}}{1} = \frac{3}{5} \cdot \frac{7}{2}$$

So, $\frac{3}{5} \div \frac{2}{7} = \frac{3}{5} \cdot \frac{7}{2}$.

In Example 1, we found that $\frac{3}{5} \div \frac{2}{7}$ is the same as $\frac{3}{5} \cdot \frac{7}{2}$. $\frac{7}{2}$ is the reciprocal of $\frac{2}{7}$.

Example 2

Divide. $\frac{1}{12} \div \frac{7}{18}$

$$\frac{1}{12} \div \frac{7}{18}$$

To divide by $\frac{7}{18}$, multiply by its reciprocal, $\frac{18}{7}$.

$$\frac{1}{12} \cdot \frac{18}{7}$$

$$\frac{1 \cdot 18}{12 \cdot 7}$$

$18 = 9 \cdot 2 = 3 \cdot 3 \cdot 2$
$12 = 4 \cdot 3 = 2 \cdot 2 \cdot 3$
Divide out like factors.

$$\frac{1 \cdot 3 \cdot 3 \cdot 2}{2 \cdot 2 \cdot 3 \cdot 7} = \frac{1 \cdot \cancel{3} \cdot 3 \cdot \overset{1}{\cancel{2}}}{\underset{1}{\cancel{2}} \cdot 2 \cdot \underset{1}{\cancel{3}} \cdot 7} = \frac{3}{14}$$

So, $\frac{1}{12} \div \frac{7}{18} = \frac{3}{14}$.

practice ▷ **Divide.**

1. $\frac{6}{7} \div \frac{10}{7}$ $\frac{3}{5}$ **2.** $\frac{3}{4} \div \frac{3}{2}$ $\frac{1}{2}$ **3.** $\frac{7}{15} \div \frac{14}{10}$ $\frac{1}{3}$ **4.** $\frac{3}{5} \div \frac{6}{5}$ $\frac{1}{2}$

Example 3   Divide.

$$4\frac{1}{2} \div 3 \qquad\qquad\qquad 5 \div 1\frac{1}{2}$$

$$3 = \frac{3}{1} \qquad 4\frac{1}{2} \div \frac{3}{1} \qquad\qquad \frac{5}{1} \div 1\frac{1}{2}$$

$$4\frac{1}{2} = \frac{(2 \cdot 4) + 1}{2} = \frac{8+1}{2} = \frac{9}{2} \qquad \frac{9}{2} \div \frac{3}{1} \qquad\qquad \frac{5}{1} \div \frac{3}{2}$$

$$\text{To divide by } \frac{3}{1}, \text{ multiply by } \frac{1}{3}. \qquad \frac{9}{2} \cdot \frac{1}{3} \qquad\qquad \frac{5}{1} \cdot \frac{2}{3}$$

$$\frac{9 \cdot 1}{2 \cdot 3} \qquad\qquad \frac{5 \cdot 2}{1 \cdot 3}$$

$$\frac{\overset{1}{\cancel{3}} \cdot 3 \cdot 1}{2 \cdot \underset{1}{\cancel{3}}} = \frac{3}{2}, \text{ or } 1\frac{1}{2} \qquad \frac{10}{3}, \text{ or } 3\frac{1}{3}$$

$$\text{So, } 4\frac{1}{2} \div 3 = 1\frac{1}{2} \text{ and } 5 \div 1\frac{1}{2} = 3\frac{1}{3}.$$

---

practice ▷ **Divide.**

5. $2\frac{1}{4} \div 6$ $\frac{3}{8}$ **6.** $14 \div 3\frac{1}{2}$ 4   **7.** $7\frac{1}{2} \div 3$ $2\frac{1}{2}$ **8.** $5\frac{1}{2} \div 22$ $\frac{1}{4}$

◇ EXERCISES ◇

Divide.

1. $\frac{3}{4} \div \frac{5}{6}$ $\frac{9}{10}$ **2.** $\frac{2}{3} \div \frac{8}{9}$ $\frac{3}{4}$ **3.** $\frac{5}{14} \div \frac{10}{7}$ $\frac{1}{4}$ **4.** $\frac{1}{18} \div \frac{8}{12}$ $\frac{1}{12}$ **5.** $\frac{3}{20} \div \frac{15}{8}$ $\frac{2}{25}$

6. $2\frac{1}{2} \div 5$ $\frac{1}{2}$ **7.** $1\frac{1}{4} \div 5$ $\frac{1}{4}$ **8.** $15 \div 3\frac{1}{3}$ $4\frac{1}{2}$ **9.** $18 \div 4\frac{1}{2}$ 4   **10.** $15 \div 2\frac{1}{2}$ 6

11. $\frac{8}{5} \div \frac{4}{15}$ 6   **12.** $\frac{3}{20} \div \frac{6}{5}$ $\frac{1}{8}$ **13.** $\frac{3}{5} \div \frac{9}{10}$ $\frac{2}{3}$ **14.** $\frac{5}{7} \div \frac{15}{14}$ $\frac{2}{3}$ **15.** $\frac{10}{8} \div \frac{15}{4}$ $\frac{1}{3}$

16. $\frac{4}{6} \div \frac{6}{21}$ $2\frac{1}{3}$ **17.** $1\frac{1}{3} \div \frac{8}{9}$ $1\frac{1}{2}$ **18.** $2\frac{6}{15} \div 1\frac{1}{3}$ $1\frac{4}{5}$ **19.** $7\frac{7}{12} \div 3\frac{1}{4}$ $2\frac{1}{3}$ **20.** $11\frac{1}{5} \div 1\frac{2}{5}$ 8

★ **21.** $\left(\frac{2}{3} \cdot \frac{9}{8}\right) \div 4\frac{1}{2}$ $\frac{1}{6}$ ★ **22.** $3\frac{1}{3} \div \left(2\frac{1}{2} \cdot \frac{4}{5}\right)$ $1\frac{2}{3}$ ★ **23.** $\left(3\frac{1}{2} \cdot 1\frac{1}{3}\right) \div \left(\frac{8}{9} \cdot \frac{3}{4}\right)$ 7

24. Donna mowed Mr. Weed's lawn for $7. It took $3\frac{1}{2}$ hours. How much did she make an hour? $2/h

25. José can run $3\frac{1}{2}$ laps in 7 minutes. How many laps can he run in 1 minute? $\frac{1}{2}$ lap

Evaluating Expressions

─── ◇ **OBJECTIVE** ◇ ───   ─────── ◇ **RECALL** ◇ ───────

To evaluate expressions like

$\dfrac{4x-2}{12}$ if $x = 3$

Evaluate $2x + 7$ if $x = 4$.

$$2 \cdot 4 + 7 \qquad \text{Substitute}$$
$$8 + 7 \qquad 4 \text{ for } x.$$
$$15$$

Example 1   Evaluate $\dfrac{4x-2}{12}$ if $x = 3$.

Substitute 3 for x. ──────→ $\dfrac{4x-2}{12} = \dfrac{4\cdot 3 - 2}{12}$

$\left.\begin{array}{c} \dfrac{4\cdot 3 - 2}{12 - 2 \text{ or } 10} \end{array}\right\}$ ──────→ $= \dfrac{10}{12}$

Factor 10 into primes. $10 = 5 \cdot 2$ ──→
Factor 12 into primes. $12 = 2 \cdot 2 \cdot 3$ ──→ $= \dfrac{5 \cdot 2}{2 \cdot 2 \cdot 3}$

Divide out like factors. ──────→ $= \dfrac{5 \cdot \overset{1}{\cancel{2}}}{2 \cdot \underset{1}{\cancel{2}} \cdot 3}$

$= \dfrac{5 \cdot 1}{2 \cdot 1 \cdot 3}$ or $\dfrac{5}{6}$

So, $\dfrac{4x-2}{12} = \dfrac{5}{6}$ if $x = 3$.

◇ **EXERCISES** ◇

Evaluate.

1. $\dfrac{3x-1}{10}$ if $x = 2$ $\quad \frac{1}{2}$

2. $\dfrac{2x+3}{15}$ if $x = 3$ $\quad \frac{3}{5}$

3. $\dfrac{5x+4}{21}$ if $x = 2$ $\quad \frac{2}{3}$

4. $\dfrac{6+2x}{20}$ if $x = 2$ $\quad \frac{1}{2}$

5. $\dfrac{5x-10}{35}$ if $x = 3$ $\quad \frac{1}{7}$

6. $\dfrac{8+2x}{18}$ if $x = 4$ $\quad \frac{8}{9}$

7. $\dfrac{3}{2x+5}$ if $x = 2$ $\quad \frac{1}{3}$

8. $\dfrac{2}{2+4x}$ if $x = 2$ $\quad \frac{1}{5}$

9. $\dfrac{10}{6x+3}$ if $x = 2$ $\quad \frac{2}{3}$

★ 10. $\dfrac{6x-6}{7y-2}$ if $x = 5, y = 6$ $\quad \frac{3}{5}$

★ 11. $\dfrac{3a+4}{2ab+a}$ if $a = 2, b = 3$ $\quad \frac{5}{7}$

Equations

───◇ **OBJECTIVE** ◇───   ───◇ **RECALL** ◇───

To solve equations like

$\frac{3}{4}x = 10$ and $\frac{3}{5}x = \frac{2}{3}$

$$\frac{2}{3} \cdot \frac{3}{2} = 1$$
$$\downarrow \quad \downarrow$$
reciprocals

$\frac{2}{3}$ times a number is 18.

Let x be the number. $\frac{2}{3} \quad \cdot \quad x \quad = 18$, or $\frac{2}{3}x = 18$

Example 1   Find the value of x that makes $\frac{2}{3}x = 18$ true.

$\frac{2}{3}x$ means $\frac{2}{3} \cdot x$. $\frac{2}{3}x = 18$

Multiply each side of the equation by $\frac{3}{2}$, $\frac{3}{2} \cdot \frac{2}{3} \cdot x = \frac{3}{2} \cdot 18$

the reciprocal of $\frac{2}{3}$. $1 \cdot x = \frac{3}{2} \cdot \frac{18}{1}$

$$x = \frac{3 \cdot 3 \cdot 3 \cdot \cancel{2}^{1}}{\cancel{2} \cdot 1}$$

$$x = 27$$

Check. $\frac{2}{3}x$ | 18

Substitute 27 for x. $\frac{2}{3} \cdot 27$ | 18

$\frac{2}{3} \cdot \frac{27}{1}$

$\frac{2 \cdot \cancel{3}^{1} \cdot 3 \cdot 3}{\cancel{3}_{1} \cdot 1}$

18

So, 27 is the value of x that makes $\frac{2}{3}x = 18$ true.

---

practice ▷   **Solve.**

1. $\frac{3}{4}x = 15$ 20   **2.** $\frac{3}{5}x = 9$ 15   **3.** $\frac{4}{7}x = 8$ 14   **4.** $\frac{5}{9}x = 20$ 36

Example 2

$\frac{1}{5}x$ means $\frac{1}{5} \cdot x$.

Multiply each side by the reciprocal of $\frac{1}{5}$.

Solve. $\frac{1}{5}x = \frac{8}{15}$

$$\frac{1}{5}x = \frac{8}{15}$$

$$\frac{5}{1} \cdot \frac{1}{5} \cdot x = \frac{5}{1} \cdot \frac{8}{15}$$

$$1 \cdot x = \frac{5 \cdot 8}{15}$$

$$x = \frac{\overset{1}{\cancel{5}} \cdot 2 \cdot 2 \cdot 2}{\underset{1}{\cancel{5}} \cdot 3} = \frac{8}{3}$$

$$x = \frac{8}{3}, \text{ or } 2\frac{2}{3}.$$

So, $2\frac{2}{3}$ is the solution of $\frac{1}{5}x = \frac{8}{15}$.

practice ▷ Solve.

5. $\frac{1}{7}x = \frac{10}{21}$ $3\frac{1}{3}$ **6.** $\frac{1}{10}x = \frac{5}{6}$ $8\frac{1}{3}$ **7.** $\frac{1}{8}x = \frac{3}{14}$ $1\frac{5}{7}$ **8.** $\frac{3}{7}x = \frac{9}{14}$ $1\frac{1}{2}$

◆ ORAL EXERCISES ◆

By what number do you multiply each side of the equation to solve?

1. $\frac{3}{7}x = 15$ $\frac{7}{3}$ **2.** $\frac{14}{3}x = 7$ $\frac{3}{14}$ **3.** $\frac{5}{6}x = \frac{11}{15}$ $\frac{6}{5}$ **4.** $\frac{2}{3}x = 8$ $\frac{3}{2}$ **5.** $\frac{4}{5}x = \frac{6}{25}$ $\frac{5}{4}$

6. $\frac{3}{5}x = \frac{9}{10}$ $\frac{5}{3}$ **7.** $\frac{4}{5}x = \frac{2}{3}$ $\frac{5}{4}$ **8.** $\frac{3}{7}x = \frac{9}{14}$ $\frac{7}{3}$ **9.** $\frac{10}{3}x = \frac{5}{7}$ $\frac{3}{10}$ **10.** $\frac{4}{5}x = 12$ $\frac{5}{4}$

◆ EXERCISES ◆

Solve.

1. $\frac{2}{3}x = 6$ 9 **2.** $\frac{4}{5}x = 12$ 15 **3.** $\frac{5}{7}x = 10$ 14 **4.** $\frac{3}{4}x = 15$ 20 **5.** $\frac{3}{5}x = 9$ 15

6. $\frac{1}{3}x = \frac{7}{15}$ $1\frac{2}{5}$ **7.** $\frac{1}{2}x = \frac{9}{4}$ $4\frac{1}{2}$ **8.** $\frac{3}{7}x = \frac{5}{14}$ $\frac{5}{6}$ **9.** $\frac{2}{5}x = \frac{7}{10}$ $1\frac{3}{4}$ **10.** $\frac{2}{3}x = \frac{5}{6}$ $1\frac{1}{4}$

11. $\frac{3}{7}x = 9$ 21 **12.** $\frac{2}{3}x = 18$ 27 **13.** $\frac{3}{4}x = 8$ $10\frac{2}{3}$ **14.** $\frac{3}{2}x = 10$ $6\frac{2}{3}$ **15.** $\frac{1}{5}x = 3$ 15

16. $\frac{4}{5}x = 8$ 10 **17.** $\frac{2}{3}x = 24$ 36 **18.** $\frac{3}{5}x = 24$ 40 **19.** $\frac{6}{5}x = 10$ $8\frac{1}{3}$ **20.** $\frac{3}{4}x = 20$ $26\frac{2}{3}$

21. $\frac{6}{5}x = \frac{2}{3}$ $\frac{5}{9}$ **22.** $\frac{4}{5}x = \frac{8}{3}$ $3\frac{1}{3}$ **23.** $\frac{2}{3}x = \frac{15}{6}$ $3\frac{3}{4}$ **24.** $\frac{4}{9}x = \frac{2}{3}$ $1\frac{1}{2}$ **25.** $\frac{3}{4}x = \frac{15}{6}$ $3\frac{1}{3}$

★ 26. $\frac{7}{2}x = \frac{14}{3}$ $1\frac{1}{3}$ **★ 27.** $\frac{27}{2}x = \frac{41}{4}$ $\frac{41}{54}$ **★ 28.** $16 = \frac{2}{3}x$ 24 **★ 29.** $\frac{3}{8} = \frac{1}{2}x$ $\frac{3}{4}$ **★ 30.** $\frac{25}{3} = \frac{5}{6}x$ 10

EQUATIONS

Evaluating Formulas

To evaluate formulas involving fractions

6^2 means $6 \cdot 6$, or 36 $\bigm|$ $\frac{12}{3}$ means $12 \div 3$, or 4

See Geometry Supplement on page 420 for more information.

Example 1

Substitute 6 for b and $3\frac{1}{2}$ for h.

$6 = \frac{6}{1}; 3\frac{1}{2} = \frac{(2 \cdot 3) + 1}{2} = \frac{6 + 1}{2} = \frac{7}{2}$

A formula for the area of a triangle is $A = \frac{1}{2}bh$.
Find A if b is 6 and h is $3\frac{1}{2}$.

$$A = \frac{1}{2}bh$$
$$= \frac{1}{2} \cdot 6 \cdot 3\frac{1}{2}$$
$$= \frac{1}{2} \cdot \frac{6}{1} \cdot \frac{7}{2}$$
$$= \frac{1 \cdot 6 \cdot 7}{2 \cdot 1 \cdot 2} = \frac{1 \cdot 3 \cdot 2 \cdot 7}{2 \cdot 1 \cdot 2}$$
$$= \frac{1 \cdot 3 \cdot \overset{1}{\cancel{2}} \cdot 7}{\underset{1}{\cancel{2}} \cdot 1 \cdot 2} = \frac{21}{2}, \text{ or } 10\frac{1}{2}$$

So, A is $10\frac{1}{2}$ when b is 6 and h is $3\frac{1}{2}$.

Example 2

$$\frac{22}{7} \doteq \pi$$
$$\uparrow$$
approximately equal to

r^2 means $r \cdot r$.

Substitute $2\frac{1}{3}$ for r.

$2\frac{1}{3} = \frac{(3 \cdot 2) + 1}{3} = \frac{6 + 1}{3} = \frac{7}{3}$

A formula for the area of a circle is $A = \pi r^2$.
Use $\frac{22}{7}$ for π.
Find A if r is $2\frac{1}{3}$.

$$A = \pi r^2 \doteq \frac{22}{7} \cdot r \cdot r$$
$$\doteq \frac{22}{7} \cdot 2\frac{1}{3} \cdot 2\frac{1}{3}$$
$$\doteq \frac{22}{7} \cdot \frac{7}{3} \cdot \frac{7}{3} = \frac{22 \cdot \overset{1}{\cancel{7}} \cdot 7}{\underset{1}{\cancel{7}} \cdot 3 \cdot 3}$$
$$\doteq \frac{154}{9}, \text{ or } 17\frac{1}{9}$$

So, A is about $17\frac{1}{9}$ when r is $2\frac{1}{3}$.

practice ▷ **1.** $A = \frac{1}{2}bh$. Find A if b is 5 and   **2.** $A \doteq \frac{22}{7}r^2$. Find A if r is $3\frac{1}{2}$.

h is $2\frac{2}{3}$. $6\frac{2}{3}$ $38\frac{1}{2}$

See Geometry Supplement on page 420 for more information.

Example 3

A formula for the volume of a cone is $V = \frac{1}{3}\pi r^2 h$. Find V in terms of π if r is $2\frac{1}{2}$ and h is 4.

$$V = \frac{1}{3}\pi r^2 h, \text{ or } \frac{1}{3}\pi \cdot r \cdot r \cdot h$$

$$2\frac{1}{2} = \frac{(2\cdot 2)+1}{2} = \frac{4+1}{2} = \frac{5}{2}$$

$$\frac{1}{3}\pi \cdot 2\frac{1}{2} \cdot 2\frac{1}{2} \cdot 4 = \frac{1}{3} \cdot \frac{5}{2} \cdot \frac{5}{2} \cdot \frac{4}{1} \cdot \pi$$

$$\frac{1 \cdot 5 \cdot 5 \cdot 2 \cdot 2}{3 \cdot 2 \cdot 2 \cdot 1}\pi = \frac{1 \cdot 5 \cdot 5 \cdot \overset{1}{\cancel{2}} \cdot \overset{1}{\cancel{2}}}{3 \cdot \underset{1}{\cancel{2}} \cdot \underset{1}{\cancel{2}} \cdot 1}\pi, \text{ or } \frac{25}{3}\pi$$

So, V is $8\frac{1}{3}\pi$ when r is $2\frac{1}{2}$ and h is 4.

---

practice ▷ $V = \frac{1}{3}\pi r^2 h$. **Find V in terms of π.**

3. r is $1\frac{1}{3}$, h is 3 $1\frac{7}{9}\pi$ **4.** r is $2\frac{1}{4}$, h is 8 $13\frac{1}{2}\pi$

◇ **EXERCISES** ◇

Find the value. $12\frac{1}{2}$ 10

1. $A = \frac{1}{2}bh$ if b is 10, h is $\frac{5}{2}$ **2.** $A = \frac{1}{2}bh$ if b is 6, h is $\frac{10}{3}$ **3.** $M = \frac{2}{3}p^2$ if p is 6 24

4. $A \doteq \frac{22}{7}r^2$ if r is $1\frac{3}{4}$ $9\frac{5}{8}$ **5.** $A \doteq \frac{22}{7}r^2$ if r is $1\frac{1}{6}$ $4\frac{5}{18}$ **6.** $G = \frac{3}{4}a^2$ if a is $3\frac{1}{3}$ $8\frac{1}{3}$

7. $R = \frac{2}{3}kb$ if k is $\frac{3}{5}$, b is 7 $2\frac{4}{5}$ **8.** $M = \frac{3}{5}p^2$ if p is $2\frac{1}{2}$ $3\frac{3}{4}$ **9.** $C \doteq \frac{22}{7}d$ if d is $1\frac{3}{4}$ $5\frac{1}{2}$

Find the value. Leave answers in terms of π.

10. $V = \frac{1}{3}\pi r^2 h$ if r is $3\frac{1}{2}$, h is 6 $24\frac{1}{2}\pi$ **11.** $V = \frac{1}{3}\pi r^2 h$ if r is $1\frac{1}{2}$, h is 9 $6\frac{3}{4}\pi$

Find the value.

★ **12.** $V = \frac{4}{3}\pi r^3$ if $\pi \doteq \frac{22}{7}$, r is $1\frac{1}{2}$ $14\frac{1}{7}$ ★ **13.** $V = \frac{1}{3}\pi r^2 h$ if $\pi \doteq \frac{22}{7}$, r is $10\frac{1}{2}$, h is 6 693

EVALUATING FORMULAS

A fraction such as $\frac{28}{12}$ can be simplified by first finding the GCF of 12 and 28.

The computer program below does this.

Example   Type this program for simplifying any fraction of the form $\frac{A}{B}$, $B \neq 0$.

   RUN the program to simplify $\frac{28}{12}$.

Clear the screen.	`100  HOME`
	`110  INPUT "TYPE IN NUMERATOR ";A`
	`120  INPUT "TYPE IN DENOMINATOR ";B`
Computer decides which is smaller, A or B. Trial divisors are decreased by 1.	`130  IF A - B < 0 THEN  LET X = A: GOTO 150`
	`140  LET X = B`
	`150  FOR I = X TO 1 STEP  - 1`
	`160  LET N = A / I`
	`170  LET D = B / I`
Computer tests whether I is a common divisor of A and B. Prints two lines of space above answer.	`180  IF N =  INT (N) AND D =  INT (D) THEN 200`
	`190  NEXT I`
	`200  PRINT : PRINT`
	`210  PRINT A"/"B" = "N"/"D`
	`220  IF D = 1 THEN  PRINT "OR "N`
	`230  END`
	`]RUN`
You type 28.	`TYPE IN NUMERATOR 28`
You type 12.	`TYPE IN DENOMINATOR 12`
Computer displays the result.	`28/12 = 7/3`

See the Computer Section beginning on page 420 for more information.

Exercises

Simplify each fraction. Then **RUN** the program to check your answer.

1. $\frac{24}{72}$ $\frac{1}{3}$ 2. $\frac{18}{12}$ $\frac{3}{2}$ 3. $\frac{48}{24}$ 2   4. $\frac{1,440}{2,160}$ $\frac{2}{3}$

5. Why does the computer seem to take so long to simplify $\frac{1,440}{2,160}$?

   because of the number of calculations it must do to find the GCF of 1,440 and 2,160

6. Change the program to find the product of any two fractions such as X/Y and Z/W. (Hint: A = X * Z, B = Y * W.) **RUN** the program to simplify (2 / 5) * (4 / 8).   See TE Answer Section for program and results.

What number corresponds to point *A*? [79]

1. 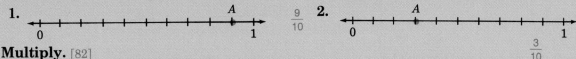 $\frac{9}{10}$ 2.   *A* $\frac{3}{10}$

Multiply. [82]

2. $4 \cdot \frac{1}{9}$ $\frac{4}{9}$

4. $\frac{1}{10} \cdot 7$ $\frac{7}{10}$

5. $6 \cdot \frac{2}{5}$ $2\frac{2}{5}$

6. $10 \cdot \frac{2}{3}$ $6\frac{2}{3}$

Tell whether each number is prime, composite, or neither. [88]

7. 18 composite

8. 24 composite

9. 1 neither

10. 19 prime

Factor into primes. Write using exponents. [88]

11. 12 $2^2 \cdot 3$

12. 28 $2^2 \cdot 7$

13. 45 $5 \cdot 3^2$

14. 42 $2 \cdot 3 \cdot 7$

Find the greatest common factor (GCF). [90]

15. 20; 24 4

16. 27; 18 9

17. 81; 108 27

18. 32; 98 2

Simplify. [92]

19. $\frac{6}{10}$ $\frac{3}{5}$

20. $\frac{20}{6}$ $3\frac{1}{3}$

21. $\frac{14}{18}$ $\frac{7}{9}$

22. $\frac{18}{12}$ $1\frac{1}{2}$

Simplify. [95]

23. $\frac{5}{14} \cdot \frac{7}{15}$ $\frac{1}{6}$

24. $1\frac{3}{8} \cdot \frac{1}{8}$ $\frac{11}{64}$

25. $7 \cdot \frac{3}{14}$ $1\frac{1}{2}$

26. $\frac{3}{4}$ of 14 $10\frac{1}{2}$

Divide. [99]

27. $\frac{2}{3} \div \frac{10}{9}$ $\frac{3}{5}$

28. $4\frac{1}{2} \div 6$ $\frac{3}{4}$

29. $15 \div 2\frac{1}{2}$ 6

30. $2\frac{1}{3} \div \frac{4}{9}$ $5\frac{1}{4}$

Evaluate. [101]

31. $\frac{4x - 2}{15}$ if $x = 2$ $\frac{2}{5}$

32. $\frac{4x + 3}{30}$ if $x = 3$ $\frac{1}{2}$

33. $\frac{2y + 5}{45}$ if $y = 5$ $\frac{1}{3}$

Solve. [102]

34. $\frac{3}{5}x = 21$ 35

35. $\frac{7}{10}x = \frac{4}{15}$ $\frac{8}{21}$

36. $\frac{2}{3}x = \frac{5}{6}$ $1\frac{1}{4}$

Find the value. [104]

37. $A = \frac{1}{2}bh$ if b is 8, h is $1\frac{1}{2}$ 6

38. $A = \frac{22}{7}r^2$ if r is $1\frac{2}{3}$ $8\frac{46}{63}$

39. $V = \frac{1}{3}\pi r^2 h$. Find V in terms of π if r is $2\frac{1}{3}$ and h is 9. $V = 16\frac{1}{3}\pi$

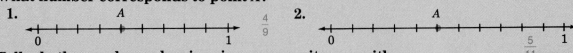

Chapter Test

What number corresponds to point *A*?

1.

$\frac{4}{9}$

2.

$\frac{5}{11}$

Tell whether each number is prime, composite, or neither.

3. 15 composite 4. 1 neither 5. 37 prime 6. 20 composite 7. 39 composite

Factor into primes. Write using exponents.

8. 24 $2^3 \cdot 3$ 9. 40 $2^3 \cdot 5$ 10. 18 $2 \cdot 3^2$ 11. 32 2^5 12. 48 $2^4 \cdot 3$

Find the greatest common factor (GCF).

13. 12; 20 4 14. 18; 36 18 15. 24; 36 12

Simplify.

16. $\frac{4}{10}$ $\frac{2}{5}$ 17. $\frac{14}{6}$ $2\frac{1}{3}$ 18. $\frac{15}{18}$ $\frac{5}{6}$

Simplify.

19. $\frac{4}{21} \cdot \frac{7}{6}$ $\frac{2}{9}$ 20. $1\frac{1}{2} \cdot \frac{1}{9}$ $\frac{1}{6}$ 21. $1\frac{1}{3} \cdot 4\frac{1}{2}$ 6

Simplify.

22. $\frac{4}{5}$ of 15 12 23. $\frac{3}{5}$ of 7 $4\frac{1}{5}$

Divide.

24. $\frac{3}{10} \div \frac{12}{5}$ $\frac{1}{8}$ 25. $3\frac{1}{2} \div 14$ $\frac{1}{4}$

Solve.

26. $\frac{2}{3}x = 8$ 12 27. $\frac{8}{5}x = \frac{4}{15}$ $\frac{1}{6}$

28. $A = \frac{1}{2}bh$. Find A if $b = 10$ and $h = 4\frac{1}{4}$. $21\frac{1}{4}$

29. $V = \frac{1}{3}\pi r^2 h$. Find V in terms of π if $r = 2\frac{3}{4}$ and $h = 8$. $V = 20\frac{1}{6}\pi$

Evaluate.

30. $\frac{7x - 1}{25}$ if $x = 3$ $\frac{4}{5}$ 31. $\frac{4x + 2}{24}$ if $x = 4$ $\frac{3}{4}$

Cumulative Review

Evaluate if $a = 5$, $x = 3$, and $t = 7$.

1. $a + 16$ 21
2. $t + 4a - 19$ 8
3. $3 + 2xa$ 33
4. $2tx + 6$ 48
5. x^4 81
6. $4a^3$ 500

Simplify.

7. $3y + 6 + 8y$ $11y + 6$
8. $3(10m - 4)$ $30m - 12$
9. $3a + 4 - 2a + 5$ $a + 9$
10. $\frac{6}{14}$ $\frac{3}{7}$
11. $\frac{6}{27} \cdot \frac{9}{21}$ $\frac{2}{21}$
12. $3\frac{1}{4} \cdot 2\frac{2}{3}$ $8\frac{2}{3}$

Solve. Check.

13. $x + 5 = 13$ 8
14. $y - 4 = 4$ 8
15. $12 = z - 7$ 19
16. $36 = 12m$ 3
17. $\frac{t}{3} = 7$ 21
18. $5 = \frac{n}{9}$ 45
19. $4t + 3 = 11$ 2
20. $5 = 3x - 13$ 6
21. $\frac{3}{4}m = \frac{3}{5}$ $\frac{4}{5}$

Write in mathematical form.

22. 17 times n $17n$
23. x divided by 4 $\frac{x}{4}$
24. 9 divided by r $\frac{9}{r}$
25. 7 times x, increased by 7 $7x + 7$
26. 3 less than 4 times a number $4n - 3$

Which property is illustrated?

27. $41 + 0 = 41$ Add. Ident.
28. $(3 \cdot 5) \cdot 12 = 3 \cdot (5 \cdot 12)$
29. $32 \cdot 1 = 32$ Mult. Ident.
30. $32 + 216 = 216 + 32$
31. $32 \cdot 9 = 9 \cdot 32$ Comm. Prop. Mult.
32. $32 \cdot 9 + 32 \cdot 13 = 32(9 + 13)$

28. Assoc. Prop. Mult.
30. Comm. Prop. Add.
32. Distributive

33. $P = 2l + 2w$. Find P if $l = 3\frac{1}{2}$ in. and $w = 1\frac{1}{4}$ in. $9\frac{1}{2}$ in.
34. $A = \frac{bh}{2}$. Find A if $b = 7$ in. and $h = 5$ in. $17\frac{1}{2}$ in.^{2}
35. $A = \pi r^2$. Find A in terms of π if $r = 9$. 81π
36. $V = \pi r^2 h$. Find V in terms of π if $r = 5$ and $h = 4$. 100π

Factor into primes.

37. 60 $2 \cdot 2 \cdot 3 \cdot 5$
38. 42 $2 \cdot 3 \cdot 7$
39. 120 $2 \cdot 2 \cdot 2 \cdot 3 \cdot 5$

Solve these problems.

40. \$35 less than Susan's salary is \$145. What is Susan's salary?   \$180
41. The sum of a number and 26 is 42. What is the number?   16
42. Four times Alfonso's age is 64. How old is Alfonso?   16
43. Olga's score of 9 is the same as Tanya's score divided by 3. What is Tanya's score?   27

Adding and Subtracting Fractions

5

The Gateway Arch in St. Louis, Missouri, has a span and height of 630 ft.

Fractions with the Same Denominator

$\diamond$ **OBJECTIVE** $\diamond$

To add and subtract fractions
with the same denominator

$\diamond$ **RECALL** $\diamond$

Show that $6 \cdot 5 + 3 \cdot 5 = (6 + 3)5$.

$6 \cdot 5 + 3 \cdot 5$	$(6 + 3)5$
$30 + 15$	$9 \cdot 5$
45	45

So, $6 \cdot 5 + 3 \cdot 5 = (6 + 3)5$.

Example 1

Show that $\frac{4}{7} + \frac{2}{7} = \frac{6}{7}$ and $\frac{5}{9} - \frac{4}{9} = \frac{1}{9}$.

$\frac{4}{7} + \frac{2}{7}$	$\frac{6}{7}$	$\frac{5}{9} - \frac{4}{9}$	$\frac{1}{9}$

$\frac{4}{7} = 4 \cdot \frac{1}{7}$ $\frac{2}{7} = 2 \cdot \frac{1}{7}$

$4 \cdot \frac{1}{7} + 2 \cdot \frac{1}{7}$	$\frac{6}{7}$	$5 \cdot \frac{1}{9} - 4 \cdot \frac{1}{9}$	$\frac{1}{9}$

Distributive property $(4 + 2)\frac{1}{7}$ $(5 - 4)\frac{1}{9}$

$6 \cdot \frac{1}{7}$ $1 \cdot \frac{1}{9}$

$\frac{6}{7}$ $\frac{1}{9}$

So, $\frac{4}{7} + \frac{2}{7} = \frac{6}{7}$ and $\frac{5}{9} - \frac{4}{9} = \frac{1}{9}$.

$\frac{4}{7} + \frac{2}{7} = \frac{4 + 2}{7} = \frac{6}{7}$ To add fractions with the same denominator: Add numerators. Keep the denominator.

$\frac{5}{9} - \frac{1}{9} = \frac{5 - 1}{9} = \frac{4}{9}$ To subtract fractions with the same denominator: Subtract numerators. Keep the denominator.

Example 2

Add. $\frac{2}{9} + \frac{5}{9}$ Subtract. $\frac{5}{7} - \frac{2}{7}$

$\frac{2}{9} + \frac{5}{9} = \frac{2 + 5}{9} = \frac{7}{9}$ $\frac{5}{7} - \frac{2}{7} = \frac{5 - 2}{7} = \frac{3}{7}$

So, $\frac{2}{9} + \frac{5}{9} = \frac{7}{9}$ and $\frac{5}{7} - \frac{2}{7} = \frac{3}{7}$.

Add or subtract.

1. $\dfrac{5}{7} + \dfrac{1}{7}$ $\dfrac{6}{7}$ **2.** $\dfrac{10}{11} - \dfrac{3}{11}$ $\dfrac{7}{11}$ **3.** $\dfrac{2}{5} + \dfrac{2}{5}$ $\dfrac{4}{5}$ **4.** $\dfrac{8}{9} - \dfrac{4}{9}$ $\dfrac{4}{9}$

Example 3   Add or subtract. Simplify if possible.

$$\dfrac{5}{8} + \dfrac{1}{8} \qquad\qquad \dfrac{9}{10} - \dfrac{3}{10}$$

$$\dfrac{5}{8} + \dfrac{1}{8} = \dfrac{5+1}{8} = \dfrac{6}{8} \qquad \dfrac{9}{10} - \dfrac{3}{10} = \dfrac{9-3}{10} = \dfrac{6}{10}$$

Factor the numerator. $\dfrac{6}{8} = \dfrac{2 \cdot 3}{2 \cdot 2 \cdot 2} \qquad\qquad \dfrac{6}{10} = \dfrac{2 \cdot 3}{2 \cdot 5}$
Factor the denominator.

Divide out like factors. $\dfrac{\overset{1}{\cancel{2}} \cdot 3}{\underset{1}{\cancel{2}} \cdot 2 \cdot 2}$, or $\dfrac{3}{4}$ $\dfrac{\overset{1}{\cancel{2}} \cdot 3}{\underset{1}{\cancel{2}} \cdot 5}$, or $\dfrac{3}{5}$

So, $\dfrac{5}{8} + \dfrac{1}{8} = \dfrac{3}{4}$ and $\dfrac{9}{10} - \dfrac{3}{10} = \dfrac{3}{5}$.

---

practice ▷   **Add or subtract. Simplify if possible.**

5. $\dfrac{5}{9} + \dfrac{1}{9}$ $\dfrac{2}{3}$ **6.** $\dfrac{5}{6} - \dfrac{1}{6}$ $\dfrac{2}{3}$ **7.** $\dfrac{1}{8} + \dfrac{3}{8}$ $\dfrac{1}{2}$ **8.** $\dfrac{5}{12} - \dfrac{1}{12}$ $\dfrac{1}{3}$

Example 4   Add or subtract. Simplify if possible.

$$\begin{array}{r} \dfrac{5}{6} \\ +\dfrac{9}{6} \\ \hline \end{array} \qquad\qquad \begin{array}{r} \dfrac{11}{8} \\ -\dfrac{1}{8} \\ \hline \end{array}$$

$\dfrac{5}{6} + \dfrac{9}{6} = \dfrac{5+9}{6} = \dfrac{14}{6}$ $\dfrac{14}{6}$ $\dfrac{10}{8}$

$$\text{Divide } 6\overline{)14} = 2\dfrac{2}{6} \qquad 2\dfrac{2}{6} \qquad\qquad 1\dfrac{2}{8}$$
$$\begin{array}{r} 12 \\ \hline 2 \end{array} \qquad\qquad 2\dfrac{1}{3} \qquad\qquad 1\dfrac{1}{4}$$

$\dfrac{2}{6} = \dfrac{\overset{1}{\cancel{2}}}{\underset{1}{\cancel{2}} \cdot 3} = \dfrac{1}{3}$ So, $\dfrac{5}{6} + \dfrac{9}{6} = 2\dfrac{1}{3}$ and $\dfrac{11}{8} - \dfrac{1}{8} = 1\dfrac{1}{4}$.

---

practice ▷   **Add or subtract. Simplify if possible.**

9. $\begin{array}{r} \dfrac{5}{4} \\ +\dfrac{1}{4} \\ \hline 1\dfrac{1}{2} \end{array}$ **10.** $\begin{array}{r} \dfrac{7}{9} \\ +\dfrac{5}{9} \\ \hline 1\dfrac{1}{3} \end{array}$ **11.** $\begin{array}{r} \dfrac{1}{3} \\ +\dfrac{7}{3} \\ \hline 2\dfrac{2}{3} \end{array}$ **12.** $\begin{array}{r} \dfrac{13}{10} \\ -\dfrac{1}{10} \\ \hline 1\dfrac{1}{5} \end{array}$

Add or subtract.

1. $\frac{2}{7} + \frac{1}{7}$ $\frac{3}{7}$ 2. $\frac{3}{10} + \frac{6}{10}$ $\frac{9}{10}$ 3. $\frac{1}{5} + \frac{3}{5}$ $\frac{4}{5}$ 4. $\frac{7}{9} + \frac{1}{9}$ $\frac{8}{9}$ 5. $\frac{3}{8} + \frac{4}{8}$ $\frac{7}{8}$ 6. $\frac{4}{7} + \frac{2}{7}$ $\frac{6}{7}$

7. $\frac{5}{9} - \frac{4}{9}$ $\frac{1}{9}$ 8. $\frac{7}{12} - \frac{6}{12}$ $\frac{1}{12}$ 9. $\frac{4}{10} - \frac{1}{10}$ $\frac{3}{10}$ 10. $\frac{5}{14} - \frac{2}{14}$ $\frac{3}{14}$ 11. $\frac{7}{8} - \frac{2}{8}$ $\frac{5}{8}$ 12. $\frac{5}{6} - \frac{4}{6}$ $\frac{1}{6}$

13. $\frac{5}{11}$ 14. $\frac{7}{11}$ 15. $\frac{2}{15}$ 16. $\frac{5}{6}$ 17. $\frac{5}{9}$ 18. $\frac{8}{13}$
$+\frac{3}{11}$ $-\frac{3}{11}$ $+\frac{5}{15}$ $-\frac{4}{6}$ $+\frac{2}{9}$ $-\frac{6}{13}$
$\frac{8}{11}$ $\frac{4}{11}$ $\frac{7}{15}$ $\frac{1}{6}$ $\frac{7}{9}$ $\frac{2}{13}$

Add or subtract. Simplify if possible.

1. $\frac{3}{9} + \frac{5}{9}$ $\frac{8}{9}$ 2. $\frac{1}{11} + \frac{5}{11}$ $\frac{6}{11}$ 3. $\frac{7}{10} + \frac{2}{10}$ $\frac{9}{10}$ 4. $\frac{3}{8} + \frac{1}{8}$ $\frac{1}{2}$ 5. $\frac{3}{6} + \frac{1}{6}$ $\frac{2}{3}$

6. $\frac{5}{9} - \frac{2}{9}$ $\frac{1}{3}$ 7. $\frac{5}{9} - \frac{1}{9}$ $\frac{4}{9}$ 8. $\frac{8}{12} - \frac{2}{12}$ $\frac{1}{2}$ 9. $\frac{7}{10} - \frac{2}{10}$ $\frac{1}{2}$ 10. $\frac{7}{8} - \frac{1}{8}$ $\frac{3}{4}$

11. $\frac{2}{9} + \frac{4}{9}$ $\frac{2}{3}$ 12. $\frac{7}{12} + \frac{2}{12}$ $\frac{3}{4}$ 13. $\frac{3}{8} + \frac{5}{8}$ 1 14. $\frac{6}{5} + \frac{2}{5}$ $1\frac{3}{5}$ 15. $\frac{9}{8} + \frac{3}{8}$ $1\frac{1}{2}$

16. $\frac{15}{8} - \frac{3}{8}$ $1\frac{1}{2}$ 17. $\frac{13}{12} - \frac{1}{12}$ 1 18. $\frac{17}{10} - \frac{3}{10}$ $1\frac{2}{5}$ 19. $\frac{11}{6} - \frac{1}{6}$ $1\frac{2}{3}$ 20. $\frac{9}{4} - \frac{3}{4}$ $1\frac{1}{2}$

21. $\frac{1}{9}$ $\frac{2}{3}$ 22. $\frac{5}{8}$ $1\frac{1}{2}$ 23. $\frac{5}{12}$ $1\frac{1}{3}$ 24. $\frac{4}{6}$ 1 25. $\frac{3}{4}$ $2\frac{1}{2}$ 26. $\frac{11}{4}$ 3
$+\frac{5}{9}$ $+\frac{7}{8}$ $+\frac{11}{12}$ $+\frac{2}{6}$ $+\frac{7}{4}$ $+\frac{1}{4}$

27. $\frac{11}{12}$ $\frac{2}{3}$ 28. $\frac{19}{6}$ $2\frac{1}{3}$ 29. $\frac{15}{4}$ $3\frac{1}{2}$ 30. $\frac{5}{18}$ $\frac{1}{6}$ 31. $\frac{7}{12}$ $\frac{1}{6}$ 32. $\frac{63}{15}$ $1\frac{1}{3}$
$-\frac{3}{12}$ $-\frac{5}{6}$ $-\frac{1}{4}$ $-\frac{2}{18}$ $-\frac{5}{12}$ $-\frac{43}{15}$

★ 33. $\frac{8}{12} + \frac{7}{12} - \frac{13}{12}$ $\frac{1}{6}$ ★ 34. $\frac{17}{15} - \frac{4}{15} + \frac{7}{15}$ $1\frac{1}{3}$ ★ 35. $\frac{20}{21} - \frac{9}{21} - \frac{4}{21}$ $\frac{1}{3}$

Solve these problems.

36. John spent $\frac{3}{4}$ hour before supper on homework and $\frac{1}{4}$ hour after supper on homework. How long did he work altogether? 1 h

37. Maria sold $\frac{3}{4}$ case of applesauce in the morning and $\frac{1}{4}$ case in the afternoon. How much more applesauce did she sell in the morning? $\frac{1}{2}$ case

FRACTIONS WITH THE SAME DENOMINATOR

The Least Common Multiple (LCM)

To find the least common
multiple (LCM) of two or
three numbers

Factor 12 into primes.

$$12 = 4 \cdot 3$$

$$= 2 \cdot 2 \cdot 3 \leftarrow \text{all primes}$$

A *multiple* of a number is the product of that
number and any whole number.

Note: 0 is a multiple of every whole
number. $n \cdot 0 = 0$ for any number n.

$2 \cdot 0 = 0$	0 is a multiple of 2.
$2 \cdot 1 = 2$	2 is a multiple of 2.
$2 \cdot 2 = 4$	4 is a multiple of 2.
$2 \cdot 3 = 6$	6 is a multiple of 2.

Definition of least common multiple
(LCM)

The *least common multiple (LCM)* of two or more
numbers is the least nonzero whole number that is a
multiple of the numbers.

Example 1

Find the least common multiple (LCM) of 4 and 5.

Find some multiples of 4.
Do not include 0.

Multiples of 4: 4, 8, 12, 16, 20, 24, 28, 32,
36, 40, 44, 48, 52, 56, 60, . . .

Find some multiples of 5.
Do not include 0.

Multiples of 5: 5, 10, 15, 20, 25, 30, 35, 40,
45, 50, 55, 60, 65, . . .

Find some multiples that are the
same (common) for both 4 and 5.

Common multiples: 20, 40, 60, . . .
↑
least common multiple

So, 20 is the least common multiple (LCM)
of 4 and 5.

Writing several common multiples can be time
consuming. A shorter way to find the LCM is shown
in the next example.

Example 2

Find the LCM of 6 and 20. Use prime factorization.

Factor 20 into primes. $20 = 10 \cdot 2 = 5 \cdot 2 \cdot 2$

Factor 6 into primes. $6 = \qquad\qquad 3 \cdot 2$

At *most* one 5, one 3, two 2's occur in either 6 or 20.   The LCM is $5 \cdot 3 \cdot 2 \cdot 2$.

$5 \cdot 3 \cdot 2 \cdot 2 = 60$ So, the LCM of 6 and 20 is 60.

Example 3

Find the LCM of 2, 4, and 6.
Use prime factorization.

Two is prime.	$2 = 2 \cdot 1 = 2$	one 2
Factor 4 into primes.	$4 = 2 \cdot 2$	two 2's
Factor 6 into primes.	$6 = 3 \cdot 2$	one 3, one 2

At *most* two 2's, one 3 occur in either 2, 4, or 6.   So, the LCM of 2, 4, and 6 is $2 \cdot 2 \cdot 3$, or 12.

---

practice ▷ **Find the LCM using prime factorization.**

1. 3 and 4   12   **2.** 5 and 6   30   **3.** 4, 5, and 6   60   **4.** 5, 10, and 15   30

You will use the concept of LCM later in adding fractions with unlike denominators. You will have to find the LCM of the denominators. The LCM will then be referred to as the LCD (lowest common denominator).

◇ EXERCISES ◇

Find the first three multiples excluding zero.

1. 5   5, 10, 15   **2.** 10   10, 20, 30   **3.** 9   9, 18, 27   **4.** 8   8, 16, 24   **5.** 11   11, 22, 33   **6.** 12   12, 24, 36

Find the LCM. (Use prime factorization.)

7. 4; 7   28   **8.** 9; 5   45   **9.** 4; 6   12   **10.** 2; 3   6

11. 3; 8   24   **12.** 10; 20   20   **13.** 5; 15   15   **14.** 6; 9   18

15. 8; 16   16   **16.** 2; 7   14   **17.** 7; 8   56   **18.** 9; 4   36

19. 4; 6; 8   24   **20.** 12; 18; 30   180   **21.** 2; 3; 5   30   **22.** 4; 6; 10   60

23. 5; 6; 10   30   **24.** 20; 24; 32   480   ★ **25.** 15; 24; 12   120   ★ **26.** 36; 54; 81   324

Fractions with Unlike Denominators

— ◇ **OBJECTIVE** ◇ —

To add and subtract fractions
with unlike denominators

— ◇ **RECALL** ◇ —

$$\frac{4}{7} + \frac{2}{7} = \frac{4+2}{7}, \text{ or } \frac{6}{7}$$

$$\frac{4}{7} - \frac{2}{7} = \frac{4-2}{7}, \text{ or } \frac{2}{7}$$

Example 1

Add. Simplify if possible. $\frac{1}{5} + \frac{2}{15}$

The denominators are not the same.
Factor 15 into primes. $15 = 5 \cdot 3$

$$\frac{1}{5} + \frac{2}{15} = \frac{1}{5} + \frac{2}{5 \cdot 3}$$

5 and 3 are the only factors present.

To have a common denominator, each denominator
must have exactly the same factors.

$$\frac{1}{5} + \frac{2}{5 \cdot 3}$$

needs 3   has 5 and 3

LCD is the same as LCM.

The least common denominator (LCD) is $5 \cdot 3$,
or 15.

Multiply the numerator and
denominator of $\frac{1}{5}$ by 3.

The denominators are now the same.

$$\frac{3}{15} + \frac{2}{15} = \frac{3+2}{15} = \frac{5}{15}$$

$$\frac{5}{15} = \frac{\cancel{5} \cdot 1}{\cancel{5} \cdot 3} = \frac{1}{3}$$

$$\frac{1 \cdot 3}{5 \cdot 3} + \frac{2}{5 \cdot 3}$$

$$\frac{3}{15} + \frac{2}{15}$$

$$\frac{5}{15}$$

$$\frac{1}{3}$$

So, $\frac{1}{5} + \frac{2}{15} = \frac{1}{3}$.

---

practice ▷ **Add or subtract. Simplify if possible.**

1. $\frac{2}{7} + \frac{3}{14}$ $\frac{1}{2}$

2. $\frac{1}{6} + \frac{2}{3}$ $\frac{5}{6}$

3. $\frac{7}{15} - \frac{1}{5}$ $\frac{4}{15}$

4. $\frac{1}{2} - \frac{3}{10}$ $\frac{1}{5}$

Example 2

Subtract. Simplify if possible. $\frac{13}{6} - \frac{2}{3}$

The denominators are not the same.
Factor 6 into primes. $6 = 3 \cdot 2$.

$$\begin{array}{cc} \frac{13}{6} & \frac{13}{3\cdot 2} \leftarrow \text{ has } 3\cdot 2 \\ -\frac{2}{3} & -\frac{2}{3} \leftarrow \text{ needs } 2 \end{array}$$

The least common denominator (LCD)
is $3 \cdot 2$, or 6.
Multiply the numerator and
denominator of $\frac{2}{3}$ by 2.

$$\begin{array}{cc} \frac{13}{3\cdot 2} & \frac{13}{6} \\ -\frac{2\cdot 2}{3\cdot 2} & -\frac{4}{6} \\ \hline & \frac{9}{6} = 1\frac{3}{6}, \text{ or } 1\frac{1}{2} \end{array}$$

So, $\frac{13}{6} - \frac{2}{3} = 1\frac{1}{2}$.

practice ▷ **Add or subtract. Simplify if possible.**

5. $\frac{7}{6}$
$+\frac{1}{3}$ $1\frac{1}{2}$

6. $\frac{4}{5}$
$+\frac{8}{15}$ $1\frac{1}{3}$

7. $\frac{17}{10}$
$-\frac{1}{2}$ $1\frac{1}{5}$

8. $\frac{13}{6}$
$-\frac{1}{2}$ $1\frac{2}{3}$

Example 3

Add or subtract. Simplify if possible.

Each denominator is prime.
Neither 3 nor 2 can be factored.

$$\frac{1}{3} + \frac{1}{2} \qquad\qquad \frac{1}{2} - \frac{1}{5}$$

$$\frac{1}{3} + \frac{1}{2} \qquad\qquad \frac{1}{2} - \frac{1}{5}$$

needs 2   needs 3   needs 5   needs 2

$$\frac{1\cdot 2}{3\cdot 2} + \frac{1\cdot 3}{2\cdot 3} \qquad\qquad \frac{1\cdot 5}{2\cdot 5} - \frac{1\cdot 2}{5\cdot 2}$$

The LCD is $3 \cdot 2$, or 6.

$$\frac{2}{6} + \frac{3}{6} \qquad\qquad \frac{5}{10} - \frac{2}{10}$$

$$\frac{5}{6} \qquad\qquad \frac{3}{10}$$

So, $\frac{1}{3} + \frac{1}{2} = \frac{5}{6}$ and $\frac{1}{2} - \frac{1}{5} = \frac{3}{10}$.

practice ▷ **Add or subtract. Simplify if possible.**

9. $\frac{1}{5} + \frac{1}{3}$ $\frac{8}{15}$ **10.** $\frac{1}{3} + \frac{3}{7}$ $\frac{16}{21}$ **11.** $\frac{1}{2} - \frac{1}{7}$ $\frac{5}{14}$ **12.** $\frac{1}{2} - \frac{2}{5}$ $\frac{1}{10}$

FRACTIONS WITH UNLIKE DENOMINATORS

Substitute for *n* the factor needed to make the denominators the same.

1. $\dfrac{1n}{3n} + \dfrac{5}{3 \cdot 2}$ 2
2. $\dfrac{1}{7 \cdot 2} + \dfrac{3n}{7n}$ 2
3. $\dfrac{1n}{5n} + \dfrac{2}{5 \cdot 3}$ 3
4. $\dfrac{1}{7 \cdot 2} + \dfrac{3n}{7n}$ 2

5. $\dfrac{4n}{2n} - \dfrac{3}{2 \cdot 5}$ 5
6. $\dfrac{15}{7 \cdot 3} - \dfrac{2n}{7n}$ 3
7. $\dfrac{13}{5 \cdot 7} - \dfrac{4n}{5n}$ 7
8. $\dfrac{3n}{5n} - \dfrac{4}{5 \cdot 7}$ 7

9. $\dfrac{4}{5 \cdot 3} + \dfrac{9n}{5n}$ 3
10. $\dfrac{6n}{2n} + \dfrac{7}{2 \cdot 3}$ 3
11. $\dfrac{5}{7 \cdot 3} + \dfrac{6n}{7n}$ 3
12. $\dfrac{8}{5 \cdot 5} + \dfrac{7n}{5n}$ 5

◇ *EXERCISES* ◇

Add or subtract. Simplify if possible.

1. $\dfrac{1}{6} + \dfrac{1}{2}$ $\dfrac{2}{3}$
2. $\dfrac{3}{7} + \dfrac{1}{14}$ $\dfrac{1}{2}$
3. $\dfrac{2}{9} + \dfrac{1}{3}$ $\dfrac{5}{9}$
4. $\dfrac{1}{3} + \dfrac{7}{15}$ $\dfrac{4}{5}$
5. $\dfrac{3}{14} + \dfrac{2}{7}$ $\dfrac{1}{2}$

6. $\dfrac{7}{10} - \dfrac{1}{2}$ $\dfrac{1}{5}$
7. $\dfrac{2}{3} - \dfrac{1}{6}$ $\dfrac{1}{2}$
8. $\dfrac{11}{15} - \dfrac{2}{5}$ $\dfrac{1}{3}$
9. $\dfrac{7}{9} - \dfrac{2}{3}$ $\dfrac{1}{9}$
10. $\dfrac{5}{4} - \dfrac{1}{2}$ $\dfrac{3}{4}$

11. $\dfrac{5}{6}$ $+\dfrac{1}{2}$ $1\dfrac{1}{3}$
12. $\dfrac{8}{9}$ $+\dfrac{2}{3}$ $1\dfrac{5}{9}$
13. $\dfrac{7}{5}$ $+\dfrac{3}{10}$ $1\dfrac{7}{10}$
14. $\dfrac{19}{10}$ $-\dfrac{1}{2}$ $1\dfrac{2}{5}$
15. $\dfrac{27}{14}$ $-\dfrac{3}{7}$ $1\dfrac{1}{2}$
16. $\dfrac{15}{4}$ $-\dfrac{1}{2}$ 3

17. $\dfrac{1}{7} + \dfrac{1}{5}$ $\dfrac{12}{35}$
18. $\dfrac{1}{3} + \dfrac{1}{7}$ $\dfrac{10}{21}$
19. $\dfrac{2}{3} + \dfrac{1}{2}$ $1\dfrac{1}{6}$
20. $\dfrac{1}{3} + \dfrac{2}{5}$ $\dfrac{11}{15}$
21. $\dfrac{2}{3} + \dfrac{3}{5}$ $1\dfrac{4}{15}$

22. $\dfrac{1}{2} - \dfrac{2}{5}$ $\dfrac{1}{10}$
23. $\dfrac{2}{3} - \dfrac{1}{2}$ $\dfrac{1}{6}$
24. $\dfrac{4}{5} - \dfrac{1}{3}$ $\dfrac{7}{15}$
25. $\dfrac{8}{5} - \dfrac{1}{3}$ $1\dfrac{4}{15}$
26. $\dfrac{5}{2} - \dfrac{3}{7}$ $2\dfrac{1}{14}$

27. $\dfrac{1}{3}$ $+\dfrac{2}{15}$ $\dfrac{7}{15}$
28. $\dfrac{6}{7}$ $+\dfrac{5}{14}$ $1\dfrac{3}{14}$
29. $\dfrac{19}{14}$ $-\dfrac{1}{2}$ $\dfrac{6}{7}$
30. $\dfrac{3}{5}$ $+\dfrac{1}{2}$ $1\dfrac{1}{10}$
★ 31. $\dfrac{13}{38}$ $+\dfrac{3}{19}$ $\dfrac{1}{2}$
★ 32. $\dfrac{22}{35}$ $-\dfrac{3}{7}$ $\dfrac{1}{5}$

Solve these problems.

33. David practices the flute $\dfrac{1}{2}$ h in the morning and $\dfrac{3}{4}$ h in the evening. How much longer does he practice in the evening?

$\dfrac{1}{4}$ h

34. Myra read $\dfrac{3}{10}$ of a book on Saturday and $\dfrac{1}{5}$ of the book on Sunday. How much of the book did she read?

$\dfrac{1}{2}$ of the book

Addition and Subtraction

To add and subtract fractions
with no denominator prime

Example 1

Add. Simplify if possible.
$$\begin{aligned} &\frac{5}{6} \\ +&\frac{3}{4} \end{aligned}$$

$$\frac{5}{6} + \frac{3}{4}$$

$6 = 3 \cdot 2 \qquad 4 = 2 \cdot 2$
$$\frac{5}{3 \cdot 2} + \frac{3}{2 \cdot 2}$$

At most there are one 3 and two 2's in any denominator. The LCD is $2 \cdot 2 \cdot 3$.

Each denominator needs $2 \cdot 2 \cdot 3$ to be the same.
$$\frac{5}{3 \cdot 2} + \frac{3}{2 \cdot 2}$$
$$\quad\uparrow \qquad\quad \uparrow$$
$$\text{needs } 2 \quad \text{needs } 3$$

$$\frac{5 \cdot 2}{3 \cdot 2 \cdot 2} + \frac{3 \cdot 3}{2 \cdot 2 \cdot 3}$$

$$\frac{10}{12} + \frac{9}{12}$$

$\dfrac{10}{12} + \dfrac{9}{12} = \dfrac{10 + 9}{12} = \dfrac{19}{12}$
$$\frac{19}{12}$$

$$1\frac{7}{12}$$

$\dfrac{7}{12}$ is in simplest form. So, $\dfrac{5}{6} + \dfrac{3}{4} = 1\dfrac{7}{12}$.

---

practice ▷ **Add. Simplify if possible.**

1. $\begin{aligned} &\frac{5}{6} \\ +&\frac{2}{9} \\ \hline &1\frac{1}{18} \end{aligned}$

2. $\begin{aligned} &\frac{3}{10} \\ +&\frac{3}{4} \\ \hline &1\frac{1}{20} \end{aligned}$

3. $\begin{aligned} &\frac{1}{6} \\ +&\frac{1}{4} \\ \hline &\frac{5}{12} \end{aligned}$

4. $\begin{aligned} &\frac{4}{9} \\ +&\frac{5}{6} \\ \hline &1\frac{5}{18} \end{aligned}$

Example 2

Subtract. Simplify if possible. $\dfrac{7}{10} - \dfrac{1}{4}$

$$\dfrac{7}{10} - \dfrac{1}{4}$$

$10 = 5 \cdot 2 \qquad 4 = 2 \cdot 2$

$$\dfrac{7}{5 \cdot 2} - \dfrac{1}{2 \cdot 2}$$

Each denominator needs $2 \cdot 2 \cdot 5$ to be the same.

needs 2   needs 5

$$\dfrac{7 \cdot 2}{5 \cdot 2 \cdot 2} - \dfrac{1 \cdot 5}{2 \cdot 2 \cdot 5}$$

$$\dfrac{14}{20} - \dfrac{5}{20}$$

$\dfrac{9}{20}$ is in simplest form.

$$\dfrac{9}{20}$$

So, $\dfrac{7}{10} - \dfrac{1}{4} = \dfrac{9}{20}$.

practice ▷ **Subtract. Simplify if possible.**

5. $\dfrac{5}{6} - \dfrac{1}{4}$ $\dfrac{7}{12}$

6. $\dfrac{9}{10} - \dfrac{1}{4}$ $\dfrac{13}{20}$

7. $\dfrac{5}{9} - \dfrac{1}{6}$ $\dfrac{7}{18}$

8. $\dfrac{5}{6} - \dfrac{1}{10}$ $\dfrac{11}{15}$

Example 3

Add. Simplify if possible. $\dfrac{5}{6} + \dfrac{7}{12} + \dfrac{1}{3}$

$6 = 3 \cdot 2$
$12 = 2 \cdot 2 \cdot 3 \qquad$ 3 is prime.

At most there are two 2's and one 3 in any denominator. The LCD is $2 \cdot 2 \cdot 3$.

$$\dfrac{5}{6} + \dfrac{7}{12} + \dfrac{1}{3}$$

$$\dfrac{5}{3 \cdot 2} + \dfrac{7}{2 \cdot 2 \cdot 3} + \dfrac{1}{3}$$

needs 2   has $2 \cdot 2 \cdot 3$ needs $2 \cdot 2$

$$\dfrac{5 \cdot 2}{3 \cdot 2 \cdot 2} + \dfrac{7}{3 \cdot 2 \cdot 2} + \dfrac{1 \cdot 2 \cdot 2}{3 \cdot 2 \cdot 2}$$

$$\dfrac{10}{12} + \dfrac{7}{12} + \dfrac{4}{12}$$

$$\dfrac{10}{12} + \dfrac{7}{12} + \dfrac{4}{12} = \dfrac{10 + 7 + 4}{12} = \dfrac{21}{12}$$

$$\dfrac{21}{12} = 1\dfrac{9}{12} = 1\dfrac{3}{4}$$

$$\dfrac{9}{12} = \dfrac{3 \cdot \cancel{3}}{2 \cdot 2 \cdot \cancel{3}} = \dfrac{3}{4} \qquad \text{So, } \dfrac{5}{6} + \dfrac{7}{12} + \dfrac{1}{3} = 1\dfrac{3}{4}.$$

practice ▷ **Add. Simplify if possible.**

9. $\dfrac{5}{12} + \dfrac{3}{4} + \dfrac{1}{2}$ $1\dfrac{2}{3}$

10. $\dfrac{1}{6} + \dfrac{3}{4} + \dfrac{1}{2}$ $1\dfrac{5}{12}$

11. $\dfrac{1}{2} + \dfrac{1}{4} + \dfrac{9}{20}$ $1\dfrac{1}{5}$

Add or subtract. Simplify if possible.

1. $\dfrac{5}{6}$
   $+\dfrac{1}{4}$ $1\dfrac{1}{12}$

2. $\dfrac{1}{6}$
   $+\dfrac{4}{9}$ $\dfrac{11}{18}$

3. $\dfrac{1}{4}$
   $+\dfrac{9}{10}$ $1\dfrac{3}{20}$

4. $\dfrac{3}{10}$
   $+\dfrac{5}{6}$ $1\dfrac{2}{15}$

5. $\dfrac{7}{15}$
   $+\dfrac{1}{6}$ $\dfrac{19}{30}$

6. $\dfrac{7}{10}$
   $+\dfrac{2}{15}$ $\dfrac{5}{6}$

7. $\dfrac{3}{4}-\dfrac{1}{6}$ $\dfrac{7}{12}$
8. $\dfrac{7}{9}-\dfrac{1}{6}$ $\dfrac{11}{18}$
9. $\dfrac{5}{6}-\dfrac{3}{4}$ $\dfrac{1}{12}$
10. $\dfrac{3}{10}-\dfrac{1}{4}$ $\dfrac{1}{20}$
11. $\dfrac{17}{18}-\dfrac{5}{6}$ $\dfrac{1}{9}$
12. $\dfrac{7}{12}-\dfrac{1}{4}$ $\dfrac{1}{3}$

13. $\dfrac{2}{3}+\dfrac{5}{12}+\dfrac{1}{6}$ $1\dfrac{1}{4}$
14. $\dfrac{3}{4}+\dfrac{1}{6}+\dfrac{7}{12}$ $1\dfrac{1}{2}$
15. $\dfrac{5}{6}+\dfrac{3}{4}+\dfrac{1}{2}$ $2\dfrac{1}{12}$
16. $\dfrac{5}{8}+\dfrac{3}{4}+\dfrac{1}{2}$ $1\dfrac{7}{8}$

17. $\dfrac{7}{12}+\dfrac{1}{6}+\dfrac{1}{2}$ $1\dfrac{1}{4}$
18. $\dfrac{7}{20}+\dfrac{1}{2}+\dfrac{3}{4}$ $1\dfrac{3}{5}$
19. $\dfrac{5}{8}+\dfrac{1}{12}+\dfrac{3}{4}$ $1\dfrac{11}{24}$
20. $\dfrac{8}{15}+\dfrac{3}{10}+\dfrac{1}{4}$ $1\dfrac{1}{12}$

Solve these problems.

21. Tom babysat for $\dfrac{3}{4}$ h on Friday, $\dfrac{1}{2}$ h on Saturday, and $\dfrac{3}{4}$ h on Sunday. For how many hours did he babysit? 2 h

22. A pharmacist wants to reduce a compound of $\dfrac{7}{8}$ oz by $\dfrac{1}{4}$ oz. How much will be left in the compound? $\dfrac{5}{8}$ oz

/ **Reading in Math** /

In each group of four mathematical terms below, one does not belong. Identify the term that does not belong.

1. variable   constant   (denominator)   expression

2. open sentence   equation   (fraction)   expression

3. multiply   (sum)   add   subtract

4. commutative   (add)   associative   distributive

5. (multiply)   formula   variable   value

6. LCM   (subtract)   LCD   multiple

Mixed Numbers

To add and subtract mixed
  numbers

$$\frac{10}{6} = 1\frac{4}{6}, \text{ or } 1\frac{2}{3}$$

Example 1

The denominators are the same.

Add or subtract. Simplify if possible.

$$3\frac{1}{8} + 6\frac{5}{8} \qquad\qquad 9\frac{5}{6} - 7\frac{1}{6}$$

Add fractions.	Subtract fractions.
Add whole numbers.	Subtract whole numbers.

$$\frac{6}{8} = \frac{3 \cdot \overset{1}{\cancel{2}}}{\underset{1}{\cancel{2}} \cdot 2 \cdot 2} \qquad \frac{4}{6} = \frac{2 \cdot \overset{1}{\cancel{2}}}{\underset{1}{\cancel{2}} \cdot 3}$$

$$\frac{3}{4} \qquad\qquad \frac{2}{3}$$

$$\begin{array}{r} 3\frac{1}{8} \\ +6\frac{5}{8} \\ \hline \frac{6}{8} \end{array} \qquad \begin{array}{r} 3\frac{1}{8} \\ +6\frac{5}{8} \\ \hline 9\frac{6}{8}, \text{ or } 9\frac{3}{4} \end{array}$$

$$\begin{array}{r} 9\frac{5}{6} \\ -7\frac{1}{6} \\ \hline \frac{4}{6} \end{array} \qquad \begin{array}{r} 9\frac{5}{6} \\ -7\frac{1}{6} \\ \hline 2\frac{4}{6}, \text{ or } 2\frac{2}{3} \end{array}$$

So, $3\frac{1}{8} + 6\frac{5}{8} = 9\frac{3}{4}$ and $9\frac{5}{6} - 7\frac{1}{6} = 2\frac{2}{3}.$

---

practice ▷ **Add or subtract. Simplify if possible.**

1. $5\frac{1}{10} + 3\frac{1}{10}$ $8\frac{1}{5}$ **2.** $6\frac{1}{8} + 7\frac{3}{8}$ $13\frac{1}{2}$ **3.** $9\frac{11}{12} - 4\frac{5}{12}$ $5\frac{1}{2}$ **4.** $8\frac{7}{8} - 1\frac{1}{8}$ $7\frac{3}{4}$

Example 2

Subtract. Simplify if possible. $8\frac{3}{4} - 2\frac{1}{6}$

$$\begin{array}{r} 8\frac{3}{4} \\ -2\frac{1}{6} \\ \hline \end{array} \qquad \begin{array}{r} 8\frac{3}{2 \cdot 2} \\ -2\frac{1}{3 \cdot 2} \\ \hline \end{array}$$

Each denominator needs $2 \cdot 2 \cdot 3$ to be the same.

The LCD is $2 \cdot 2 \cdot 3$.

$$\begin{array}{r} 8\frac{3}{2 \cdot 2} \quad \leftarrow \text{needs } 3 \\ -2\frac{1}{3 \cdot 2} \quad \leftarrow \text{needs } 2 \\ \hline \end{array}$$

$$\begin{array}{r} 8\frac{3 \cdot 3}{2 \cdot 2 \cdot 3} \\ -2\frac{1 \cdot 2}{3 \cdot 2 \cdot 2} \\ \hline \end{array} \qquad \begin{array}{r} 8\frac{9}{12} \\ -2\frac{2}{12} \\ \hline 6\frac{7}{12} \end{array}$$

So, $8\frac{3}{4} - 2\frac{1}{6} = 6\frac{7}{12}.$

---

practice ▷ **Add or subtract. Simplify if possible.**

5. $3\frac{1}{4} + 2\frac{1}{6}$ $5\frac{5}{12}$ **6.** $6\frac{1}{5} + 3\frac{1}{15}$ $9\frac{4}{15}$ **7.** $7\frac{5}{6} - 3\frac{3}{4}$ $4\frac{1}{12}$ **8.** $2\frac{3}{4} - \frac{1}{8}$ $2\frac{5}{8}$

Example 3

Add. $6 + 4\frac{1}{2}$

Subtract. $7\frac{5}{6} - 4$

$6 + 4\frac{1}{2} = (6 + 4) + \frac{1}{2} = 10 + \frac{1}{2}$, or $10\frac{1}{2}$ $6 + 4\frac{1}{2} = 10\frac{1}{2}$

$7\frac{5}{6} - 4 = 3\frac{5}{6}$

$7\frac{5}{6} - 4 = 7 + \frac{5}{6} - 4 = 3 + \frac{5}{6}$, or $3\frac{5}{6}$ So, $6 + 4\frac{1}{2} = 10\frac{1}{2}$ and $7\frac{5}{6} - 4 = 3\frac{5}{6}$.

---

practice ▷ **Add or subtract.**

9. $4\frac{3}{4} + 7$ $11\frac{3}{4}$ **10.** $3 + 5\frac{7}{8}$ $8\frac{7}{8}$ **11.** $9\frac{3}{5} - 6$ $3\frac{3}{5}$ **12.** $8\frac{9}{10} - 1$ $7\frac{9}{10}$

Example 4

Add and simplify. $2\frac{1}{3} + 3\frac{5}{6} + 4\frac{1}{2}$

Factor the denominators.
3 is prime.

$2\frac{1}{3}$ $2\frac{1}{3}$

$3\frac{5}{6}$ $3\frac{5}{3 \cdot 2}$

$6 = 3 \cdot 2$

$+ 4\frac{1}{2}$ $+ 4\frac{1}{2}$

2 is prime.

Each denominator needs $3 \cdot 2$ to be the same. The LCD is $3 \cdot 2$.

Multiply the numerator and denominator by 2.

$2\frac{1}{3}$ ← needs 2

$3\frac{5}{3 \cdot 2}$ ← has $3 \cdot 2$

$2\frac{1 \cdot 2}{3 \cdot 2}$ $2\frac{2}{6}$

$3\frac{5}{3 \cdot 2}$ $3\frac{5}{6}$

Multiply the numerator and denominator by 3.

$+ 4\frac{1}{2}$ ← needs 3

$+ 4\frac{1 \cdot 3}{2 \cdot 3}$ $+ 4\frac{3}{6}$

$9\frac{10}{6}$

$9\frac{10}{6}$ means $9 + \frac{10}{6} = 9 + 1\frac{4}{6} = 10\frac{4}{6}$, or $10\frac{2}{3}$.

So, $2\frac{1}{3} + 3\frac{5}{6} + 4\frac{1}{2} = 10\frac{2}{3}$.

---

practice ▷ **Add. Simplify if possible.**

$13\frac{3}{8}$

13. $3\frac{1}{5} + 2\frac{7}{10} + 5\frac{1}{2}$ $11\frac{2}{5}$ **14.** $\frac{2}{5} + 2\frac{7}{15} + 6\frac{1}{3}$ $9\frac{1}{5}$ **15.** $4\frac{1}{2} + 7\frac{1}{8} + 1\frac{3}{4}$

MIXED NUMBERS

◇ EXERCISES ◇

Add or subtract. Simplify if possible. See TE Answer Section.

1. $7\frac{1}{10} + 3\frac{3}{10}$ 2. $8\frac{1}{9} + 6\frac{5}{9}$ 3. $7\frac{1}{8} + 6\frac{3}{8}$ 4. $9\frac{7}{15} - 4\frac{2}{15}$ 5. $8\frac{3}{4} - 7\frac{1}{4}$

6. $8\frac{3}{10} + \frac{1}{2}$ 7. $8\frac{7}{10} + 4\frac{1}{4}$ 8. $4\frac{3}{5} + \frac{1}{3}$ 9. $3\frac{3}{8} + 2\frac{1}{4}$ 10. $7\frac{1}{6} + 3\frac{5}{9}$

11. $7\frac{1}{2} - 2\frac{1}{4}$ 12. $8\frac{2}{3} - 5\frac{1}{6}$ 13. $7\frac{7}{8} - \frac{3}{4}$ 14. $9\frac{4}{5} - 3\frac{1}{2}$ 15. $4\frac{5}{6} - 3\frac{1}{4}$

16. $9 + 3\frac{1}{2}$ 17. $7\frac{1}{4} - 2$ 18. $6\frac{5}{8} + 4$ 19. $9 + 6\frac{1}{2}$ 20. $7\frac{3}{4} - 2$

21. $2\frac{3}{5} + 5\frac{9}{10}$ 22. $4\frac{1}{2} + 7\frac{5}{6}$ 23. $7\frac{1}{4} + 9\frac{2}{3}$ 24. $5\frac{2}{3} + 4\frac{5}{9}$ 25. $8\frac{1}{2} + 6\frac{7}{8}$

26. $8\frac{2}{3} + 3\frac{1}{6} + 4\frac{1}{2}$ 27. $\frac{3}{10} + 8\frac{1}{2} + 9\frac{2}{5}$ 28. $6\frac{2}{3} + 5\frac{1}{2} + 3\frac{5}{6}$ 29. $3\frac{13}{15} + 2\frac{1}{5} + 5\frac{1}{3}$

30. $7\frac{1}{8} + 6\frac{3}{4} + \frac{1}{2}$ 31. $7\frac{9}{10} + 4\frac{3}{5} + 1\frac{1}{2}$ 32. $3\frac{3}{4} + 2\frac{1}{3} + 4\frac{5}{6}$ 33. $2\frac{5}{8} + 8\frac{2}{3} + 6\frac{1}{4}$

Solve these problems.

34. At birth, a baby weighed $7\frac{3}{10}$ pounds. Two months later, it weighed $8\frac{3}{4}$ pounds. How much weight did it gain? $1\frac{9}{20}$ lb

35. Marni's punchbowl holds $4\frac{5}{8}$ gal of punch. Gerard's punchbowl holds $1\frac{1}{4}$ gal more. How much punch does Gerard's bowl hold? $5\frac{7}{8}$ gal

36. A grocer ordered $3\frac{1}{2}$ cases of soup on Monday and $7\frac{3}{4}$ cases of soup on Tuesday. How many cases of soup were ordered? $11\frac{1}{4}$ cases

37. Abdul ran 3 kilometers in $9\frac{3}{4}$ minutes last week. This week, he ran 3 kilometers in $8\frac{1}{2}$ minutes. How many minutes less is this? $1\frac{1}{4}$ min

Find the value of n in each case.

1. $\frac{1}{2} + \frac{1}{3} = \frac{35}{n}$ 42

2. $1\frac{2}{3} + 3\frac{3}{4} = 4\frac{n}{24}$ 34

3. $\frac{5}{6} - \frac{1}{2} = \frac{n}{15}$ 5

4. $7\frac{1}{2} - 2\frac{3}{4} = 3\frac{28}{n}$ 16

      CHAPTER FIVE

Problem Solving – Applications
Jobs for Teenagers

1. Kim has a new job as a lifeguard. To get in shape, she swam $8\frac{3}{4}$ laps on Monday. She swam $7\frac{3}{4}$ laps on Tuesday. Find the total laps she swam. $16\frac{1}{2}$ laps

2. José's swimming coach told him to swim 65 laps this weekend. He swam $15\frac{1}{4}$ laps on Friday and $40\frac{1}{4}$ laps on Saturday. How many laps must he swim on Sunday to make his goal? $9\frac{1}{2}$ laps

3. Sue's softball coach told her to run 30 laps this weekend. She ran $14\frac{1}{2}$ laps on Saturday morning and $10\frac{1}{4}$ laps on Saturday afternoon. How many laps must she run on Sunday to make the 30-lap goal? $5\frac{1}{4}$ laps

4. Rudy is manager of the track team. He needed to lose weight. He lost $1\frac{1}{2}$ lb the first week, 2 lb the second week, and $1\frac{1}{4}$ lb the third week. He now weighs $171\frac{1}{4}$ lb. How much did he weigh before? 176 lb

5. Martha weighed 120 lb. Her soccer coach wanted her to reduce her weight. She lost $1\frac{1}{2}$ lb each week for 3 weeks. How much does she weigh now? $115\frac{1}{2}$ lb

6. Peter is supposed to practice 10 h this weekend for a tennis match. He practiced $2\frac{1}{2}$ h on Saturday morning, $1\frac{1}{4}$ h that afternoon, and 3 h on Sunday morning. How many more hours must he practice on Sunday? $3\frac{1}{4}$ h

Renaming in Subtraction

To subtract mixed numbers
using renaming

Regroup
1 ten as
10 ones.

$$483$$
$$-219$$

$$\overset{7\ 13}{4\cancel{8}\cancel{3}}$$
$$-219$$
$$\overline{264}$$

Example 1

Rename $6\frac{3}{10}$ as $5\frac{n}{10}$ by replacing n.

$$6\frac{3}{10} = \quad 6 \quad + \frac{3}{10}$$

Rename 6 as $5 + 1$. $\qquad 5 + 1 + \frac{3}{10}$

Rename 1 as $\frac{10}{10}$. $\qquad 5 + \frac{10}{10} + \frac{3}{10}$

$\frac{10}{10} + \frac{3}{10} = \frac{13}{10}$ $\qquad 5 + \frac{13}{10}$

So, $6\frac{3}{10} = 5\frac{13}{10}$.

---

practice ▷ **Rename by replacing n.**

1. $8\frac{1}{3} = 7\frac{n}{3}$ ⠀4

2. $7\frac{4}{5} = 6\frac{n}{5}$ ⠀9

3. $9\frac{1}{6} = 8\frac{n}{6}$ ⠀7

Example 2

Subtract. $9\frac{2}{5} - 2\frac{4}{5}$

Write in vertical form. $\quad 9\frac{2}{5}$ Think: Cannot subtract $\frac{4}{5}$ from $\frac{2}{5}$;

$\quad -2\frac{4}{5}$ Rename $9\frac{2}{5}$.

$\quad 8\frac{7}{5} \quad 9\frac{2}{5} = 8 + 1 + \frac{2}{5} = 8 + \frac{5}{5} + \frac{2}{5} = 8 + \frac{7}{5} = 8\frac{7}{5}$

$\quad -2\frac{4}{5}$

$\quad \overline{6\frac{3}{5}}$

So, $9\frac{2}{5} - 2\frac{4}{5} = 6\frac{3}{5}$.

Example 3

Subtract and simplify. $6\frac{1}{3} - 2\frac{5}{6}$

Factor denominators; 3 is prime.
The LCD is $3 \cdot 2$.

	Step 1	Step 2
$6\frac{1}{3}$	$6\frac{1}{3}$	$6\frac{1 \cdot 2}{3 \cdot 2}$
$-2\frac{5}{6}$	$-2\frac{5}{3 \cdot 2}$	$-2\frac{5}{3 \cdot 2}$

$6\frac{2}{6} = 6 + \frac{2}{6}$

$5 + 1 + \frac{2}{6}$

$5 + \frac{6}{6} + \frac{2}{6} = 5\frac{8}{6}$

	Step 3	Step 4
	$6\frac{2}{6}$	$5\frac{8}{6}$
	$-2\frac{5}{6}$	$-2\frac{5}{6}$
		$3\frac{3}{6}$, or $3\frac{1}{2}$

So, $6\frac{1}{3} - 2\frac{5}{6} = 3\frac{1}{2}$

practice ▷ **Subtract. Simplify if possible.**

$2\frac{9}{10}$

4. $7\frac{1}{10} - 3\frac{3}{5}$ $3\frac{1}{2}$ **5.** $4\frac{1}{3} - 2\frac{5}{6}$ $1\frac{1}{2}$ **6.** $7\frac{1}{4} - 4\frac{1}{2}$ $2\frac{3}{4}$ **7.** $5\frac{3}{5} - 2\frac{7}{10}$

Example 4

Subtract. $8 - 5\frac{2}{3}$

8 Think: Rename 8 as $7\frac{3}{3}$.

$-5\frac{2}{3}$

$7\frac{3}{3}$

$-5\frac{2}{3}$

$2\frac{1}{3}$ So, $8 - 5\frac{2}{3} = 2\frac{1}{3}$.

practice ▷ **Subtract.**

$4\frac{1}{2}$

8. $7 - 3\frac{1}{5}$ $3\frac{4}{5}$ **9.** $9 - 4\frac{3}{4}$ $4\frac{1}{4}$ **10.** $7 - 6\frac{1}{3}$ $\frac{2}{3}$ **11.** $8 - 3\frac{1}{2}$

RENAMING IN SUBTRACTION

◇ ORAL EXERCISES ◇

Rename by replacing *n*.

1. $4\frac{1}{5} = 3\frac{n}{5}$ 6   2. $5\frac{1}{8} = 4\frac{n}{8}$ 9   3. $9\frac{2}{3} = 8\frac{n}{3}$ 5   4. $6\frac{1}{2} = 5\frac{n}{2}$ 3   5. $7\frac{3}{4} = 6\frac{n}{4}$ 7

6. $9\frac{1}{6} = 8\frac{n}{6}$ 7   7. $8\frac{1}{4} = 7\frac{n}{4}$ 5   8. $5\frac{1}{3} = 4\frac{n}{3}$ 4   9. $9\frac{3}{4} = 8\frac{n}{4}$ 7   10. $7\frac{1}{9} = 6\frac{n}{9}$ 10

◇ EXERCISES ◇

Subtract. Simplify if possible.

1. $3\frac{1}{4}$ $-1\frac{3}{4}$ $1\frac{1}{2}$ 2. $5\frac{1}{6}$ $-2\frac{5}{6}$ $2\frac{1}{3}$ 3. $7\frac{1}{9}$ $-3\frac{4}{9}$ $3\frac{2}{3}$ 4. $6\frac{1}{10}$ $-4\frac{7}{10}$ $1\frac{2}{5}$ 5. $3\frac{3}{8}$ $-1\frac{5}{8}$ $1\frac{3}{4}$ 6. $8\frac{1}{12}$ $-3\frac{5}{12}$ 4

7. $8\frac{1}{6} - 4\frac{1}{2}$ $3\frac{2}{3}$ 8. $5\frac{3}{5} - 2\frac{9}{10}$ $2\frac{7}{10}$ 9. $7\frac{1}{6} - 4\frac{2}{3}$ $2\frac{1}{2}$ 10. $9\frac{7}{10} - 3\frac{4}{5}$ $5\frac{9}{10}$ 11. $8\frac{1}{4} - 3\frac{1}{2}$ $4\frac{3}{4}$

12. $5\frac{1}{6} - 2\frac{1}{3}$ $2\frac{5}{6}$ 13. $3\frac{1}{8} - 2\frac{3}{4}$ $\frac{3}{8}$ 14. $7\frac{1}{9} - 3\frac{2}{3}$ $3\frac{4}{9}$ 15. $8\frac{1}{5} - \frac{3}{10}$ $7\frac{9}{10}$ 16. $8\frac{1}{3} - 2\frac{5}{6}$ $5\frac{1}{2}$

17. $6 - 2\frac{1}{3}$ $3\frac{2}{3}$ 18. $6 - 3\frac{2}{5}$ $2\frac{3}{5}$ 19. $8 - 7\frac{1}{2}$ $\frac{1}{2}$ 20. $6 - 5\frac{1}{4}$ $\frac{3}{4}$ 21. $9 - 4\frac{5}{6}$ $4\frac{1}{6}$

Solve these problems.

22. Josh is on a diet. One Monday he weighed 170 lb. The following Monday, he weighed $168\frac{1}{4}$ lb. How much weight did he lose? $1\frac{3}{4}$ lb

23. A pair of slacks is $40\frac{1}{4}$ in long. They must be shortened by $8\frac{1}{2}$ in. How long will they be then? $31\frac{3}{4}$ in

24. Elke had 5 lb of potatoes. She cooked $2\frac{3}{8}$ lb of potatoes. How many lb of potatoes are left? $2\frac{5}{8}$ lb

25. The Wilsons are on a trip of $85\frac{3}{10}$ mi. They have gone $67\frac{1}{2}$ mi. How much farther must they go? $17\frac{4}{5}$ mi

Calculator

Estimate the number of digits in the answer. Check with a calculator.

1. $4,628 \div 52$ 2 digits

2. $102,764 \div 46$ 4 digits

3. $29,880 \div 72$ 3 digits

Evaluating Expressions

──◇ *OBJECTIVE* ◇──   ──────◇ *RECALL* ◇──────

To evaluate expressions like
$\frac{3x-5}{5} + \frac{2}{15}$ if $x = 2$

Evaluate $3a + 4$ if $a = 5$.
$$3 \cdot 5 + 4 \quad \text{Substitute 5 for } a.$$
$$15 + 4 \text{ or } 19$$

Example 1

Evaluate $\frac{2a}{9} + \frac{2}{9}$ if $a = 3$.

Substitute 3 for a. $\dfrac{2 \cdot 3}{9} + \dfrac{2}{9}$

$\dfrac{6}{9} + \dfrac{2}{9}$

Denominators are the same;
add the numerators. $\dfrac{8}{9}$

So, the value of $\frac{2a}{9} + \frac{2}{9}$ if $a = 3$ is $\frac{8}{9}$.

Example 2

Evaluate $\frac{2x+1}{21} + \frac{5}{21}$ if $x = 4$. Simplify if possible.

Substitute 4 for x. $\dfrac{2 \cdot 4 + 1}{21} + \dfrac{5}{21}$

$2 \cdot 4 + 1 = 9$ $\dfrac{9}{21} + \dfrac{5}{21}$

Denominators are the same;
add the numerators. $\dfrac{14}{21}$

Factor 14 into primes.
Factor 21 into primes. $\dfrac{7 \cdot 2}{7 \cdot 3}$

Divide out the like factors. $\dfrac{\overset{1}{\cancel{7}} \cdot 2}{\underset{1}{\cancel{7}} \cdot 3}$ or $\dfrac{2}{3}$

So, the value of $\frac{2x+1}{21} + \frac{5}{21}$ if $x = 4$ is $\frac{2}{3}$.

---

practice ▷ **Evaluate. Simplify if possible.**

1. $\frac{2a}{11} + \frac{4}{11}$ if $a = 3$ $\frac{10}{11}$ **2.** $\frac{5x}{20} + \frac{1}{20}$ if $x = 3$ $\frac{4}{5}$ **3.** $\frac{1}{30} + \frac{3m+5}{30}$ if $m = 2$ $\frac{2}{5}$

Example 3

Evaluate $\frac{2}{15} + \frac{3x - 5}{5}$ if $x = 4$. Simplify if possible.

Substitute 4 for x.

$$\frac{2}{15} + \frac{3 \cdot 4 - 5}{5}$$

$3 \cdot 4 - 5 = 12 - 5 = 7$.

$$\frac{2}{15} + \frac{7}{5}$$

Factor 15 into primes.

$$\frac{2}{5 \cdot 3} + \frac{7}{5}$$

Multiply numerator and denominator of $\frac{7}{5}$ by 3; the LCD is $5 \cdot 3$.

$$\frac{2}{5 \cdot 3} + \frac{7 \cdot 3}{5 \cdot 3}$$

$$\frac{2}{15} + \frac{21}{15}$$

$$1\frac{8}{15}$$
$$15\overline{)23}$$

$$\frac{23}{15} = 1\frac{8}{15}$$

So, the value of $\frac{2}{15} + \frac{3x - 5}{5}$ if $x = 4$ is $1\frac{8}{15}$.

practice ▷ **Evaluate. Simplify if possible.**

4. $\frac{1}{4} + \frac{2x + 1}{12}$ if $x = 2$ $2\frac{2}{3}$

5. $\frac{2x - 5}{24} + \frac{3}{8}$ if $x = 5$ $\frac{7}{12}$

◇ ORAL EXERCISES ◇

Evaluate.

1. $\frac{3x}{4}$ if $x = 4$ 3

2. $\frac{3y}{5}$ if $y = 10$ 6

3. $\frac{4n}{7}$ if $n = 14$ 8

◇ EXERCISES ◇

Evaluate. Simplify if possible.

1. $\frac{4x}{13} + \frac{5}{13}$ if $x = 1$ $\frac{9}{13}$

2. $\frac{7}{15} + \frac{2a}{15}$ if $a = 3$ $\frac{13}{15}$

3. $\frac{2a - 1}{11} + \frac{1}{11}$ if $a = 4$ $\frac{8}{11}$

4. $\frac{2x - 3}{20} + \frac{3}{20}$ if $x = 5$ $\frac{1}{2}$

5. $\frac{7y - 10}{9} + \frac{2}{9}$ if $y = 2$ $\frac{2}{3}$

6. $\frac{7}{30} + \frac{5x - 14}{30}$ if $x = 3$ $\frac{4}{15}$

7. $\frac{7t - 13}{6} + \frac{5}{24}$ if $t = 2$ $\frac{3}{8}$

8. $\frac{7}{9} + \frac{2k - 11}{3}$ if $k = 6$ $1\frac{1}{9}$

9. $\frac{7}{18} + \frac{3y - 14}{9}$ if $y = 5$ $\frac{1}{2}$

10. $\frac{1}{2} + \frac{7a - 25}{10}$ if $a = 4$ $\frac{4}{5}$

11. $\frac{3x - 4}{3} + \frac{2}{15}$ if $x = 2$ $\frac{4}{5}$

12. $\frac{7}{30} + \frac{4a - 23}{10}$ if $a = 6$ $\frac{1}{3}$

★ 13. $\frac{3x + 1}{40} + \frac{5x - 9}{5}$ if $x = 2$ $\frac{3}{8}$

★ 14. $\frac{7m - 9}{32} + \frac{9m - 26}{8}$ if $m = 3$ $\frac{1}{2}$

Problem Solving – Careers
Bookkeepers

A bookkeeper prepares the payroll. Find the total number of hours the bookkeeper should record.

1. John Hendrickson

Mon.	Tues.	Wed.	Thurs.	Fri.
$8\frac{1}{2}$	$9\frac{1}{4}$	$7\frac{3}{4}$	—	—

$25\frac{1}{2}$ h

2. Joanne Romano

Mon.	Tues.	Wed.	Thurs.	Fri.
$8\frac{1}{2}$	$8\frac{3}{4}$	9	$8\frac{1}{4}$	$8\frac{1}{2}$

43 h

A bookkeeper checks for mistakes in bills sent to a business. Is the total correct?

3. From: Hogg's Meats
To: Jan's Deli
For: Purchase of frankfurters

July 15	Aug. 1	Aug. 15	Total
$17\frac{1}{2}$ lb	$24\frac{1}{4}$ lb	$15\frac{3}{4}$ lb	$60\frac{1}{2}$ lb

no; $57\frac{1}{2}$ lb

4. From: ABC Textiles
To: Fabric Yard
For: Purchase of cotton fabric

Blue	Orange	Aqua	Total
$10\frac{1}{2}$ yd	$20\frac{3}{8}$ yd	$17\frac{1}{4}$ yd	$48\frac{1}{8}$ yd

yes

A firm pays overtime for all hours over 40 hours per week. How many hours overtime should Mary record for each person?

5. Bill O'Neill

Mon.	Tues.	Wed.	Thurs.	Fri.
$9\frac{3}{4}$	$8\frac{1}{2}$	$7\frac{3}{4}$	$9\frac{1}{2}$	8

$3\frac{1}{2}$ h overtime

6. Margarita Juarez

Mon.	Tues.	Wed.	Thurs.	Fri.
$9\frac{1}{4}$	8	$8\frac{1}{2}$	$8\frac{3}{4}$	$9\frac{1}{2}$

4 h overtime

PROBLEM SOLVING

Equations

─── ◇ **OBJECTIVES** ◇ ───   ─── ◇ **RECALL** ◇ ───

To solve equations whose
solutions are fractions or
mixed numbers

To solve equations that contain
fractions

Solve	$x + 3 = 9.$
Subtract 3 from	$x + 3 - 3 = 9 - 3$
each side.	
So, the solution is 6.	$x = 6$

Example 1

Solve $3x + 5 = 12$. Check the solution.

Subtract 5 from each side. $\quad 3x + 5 - 5 = 12 - 5$

$$3x = 7$$

Divide each side by 3. $\quad\quad \dfrac{3x}{3} = \dfrac{7}{3}$

$$x = \frac{7}{3}, \text{ or } 2\frac{1}{3}.$$

Check: $\quad \underline{3x + 5 = 12}$

$\dfrac{7}{3}$ is a convenient form for checking.

$$3 \cdot \frac{7}{3} + 5 \;\bigg|\; 12$$

$$7 + 5 \;\bigg|$$

$$12 = 12$$

So, $\dfrac{7}{3}$, or $2\dfrac{1}{3}$ is the solution of

$3x + 5 = 12.$

Example 2

Solve $5y - 6 = 5$. Check the solution.

Add 6 to each side. $\quad 5y - 6 + 6 = 5 + 6$

$$5y = 11$$

Divide each side by 5. $\quad\quad \dfrac{5y}{5} = \dfrac{11}{5}$

$$y = \frac{11}{5}, \text{ or } 2\frac{1}{5}$$

Check: $\quad \underline{5y - 6 = 5}$

$\dfrac{11}{5}$ is a convenient form for checking.

$$5 \cdot \frac{11}{5} - 6 \;\bigg|\; 5$$

$$11 - 6 \;\bigg|$$

$$5 = 5$$

So, $\dfrac{11}{5}$, or $2\dfrac{1}{5}$ is the solution of $5y - 6 = 5.$

Solve. Check the solution.

1. $2x + 7 = 16$ $\quad 4\frac{1}{2}$
2. $5y + 9 = 11$ $\quad \frac{2}{5}$
3. $3z + 4 = 12$ $\quad 2\frac{2}{3}$
4. $3x - 2 = 8$ $\quad 3\frac{1}{3}$
5. $6y - 3 = 12$ $\quad 2\frac{1}{2}$
6. $4z - 5 = 13$ $\quad 4\frac{1}{2}$

Example 3

Solve $\frac{2}{3}x + 6 = 15$. Check the solution.

Subtract 6 from each side of the equation.

$$\frac{2}{3}x + 6 - 6 = 15 - 6$$

$$\frac{2}{3}x = 9$$

Multiply by the reciprocal of the coefficient of x.

$$\frac{3}{2} \cdot \frac{2}{3} = 1$$

$$\frac{3}{2} \cdot \frac{2}{3}x = \frac{3}{2} \cdot 9$$

$$x = \frac{27}{2}, \text{ or } 13\frac{1}{2}$$

$\frac{27}{2}$ is a convenient form for checking.

Check: $\quad \frac{2}{3}x + 6 = 15$

$$\begin{array}{c|c} \frac{2}{3} \cdot \frac{27}{2} + 6 & 15 \\ 9 + 6 & \\ 15 & = 15 \end{array}$$

So, $\frac{27}{2}$, or $13\frac{1}{2}$ is the solution of $\frac{2}{3}x + 6 = 15$.

---

practice ▷ **Solve. Check the solution.**

7. $\frac{4}{5}x + 7 = 9$ $\quad 2\frac{1}{2}$
8. $\frac{3}{4}y + 1 = 8$ $\quad 9\frac{1}{3}$
9. $\frac{1}{5}z + 7 = 11$ $\quad 20$

Example 4

Solve $\frac{4}{7}x - 2 = 9$. Check the solution.

Add 2 to each side. $\quad \frac{4}{7}x - 2 + 2 = 9 + 2$

$$\frac{4}{7}x = 11$$

Multiply each side by $\frac{7}{4}$.

$$\frac{7}{4} \cdot \frac{4}{7}x = \frac{7}{4} \cdot 11$$

$$x = \frac{77}{4}, \text{ or } 19\frac{1}{4}$$

$\frac{77}{4}$ is a convenient form for checking.

Check: $\quad \frac{4}{7}x - 2 = 9$

$$\begin{array}{c|c} \frac{4}{7} \cdot \frac{77}{4} - 2 & 9 \\ 11 - 2 & \\ 9 & = 9 \end{array}$$

So, $\frac{77}{4}$, or $19\frac{1}{4}$ is the solution of $\frac{4}{7}x - 2 = 9$.

EQUATIONS

Solve. Check the solution.

10. $\frac{2}{3}x - 1 = 2$ $4\frac{1}{2}$ **11.** $\frac{5}{6}y - 3 = 7$ 12   **12.** $\frac{3}{5}z - 5 = 2$ $11\frac{2}{3}$

◇ ORAL EXERCISES ◇

Solve.

1. $x + \frac{3}{5} = \frac{4}{5}$ $\frac{1}{5}$ **2.** $x + \frac{1}{7} = \frac{6}{7}$ $\frac{5}{7}$ **3.** $x - \frac{1}{9} = \frac{4}{9}$ $\frac{5}{9}$ **4.** $x - \frac{1}{7} = \frac{2}{7}$ $\frac{3}{7}$

5. $x + \frac{1}{11} = \frac{7}{11}$ $\frac{6}{11}$ **6.** $x - \frac{1}{5} = \frac{2}{5}$ $\frac{3}{5}$ **7.** $x + \frac{2}{9} = \frac{7}{9}$ $\frac{5}{9}$ **8.** $x - \frac{1}{13} = \frac{3}{13}$ $\frac{4}{13}$

9. $x - \frac{1}{15} = \frac{7}{15}$ $\frac{8}{15}$ **10.** $x + \frac{1}{9} = \frac{8}{9}$ $\frac{7}{9}$ **11.** $x - \frac{3}{7} = \frac{2}{7}$ $\frac{5}{7}$ **12.** $x + \frac{2}{11} = \frac{9}{11}$ $\frac{7}{11}$

◇ EXERCISES ◇

Solve. Check the solution.

1. $4x + 7 = 9$ $\frac{1}{2}$ **2.** $3x + 6 = 7$ $\frac{1}{3}$ **3.** $5x + 9 = 15$ $1\frac{1}{5}$

4. $4x + 1 = 21$ 5   **5.** $5x + 3 = 12$ $1\frac{4}{5}$ **6.** $9x + 2 = 4$ $\frac{2}{9}$

7. $6x - 2 = 1$ $\frac{1}{2}$ **8.** $7x - 4 = 2$ $\frac{6}{7}$ **9.** $8x - 5 = 12$ $2\frac{1}{8}$

10. $9x - 3 = 10$ $1\frac{4}{9}$ **11.** $5x - 2 = 4$ $1\frac{1}{5}$ **12.** $12x - 4 = 7$ $\frac{11}{12}$

13. $\frac{1}{2}x + 5 = 8$ 6   **14.** $\frac{3}{5}x + 9 = 14$ $8\frac{1}{3}$ **15.** $\frac{7}{2}x + 7 = 9$ $\frac{4}{7}$

16. $\frac{4}{7}x - 1 = 2$ $5\frac{1}{4}$ **17.** $\frac{9}{2}x - 3 = 15$ 4   **18.** $\frac{3}{7}x - 6 = 11$ $39\frac{2}{3}$

★ **19.** $2x + 9\frac{1}{3} = 17\frac{2}{3}$ $4\frac{1}{6}$ ★ **20.** $\frac{2}{3}x - 2\frac{3}{4} = 1\frac{1}{2}$ $6\frac{3}{8}$ ★ **21.** $1\frac{1}{2}x + 4\frac{1}{3} = 9\frac{1}{4}$

  $3\frac{5}{18}$

Challenge

To what single fraction does this simplify?

$$\cfrac{1}{1 + \cfrac{1}{2 + \cfrac{1}{3 + \cfrac{1}{4 + \cfrac{1}{5 + \frac{1}{6}}}}}}$$
  $\frac{972}{1393}$

(*Hint:* Start at the bottom.)

$$\frac{1}{5 + \frac{1}{6}} = \frac{1}{5\frac{1}{6}} = \frac{1}{\frac{31}{6}} = 1 \div \frac{31}{6} = \frac{1}{1} \cdot \frac{6}{31} = \frac{6}{31}$$

     CHAPTER FIVE

Non-Routine Problems

There is no simple method used to solve these problems. Use whatever methods you can. Do not expect to solve these easily.
See TE Answer Section for sample solutions to these problems.

1. Your friend challenges you to use four 4's to make numbers, and she allows you to use any operations you wish. She gives you this beginning to get you going: $1 = \dfrac{4 + 4}{4 + 4}$, $2 = \dfrac{4 + 4}{\sqrt{4} \cdot \sqrt{4}}$. Recall that $\sqrt{4} = 2$ because $2 \cdot 2 = 4$. Now complete the work writing numbers 3 through 10. You must use *exactly* four 4's, no more no less. See TE Answer Section.

2. Now you have the challenge of writing numbers using exactly three 9's. To make 10, you can write $9 + \dfrac{9}{9}$. To make 0, you can write $\dfrac{9 - 9}{9}$. Write the following numbers using exactly three 9's.

   a. 6 $9 - \dfrac{9}{\sqrt{9}}$ **b.** 18 $\sqrt{9} \cdot \sqrt{9} + 9$ **c.** 81 $9 \cdot \sqrt{9} \cdot \sqrt{9}$ **d.** 27 $\sqrt{9} \cdot \sqrt{9} \cdot \sqrt{9}$

   e. 4 $\dfrac{\sqrt{9} + 9}{\sqrt{9}}$ **f.** 24 $9 \cdot \sqrt{9} - \sqrt{9}$ **g.** 12 $\sqrt{9} \cdot \sqrt{9} + \sqrt{9}$ **h.** 10 $9 + \dfrac{9}{9}$

3. Write the following numbers using exactly three 6's. For example, to make 6, you can write $\dfrac{6 \cdot 6}{6}$. To make 2, you can write $\dfrac{6 + 6}{6}$.

   a. 0 $\dfrac{6 - 6}{6}$ **b.** 6 $6 + 6 - 6$ **c.** 60 $66 - 6$ **d.** 18 $6 + 6 + 6$

   e. 11 $\dfrac{66}{6}$ **f.** 30 $6 \cdot 6 - 6$ **g.** 7 $6\dfrac{6}{6}$ **h.** 5 $6 - \dfrac{6}{6}$

4. You made the following transaction with your friend: You bought a pen from him for $4, sold it for $5, then bought it back for $6, and finally sold it for $7. Did you make any money? Yes Did you lose? You made $2.

5. The Carters have two children. The product of their ages is 32. The sum is 12. What are their ages? 4 yr and 8 yr

6. I am thinking of a number. If the number is multiplied by 6, then the product divided by 9, and this result increased by 7, the final result is 15. What number am I thinking of? 12

7. Sandy has $8 in her savings account. Maria has $12. From now on, Sandy will add $1 to her account each week, and Maria will add $3. After how many weeks will Maria's account be twice as big as Sandy's? After 4 wk Maria's account will be twice as big as Sandy's.

The first step in writing a program to add any two fractions is to come up with a general formula. This formula must tell the computer how to add two fractions, given their numerators and denominators.

Example 1 Write a formula for the sum of any two fractions $\frac{X}{Y}$ and $\frac{Z}{W}$.

$$S = \frac{X}{Y} + \frac{Z}{W}$$

First, find the LCD: $Y \cdot W$ $\qquad$ $S = \frac{X \cdot W}{Y \cdot W} + \frac{Z \cdot Y}{W \cdot Y}$ $\qquad$ $S = \frac{X \cdot W + Z \cdot Y}{Y \cdot W}$

This formula for finding the sum of any two fractions is used in the next example to write a program for adding fractions.

Example 2 Write a program to add any two fractions $\frac{X}{Y}$ and $\frac{Z}{W}$, $Z \neq 0$, $W \neq 0$.

Then **RUN** the program to find the sum $\frac{2}{3} + \frac{1}{10}$.

| A is the numerator and B is the denominator of the sum in the formula from **Example 1**. | ```
10   HOME
20   INPUT "WRITE 1ST NUMERATOR ";X
30   INPUT "WRITE 1ST DENOMINATOR ";Y
40   INPUT "WRITE 2ND NUMERATOR ";Z
50   INPUT "WRITE 2ND DENOMINATOR ";W
60   LET A = X * W + Y * Z
70   LET B = Y * W
80   PRINT : PRINT
90   PRINT X"/"Y" + "Z"/"W" = "A"/"B
100  END
``` |

See the Computer Section beginning on page 420 for more information.

Exercises

Find each sum on your own. Then use the program above to check.

1. $\frac{4}{7} + \frac{1}{3}$ $\frac{19}{21}$ $\qquad$ **2.** $\frac{2}{9} + \frac{3}{5}$ $\frac{37}{45}$ $\qquad$ **3.** $\frac{4}{5} + \frac{1}{3}$ $\frac{17}{15}$ $\qquad$ **4.** $\frac{3}{8} + \frac{1}{2}$ $\frac{14}{16}$

Notice that in Exercise 4 above, the result is not simplified. Modify the program in Example 1 above by adding the program from Chapter 4 on page 106. Use this new program to add each of the following and simplify the results.

See TE Answer Section for program.

5. $\frac{3}{8} + \frac{1}{2}$ $\frac{7}{8}$ $\qquad$ **6.** $\frac{1}{10} + \frac{1}{15}$ $\frac{1}{6}$ $\qquad$ **7.** $\frac{5}{12} + \frac{1}{4}$ $\frac{2}{3}$ $\qquad$ **8.** $\frac{3}{20} + \frac{3}{5}$ $\frac{3}{4}$
```

Find the least common multiple (LCM). [114]

1. 6; 15   30

2. 4; 10   20

3. 4; 6; 8   24

4. 3; 5; 30   30

Add or subtract. Simplify if possible.

5. [111] $\frac{3}{10} + \frac{1}{10}$ $\frac{2}{5}$

6. [111] $\frac{7}{12} - \frac{5}{12}$ $\frac{1}{6}$

7. [116] $\frac{3}{5} + \frac{5}{15}$ $\frac{14}{15}$

8. [116] $\frac{3}{5} - \frac{1}{3}$ $\frac{4}{15}$

9. [116]
$$\frac{11}{6}$$
$$-\frac{2}{3}$$ $1\frac{1}{6}$

10. [116]
$$\frac{3}{10}$$
$$+\frac{4}{5}$$ $1\frac{1}{10}$

11. [116]
$$\frac{17}{6}$$
$$-\frac{1}{2}$$ $2\frac{1}{3}$

12. [116]
$$\frac{5}{14}$$
$$+\frac{1}{7}$$ $\frac{1}{2}$

13. [119]
$$\frac{5}{6}$$
$$+\frac{5}{9}$$ $1\frac{7}{18}$

14. [119]
$$\frac{7}{10}$$
$$+\frac{3}{4}$$ $1\frac{9}{20}$

15. [119]
$$\frac{7}{10}$$
$$-\frac{1}{4}$$ $\frac{9}{20}$

16. [119]
$$\frac{5}{6}$$
$$-\frac{3}{4}$$ $\frac{1}{12}$

17. [122]
$$7\frac{3}{10}$$
$$+3\frac{1}{10}$$ $10\frac{2}{5}$

18. [122]
$$6\frac{3}{4}$$
$$-4\frac{1}{6}$$ $2\frac{7}{12}$

19. [122]
$$7\frac{1}{8}$$
$$-5$$ $2\frac{1}{8}$

20. [122]
$$9$$
$$+4\frac{1}{3}$$ $13\frac{1}{3}$

21. [122] $9\frac{1}{3} - 6$ $3\frac{1}{3}$

22. [122] $7\frac{2}{5} + 1\frac{3}{10} + 4\frac{1}{2}$ $13\frac{1}{5}$

23. [126] $9\frac{1}{10} - 4\frac{3}{5}$ $4\frac{1}{2}$

Evaluate. Simplify if possible. [129]

24. $\frac{5}{12} + \frac{6x - 8}{12}$ if $x = 2$ $\frac{3}{4}$

25. $\frac{2t + 1}{20} + \frac{1}{5}$ if $t = 5$ $\frac{3}{4}$

Solve. Simplify if possible. [132]

26. $\frac{3}{2}x + 5 = 9$ $2\frac{2}{3}$

27. $4x - 6 = 12$ $4\frac{1}{2}$

28. $\frac{4}{9}x - 5 = 1$ $13\frac{1}{2}$

Solve these problems. [125]

29. John jogs $5\frac{1}{2}$ laps, rests, then jogs $3\frac{1}{4}$ laps. How many laps does he jog altogether? $8\frac{3}{4}$ laps

30. Mary weighs $170\frac{1}{2}$ lb. How many pounds must she lose in order to weigh $167\frac{1}{4}$ lb? $3\frac{1}{4}$ kg

Chapter Test

Find the least common multiple (LCM).

1. 4; 6 12

2. 3; 5 15

3. 6; 10; 15 30

Add or subtract. Simplify if possible.

4. $\frac{5}{8} - \frac{1}{8}$ $\frac{1}{2}$

5. $\frac{4}{9} + \frac{5}{9}$ 1

6. $\frac{1}{5} + \frac{2}{3}$ $\frac{13}{15}$

7. $\begin{array}{r} \frac{11}{6} \\ -\frac{1}{3} \end{array}$ $1\frac{1}{2}$

8. $\begin{array}{r} \frac{3}{14} \\ +\frac{1}{2} \end{array}$ $\frac{5}{7}$

9. $\begin{array}{r} \frac{1}{6} \\ +\frac{2}{9} \end{array}$ $\frac{7}{18}$

10. $\begin{array}{r} \frac{3}{4} \\ -\frac{1}{6} \\ \hline \frac{7}{12} \end{array}$

11. $\frac{7}{12} + \frac{2}{3} + \frac{1}{2}$ $1\frac{3}{4}$

12. $\frac{1}{10} + \frac{1}{5} + \frac{1}{2}$ $\frac{4}{5}$

13. $\begin{array}{r} 9\frac{7}{10} \\ +2\frac{1}{10} \end{array}$ $11\frac{4}{5}$

14. $\begin{array}{r} 6\frac{3}{4} \\ -5\frac{1}{6} \end{array}$ $1\frac{7}{12}$

15. $\begin{array}{r} 8\frac{2}{3} \\ -5 \end{array}$ $3\frac{2}{3}$

16. $\begin{array}{r} 7 \\ +5\frac{3}{4} \\ \hline 12\frac{3}{4} \end{array}$

17. $8\frac{1}{2} + 1\frac{7}{10} + 6\frac{1}{5}$ $16\frac{2}{5}$

18. $8\frac{2}{5} + 4\frac{7}{15} + 7\frac{1}{3}$ $20\frac{1}{5}$

Evaluate. Simplify if possible.

19. $\frac{2x+1}{7} + \frac{3}{7}$ if $x = 4$ $1\frac{5}{7}$

20. $\frac{2m-1}{28} + \frac{1}{4}$ if $m = 3$ $\frac{3}{7}$

Solve. Simplify if possible.

21. $\frac{3}{4}x + 2 = 8$ 8

22. $\frac{4}{5}x - 2 = 3$ $6\frac{1}{4}$

23. $5x - 2 = 7$ $1\frac{4}{5}$

Solve these problems.

24. Mona jogs $4\frac{1}{4}$ laps, rests, then jogs $8\frac{1}{2}$ laps. How many laps does she jog altogether? $12\frac{3}{4}$ laps

25. Martin weighs $175\frac{3}{4}$ lb. How much will he weigh if he loses $9\frac{1}{2}$ lb? $166\frac{1}{4}$ lb

Organizing Data

6

The Statue of Freedom tops the 287 ft Capitol Dome in Washington, D.C.

Reading and Making Tables

To read a table
To make a table

TEST SCORE	
Eric	97
Mort	88
Dawn	99 ←——Highest Test Score

Example 1

What tax would you pay if your taxable income was between $22,250 and $22,300 and you are single? Between $22,700 and $22,750 and are married filing jointly?

Single

Taxable Income: $22,250–$22,300

1. Find $22,250–$22,300 in the left column.
2. Find *Single* at the top.
3. Find the intersection of the horizontal line from the left column and the vertical line from the top.
4. Your tax is $4,889.

Married filing jointly

Taxable Income: $22,700–$22,750

1. Find $22,700–$22,750 in the left column.
2. Find *Married filing jointly* at the top.
3. Find the intersection of the horizontal and vertical lines.
4. Your tax is $3,930.

TAX TABLE

If line 34 (taxable income) is—		And you are—			
At least	But less than	Single	Married filing jointly	Married filing separately	Head of a household
			Your tax is—		
22,000	22,050	4,805	3,737	5,909	4,438
22,050	22,100	4,821	3,751	5,930	4,453
22,100	22,150	4,838	3,764	5,951	4,468
22,150	22,200	4,855	3,778	5,973	4,483
22,200	22,250	4,872	3,792	5,994	4,499
22,250	22,300	4,889	3,806	6,015	4,514
22,300	22,350	4,905	3,820	6,036	4,529
22,350	22,400	4,922	3,833	6,058	4,545
22,400	22,450	4,939	3,847	6,079	4,560
22,450	22,500	4,956	3,861	6,100	4,575
22,500	22,550	4,973	3,875	6,121	4,591
22,550	22,600	4,989	3,889	6,142	4,606
22,600	22,650	5,006	3,903	6,164	4,621
22,650	22,700	5,023	3,916	6,185	4,637
22,700	22,750	5,040	3,930	6,206	4,652
22,750	22,800	5,056	3,944	6,227	4,667
22,800	22,850	5,073	3,958	6,249	4,682
22,850	22,900	5,090	3,972	6,270	4,698
22,900	22,950	5,107	3,986	6,293	4,713
22,950	23,000	5,124	3,999	6,317	4,728

So, the tax on taxable income between $22,250 and $22,300 and Single is $4,889. The tax on taxable income between $22,700 and $22,750 and Married filing jointly is $3,930.

practice ▷ **Determine the tax using the tax table in Example 1.**

1. Single
   Taxable income:
   $22,500–$22,550 $4,973

2. Married filing separately
   Taxable income:
   $22,850–$22,900 $6,270

It is often useful to make a table to display information.

Example 2

Make a table for the following information. Assume each fruit contains 100 edible grams. An apple contains 80 calories, 20 grams of carbohydrates, and 120 units of vitamin A. A banana contains 100 calories, 26 grams of carbohydrates, and 230 units of vitamin A. A $\frac{1}{2}$ cantaloupe contains 80 calories, 20 grams of carbohydrates, and 9,240 units of vitamin A.

1. Label the vertical scale with the names of the fruits.

2. Label the horizontal scale with the composition of the fruits.

3. Draw vertical and horizontal lines separating fruits and their composition.

4. Put data in the appropriate boxes.

5. Give the table a title.

COMPOSITION OF FOODS

	Calories	Carbohydrates (grams)	Vitamin A (units)
Apple	80	20	120
Banana	100	26	230
$\frac{1}{2}$ Cantaloupe	80	20	9,240

practice ▷ **Make a table from the following information.**   Check students' tables.

3. At 5% interest, $1,000 would grow to $1,051 at the end of 1 year; $1,105 in 2 years, $1,161 in 3 years; and $1,221 in 4 years. At 6%, it would grow to $1,061; $1,127; $1,197; and $1,271 respectively.

READING AND MAKING TABLES

◇ *ORAL EXERCISES* ◇

Answer the exercises below. Use the bus schedule.

1. When will the 6:20 Court St. bus arrive at Morgan Rd. and Ross Park? 6:34
2. When will the 7:50 Court St. 8:20 bus arrive back at Court St.?
3. A passenger leaves Telegraph St. and Conklin Ave. at 7:25. When will he get to Vestal Ave. and Park Ave.? 7:40
4. A passenger leaves James St. and Washington St. at 6:03. 6:20 When will she get to Court St.?

Leave Court St.	Telegraph St. and Conklin Ave.	James St. and Washington St.	Arrive Morgan Rd. and Ross Park	Leave Morgan Rd. and Ross Park	Vestal Ave. and Park Ave.	Arrive Court St.
5:55	5:59	6:03	6:07	6:07	6:15	6:20
6:20	6:25	6:29	6:34	6:35	6:40	6:50
6:50	6:55	6:59	7:04	7:05	7:10	7:20
7:20	7:25	7:29	7:34	7:35	7:40	7:50
7:50	7:55	7:59	8:04	8:05	8:10	8:20

◇ *EXERCISES* ◇

Answer Exercises 1–4 below. Use the table.

1. What tax will you pay if your taxable income is $19,225 and you are single? $3,865

2. What tax will you pay if your taxable income is $19,455 and you are married filing jointly? $3,060

3. What tax will you pay if your taxable income is $19,048 and you are the head of a household? $3,519

4. What tax will you pay if your taxable income is $19,275 and you are married filing separately? $4,741

If line 34 (taxable income) is—		And you are—			
At least	But less than	Single	Married filing jointly	Married filing separately	Head of a household
			Your tax is—		
19,000					
19,000	19,050	3,797	2,954	4,635	3,519
19,050	19,100	3,814	2,965	4,656	3,535
19,100	19,150	3,831	2,977	4,678	3,550
19,150	19,200	3,848	2,989	4,699	3,565
19,200	19,250	3,865	3,001	4,720	3,580
19,250	19,300	3,881	3,013	4,741	3,596
19,300	19,350	3,898	3,025	4,762	3,611
19,350	19,400	3,915	3,037	4,784	3,626
19,400	19,450	3,932	3,048	4,805	3,642
19,450	19,500	3,949	3,060	4,826	3,657

Algebra Maintenance

Solve.

1. $3x = 48$ 16
2. $x - 12 = 7$ 19
3. $24 = 5z - 1$ 5
4. $x + \frac{1}{2} = \frac{3}{4}$ $\frac{1}{4}$
5. $3 = k - 2\frac{1}{2}$ $5\frac{1}{2}$
6. $3x - 1 = 4$ $1\frac{2}{3}$
7. $\frac{3}{4}r + 4 = 5$ $1\frac{1}{3}$
8. $\frac{5}{6}t - 4 = 1$ 6
9. $5k + 9 = 17$ $1\frac{3}{5}$

Pictographs

◇ **OBJECTIVE** ◇

To read and make pictographs

◇ **RECALL** ◇

$$4\frac{1}{2} \cdot 60 = \frac{9}{2} \cdot \frac{60}{1}, \text{ or } 270$$

Example 1

Pictures are used to show
student enrollment.

The pictograph shows student enrollment in grades 10–12. How many are in each grade?

ENROLLMENT IN MADISON HIGH SCHOOL

Grade	Number of Students
10	⚲ ⚲ ⚲ ⚲ ⚲
11	⚲ ⚲ ⚲ ⚲ ⚲
12	⚲ ⚲ ⚲ ⚲

Key: Each ⚲ represents 100 students.

$4\frac{1}{2} \cdot 100 = \frac{9}{2} \cdot 100, \text{ or } 450$

Grade 10: 5 ⚲ means $5 \cdot 100$, or 500

Grade 11: $4\frac{1}{2}$ ⚲ means $4\frac{1}{2} \cdot 100$, or 450

Grade 12: 4 ⚲ means $4 \cdot 100$, or 400

So, there are 500 students in grade 10; 450 in grade 11; and 400 in grade 12.

Example 2

Make a pictograph for the following table.

NEW HOUSES

Jan. – Mar.	Apr. – June	July – Sept.	Oct. – Dec.
60	45	55	30

Title the pictograph.
Put heads on the columns.

For 60 houses, use 6 🏠 .

For 45 houses, use $4\frac{1}{2}$ 🏠.

For 55 houses, use $5\frac{1}{2}$ 🏠 .

For 30 houses, use 3 🏠 .
Determine key.

Choose a symbol to represent houses.
Use 🏠 to represent 10 new houses.

NEW HOUSES

Months	Number of New Houses
Jan.-Mar.	🏠 🏠 🏠 🏠 🏠 🏠
Apr.-June	🏠 🏠 🏠 🏠 🏠
July-Sept.	🏠 🏠 🏠 🏠 🏠 🏠
Oct.-Dec.	🏠 🏠 🏠

Key: Each 🏠 represents 10 new houses.

Make a pictograph.
Check students' graphs.

ATTENDANCE AT SOCCER GAMES

Game	Number of Fans
1	6,000
2	7,500
3	7,000
4	4,000

◇ *ORAL EXERCISES* ◇

ꝰ **represents 100 people. How many** ꝰ **would you need to show each?**

1. 400 people 4
2. 350 people $3\frac{1}{2}$
3. 1,000 people 10
4. 150 people $1\frac{1}{2}$

◇ *EXERCISES* ◇

1. How many birthdays in Jan.? 50
2. How many birthdays in Feb.? 45
3. How many birthdays in Mar.? 30
4. How many birthdays in Apr.? 40
5. What was the total number of birthdays for the 4 months? 165

BIRTHDAYS IN CHEROKEE GRADE SCHOOL

Month	Number of Birthdays
Jan.	
Feb.	
Mar.	
Apr.	

Key: Each 🎂 represents 10 birthdays.

6. How many boxes sold by the ninth grade? 100
7. How many boxes sold by the tenth grade? 90
8. How many boxes sold by the eleventh grade? 110
9. How many boxes sold by the twelfth grade? 120

SCHOOL STATIONERY SALE

Grade	Number of Boxes Sold
9	
10	
11	
12	

Key: Each 📦 represents 20 boxes.

Make a pictograph. Check students' graphs.

10.
WORLD SERIES ATTENDANCE

Day	Number of Fans
Friday	50,000
Saturday	65,000
Monday	70,000
Tuesday	60,000
Wednesday	75,000

★ 11.
STAMP COLLECTIONS

Name	Number of Stamps
Josh	175
Randi	225
Mario	150
Carol	100

Problem Solving – Applications
Jobs for Teenagers

Example 1

Jason worked 2 hours cutting a lawn. He charged $3.75 per hour. His expenses were $0.50 per hour for gas and oil and $0.90 for equipment costs and repairs. How much was his net profit?

Net Profit = Amount Charged − Expenses

$2 \cdot \$3.75 = \$7.50 \qquad 2 \cdot \$0.50 + \$0.90 = \$1.90$

Net Profit = $7.50 − $1.90 = $5.60

So, Jason's net profit was $5.60.

$$\begin{array}{r} \$7.50 \\ -1.90 \\ \hline \$5.60 \end{array}$$

Example 2

One year, Wanda used 6 bags of fertilizer, $1\frac{1}{2}$ bags of seed, and 4 bags of limestone. The prices follow: fertilizer $8.95 per bag, seed $8.50 per bag, limestone $1.25 per bag. What was the total cost for the actual amounts used?

Total cost = fertilizer + seed + limestone.

Fertilizer $8.95 × 6 = $53.70

Seed $8.50 × $1\frac{1}{2}$ = $8.50 \times \frac{3}{2} = \frac{25.50}{2}$, or $12.75

Limestone $1.25 × 4 = $5.00

$53.70 + $12.75 + $5.00 = $71.45

So, the total cost for the amounts used was $71.45.

$$\begin{array}{r} \$53.70 \\ 12.75 \\ +5.00 \\ \hline \$71.45 \end{array}$$

Solve these problems.

1. Gale took 3 hours to cut the O'Neals' lawn. She charged $3.90 per hour. Her expenses were 60 cents per hour for gas and oil and 50 cents per hour for equipment costs and repairs. How much did she net? $8.40

2. Keith took 4 hours to cut the Issas' lawn. He charged $2.60 per hour. He used their lawnmower and gas. How much did he earn? $10.40

3. Carol purchased fertilizer at $9 per bag, limestone at $2 per bag, seed at $6.50 per bag, and weed killer at $4 per bottle. In one year she used 3 bags of fertilizer, $\frac{1}{2}$ bag of seed, $2\frac{1}{2}$ bags of limestone, and $\frac{1}{4}$ bottle of weed killer. What was the total cost for the actual amounts used? $36.25

4. Miguel purchased fertilizer at $12 per bag, seed at $8.50 per bag, limestone at $2.50 per bag, and weed killer at $4.20 per bottle. In one year he used bags of fertilizer, 1 bag of seed, $4\frac{1}{2}$ bags of limestone, and $\frac{1}{4}$ bottle of weed killer. What was the total cost for the actual amounts used? $44.80

5. Randy spent 8 hours working with a rototiller. He charged $6.50 per hour. His expenses were $20 for renting the rototiller and 50 cents per hour for gas. How much was his net profit? $28.00

6. Reiko worked 6 hours digging gardens with a rototiller. She charged $6.50 per hour. Her expenses were $4 per hour for the rototiller and 40 cents per hour for gas. How much was her net profit? $12.60

7. Karen took 3 hours to cut a lawn. She charged $4.10 per hour. Her expenses were 50 cents per hour for gas and oil and 40 cents per hour for equipment costs and repairs. How much did she net? $9.60

8. Ralph took $1\frac{1}{2}$ hours to cut the Burns' lawn. He charged $3 per hour. He had no expenses. How much did he earn? $4.50

9. Wally purchased fertilizer at $9.50 per bag, limestone at $2.25 per bag, seed at $6 per bag, and weed killer at $4.50 per bottle. In one year he used $5\frac{1}{2}$ bags of fertilizer, $1\frac{1}{2}$ bags of seed, 3 bags of limestone, and $\frac{1}{2}$ bottle of weed killer. Find the total cost for the actual amounts used. $70.25

10. Roz purchased fertilizer at $10 per bag, limestone at $1.75 per bag, seed at $7.25 per bag, and weed killer at $3.50 per bottle. In one year she used $4\frac{1}{2}$ bags of fertilizer, 1 bag of seed, 2 bags of limestone, and $\frac{1}{5}$ bottle of weed killer. What was the total cost of the actual amounts used? $56.45

11. Carmen spent 4 hours working with a rototiller. She charged $6 per hour. Her expenses were $10 for renting the rototiller and 40 cents per hour for gas. How much was her net profit? $12.40

12. Michael worked $7\frac{1}{2}$ hours landscaping the Paisley's lawn. He charged $10.50 per hour, plus bus fare. If his total bus fare was $2.95, how much must the Paisley's pay him? $81.70

Bar Graphs

To read and make vertical bar
  graphs
To read and make horizontal bar
  graphs

$$25,000 = 25 \text{ thousand}$$
$$110,000,000 = 110 \text{ million}$$

Example 1

The graph is a vertical bar graph.

Number of babies born:
  in Sept.: 40
    Oct.: 30
    Nov.: 15
    Dec.: 35

How many babies were born each month at
Northside Hospital?

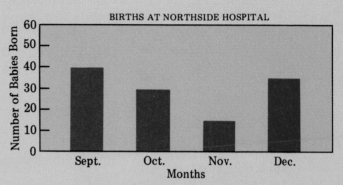

So, there were 40 babies born in Sept., 30 in Oct.,
15 in Nov., and 35 in Dec.

Example 2

The graph is a horizontal bar graph.

Number of yearbooks sold:
  grade 10: 80
  grade 11: 130
  grade 12: 200

How many yearbooks were sold by each grade?

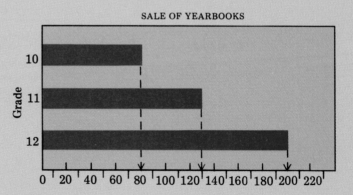

So, 80 yearbooks were sold by the tenth grade, 130
by the eleventh grade, and 200 by the twelfth
grade.

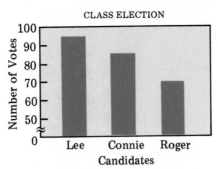

CLASS ELECTION

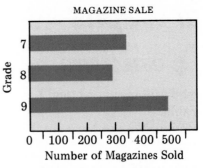

MAGAZINE SALE

1. How many votes did each candidate get? Lee 95; Connie 85; Roger 70

2. How many magazines were sold by each grade? 7th: 350; 8th: 300; 9th: 500

Example 3

Make a bar graph to show the approximate U.S. population.
1940, 130,000,000; 1950, 150,000,000;
1960, 180,000,000; 1970, 200,000,000

This is a vertical bar graph. Give the graph a title.

Label the vertical scale with population in millions. 130,000,000 = 130 million

Label the horizontal scale with the years. Draw the bars, and give the graph a title.

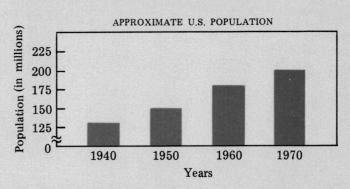

This is a horizontal bar graph. Give the graph a title.

Label the vertical scale with the years.

Label the horizontal scale with the population in millions. Draw the bars, and give the graph a title.

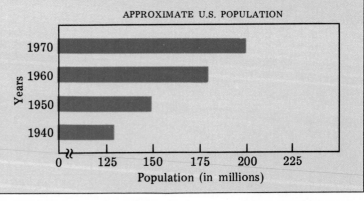

practice ▷ 3. Make a vertical bar graph and a horizontal bar graph to show the average yearly rainfall.
Atlanta, 130 cm; Los Angeles, 30 cm; Philadelphia, 100 cm; Minneapolis, 50 cm   Check students' graphs

How many fans attended the games on each day?

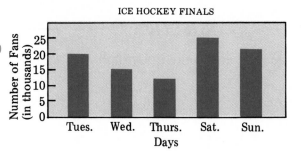

ICE HOCKEY FINALS

1. on Tues. 20,000
2. on Wed. 15,000
3. on Thurs. 12,500
4. on Sat. 25,000
5. on Sun. 22,500
6. On which day did the greatest number of fans attend? Saturday
7. On which day did the fewest number of fans attend? Thursday

◇ EXERCISES ◇

What is the life expectancy of each?

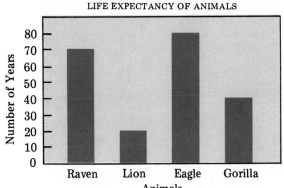

LIFE EXPECTANCY OF ANIMALS

1. the raven 70 yr
2. the lion 20 yr
3. the eagle 80 yr
4. the gorilla 40 yr
5. Which animal has the longest life expectancy? the eagle
6. Which animal has the shortest life expectancy? the lion

Answer the exercises below. Use the graph at the right.

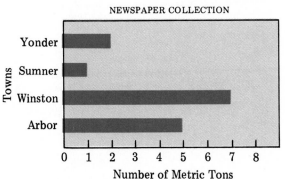

NEWSPAPER COLLECTION

Yonder 2t; Sumner 1t,
7. How much newspaper did each town collect? Winston 7t, Arbor 5t
8. Which town collected the most newspaper? Winston
9. Which town collected the least newspaper? Sumner
10. How much did all four towns collect together? 15t
11. Make a vertical bar graph to show these basketball scores.
    Game 1, 90 points; Game 2, 85 points; Game 3, 95 points;
    Game 4, 80 points
12. Make a horizontal bar graph to show the average yearly rainfall.
    Buffalo, 100 cm; Salt Lake City, 40 cm; Miami, 125 cm;
    Juneau, 160 cm   Check students' graphs for 11 and 12.

Line Graphs

<table>
<tr><td colspan="5">◇ OBJECTIVE ◇   ◇ RECALL ◇</td></tr>
</table>

To read and make line graphs

Date	Withdrawal	Deposit	Balance	
Jan. 7			$12.50	}increase
Jan. 15		$2.25	$14.75	}decrease
Jan. 24	$1.25		$13.50	}increase
Jan. 30		$2.50	$16.00	

Example 1

Answer the exercises below. Use the line graph.

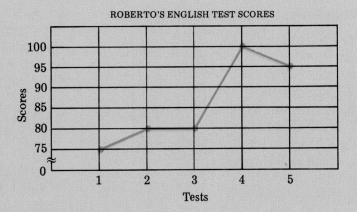

ROBERTO'S ENGLISH TEST SCORES

To find the score on a test, read up from the test number to the dot, then read across to the score.

Highest score: 100
Lowest score: − 75
Difference: 25

Test 3: 80 }increase
Test 4: 100

1. On which test did Roberto get the highest score?
   Roberto scored the highest on Test 4.
2. What is the difference between the highest and lowest scores?
   The difference between the highest and lowest scores is 25.
3. Did the scores increase, decrease, or stay the same from Test 3 to Test 4?
   The score increased from Test 3 to Test 4.

---

practice ▷   **Use the graph in Example 1. Did the scores increase, decrease, or stay the same?**

1. From Test 1 to Test 2   increased
2. From Test 2 to Test 3   stayed the same
3. From Test 4 to Test 5   decreased

Example 2

Make a line graph to show the average daily high temperatures. Mon., 12°; Tues., 15°; Wed., 12°; Thurs., 14°; Fri., 16°

Use graph paper or make a grid.
Give the graph a title.

Label the vertical scale with temperature in degrees.
Label the horizontal scale with days.
Draw dots to show the daily high temperatures.

Connect the dots.

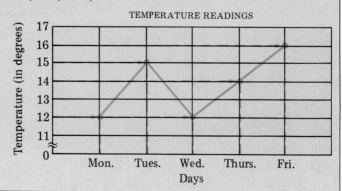

practice ▷ **4.** Make a line graph to show how many books were sold each day.
Mon., 25; Tues., 20; Wed., 15; Thurs., 20; Fri., 30   Check students' graphs.

◇ ORAL EXERCISES ◇

What was the time for the men's 100-meter dash for each year?

1. 1964 10.0 sec
2. 1968 9.9 sec
3. 1972 10.1 sec
4. 1976 10.1 sec
5. 1980 10.2 sec

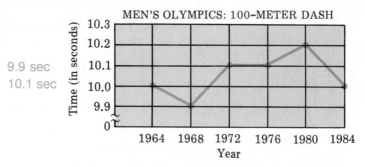

◇ EXERCISES ◇

Answer the exercises below. Use the line graph at the right.

1. In what month did Marcy read the greatest number of books? July

2. In what month did Marcy read the least number of books? October

3. In what months did Marcy read the same number of books? June and Sept.

4. How many books did Marcy read during the five months? 28

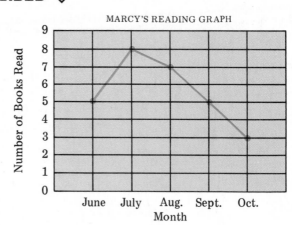

What was the approximate time for women's downhill skiing for each year?

WOMEN'S OLYMPICS: DOWNHILL SKIING

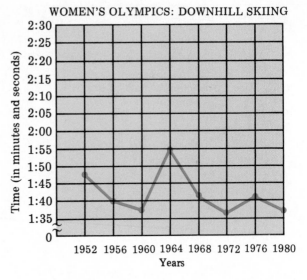

5. 1952   1:48
6. 1956   1:40
7. 1960   1:38
8. 1964   1:55
9. 1968   1:41
10. 1972   1:37
11. 1976   1:46
12. 1980   1:38
13. What is the difference in the times of 1964 and 1972?   0:18
14. Make a line graph to show the average monthly rainfall.

    July, 18 cm; Aug., 16 cm;

    Sept., 23 cm; Oct., 20 cm; Nov., 8 cm;

    Dec., 4 cm   Check students' graphs.

Does the amount of rainfall increase or decrease? Use the graph you made in Exercise 14.

15. From July to Aug.?   decrease
16. From Aug. to Sept.?   increase
17. From Oct. to Nov.?   decrease

18. Make a line graph to show the price of a stock during a one-week period.   Check students' graphs.

    Mon., $4\frac{1}{2}$; Tues., $4\frac{3}{8}$; Wed., $4\frac{5}{8}$; Thurs., $5\frac{1}{8}$; Fri., $4\frac{3}{4}$

Did the price of the stock increase or decrease? Use the graph you made in Exercise 18.

19. From Mon. to Tues.?   decrease
20. From Wed. to Thurs.?   increase
21. From Thurs. to Fri.?   decrease

Challenge

$$1 \cdot 91 = \ \ 91$$
$$2 \cdot 91 = 182$$
$$3 \cdot 91 = 273$$
$$4 \cdot 91 = 364$$

Do you see a pattern? Use the pattern to find these products.

$$5 \cdot 91 = \underline{\ \ 455\ \ }$$
$$6 \cdot 91 = \underline{\ \ 546\ \ }$$
$$7 \cdot 91 = \underline{\ \ 637\ \ }$$
$$8 \cdot 91 = \underline{\ \ 728\ \ }$$
$$9 \cdot 91 = \underline{\ \ 819\ \ }$$

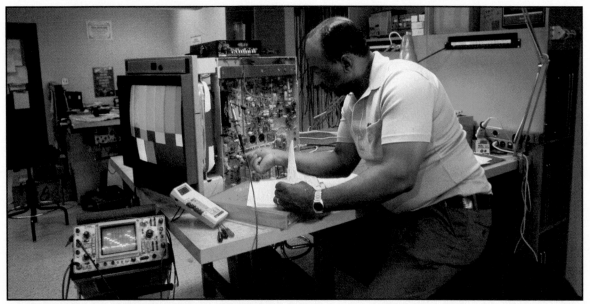

Problem Solving – Careers
Appliance-Repair Specialists

1. Marco's Appliance Repair specializes in fixing small appliances. Last year, Marco's repaired 4,721 appliances. About how many appliances were repaired per week? about 91 appliances/wk
2. Chin's Major Appliance Service repairs refrigerators, washers, and dryers. Lucy Chin charges $25 for a house call and $16 per hour for labor. It took her 45 minutes to repair the Sragows' washer. How much did she charge if the new part cost $8.39? $45.39
3. Ira Kaplan repairs small appliances. He repairs about 50 appliances per week. He makes $250 profit for each appliance. About how much does he earn in a year? about $19,500/yr

Answer the exercises below. Use the paycheck stub.

Fitzpatrick Appliance Repair Co.						
Payroll statement for George M. Manitzas						Weekly Statement
1. 25 h Regular Hours	2. 2h 15 min Overtime Hours	3. 293.92 Gross Pay	4. 28.46 Withholding Tax	5. 10.34 State Tax	6. 4.83 City Tax	7. 18.02 Social Security
8. Bonds Savings 6.50	9. Credit Union Savings 2.00	10. Dues 4.00	11. United Fund 2.00	12. Insurance 2.40	13. Net Pay	14. End of Pay Period 10/24/86
Detach stub before cashing.						Retain this record for tax purposes.

4. What is the net pay for the pay period ending on 10/24/86? $215.37
5. What is the approximate net pay for the year? approx. $11,199.24
6. What are the credit union savings for the year? $104
7. What are the approximate taxes for the year? approx. $2,268.76

Using Graphs

To solve problems by using graphs

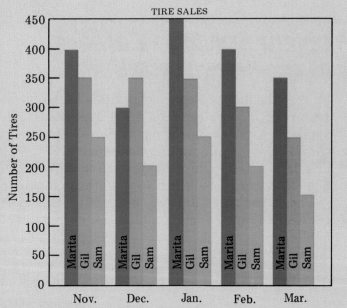

SNOWFALL

Day	Number of Centimeters
Mon.	4
Tues.	2
Wed.	0
Thurs.	6
Fri.	5

Example 1

Add across to find the number of tires sold by each.
Add down to find the number of tires sold each month.

Copy and complete the table below. Use the bar graph.

TIRE SALES

	Nov.	Dec.	Jan.	Feb.	Mar.	Total
Marita	400	300	450	400	350	1,900
Gil	350	350	350	300	250	1,600
Sam	250	200	250	200	150	1,050
Total	1,000	850	1,050	900	750	4,550

Example 2

Marita sold 1,900 tires.

Sam sold 1,050 tires.

1,050 tires were sold in Jan.

Gil sold 350 tires in Dec.

Answer the questions. Use the table in Example 1.
1. Who sold the most tires?
   Marita sold the most tires.
2. Who sold the fewest tires?
   Sam sold the fewest tires.
3. In which month were the most tires sold?
   The most tires were sold in January.
4. Who sold the most tires in December?
   Gil sold the most tires in December.

practice ▷ **Use the table in Example 1.**

1. Who sold the most tires in Mar.?   Marita

2. Who sold the fewest tires in Jan.?   Sam

Example 3

Answer the exercises below. Use the graph.

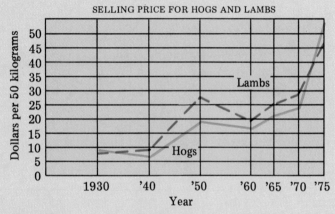

SELLING PRICE FOR HOGS AND LAMBS

Key: Hogs ——— Lambs — — — —

Look for highest point on the graph.

Look at the solid line graph.

1. In what year did the farmer get the best price for hogs and lambs?
   The farmer got the best price for hogs and lambs in 1975.
2. What was the selling price for hogs in 1930?
   The selling price for hogs in 1930 was about $9 per 50 kilograms.

practice ▷ **Use the graph in Example 3.**

3. In what year did the farmer get the poorest price for hogs and lambs?   1940, hogs; 1930, lambs

4. What was the selling price for lambs in 1950? $27/50 kg

◇ EXERCISES ◇

Use the bar graph for Exercises 1–7.

1. Copy and complete the table.

2. Which store had the greatest sales?
   Store 2

3. Which store had the least sales?
   Store 3

4. In which year were sales the greatest?
   1985

5. In which year were sales the worst?
   1982

6. Which store's sales increased each year?
   Store 3

7. What were the total sales for all three stores? $1,424,000

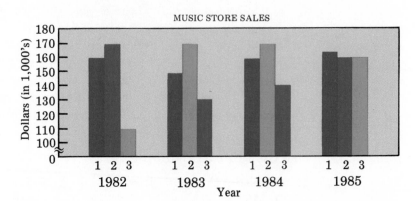

MUSIC STORE SALES

	1982	1983	1984	1985	Total
Store 1	$160,000	$150,000	$160,000	$165,000	$475,000
Store 2	$170,000	$170,000	$170,000	$179,000	$519,000
Store 3	$110,000	$130,000	$140,000	$160,000	$430,000
Total	$440,000	$450,000	$470,000	$504,000	$1,424,000

Answer the exercises below. Use the graph.

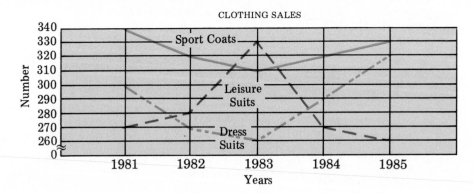

8. In what year were the most sport coats sold?   1981

9. In what year were the least number of leisure suits sold?   1985

10. What was the number of dress suits sold in 1984?   290

11. What was the total number of sport coats, dress suits, and leisure suits sold in 1982?   870

156   *CHAPTER SIX*

Applying Equations

◇ **OBJECTIVE** ◇

To solve perimeter problems
using equations

◇ **RECALL** ◇

Perimeter means the distance around.

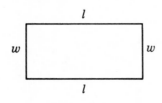

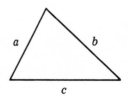

$$P = l + w + l + w$$
$$P = 2l + 2w$$

$$P = a + b + c$$

See Geometry Supplement on page 420 for more information.

Example 1

The length of a rectangle is 19 cm and the
perimeter is 62 cm. Find the width.

$$P = 2l + 2w$$

Substitute 62 for P, 19 for l. $\quad 62 = 2 \cdot 19 + 2w$

$$62 = 38 \quad + 2w$$

Subtract 38 from each side. $\quad 62 - 38 = 38 - 38 + 2w$

$$24 = 2w$$

Divide each side by 2. $\quad \dfrac{24}{2} = \dfrac{2w}{2}$

$$12 = w, \text{ or } w = 12$$

So, the width is 12 cm.

---

practice ▷

1. The length of a rectangle is
15 cm and the perimeter is 40
cm. Find the width. 5 cm

2. The width of a rectangle is 8
cm and the perimeter is 58
cm. Find the length. 21 cm

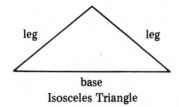

leg   leg

base
Isosceles Triangle

A triangle is *isosceles* if two of the sides have the
same length (are congruent). The two congruent
sides are called *legs,* and the third side is called the
base.

Example 2

The length of each leg of an isosceles triangle is 10 cm and the perimeter is 25 cm. Find the length of the base, b.

$$P = a + b + c$$

Substitute 10 for a, 10 for c, and 25 for P.

$$25 = 10 + b + 10$$
$$25 = 20 + b$$

Subtract 20 from each side.

$$25 - 20 = 20 - 20 + b$$
$$5 = b, \text{ or } b = 5$$

So, the length of the base is 5 cm.

practice ▷
3. The length of each leg of an isosceles triangle is 12 cm, and the perimeter is 30 cm. Find the length of the base.
6 cm

4. The base of an isosceles triangle is 18 cm, and the perimeter is 42 cm. Find the length of each leg. 12 cm

◇ ORAL EXERCISES ◇

Find the perimeter.

1.
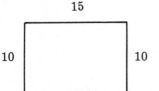
P = 50

2.
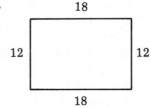
P = 60

3.
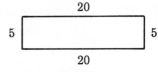
P = 50

4.
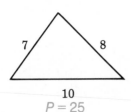
P = 25

5.
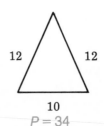
P = 34

6.
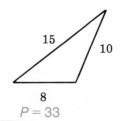
P = 33

◇ EXERCISES ◇

Find the missing side.

1. $46 = 2 \cdot 15 + 2w$ 8
3. $108 = 2l + 2 \cdot 22$ 32

2. $62 = 2l + 2 \cdot 10$ 21
4. $180 = 2 \cdot 61 + 2w$ 29

Solve these problems.

5. The width of a rectangle is 14 cm, and the perimeter is 98 cm. Find the length.   35 cm

6. The length of a rectangle is 17 cm, and the perimeter is 38 cm. Find the width.   2 cm

7. The length of each leg of an isosceles triangle is 12 cm, and the perimeter is 34 cm. Find the base, b.   10 cm

8. The base of an isosceles triangle is 20 cm, and the perimeter is 42 cm. Find the length of each leg.   11 cm

9. The base of an isosceles triangle is 15 cm, and the perimeter is 35 cm. Find the length of each leg.   10 cm

10. The length of each leg of an isosceles triangle is 17 cm, and the perimeter is 62 cm. Find the base.   28 cm

11. The length of two sides of a triangle are 16 cm and 14 cm, and the perimeter is 42 cm. Find the length of the third side.   12 cm

12. The length of two sides of a triangle are 8 cm and 7 cm, and the perimeter is 19 cm. Find the length of the third side.   4 cm

★ 13. The perimeter of an equilateral triangle (3 congruent sides) is 33 cm. Find the length of each side.   11 cm

★ 14. The perimeter of an equilateral triangle is 72 m. Find the length of each side.   24 m

★ 15. The length of a rectangle is 4 m more than 3 times the width. The perimeter is 32 m. Find the length and width. l = 13m; w = 3m

★ 16. The width of a rectangle is one half the length. The perimeter is 24 m. Find the length and width. l = 8m; w = 4m

★ 17. The first side of a triangle is 2 cm longer than the second side. The third side is 3 cm shorter than twice the second side. The perimeter is 15 cm. How long is each side?   1st side: 6 cm; 2nd side: 4cm; 3rd side: 5 cm

★ 18. One side of a triangle is 4 cm shorter then the second side. The remaining side is twice the first side. The perimeter is 48 cm. Find the length of each side.   1st side: 11 cm; 2nd side: 15 cm; 3rd side: 22 cm

Calculator

Find the length of a rectangle. Round to the nearest tenth.

P = 56.9 cm
w = 12.3 cm

Formula $P = 2l + 2w$

$$l = \frac{P - 2w}{2}$$

PRESS   2 × 12.3 ⊖ 24.6
            56.9 ⊖ 24.6 ⊖ 32.3 ⊙ 2 ⊖
DISPLAY   16.15
So, the length is 16.2 cm.

1. P = 72.3 cm and w = 11.7 cm   24.5 cm
2. P = 63.8 m and w = 9.8 m   22.1 m

This test will prepare you for taking standardized tests. Choose the best answer for each question.

1. Which of the following is greater than $\frac{1}{3}$?

   a. $\left(\frac{1}{3}\right)^2$ **b.** 0.03   **c.** $\frac{1}{4}$ **d.** None of the above.

2. If it is snowing at the rate of 3 centimeters per hour, then how many centimeters of snow will fall in n minutes?

   a. $\frac{n}{20}$ **b.** $\frac{20}{n}$ **c.** $20n$ **d.** $3n$

3. There are 880 students in a college. One half of the enrollment is women. One fourth of the women are studying psychology. One fifth of those women studying psychology are studying statistics. One half of those women studying statistics are studying computers. How many women are studying computers?

   a. 10   **b.** 11   **c.** 44   **d.** 396

4. The sum of two numbers is 10. Their difference is 4. What is the whole number quotient of the two numbers?

   a. 2   **b.** 4   **c.** 6   **d.** 8

5. How many one fourths are there in $\frac{1}{2} + 1\frac{1}{4} + 2\frac{3}{4}$?

   a. $4\frac{1}{2}$ **b.** 7   **c.** 9   **d.** 18

6. If the same number is added to both the numerator and denominator of a proper fraction, then the value of the new fraction is
   a. less than the value of the original fraction.
   b. greater than the value of the original fraction.
   c. the same as the value of the original fraction.
   d. cannot be determined.

7. If 4 times a number A is 12, and 5 times a number B is 11, then 10 times AB is?
   a. 66   **b.** 75   **c.** 132   **d.** 230

8. How many nickels are there in q quarters?

   a. $\frac{q}{5}$ **b.** $25q$ **c.** $4q$ **d.** $5q$

9. How many numbers between 1 and 100 are exactly divisible by 2 or 3 but not both?
   a. 67   **b.** 76   **c.** 77   **d.** 83

Another way to enter data into a computer is by using the **READ** and **DATA** statements. These statements are advantageous when you know ahead of time all the data you need to enter.

Example   Listed are the wages and weekly hours of four workers. Write a program to show the hours, wages, and resulting pay for each of the four workers: $5.75/h, 35 h; $6.70/h, 40 h; $5.75/h, 39 h; $7.00/h, 35 h.

First write a formula to find the pay.

Pay = Wage times Hours

P =   W   *   H

Prints heading for 3 columns.	10 PRINT "WAGE","HOURS","PAY"
There are 4 sets of **DATA.**	20 FOR K = 1 TO 4
Two memory locations are reserved for wage, hours.	30 READ W,H
Formula for finding pay.	40 LET P = W * H
Four sets of **DATA.**	50 DATA 5.75,35, 6.70,40, 5.75,39, 7,35
Compute **PRINTS** wage, hours, and resulting pay.	60 PRINT W,H,P
	70 NEXT K
	80 END

See the Computer Section beginning on page 420 for more information.

EXERCISES

1. Modify the program above to compute the pay for the following 3 sets of **DATA:** $9.75/h, 40 h; $5.89/h, 35 h; $4.85/h, 38 h. Compute the pay for each. Then **RUN** the program to check your answers. $390, $206.15, $184.30

2. Write and **RUN** a program to find P in the formula P = 2L + 2W (perimeter of a rectangle). Use the **DATA** L = 5, W = 8; L = 19, W = 23; L = 42, W = 17.   See TE Answer Section for program and results.

Chapter Review

Answer Exercises 1–3 below using the tax table. [140]

1. What tax will you pay if your taxable income is $30,190 and you are single? $7,939

2. What tax will you pay if your taxable income is $30,220 and you are married filing jointly? $6,242

3. What tax will you pay if your taxable income is $30,075 and you are married filing separately? $9,756

4. Make a pictograph for attendance at four tennis
[143] matches. Match 1: 800 fans; Match 2: 750 fans; Match 3: 950 fans; Match 4: 900 fans Check students' graphs.

If line 34 (taxable income) is—		And you are—		
At least	But less than	Single	Married filing jointly	Married filing separately
		Your tax is—		
30,000	30,050	7,873	6,169	9,729
30,050	30,100	7,895	6,187	9,756
30,100	30,150	7,917	6,206	9,783
30,150	30,200	7,939	6,224	9,809
30,200	30,250	7,960	6,242	9,836

Use the bar graph to find the approximate speed for each. [147]

5. the reindeer 33 km/h

6. the wildebeest 60 km/h

7. the quarter horse 55 km/h

8. the bear 30 km/h

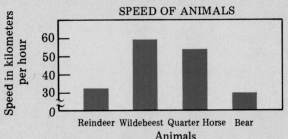

SPEED OF ANIMALS

9. Make a line graph to show the average
[150] monthly temperature. Jan., 2°C; Mar., 7°C; May, 18°C; July, 26°C; Sept., 22°C; Nov., 9°C Check students' graphs.

Use the graph to answer Exercises 10–11.

10. In what year did Brand A have the
[154] greatest sales? 1981

11. Who sold the most jeans in 1985? Brand A

Solve these problems. [157]

12. The length of each leg of an isosceles triangle is 11 cm and the perimeter is 38 cm. Find the length of the base. 16 cm

13. The base of an isosceles triangle is 16 cm and the perimeter is 58 cm. Find the length of each side. 21 cm

14. The length of a rectangle is 14 cm and the perimeter is 52 cm. Find the width. 12 cm

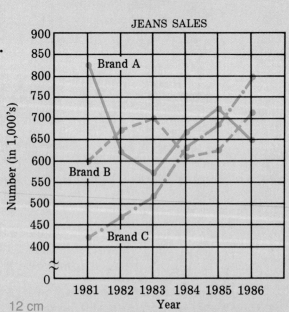

JEANS SALES

Complete Exercises 1–3 below using the tax table.

1. What tax will you pay if your taxable income is $41,240 and you are single? $11,947

2. What tax will you pay if your taxable income is $41,125 and you are married filing jointly?
$9,634

3. What tax will you pay if your taxable income is $41,080 and you are married filing separately? $14,279

If line 37 (taxable income) is—		And you are—		
At least	But less than	Single	Married filing jointly	Married filing separately
		Your tax is—		
41,000	41,050	11,859	9,595	14,255
41,050	41,100	11,881	9,614	14,279
41,100	41,150	11,903	9,634	14,304
41,150	41,200	11,925	9,653	14,328
41,200	41,250	11,947	9,673	14,353

Use the pictograph to answer these questions.

4. How many cars were sold by Dealer 1? 40 cars

5. Who sold the most cars? Dealer 3

6. How many cars were sold in November? 170 cars

NEW CAR SALES

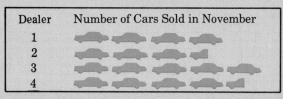

Dealer	Number of Cars Sold in November
1	
2	
3	
4	

Key: Each ⬤ represents 10 cars.

7. Make a vertical bar graph to show the approximate life expectancy of these animals. Check students' graphs.
cow, 18 years; horse, 27 years; mouse, 4 years; monkey, 7 years

8. Make a line graph to show Leah's scores on 5 French tests. Check students' graphs.
Test 1, 75; Test 2, 90; Test 3, 80; Test 4, 70; Test 5, 85

Use the bar graph to complete Exercises 9 and 10.

9. In 1986 which class had the best attendance? 10

10. In 1983 which class had the poorest attendance? 10

Solve these problems.

11. The length of each leg of an isosceles triangle is 15 cm and the perimeter is 41 cm. Find the length of the base. 11 cm

12. The length of a rectangle is 16 cm and the perimeter is 58 cm. Find the width. 13 cm

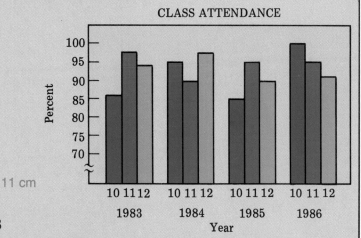

CLASS ATTENDANCE

Decimals

In the spring, Niagara Falls doubles its flow to 100,000 ft^3/s.

Introduction to Decimals

—◇ OBJECTIVES ◇—

To read and write numerals
  for word names from millions
  to millionths

To change fractions like $\frac{3}{10}$, $\frac{75}{100}$,
  and $\frac{4}{1,000}$ to decimals

To round decimals to the nearest
  tenth or hundredth

—◇ RECALL ◇—

Read $\frac{5}{10}$ as 5 tenths.

Read $\frac{7}{100}$ as 7 hundredths.

Read $\frac{95}{1,000}$ as 95 thousandths.

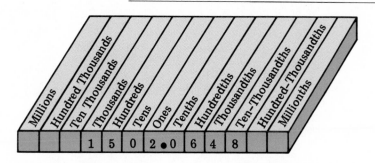

This number is read as follows:
One thousand, five hundred two, and six hundred forty-eight ten-thousandths.

Example 1

Study the table below. Then cover the right hand column, and give the word name for each decimal. Cover the left hand column, and give the decimal for each word name.

DECIMAL	WORD NAME
12.9	twelve and nine tenths
128.60	one hundred twenty-eight and sixty hundredths
0.481	four hundred eighty-one thousandths
5,063,198.0006	five million, sixty-three thousand, one hundred ninety-eight and six ten-thousandths
0.000741	seven hundred forty-one millionths

practice ▷

Write a decimal for each word name.

1. six and eight tenths   6.8
2. four hundred three millionths
   0.000403

Write a word name for each decimal.

3. 17.4   seventeen and four tenths
4. 0.0659   six-hundred fifty-nine ten-thousandths

We can use decimals to represent fractions.

2 tenths	25 hundredths	365 thousandths
$\frac{2}{10}$ or 0.2	$\frac{25}{100}$ or 0.25	$\frac{356}{1,000}$ or 0.356
one 0 one decimal place	two 0's two decimal places	three 0's three decimal places

---

Example 2

Change $\frac{4}{100}$ to a decimal.

Write 4.   4.

0.04 Move the decimal point 2 places to the left.

So, $\frac{4}{100}$ = 0.04.

Example 3

Change $\frac{66}{1,000}$ to a decimal.

Write 66.   66.

0.066 Move the decimal point 3 places to the left.

So, $\frac{66}{1,000}$ = 0.066.

---

practice ▷ **Change to decimals.**

5. $\frac{7}{10}$ 0.7   **6.** $\frac{65}{100}$ 0.65   **7.** $\frac{46}{1,000}$ 0.046   **8.** $\frac{9}{100}$ 0.09

Example 4

Round to the nearest tenth.
48.32, 48.35, and 48.37

Look at the hundredths digit.   48.3 **2**   48.3 **5**   48.3 **7**

If the hundredths digit is less than 5, round down.   tenths   less   5 or   5 or
                                                        place   than 5   more   more

If the hundredths digit is 5 or greater, round up.   48.3   48.4   48.4

So, 48.32 is 48.3, 48.35 is 48.4, and 48.37 is 48.4 to the nearest tenth.

Example 5

Round 0.3459 to the nearest hundredth.

Look at the thousandths digit.   0.34 **5** 9

The thousandths digit is 5; round up.   So, 0.3459 is 0.35 to the nearest hundredth.

practice ▷ **Round to the nearest tenth and to the nearest hundredth.**

9. 0.627 ⁰·⁶ ₀.₆₃ **10.** 0.6559 ⁰·⁷ ₀.₆₆ **11.** 0.683 ⁰·⁷ ₀.₆₈ **12.** 24.178 ²⁴·² ₂₄.₁₈

◇ ORAL EXERCISES ◇

Read each numeral. See TE Answer Section.

1. 2.4 **2.** 36.21 **3.** 42.07 **4.** 0.05
5. 271.632 **6.** 306.0145 **7.** 79.006 **8.** 9.24863
9. 1,352.060 **10.** 4,628,000.9 **11.** 0.000592 **12.** 160.00083

◇ EXERCISES ◇

Write a numeral for each word name.

1. nine and five tenths 9.5

2. sixty-three hundredths 0.63

3. eight hundred twenty-three thousandths 0.823

4. one hundred five and eight thousandths 105.008

5. eight hundred thirteen and two hundred thirty-six millionths 813.000236

6. two million, seven hundred twenty-four thousand and nine tenths 2,724,000.9

Change to decimals.

7. $\frac{76}{100}$ 0.76 **8.** $\frac{8}{10}$ 0.8 **9.** $\frac{6}{100}$ 0.06 **10.** $\frac{5}{1,000}$ 0.005 **11.** $\frac{67}{100}$ 0.67 **12.** $\frac{495}{1,000}$ 0.495

13. $\frac{763}{1,000}$ 0.763 **14.** $\frac{98}{100}$ 0.98 **15.** $\frac{1}{100}$ 0.01 **16.** $\frac{3}{1,000}$ 0.003 **17.** $\frac{13}{1,000}$ 0.013 **18.** $\frac{4}{10}$ 0.4

19. $\frac{54}{1,000}$ 0.054 **20.** $\frac{2}{100}$ 0.02 **21.** $\frac{16}{1,000}$ 0.016 **22.** $\frac{40}{1,000}$ 0.040 **23.** $\frac{11}{1,000}$ 0.011 **24.** $\frac{8}{100}$ 0.08

Round to the nearest tenth.

25. 0.617 ₀.₆ **26.** 0.657 ₀.₇ **27.** 0.6874 ⁰·⁷ **28.** 57.454 ⁵⁷·⁵ **29.** 116.93 ¹¹⁶·⁹ **30.** 17.181 ¹⁷·²
31. 14.683 **32.** 14.6237 **33.** 14.653 **34.** 114.08 **35.** 72.115 **36.** 112.0586
14.7 14.6 14.7 114.1 72.1 112.1

Round to the nearest hundredth.

37. 0.638 ₀.₆₄ **38.** 0.635 ₀.₆₄ **39.** 0.6328 ⁰·⁶³ **40.** 2.788 ₂.₇₉ **41.** 3.169 ₃.₁₇ **42.** 4.2215 ⁴·²²
43. 16.486 **44.** 16.483 **45.** 16.485 **46.** 29.057 **47.** 46.118 **48.** 192.7394
16.49 16.48 16.49 29.06 46.12 192.74

★ **49.** Change $\frac{768}{1,000}$ to a decimal and round to the nearest hundredth. 0.77

★ **50.** Change $\frac{2,651}{1,000}$ to a decimal and round to the nearest hundredth. 2.65

INTRODUCTION TO DECIMALS

Changing Fractions to Decimals

◇ OBJECTIVES ◇

To change fractions to decimals

To change a fraction like $\frac{5}{12}$ to a decimal to the nearest hundredth

◇ RECALL ◇

$\frac{15}{5}$ means $5)\overline{15}$
$\phantom{\frac{15}{5} \text{ means } 5)}\underline{15}$
$\phantom{\frac{15}{5} \text{ means } 5)}\ 0$

Example 1

Change $\frac{2}{5}$ to a decimal.

Divide. $\frac{2}{5}$ means $5)\overline{2}$.

Write a 0 after the decimal point since 20 is divisible by 5.

$$5)\overline{2.0} \rightarrow 5)\overline{2.0}$$
$$ \underline{2\ 0}$$
$$\ \ 0$$

So, $\frac{2}{5} = 0.4$.

Example 2

Change $\frac{3}{8}$ to a decimal.

$\frac{3}{8}$ means $8)\overline{3}$.

Write one 0.	Write two 0's.	Write three 0's.
0.3	0.37	0.375
8)3.0	8)3.00	8)3.000
2 4	2 4	2 4
6	60	60
	56	56
remainder	4	40
	remainder	40
		0

Divide.
Write enough 0's after the decimal point so that the remainder will be 0.

So, $\frac{3}{8} = 0.375$.

---

practice ▷ **Change to decimals.**

1. $\frac{3}{5}$ 0.6

2. $\frac{1}{4}$ 0.25

3. $\frac{5}{8}$ 0.625

4. $\frac{3}{20}$ 0.15

Example 3

Change $\frac{4}{9}$ to a decimal to the nearest tenth.

$\frac{4}{9}$ means $9\overline{)4}$.

Divide.

To find the answer to the nearest tenth, carry out the division to hundredths.

$$\begin{array}{r} 0.44 \\ 9\overline{)4.00} \\ 3\,6 \\ \hline 40 \\ 36 \\ \hline 4 \end{array}$$

0.4 4
— less than 5
Round to 0.4.

So, $\frac{4}{9}$ = 0.4 to the nearest tenth.

Notice that $\frac{4}{9}$ = 0.4444 . . . The 4's repeat forever. 0.4444 . . . is called a *repeating decimal*. It is written as $0.4\overline{4}$.

practice ▷ **Change to a decimal to the nearest tenth.**

0.7

5. $\frac{2}{3}$ 0.7

6. $\frac{7}{9}$ 0.8

7. $\frac{5}{6}$ 0.8

8. $\frac{7}{12}$

Example 4

Change $\frac{5}{12}$ to a decimal to the nearest hundredth.

Divide.

To find the answer to the nearest hundredth, carry out the division to thousandths.

$$\begin{array}{r} 0.416 \\ 12\overline{)5.000} \\ 4\,8 \\ \hline 20 \\ 12 \\ \hline 80 \\ 72 \\ \hline 8 \end{array}$$

0.41 6
— 5 or more
Round to 0.42.

So, $\frac{5}{12}$ = 0.42 to the nearest hundredth.

practice ▷ **Change to a decimal to the nearest hundredth.**

0.17

9. $\frac{4}{7}$ 0.57

10. $\frac{5}{9}$ 0.56

11. $\frac{11}{12}$ 0.92

12. $\frac{1}{6}$

CHANGING FRACTIONS TO DECIMALS

169

Change to decimals.

1. $\frac{1}{5}$ 0.2

2. $\frac{4}{5}$ 0.8

3. $\frac{3}{4}$ 0.75

4. $\frac{1}{8}$ 0.125

5. $\frac{7}{8}$ 0.875

6. $\frac{1}{2}$ 0.5

7. $\frac{7}{20}$ 0.35

8. $\frac{9}{25}$ 0.36

9. $\frac{3}{50}$ 0.06

10. $\frac{11}{40}$ 0.275

★ 11. $\frac{5}{16}$ 0.3125

★ 12. $\frac{13}{32}$ 0.40625

Change to a decimal to the nearest tenth.

13. $\frac{5}{7}$ 0.7

14. $\frac{2}{3}$ 0.7

15. $\frac{5}{9}$ 0.6

16. $\frac{6}{7}$ 0.9

17. $\frac{7}{9}$ 0.8

18. $\frac{4}{7}$ 0.6

19. $\frac{1}{3}$ 0.3

20. $\frac{1}{7}$ 0.1

21. $\frac{2}{9}$ 0.2

22. $\frac{5}{11}$ 0.5

23. $\frac{3}{14}$ 0.2

24. $\frac{4}{11}$ 0.4

Change to a decimal to the nearest hundredth.

25. $\frac{8}{9}$ 0.89

26. $\frac{1}{12}$ 0.08

27. $\frac{1}{9}$ 0.11

28. $\frac{6}{7}$ 0.86

29. $\frac{1}{7}$ 0.14

30. $\frac{7}{12}$ 0.58

31. $\frac{5}{11}$ 0.45

32. $\frac{3}{14}$ 0.21

33. $\frac{5}{6}$ 0.83

34. $\frac{5}{14}$ 0.36

★ 35. $\frac{5}{29}$ 0.17

★ 36. $\frac{3}{41}$ 0.07

37. Anthony saves $\frac{1}{5}$ of his salary. Write the part he saves as a decimal. 0.2

38. The gas tank of Edna's car is $\frac{3}{4}$ full. Write the part of the tank that is full as a decimal. 0.75

Algebra Maintenance

Simplify.

1. $3 \cdot \frac{4}{3}m$ $4m$

2. $5\left(\frac{3}{5}x - 6\right)$ $3x - 30$

3. $7\left(\frac{3}{14}y + \frac{4}{7}\right)$ $\frac{3}{2}y + 4$

4. $4a + 2 - 7b + 3a - 4b$ $7a + 2 - 11b$

5. $2x - 5 + \frac{3}{4}x + 5y + 8 - \frac{4}{5}y$ $2\frac{3}{4}x + 3 + 4\frac{1}{5}y$

Solve each equation. Check.

6. $\frac{1}{2}y = 4$ 8

7. $\frac{3}{4}t - 4 = 8$ 16

8. $8k - \frac{3}{4} = \frac{1}{2}$ $\frac{5}{32}$

Solve.

9. One-half of a number, increased by 2, is 13. What is the number? 22

10. Larry's age is 5 more than one half Frank's age. Frank is 14 years old. How old is Larry? 12

Adding and Subtracting Decimals

To add decimals
To subtract decimals

Add. 48 + 3 + 116

$$
\begin{array}{r}
48 \\
3 \\
+\,116 \\
\hline
167
\end{array}
$$

Example 1

Add. 11.02 + 0.1 + 100

Write in vertical form with decimal points in line with each other.

$$
\begin{array}{r}
11.02 \\
0.1 \\
+\,100. \\
\hline
\end{array}
\quad\longrightarrow\quad
\begin{array}{r}
11.02 \\
0.10 \\
+\,100.00 \\
\hline
111.12
\end{array}
$$

The answer has its decimal point in line with the others.

So, the sum is 111.12.

Example 2

Add. 92.68 + 0.003 + 9.378

Write in vertical form with decimal points in line with each other.

$$
\begin{array}{r}
92.68 \\
0.003 \\
+\;\;9.378 \\
\hline
\end{array}
\qquad
\begin{array}{r}
92.680 \\
0.003 \\
+\;\;9.378 \\
\hline
102.061
\end{array}
$$

The answer has its decimal point in line with the others.

So, the sum is 102.061.

---

practice ▷　**Add.**

1. 149.68 + 0.48 + 91.32
   241.48

2. 87.65 + 0.004 + 9.478
   97.132

Example 3

Subtract.　146.38 − 28.99

Decimal points are lined up. The answer has its decimal point in line with the others.

$$
\begin{array}{r}
146.38 \\
-\;\;28.99 \\
\hline
\end{array}
\qquad
\begin{array}{r}
146.38 \\
-\;\;28.99 \\
\hline
117.39
\end{array}
$$

So, the difference is 117.39.

---

practice ▷　**Subtract.**

3. 38.65 − 25.43　13.22

4. 17.146 − 9.38　7.766

Example 4

Evaluate $x + 5.8$ if $x = 0.932$.

Substitute 0.932 for x. Write in vertical form. Insert zeros. Then add.

$$\begin{array}{r} 0.932 \\ +5.800 \\ \hline 6.732 \end{array}$$

So, the sum is 6.732.

practice ▷ **Evaluate for the given value of the variable.**

5. $x + 9.8$ if $x = 12.05$ 21.85   **6.** $9.3 - y$ if $y = 0.46$ 8.84

◇ EXERCISES ◇

Add.

1. $13.06 + 0.8 + 400.0$ 413.86
3. $\$25.43 + \$170 + \$0.69$ $196.12
5. $81.43 + 2.089 + 0.6 + 20.098$ 104.217

2. $24.35 + 0.046 + 3.047$ 27.443
4. $\$112.45 + \$1.04 + \$0.89$ $114.3
6. $0.4308 + 2.85 + 300 + 47.1$
350.3808

Subtract.

7. $48.75 - 26.32$ 22.43
9. $\$412.42 - \$59.86$ $352.56
11. $17.003 - 3.68$ 13.323

8. $16.146 - 8.236$ 7.910
10. $466 - 65.89$ 400.11
12. $18.5 - 9.836$ 8.664

Evaluate for the given value of the variable.

13. $x + 4.3$ if $x = 9.8$ 14.1
15. $y - 2.3$ if $y = 17.1$ 14.8

14. $62.3 - x$ if $x = 5.9$ 56.4
16. $7.46 + z$ if $z = 8.25$ 15.71

★ **17.** Subtract 4.12 from the sum of 112.46, 2.18, and 1.06.   111.58

Calculator

1. Add. $249.6182 + 0.0083 + 12.49 + 0.0349$ 262.1514

2. Subtract. 241.346 from 3,568   3326.654

3. Add. $0.0046 + 1.0049 + 21.1678 + 1119.32$ 1141.4973

4. Subtract. 0.00584 from 12.001578   11.995738

5. Subtract the sum of $115.45, $240.65, and $19.45 from the sum of $255.69, $189.39, and $1,156.95. $1,226.48

Multiplying Decimals

To multiply decimals
To multiply whole numbers
 and decimals

$$0.03 = \frac{3}{100} \qquad 0.1 = \frac{1}{10}$$

Example 1

Write each as a fraction and multiply.
Multiply the numerators.
Multiply the denominators.
$$\frac{6}{1,000} = 0.006$$

Show that $(0.03)(0.2) = 0.006$.
$(0.03)(0.2)$
$$\frac{3}{100} \cdot \frac{2}{10}$$
$$\frac{6}{1,000}$$
0.006   So, $(0.03)(0.2) = 0.006$.

Example 1 suggests that we count the decimal places.

$0.03 \leftarrow$ 2 decimal places
$\underline{\times 0.2} \leftarrow$ 1 decimal place
$0.006 \leftarrow$ 2 + 1, or 3 decimal places

Example 2

Multiply. $(3.45)(0.014)$
$3.45 \leftarrow$ 2 decimal places
$\underline{\times 0.014} \leftarrow$ 3 decimal places
1380
$\underline{345}$
$0.04830 \leftarrow$ 2 + 3, or 5 decimal places

So, $(3.45)(0.014) = 0.04830$, or 0.0483.

---

practice ▷ **Multiply.**

0.018468

1. $(62.3)(2.4)$ 149.52   **2.** $(3.72)(0.017)$ 0.06324   **3.** $(0.243)(0.076)$

Example 3

Substitute 0.8 for x and write in vertical
form. Then multiply.
Three decimal places are in the product.

Evaluate $1.36x$ if $x = 0.8$.
1.36
$\underline{\times 0.8}$
1.088

So, the product is 1.088.

---

practice ▷ **Evaluate for the given value of the variable.**

4. $7.52x$ if $x = 0.6$ 4.512

5. $0.42y$ if $y = 1.5$ 0.63

◇ ORAL EXERCISES ◇

Where should the decimal point be placed?

1.	**2.**	**3.**	**4.**	**5.**	**6.**
3.48	12.16	0.24	2.86	1.46	0.0867
×0.2	×0.02	×0.3	×0.014	×0.028	×0.05
0.696	0.2432	0.072	0.04004	0.04088	0.004335

◇ EXERCISES ◇

Multiply.   See TE Answer Section.

1. $(2.16)(0.015)$ **2.** $(72.4)(3.2)$ **3.** $(1.06)(0.32)$ **4.** $(0.26)(0.04)$
5. $(1.1)(0.038)$ **6.** $(21.6)(0.042)$ **7.** $(2.17)(0.43)$ **8.** $(0.512)(3.4)$

9.	**10.**	**11.**	**12.**	**13.**	**14.**
215	26	0.047	0.0082	21.3	249
×0.034	×0.13	×35	×23	×17	×0.046

15.	**16.**	**17.**	**18.**	**19.**	**20.**
145.32	72.89	15.45	212.19	11.43	15.3
×0.25	×0.06	×0.32	×0.013	×0.013	×0.0015

★ **21.** $(7.1)(0.04)(1.13)$ ★ **22.** $(0.014)(0.13)(16.23)$ ★ **23.** $(30.62)(0.25)(4.6)$

Evaluate for the given value of the variable.

24. $6.32y$ if $y = 0.4$ 2.528   **25.** $17.8x$ if $x = 0.12$ 2.136
26. $23.5z$ if $z = 0.08$ 1.88   **27.** $1.53y$ if $y = 1.09$ 1.6677

Solve these problems.

28. Joanna saves $1.75 each week. How much has she saved after 48 weeks?
$84

29. Plums cost $1.40 per kilogram. What is the cost of 4.5 kilograms?
$6.30

Challenge

See if you can move the disks to one of the other pegs. Follow these rules.

1. Move no more than one disk at a time.
2. Do not put a disk on top of one smaller than itself.
What is the least possible number of moves?
See TE Answer Section.

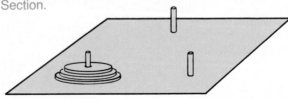

Problem Solving – Applications
Jobs for Teenagers

Example

How much should Faith charge a customer for an order of 2.5 lb of pastrami and 1.4 lb of cheese?

First find the cost of each. Then find the total.

DELICATESSEN PRICES

ITEM	PRICE PER POUND
CHEESE	3.80
TURKEY	3.84
ROAST BEEF	6.46
PASTRAMI	6.43
BOLOGNA	3.20

2.5 lb pastrami
$$\begin{array}{r} \$6.43 \leftarrow \text{pastrami } \$6.43/\text{lb} \\ \times\ 2.5 \leftarrow 2.5 \text{ lb} \\ \hline 3215 \\ 1286 \\ \hline \$16.075 \leftarrow \text{cost} \end{array}$$

1.4 lb cheese
$$\begin{array}{r} \$3.80 \\ \times\ 1.4 \\ \hline 1520 \\ 380 \\ \hline \$5.320 \end{array}$$

Round 16.0 **7** 5 one cent higher to $16.08.

$$\begin{array}{r} \$16.08 \quad \text{Add to find the total.} \\ +\ 5.32 \\ \hline \$21.40 \end{array}$$

So, the total bill is $21.40.

In business, costs are usually rounded up to the next higher cent unless there are 0's after the cents place.

$7.3241	$12.6492	$14.7800 ← only 0's after 8
rounds to $7.33	rounds to $12.65	stays at $14.78

Use the chart for the price per pound.

DELICATESSEN PRICES	
Item	Price per Pound
Cheese	$3.80
Turkey	$3.84
Roast beef	$6.46
Pastrami	$6.43
Bologna	$3.20

1. José ordered 2.5 lb of bologna and 1.7 lb of turkey. Find the cost of the order. $14.53

2. Maureen ordered 2.4 lb of roast beef and 1.5 lb of bologna. Find the cost of the order. $20.31

3. Miss Riggio called for 1.6 lb of roast beef and 2.7 lb of turkey. How much will her bill be? $20.71

4. Mr. Rothstein called for 0.6 lb of pastrami and 1.5 lb of roast beef. Find the cost of the order. $13.55

5. Janet is planning a party. She ordered 2.7 lb of cheese, 3.2 lb of turkey, and 0.6 lb of bologna. Find the cost of the order. $24.47

6. The Martinsons are planning a luncheon. They ordered 4.5 lb of cheese, 2.5 lb of pastrami, and 3.6 lb of turkey. How much did the order cost? $47.01

7. Ms. Becker is having a party. She ordered 1.4 lb of cheese and bologna and 1.7 lb of roast beef and turkey. How much will her bill be? $27.31

8. Julia bought 0.6 lb of turkey and 1.2 lb of cheese. Find the cost. How much change did she get from a $20 bill? $6.87, $13.13

9. Bill bought 1.4 lb of roast beef and 0.4 lb of pastrami. Find the cost. How much change did he get from a $20 bill? $11.63, $8.37

10. Nick bought 0.6 lb of cheese, 0.5 lb of roast beef, and 0.3 lb of pastrami. Find the total cost. How much change did he get from a $20 bill? $7.44, $12.56

11. The Serkins ordered 1.5 lb of cheese, 0.6 lb of turkey, and 1.1 lb of bologna. Find the total cost. How much change did they get from a $20 bill? $11.53, $8.47

12. Mrs. White uses 0.1 lb of roast beef on a sandwich. How many lb of roast beef should she buy for 3 sandwiches? How much will she pay?   0.3 lb/$1.94

13. Lars uses 0.08 lb of bologna and 0.08 lb of cheese on a sandwich. He bought enough bologna and cheese for 4 sandwiches. How much did he pay? $2.24

14. The Bryants bought 6 rolls at 15 cents each and 0.5 lb of turkey. They made 6 sandwiches. How much did each sandwich cost them? $0.47

15. Mr. Snow bought a pickle for $0.15, an iced tea for $0.50, 2 rolls for $0.15 each, and 0.09 lb of pastrami. How much did his order cost? $1.53

Dividing Decimals

—◇ **OBJECTIVES** ◇—

To divide decimals by decimals
To round quotients to the
  nearest tenth or hundredth

———◇ **RECALL** ◇———

$10(1.4) = 14$ $100(1.4) = 140$ $1,000(1.4) = 1,400$

$8\overline{)16}$ means $\frac{16}{8}$

Example 1

Multiply the numerator and
denominator by 100.
$100(12.36\,5) = 1,236.5$
$100(4.12\,) = 412$

The divisor 412 is a whole number.

Rewrite $4.12\overline{)12.365}$ so that the divisor is a whole number.

$4.12\overline{)12.365}$ means $\frac{12.365}{4.12}$ and $12.365 \div 4.12$

$\frac{12.365(100)}{4.12(100)}$

$\frac{1236.5}{412}$ or $412\overline{)1236.5}$

            ↑

    whole number

So, $4.12\overline{)12.365}$ can be rewritten as $412\overline{)1,236.5}$.

Example 1 suggests how to move decimal points when dividing by decimals.

$4.12\overline{)12.365}$

$4.12\,\overline{)12.36\,5}$

Move the decimal point two places to the right to get a whole number.

Also move this decimal point the same number of decimal places to the right.

Example 2

Make the divisor, 0.03, a whole number.
Move each decimal point two places to
the right.
Divide.
Line up the decimal point in the
quotient with the decimal point in the
dividend.

Divide. $2.796 \div 0.03$

$0.03\,\overline{)2.79\,6}$

two places

$\begin{array}{r} 93.2 \\ 3\overline{)279.6} \end{array}$

So, $2.796 \div 0.03 = 93.2$

---

practice ▷ **Divide.**

1. $0.04\overline{)3.684}$ (92.1)

2. $0.02\overline{)68.462}$ (3423.1)

3. $3.575 \div 0.005$ (715.)

DIVIDING DECIMALS

Example 3

Make the divisor, 0.07, a whole number. Move each decimal point two places to the right.

To find the answer to the nearest tenth, carry the division to hundredths, or two decimal places.

Divide. Round to the nearest tenth. 0.41 ÷ 0.07

0.07̸)0.41̸

```
      5.85
  7)41.00  ← Write two 0's to get two
    35          decimal places.
    ──
    60
    56
    ──
    40
    35
    ──
     5
```

5.8 **5** ← 5 or more.
Round to 5.9.

5.85 rounded to the nearest tenth is 5.9.

So, 0.41 ÷ 0.07 = 5.9 rounded to the nearest tenth.

Example 4

Make the divisor, 0.048, a whole number. Move each decimal point three places to the right.

To find the answer to the nearest hundredth, carry the division to thousandths, or three decimal places.

Divide. Round to the nearest hundredth.
0.03375 ÷ 0.048

0.048̸)0.033̸75

```
48)33.750  ← Write one 0 to get
               three decimal places.
```

48 is close to 50. 50)337̄

5)33̄   Try 7.

```
       0
  48)15   Write 0 above 5.
```

48)150̄; 48 is close to 50.
Think: 50)150̄ = 3.

```
     0.7            0.70
 48)33.750      48)33.750
    33 6            33 6
    ────            ────
      15             150 ← Bring down the 0.
```

```
    0.703
 48)33.750
    33 6
    ────
     150
     144
     ───
       6
```

0.70 **3** ← less than 5
Round to 0.70.

0.703 rounded to the nearest hundredth is 0.70.

So, 0.03375 ÷ 0.048 = 0.70 rounded to the nearest hundredth.

practice ▷ **Divide. Round to the nearest tenth and to the nearest hundredth.**

         3.8; 3.83   0.6; 0.61   0.8; 0.83

4. 0.06)0.23̄   **5.** 0.01393 ÷ 0.023   **6.** 0.46)0.3827̄

Where should the decimal point be placed?

1. $\overset{7.12}{0.4)\overline{2.848}}$
2. $\overset{712}{0.04)\overline{2.848}}$
3. $\overset{712.}{0.004)\overline{2.848}}$
4. $\overset{21}{0.06)\overline{0.126}}$

5. $\overset{911.}{0.005)\overline{4.555}}$
6. $\overset{243}{0.002)\overline{0.0486}}$
7. $\overset{0.11}{0.05)\overline{0.0055}}$
8. $\overset{0.34}{0.21)\overline{0.0714}}$

◇ *EXERCISES* ◇

Divide.

1. $\overset{123.4}{0.02)\overline{2.468}}$
2. $\overset{2231.3}{0.03)\overline{66.939}}$
3. $\overset{1962}{0.004)\overline{7.848}}$
4. $\overset{0.9}{0.05)\overline{0.045}}$

5. $0.002)\overline{8.42}$ 4,210
6. $0.03)\overline{8.43}$ 281
7. $0.03)\overline{0.3249}$ 10.83
8. $0.06)\overline{0.186}$ 3.1

9. $0.003)\overline{273.9}$ 91,300
10. $0.06)\overline{246}$ 4,100
11. $0.008)\overline{82.48}$ 10,310
12. $0.002)\overline{84.62}$ 42,310

Divide. Round to the nearest tenth.

13. $23 \div 0.07$ 328.6
14. $0.77 \div 0.03$ 25.7
15. $1.337 \div 0.5$ 2.7

16. $7.62 \div 2.4$ 3.2
17. $24.6 \div 7.1$ 3.5
18. $0.786 \div 0.34$ 2.3

19. $0.70 \div 0.32$ 2.2
20. $0.4965 \div 0.043$ 11.5
21. $0.893 \div 0.63$ 1.4

Divide. Round to the nearest hundredth.

22. $0.754 \div 2.3$ 0.33
23. $0.4935 \div 0.42$ 1.18
24. $0.497 \div 0.41$ 1.21

25. $0.7577 \div 6.8$ 0.11
★ 26. $0.7856 \div 0.0462$ 17.00
★ 27. $0.04682 \div 1.19$ 0.04

Solve these problems.

28. It takes Jim 9.2 minutes to run 2.5 kilometers. How many kilometers can he run in 1 minute? Round to the nearest tenth.   0.3 km

29. Jane traveled by car 199.7 kilometers in 4.5 hours. Find her average speed. Round to the nearest hundredth.   44.38 km/h

Challenge

A customer in Mr. Sole's shoe store bought a pair of shoes for $42 and paid with a $50 bill. Mr. Sole had no change. So he walked to Harry's Barber Shop to change the bill. Just after the customer left, Harry stormed into Mr. Sole's store calling the $50 bill counterfeit. Mr. Sole had to give him $50. Also Mr. Sole had given the crook $8 change. It looks like Mr. Sole lost $58 in all. Right or wrong?

Wrong, the loss is $100:
$42 + $50 + $8 = $100
↓   ↓   ↓
shoes   bill   change

Scientific Notation and Large Numbers

— ◇ *OBJECTIVES* ◇ —

To multiply decimals by powers of ten
To write large numbers in scientific notation

——— ◇ *RECALL* ◇ ———

Powers of 10

$$10^2 = 100, 10^3 = 1{,}000, 10^4 = 10{,}000$$

Example 1

Multiply by 10: Move the decimal point 1 place right.
Multiply by 100: Move the decimal point 2 places right.

Multiply.

3.47×10
$3.47 \times 10 = 34.7$
1 zero   1 place

5.967×100
$5.967 \times 100 = 596.7$
2 zeros   2 places

So, $3.47 \times 10 = 34.7$ and $5.967 \times 100 = 596.7$.

practice ▷ **Multiply.**

1. 2.87×10 28.7
2. 4.385×100 438.5
3. $9.876 \times 1{,}000$ 9,876

Example 2

$10^2 = 100$

Multiply.
5.876×10^2
5.876×10^2
$5.876 \times 100 = 587.6$
So, $5.876 \times 10^2 = 587.6$ and

$10^3 = 1{,}000$
4.375×10^3
4.375×10^3
$4.375 \times 1{,}000 = 4{,}375$
$4.375 \times 10^3 = 4{,}375$.

practice ▷ **Multiply.**

4. 5.325×10^3 5,325
5. 9.6257×10^4 96,257
6. 10.4156×10^2 1,041.56

Example 3

7.36 multiplied by what power of 10 is 736?

What number should be substituted for n to make the sentence true?

$736 = 7.36 \times 10^n$
$736 = 7.36 \times 100$
$736 = 7.36 \times 10^2$

So, $n = 2$.

practice ▷ **Find n to make the sentence true.**

7. $547 = 5.47 \times 10^n$ 2
8. $7{,}635.7 = 7.6357 \times 10^n$ 3

Large numbers can be written as a product of a number between 1 and 10 and a power of 10. This is called *scientific notation*.

Example 4

Write 497 and 5,000,000 in scientific notation.

To make 497 a number between 1 and 10, move the decimal point and multiply by a power of ten.

$$497 = 4.97 \times 10^n$$

2 places

$$n = 2$$
$$497 = 4.97 \times 10^2$$

So, $497 = 4.97 \times 10^2$ in scientific notation.

$$5,000,000 = 5.000000 \times 10^n$$

6 places

$$n = 6$$
$$5,000,000 = 5 \times 10^6$$

So, $5,000,000 = 5 \times 10^6$ in scientific notation.

practice ▷ Write in scientific notation.

9. 234 2.34×10^2 10. 49,000 4.9×10^4 11. 30,000,000 3×10^7

◇ EXERCISES ◇

Multiply.

1. 3.65×100 365
2. 4.155×10 41.55
3. $9.8763 \times 1,000$ 9,876.3
4. $8.1456 \times 10,000$ 81,456
5. 1.3456×100 134.56
6. $12.432158 \times 100,000$ 1,243,215.8
7. 4.872×10^2 487.2
8. 12.34125×10^3 12,341.25
9. 10.176235×10^4 101,762.35
10. 18.145×10 181.45
11. 5.449872×10^5 544,987.2
12. 27.92976×10^2 2,792.976

Write in scientific notation.

13. 487 4.87×10^2
14. 96,000 9.6×10^4
15. 3,000,000 3×10^6
16. 12,000,000 1.2×10^7
17. 582,000 5.82×10^5
18. 140,000,000 1.4×10^8
19. 750,000 7.5×10^5
20. 635,000,000 6.35×10^8
21. 8,050,000 8.05×10^6
22. 4,100,000 4.1×10^6
23. 2,000,000,000 2×10^9
24. 350,000,000,000 3.5×10^{11}

Write in scientific notation the combined populations of:

25. Philadelphia and Detroit. 9.2442×10^6

26. Boston and Honolulu. 4.6096×10^6

27. All four population areas. 1.38538×10^7

CITY	POPULATION
Philadelphia	4,809,900
Detroit	4,434,300
Boston	3,918,400
Honolulu	691,200

SCIENTIFIC NOTATION AND LARGE NUMBERS

Decimal Equations

To solve equations like
  $1.62 + x = 4.39$, $0.05x = 7.5$,

  $\frac{x}{4.6} = 9.71$, and $3x - 4.2 = 5.4$

$$(0.7)(10) = 7$$

$$(0.05)(100) = 5$$

Example 1

Subtract 1.62 from each side.

Solve and check. $1.62 + x = 4.39$
$$1.62 + x = 4.39$$
$$1.62 - 1.62 + x = 4.39 - 1.62$$
$$x = 2.77$$

Substitute 2.77 for x.

Check: $\begin{array}{c|c} 1.62 + x = 4.39 \\ \hline 1.62 + 2.77 & 4.39 \\ 4.39 = 4.39 \end{array}$

So, the solution is 2.77.

---

practice ▷ **Solve.**

1. $x + 1.8 = 4.7$ 2.9 **2.** $y - 2.63 = 5.98$ 8.61 **3.** $9.3 + z = 12.7$ 3.4

Example 2

Divide each side by 0.05.

Solve. $0.05x = 7.5$
$$0.05x = 7.5$$
$$\frac{0.05x}{0.05} = \frac{7.5}{0.05}$$
$$x = 150$$

So, the solution is 150.

Example 3

Solve. $\frac{x}{4.6} = 9.71$

$$\frac{x}{4.6} = 9.71$$

Multiply each side by 4.6.

$$(4.6)\frac{x}{4.6} = (9.71)(4.6)$$

$$x = 44.666$$

So, the solution is 44.666.

---

practice ▷ **Solve and check.**

4. $4.1x = 11.48$ 2.8   **5.** $\frac{y}{7.25} = 3.1$ 22.475 **6.** $0.12x = 0.06$ 0.5

<table>
<tr>
<td>Example 4</td>
<td colspan="2">Solve. $3x - 4.2 = 5.4$
$3x - 4.2 = 5.4$</td>
</tr>
<tr>
<td>Add 4.2 to each side.</td>
<td colspan="2">$3x - 4.2 + 4.2 = 5.4 + 4.2$
$3x = 9.6$</td>
</tr>
<tr>
<td>Divide each side by 3.</td>
<td colspan="2">$\dfrac{3x}{3} = \dfrac{9.6}{3}$
$x = 3.2$</td>
</tr>
<tr>
<td></td>
<td colspan="2">So, the solution is 3.2.</td>
</tr>
</table>

practice ▷   Solve.

     7. $2x - 1.2 = 6.4$ 3.8   **8.** $4y + 3.2 = 4.4$ 0.3   **9.** $5x - 3.8 = 2.7$ 1.3

◇ ORAL EXERCISES ◇

Tell what you would do to each side of the equation to solve it.
See TE Answer Section.

1. $x + 1.8 = 17.9$	**2.** $0.7y = 4.9$	**3.** $12.6 + y = 18.2$
4. $c - 4.2 = 16.3$	**5.** $\dfrac{x}{7.9} = 14.8$	**6.** $12.03z = 36.09$
7. $x - 3.8 = 14.9$	**8.** $\dfrac{x}{1.07} = 7.21$	**9.** $0.3y = 0.09$
10. $\dfrac{z}{8.1} = 5.9$	**11.** $r - 9.2 = 7.3$	**12.** $s + 0.09 = 0.26$

◇ EXERCISES ◇

Solve.

1. $x + 7.6 = 8.2$ 0.6	**2.** $0.3 + y = 8.9$ 8.6	**3.** $c + 0.5 = 9.1$ 8.6
4. $r + 3.21 = 8.96$ 5.75	**5.** $4.98 + z = 7.14$ 2.16	**6.** $6.8 + b = 8.0$ 1.2
7. $x + 3.5 = 12.8$ 9.3	**8.** $7.02 + z = 12.51$ 5.49	**9.** $x - 2.5 = 7.1$ 9.6
10. $y - 4.07 = 6.31$ 10.38	**11.** $z - 5.4 = 0.9$ 6.3	**12.** $c - 3.02 = 8.15$ 11.17
13. $r - 0.04 = 1.39$ 1.43	**14.** $s - 2.8 = 4.13$ 6.93	**15.** $x - 0.3 = 8.97$ 9.27
16. $z - 0.06 = 0.08$ 0.14	**17.** $0.3x = 0.06$ 0.2	**18.** $0.8y = 0.24$ 0.3
19. $0.9x = 0.54$ 0.6	**20.** $0.7y = 0.63$ 0.9	**21.** $0.06x = 0.03$ 0.5
22. $0.04z = 1.6$ 40	**23.** $4.2x = 11.34$ 2.7	**24.** $2.3x = 11.04$ 4.8
25. $\dfrac{x}{0.2} = 0.9$ 0.18	**26.** $\dfrac{y}{1.9} = 6.7$ 12.73	**27.** $\dfrac{z}{0.6} = 4.9$ 2.94
28. $\dfrac{r}{2.8} = 0.4$ 1.12	**29.** $\dfrac{a}{0.9} = 2.5$ 2.25	**30.** $\dfrac{c}{4.3} = 2.85$ 12.255
31. $2x + 1.9 = 3.7$ 0.9	**32.** $4y - 7.2 = 5.2$ 3.1	**33.** $6x - 4.2 = 7.2$ 1.9
★ **34.** $5.3 + 3z = 8.1$ $0.9\overline{3}$	★ **35.** $\dfrac{a}{2.5} + 1.6 = 3.04$ 3.6	★ **36.** $\dfrac{r}{0.7} - 0.74 = 7.26$ 5.6

Solve these problems.

37. In the formula $T = 1.12a$, find a if $T = \$196.00$. $a = \$175$

38. In the formula $c = 3.14d$, find d if $c = 12.56$. $d = 4$

DECIMAL EQUATIONS

Problem Solving – Careers
Machinists

1. A rod for a machine must be 8.97 cm long. It is now 9.01 cm. How much must be filed down?   0.04 cm

2. Three holes are to be drilled into a metal strip that is 13.6 cm long. Two of the holes are to be at either end of the strip, and the third hole is to be exactly in the middle. How far is the middle hole from the end?   6.8 cm

3. A piece of sheet metal is 42.6 cm wide. It is to be cut into 3 pieces equal in width. How wide will each be?   14.2 cm

4. On a blueprint, 1 dm represents 15 cm. The diameter of a gear on the blueprint measures 1.2 dm. Find the actual diameter of the gear.   18 cm

5. A machinist earns $5.85 an hour. How much is earned by a machinist who works 48 hours in a week? $280.80/wk

6. The length of a metal rod is 6.31 cm. It must be shortened to 6.02 cm. How much must be filed down?   0.29 cm

7. A wheel has a diameter of 8.1 centimeters. A piece of wire is to be wrapped around the wheel. Find the length of the wire to the nearest tenth of a centimeter. Use $c = 3.14d$.   25.4 cm

Non-Routine Problems

There is no simple method used to solve these problems. Use whatever methods you can. Do not expect to solve these easily.

See TE Answer Section for sample solutions to these problems.

1. Kathy wanted to find out the difference of the sum of all even numbers from 2 through 1,000 and all of the odd numbers from 1 through 999. To avoid the tedious work of finding the two sums, she arranged her work as follows:

$$
\begin{array}{r}
2 + 4 + 6 + \ldots + 996 + 998 + 1{,}000 \\
-1 + 3 + 5 + \ldots + 995 + 997 + 999 \\
\hline
1 + 1 + 1 + \ldots + 1 + 1 + 1
\end{array}
$$

Use the above approach and tell what the difference is. 500

2. Do you know that the number 96 can be divided into three parts in the ratio 1 to 3 to 8? Find these three numbers. 8, 24, 64

3. The sum of two numbers is 105. One-half of one number is equal to one-third of the other number. What are the two numbers? 42 and 63

4. A father gave some money to his three sons. To the first son he gave half the money plus one-half dollar. To the second son he gave half of what was left plus one-half dollar. To the third son he gave half of what was left plus one-half dollar. Having done that, he had no money left. How much money did the father give away in all? $7

5. The number of bacteria in a bottle triples each minute. The bottle is completely filled after 15 minutes. After how many minutes was the bottle one-third full? 14 min

6. Choose any three consecutive numbers. Find their product. Is the product divisible by 6? Now do the same with other triplets of consecutive numbers. Is the product in each case divisible by 6? Do you think it is always true? Can you find an exception? yes, yes, no, yes, 0, 1, 2

Recall that commission is a way many businesses encourage their employees to work harder in order to increase their incomes. The more you sell, the greater your earnings. The program below illustrates an application of the branching concept of computer programming taught in the computer section at the end of this text.

Example   Joan is paid a regular salary of $145 per week plus a commission of $7.00 for each T.V. she sells over 6. Write a program to find her total pay. Then type and **RUN** the program to find her total pay for 10 sales; then for 3 sales.

```
Allows for two RUNS              10   FOR K = 1 TO 2
                                 20   INPUT "TYPE NUMBER SOLD ";N
If more than 6 are sold, computer goes   30   IF N > 6 THEN 60
          to line 60 to compute total.   40   LET T = 145
                                 50   GOTO 70
                                 60   LET T = 145 + 7 * (N - 6)
                                 70   PRINT "TOTAL PAY IS "T
                                 80   NEXT K
                                 90   END

                                 ]RUN
                                 TYPE NUMBER SOLD 10
                                 TOTAL PAY IS 173
                                 TYPE NUMBER SOLD 3
Total is $145 + 7 for each sold over 6.   TOTAL PAY IS 145
```

See the computer section beginning on page 420 for more information.

Exercises

$194, $145

1. **RUN** the program above to find the total pay for 13 sales; for 5 sales.

2. Suppose the way Joan is paid is changed to a regular salary of $185 per week plus a commission of $10 for each sale she makes after 8 or more. Make the necessary changes in lines of the program. Then **RUN** the program to find the total pay for 6 sales; then for 8 sales; and for 12 sales. See TE Answer Section for program. $185, $195, $235

★ 3. José is paid a basic salary of $220 plus a commission of $6 for each sale up to 5; $8 for each sale over 5 up to 9 sales; $9 for each sale when 10 or more. Write and **RUN** a program to find his total pay for 3 sales; then for 9 sales; and for 10 sales.   See TE Answer Section for program. $238, $282, $291

Chapter Review

Write a numeral for each word name. [165]

1. sixty-eight and ninety-four ten-thousandths 68.0094

2. nine million, seven hundred fifty thousand and six tenths 9,750,000.6

Write a word name for each numeral. [165]

3. 700.96 seven hundred and ninety-six hundredths

4. 3,200,000.001 three million, two hundred thousand and one thousandth

Change to decimals. [165]

5. $\frac{85}{100}$ 0.85 **6.** $\frac{9}{10}$ 0.9 **7.** $\frac{8}{100}$ 0.08 **8.** $\frac{8}{1,000}$ 0.008 **9.** $\frac{77}{100}$ 0.77 **10.** $\frac{77}{1,000}$ 0.077

Round to the nearest tenth and to the nearest hundredth. [165]

11. 0.934 0.9 0.93 **12.** 9.758 9.8 9.76 **13.** 12.436 12.4 12.44 **14.** 10.952 11.0 10.95

Change to a decimal to the nearest tenth and to the nearest hundredth. [168]

15. $\frac{2}{5}$ 0.4 0.40 **16.** $\frac{1}{3}$ 0.3 0.33 **17.** $\frac{4}{11}$ 0.4 0.36 **18.** $\frac{3}{7}$ 0.4 0.43 **19.** $\frac{3}{11}$ 0.3 0.27 **20.** $\frac{50}{9}$ 0.6 0.56

Add. [171]

21. $149.83 + 0.52 + 88.43$ 238.78

22. $67.35 + 0.005 + 8.473$ 75.828

Subtract. [171]

23. $38.43 - 22.78$ 15.65

24. $17.14 - 8.384$ 8.756

Multiply. [173]

25. $(71.2)(3.4)$ 242.08

26. $\begin{array}{r} 41.6 \\ \times\, 0.036 \\ \hline 1.4976 \end{array}$

27. $(2.45)(0.014)$ 0.0343

28. $\begin{array}{r} 218 \\ \times\, 0.31 \\ \hline 67.58 \end{array}$

Evaluate. [171, 173]

29. $x - 5.23$ if $x = 19.1$ 13.87

30. $4.3y$ if $y = 7.01$ 30.143

Divide. Round to the nearest tenth or nearest hundredth. [177]

31. $0.04\overline{)2.448}$ 61.2

32. $0.24 \div 0.07$ 3.43

33. $0.024\overline{)0.01384}$ 0.58

34. Multiply. 3.6453×10^4 36,453 [180]

35. Write 532,000 in scientific notation. [180] 5.32×10^5

36. Solve. $x - 4.8 = 9.21$ [182] 4.41

37. Solve. $0.3x = 0.93$ [182] 3.1

38. Horace saves $4.25 each week. How [173] much has he saved after 52 weeks? $221

39. Orlando traveled 156 kilometers in [177] 2.5 hours. Find his average speed. 62.4 km/h

Chapter Test

Write a numeral for each word name.

1. five hundred twenty-eight and three thousandths 528.003

2. forty-two and three hundred one millionths 42.000301

Write a word name for the numeral.

3. 1,462.0075 one thousand, four hundred sixty-two and seventy-five ten-thousandths

Change to decimals.

4. $\frac{3}{10}$ 0.3

5. $\frac{75}{100}$ 0.75

6. $\frac{5}{1,000}$ 0.005

Round to the nearest tenth and to the nearest hundredth.

7. 0.732 0.7 0.73

8. 6.845 6.8 6.85

Change to decimals.

9. $\frac{2}{11}$ Round to the nearest tenth. 0.2

10. $\frac{5}{7}$ Round to the nearest hundredth. 0.71

Add.

11. 258.43 + 0.62 + 99.28 358.33

12. 72.36 + 0.004 + 9.416 81.78

Subtract.

13. 78.42 − 19.68 58.74

14. 19 − 3.14 15.86

Multiply.

15. (84.3)(6.2) 522.66

16. 249 × 0.012 2.988

17. 49.3567×10^3 49,356.7

Evaluate.

18. $y + 16.3$ if $y = 2.78$ 19.08

19. $2.03x$ if $x = 6.4$ 12.992

Divide.

20. $0.04\overline{)7.284}$ 182.1

21. 6.4862 ÷ 0.02 324.31

22. Divide. Round to the nearest tenth. 0.01372 ÷ 0.021 0.7

23. Write in scientific notation. 38,000,000 3.8×10^7

24. Solve. $x - 4.8 = 18.9$ 23.7

25. Solve. $\frac{x}{8.2} = 0.7$ 5.74

26. Potatoes cost $1.18 per kilogram. What is the cost of 2.5 kilograms? $2.95

27. Norma traveled 279.3 kilometers in 3.8 hours. Find her average speed. 73.5 km/h

Measurement

The Columbia *space shuttle at lift-off weighs 4.5 milllion pounds.*

Centimeters and Millimeters

To measure to the nearest
centimeter

To measure to the nearest
millimeter

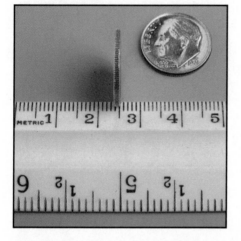

To measure the length of segment *AB*:
1. Place the end of a ruler at point *A*.
2. Read the number of units at point *B*.

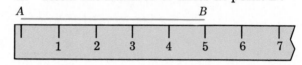

Segment *AB* measures 5 units.

The centimeter (cm) is a unit for measuring length.

1 cm

The ruler below is marked in centimeters.

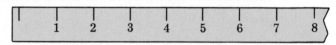

Example 1

Use a centimeter ruler to measure segment *AB*.

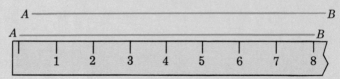

Read $\overline{AB}$ as segment *AB*.

So, $\overline{AB}$ measures 8 cm.

Example 2

Measure $\overline{CD}$ to the nearest centimeter.

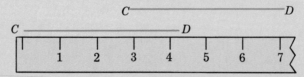

$\overline{CD}$ is between 4 cm and 5 cm but
closer to 4 cm.

Measurements are approximate.

So, $\overline{CD}$ measures 4 cm to the nearest centimeter.

---

practice ▷ **Measure to the nearest centimeter.**

1. R————————S   2. X————————————————Y
        3 cm   6 cm

The millimeter (mm) is used to measure lengths with greater accuracy. There are 10 mm in 1 cm.

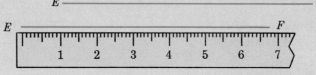

1 cm = 10 mm, or
1 mm = 0.1 cm

Example 3

Use a ruler marked in millimeters to measure $\overline{EF}$ to the nearest millimeter.

E ——————————————————— F

So, $\overline{EF}$ measures 68 mm to the nearest millimeter.

practice ▷ **Measure to the nearest millimeter.**

3. P ——————— Q
37 mm

4. M ———————————— N
42 mm

◇ EXERCISES ◇

Measure to the nearest centimeter.

1. A ——————————————— B
6 cm

2. C ——————————— D
4 cm

3. E ————— F
3 cm

4. G ———— H
2 cm

Measure to the nearest millimeter.

5. K ———————————— L
55 mm

6. M ————————————— N
44 mm

7. P ———————— Q
32 mm

8. R ———————————— S
39 mm

Measure to the nearest centimeter.   Answers will vary.

9. The length of your right foot.
11. Your height

10. The length of your left index finger
12. The length of a chalkboard eraser

Measure to the nearest millimeter.   Answers will vary.

13. The length of your right thumbnail
15. The length of your pencil

14. The length of a paper clip
16. The width of your math book

Units of Length

—◇ **OBJECTIVES** ◇—

To state the meaning of metric prefixes

To express one metric unit of length in terms of another

|←1 m→|

|←—2 dm—→|

—————— ◇ **RECALL** ◇ ——————

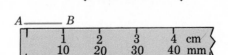

$\overline{AB}$ measures 1 cm.
$\overline{AB}$ measures 10 mm.

$$1 \text{ cm} = 10 \text{ mm, or}$$
$$1 \text{ mm} = 0.1 \text{ cm}$$

The decimeter (dm) is another unit for measuring length.

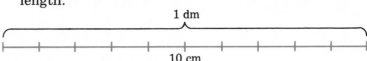

1 dm

10 cm

There are 10 cm in 1 dm.

$$1 \text{ dm} = 10 \text{ cm, or}$$
$$1 \text{ cm} = 0.1 \text{ dm}$$

Example 1

How many millimeters are there in a decimeter? A millimeter is what part of a decimeter?

$$1 \text{ dm} = 10 \text{ cm}$$
$$1 \text{ cm} = 10 \text{ mm}$$

So, 1 dm = 10 × (10 mm), or 100 mm.

$$1 \text{ cm} = 0.1 \text{ dm}$$
$$1 \text{ mm} = 0.1 \text{ cm}$$

So, 1 mm = 0.1 × (0.1 dm), or 0.01 dm.

A baseball bat is about 1 m long.

The meter (m) is the basic unit for measuring length. There are 10 dm in 1 m.

$$1 \text{ m} = 10 \text{ dm, or}$$
$$1 \text{ dm} = 0.1 \text{ m}$$

Example 2

Make true sentences.

$1 \text{ m} = \underline{\quad ? \quad} \text{ cm}$ $1 \text{ cm} = \underline{\quad ? \quad} \text{ m}$
$1 \text{ m} = 10 \text{ dm}$ $1 \text{ cm} = 0.1 \text{ dm}$
$1 \text{ dm} = 10 \text{ cm}$ $1 \text{ dm} = 0.1 \text{ m}$

So, $1 \text{ m} = 10 \times (10 \text{ cm})$, or 100 cm, and
$1 \text{ cm} = 0.1 \times (0.1 \text{ m})$, or 0.01 m.

Example 3

Make true sentences.

$1 \text{ m} = \underline{\quad ? \quad} \text{ mm}$ $1 \text{ mm} = \underline{\quad ? \quad} \text{ m}$
$1 \text{ m} = 10 \text{ dm}$ $1 \text{ mm} = 0.1 \text{ cm}$
$1 \text{ dm} = 10 \text{ cm}$ $1 \text{ cm} = 0.1 \text{ dm}$
$1 \text{ cm} = 10 \text{ mm}$ $1 \text{ dm} = 0.1 \text{ m}$

So, $1 \text{ m} = 10 \times 10 \times 10 \text{ mm}$, or $1,000 \text{ mm}$, and
$1 \text{ mm} = 0.1 \times 0.1 \times 0.1 \text{ m}$, or 0.001 m.

practice ▷ **Make true sentences.**

1. $1 \text{ cm} = \underline{\quad ? \quad} \text{ mm}$ 10

2. $1 \text{ mm} = \underline{\quad ? \quad} \text{ dm}$ 0.01

METRIC PREFIXES

kilo thousand	hecto hundred	deka ten	basic unit	deci tenth	centi hundredth	milli thousandth

The chart below shows the units for measuring length. The units that are used most often are shaded.

kilometer km 1,000 m	hectometer hm 100 m	dekameter dam 10 m	meter m	decimeter dm 0.1 m	centimeter cm 0.01 m	millim...

Example 4

Make true sentences

$1 \text{ hm} = \underline{\quad ? \quad} \text{ m}$

Use the chart. $1 \text{ hm} = 100 \text{ m}$

So, $1 \text{ hm} = 100 \text{ m}$ and

194

10. Solve
Rita's age of
Rita's age divided by
father's age divided by
quotient decreased by
father's age?
48

practice ▷ **Make true sentences.**

3. $1 \text{ dam} = \underline{\quad ? \quad} \text{ m}$ 10

UNITS OF LENGTH

◇ EXERCISES ◇

What is the meaning of each prefix?

1. deka ten
4. milli thousandth

2. centi hundredth
5. hecto hundred

3. kilo thousand
6. deci tenth

Make true sentences.

7. 1 m = __?__ cm 100
10. 1 cm = __?__ mm 10
13. 1 hm = __?__ m 100
16. 1 dm = __?__ m 0.1

8. 1 m = __?__ dam 0.1
11. 1 m = __?__ dm 10
14. 1 m = __?__ mm 1,000
17. 1 mm = __?__ cm 0.1

9. 1 km = __?__ m 1,000
12. 1 mm = __?__ m 0.00
15. 1 cm = __?__ m 0.01
18. 1 m = __?__ hm 0.01

Solve these problems.

19. The distance across Jim's thumbnail is 1 cm. How many mm is this? 10 mm

20. Juanita walked 1 km to school. How many m is this? 1,000 m

21. Marie is 1 m tall. How many cm is this? 100 cm

22. A housing complex is 1,000 m long. How many km is this? 1 km

23. A lightning bug is 10 mm long. How many cm is this? 1 cm

24. Jan's baseball bat is 100 cm long. How many m is this? 1 m

Algebra Maintenance

Evaluate.

1. $3a + 4b$ if $a = 4$ and $b = \frac{3}{4}$ 15

2. $3mr^2$ if $m = 5$ and $r = \frac{1}{5}$ $\frac{3}{5}$

3. $3.05x$ if $x = 5.6$ 17.08

4. $\frac{1}{3}c + \frac{3}{4}d$ if $c = 6$ and $d = \frac{1}{2}$ $2\frac{3}{8}$

5. $4(x + 3)$ if $x = 0.27$ 13.08

6. $0.3(x - 0.2)$ if $x = 5.1$ 1.47

Solve.

7. $2x + 5 = 13$ 4

8. $3x - 7 = 26$ 11

9. $\frac{2}{3}x - 5 = 1$ 9

these problems.

614 is the same as her
3 and the
2. What is her

11. 5 more than 6 times a number is 29. What is the number? 4

CHAPTER EIGHT

Changing Units of Length

◇ OBJECTIVE ◇

To change units of length within the metric system

◇ RECALL ◇

To multiply 5.63 by 100, move the decimal point two places to the right.

$$5.63 \times 100 = 563$$

To divide 28 by 1,000, move the decimal point three places to the left.

$$28 \div 1,000 = 0.028$$

Example 1

Use the chart of units of length.

Change 4.25 meters to centimeters.

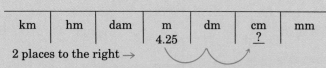

To get from m to cm, move two places to the right.

4.25

1 m = 100 cm
To multiply by 100, move the decimal point two places to the right.

So, 4.25 m = 425 cm.

Example 2

Change 5 kilometers to meters.

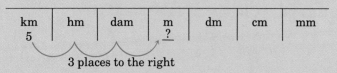

To get from km to m, move three places to the right.

5.000

1 km = 1,000 m
To multiply by 1,000, move the decimal point three places to the right.

So, 5 km = 5,000 m.

---

practice ▷ **Change as indicated.**

1. 12.5 m to cm   1,250 cm

2. 7 km to m   7,000 m

Example 3

Change 625 millimeters to meters.

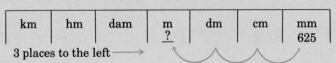

| km | hm | dam | m
? | dm | cm | mm
625 |

3 places to the left ⟶

1 mm = 0.001 m
To multiply by 0.001, move the decimal
point three places to the left.

To get from mm to m, move three places to the left.

0625.

So, 625 mm = 0.625 m.

practice ▷ **Change as indicated.**

3. 432 mm to m   0.432 m

4. 1,500 mm to m   1.5 m

Example 4

Change 4 meters to kilometers.

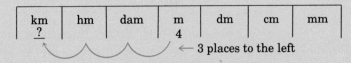

| km
? | hm | dam | m
4 | dm | cm | mm |

⟵ 3 places to the left

1 m = 0.001 km
To multiply by 0.001, move the decimal
point three places to the left.

To get from m to km, move three places to the left.

0004.

So, 4 m = 0.004 km.

practice ▷ **Change as indicated.**

5. 529 m to km   0.529 km

6. 82 m to km   0.082 km

◇ ORAL EXERCISES ◇

To change as indicated, tell how the decimal point must be moved.

1. m to cm   2 places right
2. km to m   3 places right
3. m to mm   3 places right

4. mm to cm   1 place left
5. mm to m   3 places left
6. cm to mm   1 place right

7. m to km   3 places left
8. cm to m   2 places left
9. m to dm   1 place right

◇ EXERCISES ◇

Change as indicated.

1. 8 m to cm   800 cm
2. 3,000 mm to m   3 m
3. 6.5 km to m   6,500 m

4. 4 cm to mm   40 mm
5. 7 m to mm   7,000 mm
6. 800 m to km   0.8 km

7. 55 mm to cm   5.5 cm
8. 63 cm to m   0.63 m
9. 420 mm to m   0.42 m

10. 7.5 m to cm   750 cm
11. 8.1 cm to mm   81 mm
12. 9.85 km to m   9,850 m

13. 43 m to km   0.043 km
14. 2.6 m to mm   2,600 mm
15. 359 cm to m   3.59 m

16. 12 km to m   12,000 m
17. 2 mm to cm   0.2 cm
18. 4 cm to m   0.04 m

19. 5 mm to m   0.005 m
20. 30 cm to mm   300 mm
21. 6 m to km   0.006 km

Solve these problems.

22. Pat bought 2.5 meters of fabric. How many centimeters is this?   250 cm

23. A caterpillar is 3.4 cm long. How many millimeters is this?   34 mm

24. José walked 0.75 km to the store. How many meters is this?   750 m

25. Mary won the 100-meter dash. How many kilometers did she run?   0.1 km

★ 26. John jogged 800 m on Monday, 1,300 m on Tuesday, and 1,150 m on Wednesday. How many kilometers did he jog in the 3 days?   3.25 km

★ 27. Amy is making a bookcase. She needs 3 shelves each 850 cm long and 2 end pieces each 1,200 cm long. How many meters of wood are needed?   49.5 m

Work with a partner to measure both your height and the distance between your fingertips when your arms are outstretched. Use a tape measure, or a piece of string and a meter stick. Give measurements to the nearest centimeter. What do you notice about the measurements of your height and your arm span?

Answers may vary. However, both measurements should be about the same.

Problem Solving – Applications
Jobs for Teenagers

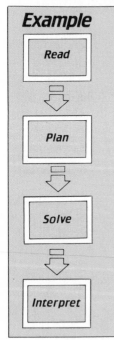

Example

Read

Plan

Solve

Interpret

When taking a road trip, the Adams family fills up the gas tank each time they stop for gas. On the last stop, it took 18.6 gal to fill up the tank. They traveled 604.5 mi since the last fill-up. How many miles per gallon did they get?

To compute mileage, use the formula:

$$mpg = \frac{m}{g}, \text{ where}$$

mpg = miles per gallon (mi/gal)
m = miles traveled
g = gallons of gas used

$mpg = \dfrac{604.5}{18.6}$ Replace m by 604.5 and g by 18.6.

     = 32.5 mi/gal.

So, the Adams got 32.5 mi/gal.

practice ▷ What is the mileage of a limousine that traveled 225.6 mi on 23.5 gal of gas? Use the formula given above to solve.   9.6 mi/gal

Solve these problems.

1. Pete's motorcycle traveled 304.5 mi on 5.8 gal of gas. What was the mileage? 52.5 mi/gal

2. On the last trip, your family got 29.5 mi/gal with your car. Getting the same mileage, how many gallons of gas will you need to travel 531 mi? 18 gal

3. The Pavuks traveled 396.8 mi averaging 24.8 mi/gal. They paid $1.179/gal of gas. How much did the gas cost for this trip? $18.86

4. The Howells traveled 592.2 mi getting 28.2 mi/gal. They paid $26.25 for the gas. What was the price of the gas? $1.25/gal

5. Ms. Lowery's car gets 18 mi/gal of gas. She paid $1.079/gal for gasoline. What is the cost of gas for a trip of 306 mi? $18.34

6. Mrs. Claire travels on the average 18 mi/day. She pays $1.099/gal for gas, and the car gets 28 mi/gal. What is the cost of gas for a 7-day week? $4.95

7. Chris works from 3:30 P.M. to 6:30 P.M. on Monday to Friday and from 8:00 A.M. to 4:00 P.M., with an hour off, on Saturday. She earns $3.25 per hour. How much does she earn in a week? $71.50

8. Chris wants to buy a stereo system with speakers that costs $119.95. How many weeks will she have to work before she can buy the system? How much money will she have left? 2 wks; $23.05 left

9. Soon, Chris will get a raise of $0.25/hr. How much more per week will she earn after the raise? How much per week will she earn then? $5.50 more; $77.00

10. Leroy changes Mr. Haley's oil filter every 10,000 mi. His odometer read 28,397 when the filter was last changed. When should it be changed again? 38,397 mi

11. Alan pumped 18 gal of gas into Anthony's car. Gas sells for $1.29/gal. How much did Anthony pay? $23.22

12. Ms. Gibson's car gets about 8 mi/gal of gas. Gas costs $1.29/gal. What is the cost for a trip of 192 mi? $30.96

13. Mrs. Rodriguez brought her car to the station for an oil change that cost $6.95, a new oil filter that cost $4.50, and a lubrication that cost $5.75. What was her total bill? $17.20

14. Jerry needed a new windshield-wiper blade and some windshield cleaning fluid. The blade cost $2.50 and the fluid cost $1.95. What was his change from a $10 bill? $5.55

15. The transmission in Jack's car has a leak. He must add a quart of transmission fluid every month. The fluid costs $2.50/qt. How much will Jack spend on transmission fluid in a year? $30

16. Eric needs to replace the water pump in his car. A new pump costs $35 plus 5% sales tax. The labor will cost $15. How much money will Eric need to replace the pump? $51.75

Area

To change units of area within
   the metric system
To compute areas of rectangles
   and squares

Area is measured in square units. The *square
centimeter* (cm^2) is a unit of area.

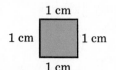

1 cm

1 cm   1 cm

1 cm

1 square centimeter
1 cm^2

See Geometry Supplement on page 420 for more information.

Example 1

Formula for the area of rectangle

To change from mm^2 to cm^2, divide
by 100.

A rectangle measures 42 mm by 16 mm. Find the
area in cm^2.

$A = l \cdot w$
   42 mm $\cdot$ 16 mm
   672 mm^2
   672 mm^2 = ___?___ cm^2
672   To divide by 100, move the decimal point two places
         to the left.
So, the area is 6.72 cm^2.

Example 2

Formula for the area of a square

To change from m^2 to cm^2, multiply
by 10,000.

A square measures 5.8 m on each side. Find the
area in cm^2.

$A = s^2$
   $(5.8 \text{ m})^2$
   33.64 m^2
   33.64 m^2 = ___?___ cm
33.6400   To multiply by 10,000, move the decimal point
             four places to the right.
So, the area is 336,400 cm^2.

1 cm
10 mm

1 cm { 10 mm

To change from cm^2 to mm^2, *multiply* by 100.
To change from mm^2 to cm^2, *divide* by 100.

1 m
100 cm

1 m { 100 cm

To change from m^2 to cm^2, *multiply* by 10,000.
To change from cm^2 to m^2, *divide* by 10,000.

practice ▷ **Make true sentences.**

1. $288 \text{ mm}^2 = \underline{} \text{ cm}^2$ 2.88
2. $36 \text{ cm}^2 = \underline{} \text{ mm}^2$ 3,600
3. $5.9 \text{ m}^2 = \underline{} \text{ cm}^2$ 59,000
4. $15,890 \text{ cm}^2 = \underline{} \text{ m}^2$ 1.589

Larger metric units of area are the are (a) and the hectare (ha). $1 \text{ a} = 100 \text{ m}^2$ $1 \text{ ha} = 10,000 \text{ m}^2$

Example 3

A rectangular lot measure 160 m by 240 m.
Find the area in ha.
$A = l \cdot w$
$ 160 \text{ m} \cdot 240 \text{ m}$
To change from m^2 to ha, divide by 10,000.
$ 38,400 \text{ m}^2$
$ 38,400 \text{ m}^2 = \underline{} \text{ ha}$
38400.   To divide by 10,000, move the decimal point four places to the left.
So, the area is 3.84 ha.

practice ▷ **Make true sentences.**

5. $460 \text{ m}^2 = \underline{} \text{ ha}$ 0.046
6. $21,900 \text{ m}^2 = \underline{} \text{ ha}$ 2.19

Make true sentences.

◇ **EXERCISES** ◇

See TE Answer Section.

1. $3 \text{ cm}^2 = \underline{} \text{ mm}^2$
2. $8 \text{ m}^2 = \underline{} \text{ cm}^2$
3. $6 \text{ ha} = \underline{} \text{ m}^2$
4. $70,000 \text{ cm}^2 = \underline{} \text{ m}^2$
5. $60,000 \text{ m}^2 = \underline{} \text{ ha}$
6. $400 \text{ mm}^2 = \underline{} \text{ cm}^2$
7. $6.1 \text{ m}^2 = \underline{} \text{ cm}^2$
8. $4.2 \text{ cm}^2 = \underline{} \text{ mm}^2$
9. $2 \text{ a} = \underline{} \text{ m}^2$
10. $7,600 \text{ m}^2 = \underline{} \text{ ha}$
11. $93,000 \text{ cm}^2 = \underline{} \text{ m}^2$
12. $39 \text{ m}^2 = \underline{} \text{ cm}^2$
13. $26 \text{ cm}^2 = \underline{} \text{ mm}^2$
14. $7,640 \text{ mm}^2 = \underline{} \text{ cm}^2$
15. $83 \text{ ha} = \underline{} \text{ m}^2$
16. $7.9 \text{ ha} = \underline{} \text{ m}^2$
17. $4,500 \text{ cm}^2 = \underline{} \text{ m}^2$
18. $380 \text{ mm}^2 = \underline{} \text{ cm}^2$
★ 19. $420 \text{ m}^2 = \underline{} \text{ a}$
★ 20. $34,000 \text{ m}^2 = \underline{} \text{ a}$
★ 21. $67 \text{ a} = \underline{} \text{ m}^2$
★ 22. $375 \text{ a} = \underline{} \text{ ha}$
★ 23. $42 \text{ km}^2 = \underline{} \text{ ha}$
★ 24. $7,100 \text{ a} = \underline{} \text{ km}^2$

Solve these problems.

25. A rectangle measures 4 cm by 8 cm. Find the area in mm^2.   3,200
26. A square measures 360 cm on each side. Find the area in m^2.   12.96
27. A square lot measures 160 m on each side. Find the area in ha.   2.56
28. A rectangle measures 50 mm by 110 mm. Find the area in cm^2.   55
29. A rectangle measures 4.2 m by 7.3 m. Find the area in cm^2.   306,600
30. A rectangular lot measures 630 m by 370 m. Find the area in ha.   23.31
31. A square deck measures 4 m on each side. Find the area in cm^2.   160,000
32. A rectangular countertop measures 230 cm by 58 cm. Find the area in m^2.   1.334
33. A rectangular park measures 170 m by 80 m. Find the area in ha.   1.36
34. A business card measures 9 cm by 6.2 cm. Find the area in mm^2.   5,580
★ 35. A square lot is 2.3 km on each side. Find the area in ha.   529
★ 36. A state park contains 5,300 ha. Express its area in km^2.   53

Capacity

To choose the best unit for
measuring the capacity of an
object

To change between milliliters
and liters

cm³ means *cubic centimeter.*

200 mm = ___?___ m

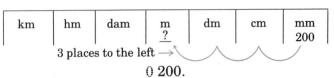

km	hm	dam	m ?	dm	cm	mm 200

3 places to the left →

0 200.

So, 200 mm = 0.2 m.

To measure the capacity of an object, we find out
how much it holds.

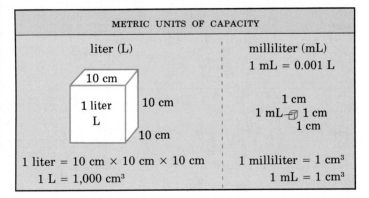

METRIC UNITS OF CAPACITY	
liter (L)	milliliter (mL) 1 mL = 0.001 L
1 liter = 10 cm × 10 cm × 10 cm 1 L = 1,000 cm³	1 milliliter = 1 cm³ 1 mL = 1 cm³

Example 1

A gas tank is greater than 1 L, so use L.
A dose of cough syrup is less than 1 L,
so use mL.

Which unit, L or mL, is better for measuring the
capacity of each?

gas tank of a car   dose of cough syrup

L is better for measuring the capacity of a tank of
gas, and mL is better for measuring the capacity of
a dose of cough syrup.

---

practice ▷ **Which unit, L or mL, is better for measuring the capacity of
each?**

1. a glass   mL

2. a fishbowl   L

The chart below shows most of the units for measuring capacity. The units that are used most often are shaded.

kiloliter kL 1,000 L	hectoliter hL 100 L	dekaliter daL 10 L	liter L	deciliter dL 0.1 L	centiliter cL 0.01 L	milliliter mL 0.001 L

Example 2

Use a chart like the meter chart.

To get from L to mL, move three places to the right.

$5 L = \underline{\ ?\ } mL$

kL	hL	daL	L 5	dL	cL	mL ?

3 places to the right →

5.000

So, 5 L = 5,000 mL.

practice ▷ **Make true sentences.**

3. $8.2 L = \underline{\ ?\ } mL$ 8,200

4. $30 L = \underline{\ ?\ } mL$ 30,000

Example 3

$425 mL = \underline{\ ?\ } L$

kL	hL	daL	L ?	dL	cL	mL 425

To get from mL to L, move three places to the left.

3 places to the left →

0 425.

So, 425 mL = 0.425 L.

practice ▷ **Make true sentences.**

5. $620 mL = \underline{\ ?\ } L$ 0.62

6. $3,800 mL = \underline{\ ?\ } L$ 3.8

Example 4

A punch recipe calls for $1\frac{1}{2}$ liters of juice. How many milliliters is this?

To get from L to mL, move three places to the right.

$1\frac{1}{2} L = 1.5 L = 1.500 mL$

So, $1\frac{1}{2}$ liters is 1,500 milliliters.

practice ▷ **Make true sentences.**

7. $0.6 L = \underline{\ ?\ } mL$ 600

8. $2.3 L = \underline{\ ?\ } mL$ 2,300

Make true sentences.

1. 1 mL = __?__ L   0.001
2. 1,000 mL = __?__ L   1
3. 1 L = __?__ mL   1,000
4. 0.001 L = __?__ mL   1

◇ EXERCISES ◇

Which unit, L or mL, is better for measuring the capacity of each?

1. a cup of hot chocolate   mL
2. bottled water for a drinking fountain   L
3. a bathtub   L
4. a baby's bottle   mL
5. a spoon   mL
6. a water tank   L
7. a glass of milk   mL
8. a bottle of cough syrup   mL

Make true sentences.

9. 8,000 mL = __?__ L   8
10. 6 L = __?__ mL   6,000
11. 1,500 mL = __?__ L   1.5
12. 12 L = __?__ mL   12,000
13. 900 mL = __?__ L   0.9
14. 8.2 L = __?__ mL   8,200
15. 365 mL = __?__ L   0.365
16. 4.25 L = __?__ mL   4,250
17. 10.4 L = __?__ mL   10,400
★ 18. 3,600 L = __?__ kL   3.6
★ 19. 2.1 kL = __?__ L   2,100
★ 20. 3 kL = __?__ mL   3,000,000

Solve these problems.

21. A recipe calls for 0.5 L of milk. How many milliliters is this?   500 mL

22. Judy drank 0.25 L of juice. How many milliliters is this?   250 mL

See Geometry Supplement on page 420 for more information.

Calculator

Find the capacity of the rectangular solid in liters. Round to the nearest tenth.

7.1 cm
9.8 cm
16.3 cm

Formula $V = l \cdot w \cdot h$

PRESS   16.3 ⊗ 9.8 ⊗ 7.1 ⊙ 1,000 ⊝

DISPLAY   1.134154

So, the capacity is 1.1 liters.

To change from cm^3 to L, divide by 1,000.

Find the capacity of the rectangular solid in liters.

1. 24 cm by 11 cm by 7 cm
   1.8 L
2. 18.1 cm by 10.5 cm by 8.4 cm
   1.6 L

Weight

◇ OBJECTIVES ◇

To choose the best unit for
  weighing an object
To change between milligrams,
  grams, kilograms, and tons

◇ RECALL ◇

1 mL — ⬚ 1 cm³

1 cm³ = 1 mL

1 L

10 cm

10 cm

10 cm

1,000 cm³ = 1L

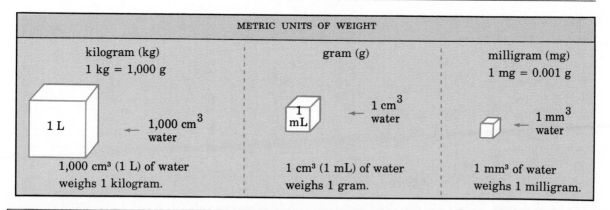

METRIC UNITS OF WEIGHT		
kilogram (kg) 1 kg = 1,000 g	gram (g)	milligram (mg) 1 mg = 0.001 g
1 L ← 1,000 cm³ water	1 mL ← 1 cm³ water	← 1 mm³ water
1,000 cm³ (1 L) of water weighs 1 kilogram.	1 cm³ (1 mL) of water weighs 1 gram.	1 mm³ of water weighs 1 milligram.

Example 1

A person weighs more than 1 kg, so use kg. A mosquito weighs more than 1 mg, but less than 1 g, so use mg.

Which unit, kg, g, or mg, is best for weighing each object?

a person   a mosquito   an apple   a cat

So, kg is the best unit for weighing a person and a cat, mg is the best unit for weighing a mosquito, and g is the best unit for weighing an apple.

practice ▷ **Which unit, kg, g, or mg, is best for weighing each object?**

1. a mouse g

2. 50 grains of sand mg

The chart below shows most of the units for measuring weight. The units used most often are shaded.

kilogram kg 1,000 g	hectogram hg 100 g	dekagram dag 10 g	gram g	decigram dg 0.1 g	centigram cg 0.01 g	milligram mg 0.001 g

WEIGHT

Example 2

Use a chart like the meter chart.

To get from g to mg, move three places to the right.

$8 \text{ g} = \underline{\ \ ?\ \ } \text{ mg}$

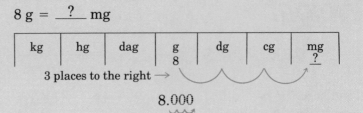

kg	hg	dag	g 8	dg	cg	mg ?

3 places to the right →

8.000

So, 8 g = 8,000 mg.

practice ▷ **Make true sentences.**

3,000

3. $5.6 \text{ g} = \underline{\ \ ?\ \ } \text{ mg}$ 5,600   **4.** $3 \text{ kg} = \underline{\ \ ?\ \ } \text{ g}$

Example 3

$375 \text{ g} = \underline{\ \ ?\ \ } \text{ kg}$

kg ?	hg	dag	g 375	dg	cg	mg

← 3 places to the left

To get from g to kg, move three places to the left.

0375.

So, 375 g = 0.375 kg.

practice ▷ **Make true sentences.**

0.4

5. $80 \text{ g} = \underline{\ \ ?\ \ } \text{ kg}$ 0.08   **6.** $400 \text{ mg} = \underline{\ \ ?\ \ } \text{ g}$

The metric ton (t) is used for weighing very heavy things, such as trucks or concrete.

1 metric ton = 1,000 kilograms

1 t = 1,000 kg, or 1 kg = 0.001 t

Example 4

To multiply by 1,000, move the decimal point three places to the right. To multiply by 0.001, move the decimal point three places to the left.

Make true sentences.

$3.2 \text{ t} = \underline{\ \ ?\ \ } \text{ kg}$ $750 \text{ kg} = \underline{\ \ ?\ \ } \text{ t}$

1 t = 1,000 kg   1 kg = 0.001 t

3.2 t = 3.2 × 1,000 kg   750 kg = 750 × 0.001 t

3.200 , or 3,200   0750, or 0.75

So, 3.2 t = 3,200 kg and 750 kg = 0.75 t.

practice ▷ **Make true sentences.**

63,000

7. $4.5 \text{ kg} = \underline{\ \ ?\ \ } \text{ t}$ 0.0045   **8.** $63 \text{ t} = \underline{\ \ ?\ \ } \text{ kg}$

Make true sentences.

1. 1 g = __?__ kg   0.001
2. 1 t = __?__ kg   1,000
3. 1 mg = __?__ g   0.001
4. 1 kg = __?__ g   1,000
5. 1 kg = __?__ t   0.001
6. 1 g = __?__ mg   1,000

◇ EXERCISES ◇

Which unit, mg, g, kg, or t, is best for weighing each object?

1. a bicycle   kg
2. a fly   mg
3. an elephant   t
4. a baby   kg
5. a quarter   g
6. a magazine   g
7. an aspirin   mg
8. a boxcar   t
9. a bus   t
10. a dog   kg
11. a pinch of salt   mg
12. a potato   g

Make true sentences.

13. 7.5 g = __?__ mg   7,500
14. 300 mg = __?__ g   0.3
15. 8.2 kg = __?__ g   8,200
16. 6.4 t = __?__ kg   6,400
17. 8,700 g = __?__ kg   8.7
18. 4,600 kg = __?__ t   4.6
19. 12 kg = __?__ g   12,000
20. 7,200 mg = __?__ g   7.2
21. 5,500 kg = __?__ t   5.5
22. 4.9 t = __?__ kg   4,900
23. 30 g = __?__ kg   0.03
24. 16 g = __?__ mg   16,000
★ 25. 6 t = __?__ g   6,000,000
★ 26. 12,000 mg = __?__ kg   0.012
★ 27. 1,000,000 g = __?__ t   1

Solve these problems.

28. A pork roast weighs 2.1 kg. How many grams is this?   2,100 g

29. A car weighs 888 kg. How many tons is this?   0.888 t

★ 30. Bill must take 4 mg of a sinus pill 3 times a day. How many grams will he take in a week?   0.084 g

★ 31. Meredith takes an iron tablet every day. Each tablet contains 63 mg of iron. How many grams of iron does she take in a year?   22.995 g

The smallest unit in our system of weight is the grain. A grain is equal to 0.0648 gram.

1. An aspirin tablet weighs 5 grains. How many milligrams is this?   324 mg

2. How many grams do 50 aspirin tablets weigh?   16.2 g

3. How many milligrams do 100 aspirin tablets weigh?   32,400 mg

4. How many kilograms do 1,000 aspirin tablets weigh?   0.324 kg

Problem Solving – Careers
Stonemasons

1. A marble floor measures 12 meters by 18 meters. How many square meters is the floor? 216 m²

2. Mrs. Spencer is a stonemason. She earns $8.75 an hour. She worked 7 hours one day. How much did she earn for the day? $61.25

3. Mr. Alvarez works as a stonemason. He builds stone exteriors of buildings. A building requires 1,200 granite stones. Each stone weighs 96 kg. How many kg of granite are needed for the building? 115,200 kg

4. Ms. Hayes is a cost estimator for a stonemasonry contractor. One job requires 430 stones to be cut. A stonemason needs about $\frac{1}{2}$ hour to cut each stone. How many hours should she estimate will be needed for the job? 215 h

5. Ms. Tsu cuts marble for marble floors. A floor of a theater lobby measures 50 m by 28 m. How many square meters of marble are needed to cover the floor? 1,400 m²

6. To repair a church Mr. Black must replace 5 damaged stones in the exterior. He needs 3 hours to replace each stone. He is paid $8.60 per hour. How much will he earn? $129

7. Ms. Newman is a stonemason's apprentice. When she completes her apprenticeship, her wages will increase by $3.00 per hour. How much more will she earn in a 40-hour week? $120 more

Temperature: Celsius and Fahrenheit

◇ OBJECTIVE ◇

To change between degrees
   Celsius and degrees
   Fahrenheit

◇ RECALL ◇

Evaluate for y if $x = 6$.

$$y = \frac{4}{3}x$$

$$y = \frac{4}{3} \cdot 6 = 8$$

The table shows a comparison between Celsius and
Fahrenheit readings.

C	0°	9.9°	15.4°	20.9°	26.4°	37°	48.2°	100°
F	32°	50°	60°	70°	80°	98.6°	120°	212°

Water freezes   Normal body   Water boils
                       temperature

Example 1

Find an approximate relationship of the Fahrenheit
(F) scale to the Celsius (C) scale.

C	Twice C	Add 30°	F	
0°	0°	30°	32°	←—exact
9.9°	19.8°	49.8°	50°	
15.4°	30.8°	60.8°	60°	
20.9°	41.8°	71.8°	70°	
26.4°	52.8°	82.8°	80°	

So, F is approximately twice C plus 30°.

Celsius to Fahrenheit formula

$$F = \frac{9}{5}C + 32°$$

Example 2

Change C = 15° and C = 20° to Fahrenheit. Use the
Celsius to Fahrenheit formula.

Substitute the Celsius degrees for C in
the formula.

$$C = 15° \qquad\qquad C = 20°$$

$$F = \frac{9}{5} \cdot 15 + 32 \qquad F = \frac{9}{5} \cdot 20 + 32$$

$$= 27 + 32 \qquad\qquad = 36 + 32$$

$$= 59° \qquad\qquad\quad = 68°$$

So, F = 59° if C = 15°, and F = 68° if C = 20°.

practice ▷ **Change Celsius to Fahrenheit.**

     1. C = 10° F = 50°   **2.** C = 25° F = 77°   **3.** C = 18° F = 6

Fahrenheit to Celsius formula
$$C = \frac{5}{9}(F - 32°)$$

Example 3

Change F = 86° and F = 40° to Celsius. Use the Fahrenheit to Celsius formula.

$$F = 86°$$

Substitute the Fahrenheit degrees for F in the formula.

$$C = \frac{5}{9}(86 - 32)$$

$$= \frac{5}{9} \cdot 54$$

$$= 30°$$

$$F = 40°$$

$$C = \frac{5}{9}(40 - 32)$$

$$= \frac{5}{9} \cdot 8$$

$4.\overline{4}$ means 4.444 . . .

$$= \frac{40}{9} \text{ or } 4.\overline{4}°$$

So, C = 30° if F = 86°, and C = $4.\overline{4}°$ if F = 40°.

---

practice ▷ **Change Fahrenheit to Celsius.**

     4. F = 80° C = $26.\overline{6}°$ **5.** F = 60° C = $15.\overline{5}°$ **6.** F = 50° C =

◇ EXERCISES ◇

Change to Fahrenheit.

1. 30°C 86°F	**2.** 12°C 53.6°F	**3.** 0°C 32°F	**4.** 24°C 75.
5. 8°C 46.4°F	**6.** 35°C 95°F	**7.** 16°C 60.8°F	**8.** 32°C 89.
9. 5°C 41°F	**10.** 27°C 80.6°F	**11.** 34°C 93.2°F	**12.** 29°C 84.

Change to Celsius.

13. 32°F 0°C	**14.** 37°F $2.\overline{7}°$C	**15.** 68°F 20°C	**16.** 49°F 9.4
17. 72°F $22.\overline{2}°$C	**18.** 41°F 5°C	**19.** 59°F 15°C	**20.** 96°F 35.
21. 54°F $12.\overline{2}°$C	**22.** 60°F $15.\overline{5}°$C	**23.** 78°F $25.\overline{5}°$C	**24.** 84°F 28.
★ **25.** 18°F $-7.\overline{7}°$C	★ **26.** 0°F $-17.\overline{7}°$C	★ **27.** 10°F $-12.\overline{2}°$C	★ **28.** 24°F -4

There are three commonly used methods for putting data into a computer.

INPUT   Allows you to change the data each time you **RUN** the program.

READ/DATA   The data is typed in as part of the program. The program can be **RUN** for only that *data*.

SEQUENTIAL INPUT   The data consists of numbers in *sequence*.

Example 1   The formula for changing from Celsius to Fahrenheit is

$F = \frac{9}{5} C + 32$. Write a program to find F for any 6 different

values of C. Decide which of the three methods described above is most appropriate. Use **INPUT** since you wish to be able to change the data with each **RUN**.

```
Allows for 6 values.    10   FOR K = 1 TO 6
                        20   INPUT "TYPE IN A VALUE FOR C ";C
Formula in BASIC.       30   LET F = 9 / 5 * C + 32
                        40   PRINT "FAHRENHEIT IS "F
                        50   NEXT K
```

Example 2   Choose the best method to write a program to find F for the following values of C: 25, 30, 35, 40, 45, 50. The data are in *sequence* 25, 30, 35, 40, 45, 50, each increased by 5.
Use **SEQUENTIAL INPUT.**

```
Data are stepped up by 5.    10   FOR C = 25 TO 50 STEP 5
No INPUT statement is needed. 20   LET F = 9 / 5 * C + 32
The variable C in line 20 is  30   PRINT "FAHRENHEIT IS "F
the same as C in line 10.     40   NEXT C
```

See the Computer Section beginning on page 420 for more information.

Exercises

1.   Write a program to find F for the following values of C: 10, 8, 26, 45, 67, 90. Why would the **READ/DATA** format be the best choice? **RUN** the program.
See TE Answer Section for the program and answer. Because a set of data is given.

Write and **RUN** a program to find F for the following values of C:

2.   Any 7 different values of C   3.   14, 16, 18, 20, 22   2.   change line 10 in Example 1's program to read **FOR K = 1 to 7.**   3.   change line 10 in Example 2's program to read **FOR C = 14 to 22 STEP 2.** 57.2°F, 60.8°F, 64.4°F, 68°F, 71.6°F

COMPUTER ACTIVITIES   211

Measure to the nearest centimeter. [190]

1. _____ 6 cm   2. _____ 4 cm

Measure to the nearest millimeter. [190]

3. _____ 53 mm   4. _____ 18 mm

Make true sentences. [192]

5. 1 cm = __?__ mm   10   6. 1 dm = __?__ m   0.1   7. 1 km = __?__ m
                                                                              1,000

Change as indicated. [195]

8. 32 mm to cm   3.2 cm   9. 4.65 km to m   4,650 m   10. 3 cm to m   0.03 m

Make true sentences. [200]

11. $9.2 \text{ m}^2 = $ __?__ cm^2 12. $460 \text{ mm}^2 = $ __?__ cm^2 13. $80,000 \text{ m}^2 = $ __?__ ha
    92,000   4.6   8

Solve this problem. [200]

14. A square measures 7.2 cm on each side. Find the area in mm^2. $5,184 \text{ mm}^2$

Which unit, L or mL, is better for measuring the capacity of each? [202]

15. a soup bowl   mL   16. the kitchen sink   L

Make true sentences. [202]

17. 4 L = __?__ mL   4,000   18. 23 mL = __?__ L   0.023   19. 5.62 L = __?__ mL
                                                                                  5,620

Solve these problems.

20. The wingspan of a moth is 42 mm.   [195] How many cm is this?   4.2 cm

21. Ramon drank 0.5 L of lemonade.   [202] How many milliliters is this?   500 mL

Which unit, mg, g, kg, or t, is best for weighing each object? [205]

22. a truck   t   23. a fox   kg   24. a glass of water   g

Make true sentences. [205]

25. 8.4 g = __?__ mg   8,400   26. 4,700 g = __?__ kg   4.7   27. 25 kg = __?__ t
                                                                                  0.025

Solve these problems. [209]

28. Change 28°C to Fahrenheit.   82.4°F   29. Change 83°F to Celsius.   28.$\overline{3}$°C

Chapter Test

Measure to the nearest centimeter.

1. _____ 3 cm

Measure to the nearest millimeter.

2. _____ 83 mm

Make true sentences.

3. 1 m = _?_ cm   100

4. 1 km = _?_ cm   100,000

5. 1 mm = _?_ m
0.001

Change as indicated.

6. 2.4 m to cm   240 cm

7. 6 mm to m   0.006 m

8. 8.42 km to m
8,420 m

Make true sentences.

9. 73 cm^2 = _?_ mm^2 7,300

10. 37,500 cm^2 = _?_ m^2 3.75

11. 91 ha = _?_ m^2
910,000

Solve this problem.

12. A rectangle measures 70 mm by 420 mm. Find its area in cm^2.   294 cm^2

Which unit, L or mL, is better for measuring the capacity of each?

13. a bucket   L

14. a soup spoon   mL

Make true sentences.

15. 6.8 L = _?_ mL   6,800

16. 420 mL = _?_ L   0.42

Solve these problems.

17. Jane rode 0.85 km on her bike. How many m is this?   850 m

18. A recipe for pancakes calls for 0.25 L of milk. How many milliliters is this?
250 mL

Which unit, mg, g, kg, or t, is best for weighing each object?

19. an orange
g

20. a baby
kg

21. a flea
mg

Make true sentences.

22. 4.3 kg = _?_ g
4,300

23. 28 mg = _?_ g
0.028

Solve these problems.

24. Change 21°C to Fahrenheit.
69.8°F

25. Change 71°F to Celsius.
21.6̄°C

Cumulative Review

Find the least common multiple (LCM).

1. 4; 14 28

2. 7; 9 63

3. 2; 5; 25 50

4. 4; 6; 10 60

Add or subtract. Simplify if possible.

5. $\frac{1}{2} + \frac{2}{7}$ $\frac{11}{14}$

6. $3\frac{3}{4} + 7\frac{5}{6}$ $11\frac{7}{12}$

7. $\frac{4}{7} - \frac{1}{3}$ $\frac{5}{21}$

8. $5\frac{1}{4} - 2\frac{7}{10}$ $2\frac{11}{20}$

Evaluate.

9. $\frac{2}{3}x + \frac{4}{5}y$ if $x = \frac{1}{2}$ and $y = \frac{5}{2}$ $2\frac{1}{3}$

10. $1.06x$ if $x = 2.5$ 2.65

11. Make a line graph to show Karen's scores on 6 mathematics tests: Test 1, 92; Test 2, 71; Test 3, 89; Test 4, 99; Test 5, 78; Test 6, 90. Check students' graphs.

Perform the indicated operations.

12. 4.78 + 2.006 + 10.4 17.186

13. 57.04 − 12.678 44.362

14. (54.8)(7.4) 405.52

15. $0.07\overline{)2.492}$ 35.6

16. Write using scientific notation. 47,000,000 4.7×10^7

17. Solve. $0.05n = 1.345$ 26.9

Make true sentences.

18. 3 m = ___?___ cm 300

19. 6 mm = ___?___ cm 0.6

20. 6.8 kg = ___?___ g 6,800

21. 45 mg = ___?___ g 0.045

22. 3.5 L = ___?___ mL 3,500

23. 680 mL = ___?___ L 0.68

24. Change 24°C to Fahrenheit. 75.2°F

25. Change 56°F to Celsius. $13.\overline{3}$°C

Solve these problems.

26. Carla worked $5\frac{2}{5}$ hours on Thursday and $6\frac{1}{4}$ hours on Friday. How many hours did she work on both days? $11\frac{13}{20}$

27. Fernando set himself a goal of working $45\frac{1}{4}$ hours during the 5-day week. He worked $38\frac{3}{4}$ hours during the 4 days. How many more hours does he have to work to meet his goal? $6\frac{1}{2}$

28. Lyle traveled 259.9 km in 4.6 hours. What was his average speed? 56.5 km/h

29. The length of a rectangle is 29 cm. Its perimeter is 97 cm. Find the width. 19.5 cm

Percents

9

Chicago's famous skyline is dominated by some of the world's tallest buildings.

Introduction to Percent

To change percents to decimals
To change decimals to percents
To change fractions to percents
To solve problems using
  percents

$0.62 = \frac{62}{100}$ Change $\frac{2}{5}$ to a decimal.

$$\begin{array}{r} 0.4 \\ 5\overline{)2.0} \end{array}$$
Divide.

42% means 42 out of 100 or 0.42.

Example 1

Change 59% to a decimal.
  Drop the % symbol.
0.59 ← Move the decimal point two places to the
  left.

59% means 59 out of 100.

So, 59% = 0.59.

---

practice ▷ **Change to decimals.**

                                                          0.85
1. 73% 0.73 2. 62% 0.62 3. 35% 0.35 4. 56% 0.56 5. 85%

Example 2

Change to decimals. 38.3%, $45\frac{1}{2}$%, 175%

| 38.3% | $45\frac{1}{2}$% | 175% |

Drop the % symbol. Move the decimal
  point two places to the left.

0.38 3 | $0.45\frac{1}{2}$ | 1.75

So, 38.3% = 0.383, $45\frac{1}{2}$% = $0.45\frac{1}{2}$, and
175% = 1.75.

---

practice ▷ **Change to decimals.**

                                                          0.713
6. 42.6%   7. $67\frac{1}{3}$%   8. 225% 2.25 9. 200% 2.00 10. 71.3%
   0.426 $0.67\frac{1}{3}$

Example 3

Drop the % symbol. Move the decimal
  point two places to the left.

Change to decimals. 6%, 0.6%, 6.2%
  6% | 0.6% | 6.2%
0.06 | 0.00 6 | 0.06 2

So, 6% = 0.06, 0.6% = 0.006, and 6.2% = 0.062.

---

practice ▷ **Change to decimals.**

11. 4% _{0.04} **12.** 0.4% _{0.004} **13.** 4.3% _{0.043} **14.** 5% _{0.05} **15.** 0.7% ^{0.007}

Example 4

Change to percents. 0.03, 0.625, $0.33\frac{1}{3}$, 4

| 0.03 | 0.625 | $0.33\frac{1}{3}$ | 4 |

To change a decimal to a percent, move the decimal point two places to the right. Then write the % symbol.

| 003. | 62.5 | $33\frac{1}{3}$ | 400. |
| 3% | 62.5% | $33\frac{1}{3}\%$ | 400% |

So, $0.03 = 3\%$, $0.625 = 62.5\%$, $0.33\frac{1}{3} = 33\frac{1}{3}\%$, and $4 = 400\%$.

---

practice ▷ **Change to percents.**

16. 0.07   **17.** 0.46   **18.** 0.667   **19.** $0.35\frac{1}{2}$ **20.** 8   **21.** 0.4

7%   46%   66.7% $35\frac{1}{2}\%$ 800%   40%

Example 5

Change $\frac{3}{5}$ to a percent. Change $\frac{2}{3}$ to a percent.

Change each to a decimal. Carry the division to hundredths.

$\frac{3}{5}$: $5\overline{)3.}$ $\frac{2}{3}$: $3\overline{)2.}$

Write two 0's for hundredths.

$$5\overline{)3.00} = 0.60$$
$$\begin{array}{r} 0.60 \\ 5\overline{)3.00} \\ \underline{3\ 0} \\ 0 \\ \underline{0} \\ 0 \end{array}$$

$$\begin{array}{r} 0.66 = 0.66\frac{2}{3} \\ 3\overline{)2.00} \\ \underline{1\ 8} \\ 20 \\ \underline{18} \\ 2 \end{array}$$

To change a decimal to a percent, move the decimal point two places to the right. Then write the % symbol.

$\frac{3}{5} = 0.60$ $\frac{2}{3} = 0.66\frac{2}{3}$

60% $66\frac{2}{3}\%$

So, $\frac{3}{5} = 60\%$ and $\frac{2}{3} = 66\frac{2}{3}\%$.

---

practice ▷ **Change to percents.** $22\frac{2}{9}\%$

22. $\frac{1}{2}$ 50%   **23.** $\frac{3}{4}$ 75%   **24.** $\frac{1}{3}$ $33\frac{1}{3}\%$ **25.** $\frac{2}{7}$ $28\frac{4}{7}\%$ **26.** $\frac{2}{9}$

Change to decimals.

1. 48% 0.48 2. 59% 0.59 3. 79% 0.79 4. 66% 0.66 5. 88% 0.88 6. 43% 0.43

7. 45.3% 0.453 8. 37.9% 0.379 9. $52\frac{1}{4}$% $0.52\frac{1}{4}$ 10. $68\frac{1}{3}$% $0.68\frac{1}{3}$ 11. 180% 1.8 12. 400% 4.00

13. 5% 0.05 14. 2% 0.02 15. 0.1% 0.001 16. 0.3% 0.003 17. 7.9% 0.079 18. 6.3% 0.063

Change to percents.

19. 0.77 77% 20. 0.63 63% 21. 0.48 48% 22. 0.66 66% 23. 0.38 38% 24. 0.51 51%

25. 0.02 2% 26. 0.08 8% 27. 0.05 5% 28. 0.449 44.9% 29. 0.912 91.2% 30. 0.551 55.1%

31. $0.45\frac{1}{3}$ $45\frac{1}{3}$% 32. $0.66\frac{1}{2}$ $66\frac{1}{2}$% 33. $0.22\frac{2}{9}$ $22\frac{2}{9}$% 34. 0.046 4.6% 35. 0.073 7.3% 36. 0.018 1.8%

37. 0.1 10% 38. 0.5 50% 39. 0.7 70% 40. 6 600% 41. 1 100% 42. 2 200%

43. $\frac{4}{5}$ 80% 44. $\frac{3}{4}$ 75% 45. $\frac{2}{5}$ 40% 46. $\frac{4}{9}$ $44\frac{4}{9}$% 47. $\frac{3}{7}$ $42\frac{6}{7}$% 48. $\frac{5}{6}$ $83\frac{1}{3}$%

Solve these problems.

49. Jake got 32 out of 100 correct on a test. What percent did he get correct? 32%

50. Mary got 7 out of 10 correct on a quiz. What percent did she get correct? 70%

★ 51. Find $\frac{1}{4} + \frac{1}{2}$. Change the answer to a percent. 75%

★ 52. Find $\frac{1}{3} + \frac{1}{6}$. Change the answer to a percent. 50%

Algebra Maintenance

Solve. Check.

1. $a + 3 = 21$ 18
2. $x - 3 = 6$ 9
3. $12 = 4z$ 3
4. $\frac{t}{5} = 7$ 35
5. $2x + 9 = 13$ 2
6. $13 = 5s - 7$ 4
7. $0.5m = 2.5$ 5
8. $0.04k = 1.6$ 40
9. $9 = 13 - \frac{1}{2}y$ 8

Evaluate for $l = 13$ and $w = 9$.

10. $2l + 3w$ 53
11. $7(l + w)$ 154

Solve.

12. One-tenth of the cost of a book increased by $2 is $3.90. What is the cost of the book? $19

13. If you multiply the cost of a football ticket by 5 and add $2.50, you get $50. What is the cost of the ticket? $9.50

Percent of a Number

To find a percent of a number

Change 46.3% to a decimal.

46.3%

0.46 3   Drop the % symbol.
          Move the decimal point
          two places to the left.

So, 46.3% = 0.463.

Example 1

26% of 246 is what number?

26% of 246 is what number?

Write 26% as a decimal.
Of means *multiply*.
Multiply.

$0.26 \times 246 = n$

$$
\begin{array}{r}
246 \\
\times 0.26 \\
\hline
14\,76 \\
49\,2 \\
\hline
63.96
\end{array}
$$

246 ← 0 decimal places
×0.26 ← 2 decimal places
63.96 ← 2 decimal places

$63.96 = n$

So, 26% of 246 is 63.96.

---

practice ▷   **1.**   35% of 146 is what number?   51.1   **2.**   8% of 216 is what number?

17.28

Example 2

3.5% of 118.43 is what number?

3.5% of 118.43 is what number?

Write 3.5% as a decimal.
Of means *multiply*.
Multiply.

$0.03\,5 \times 118.43 = n$

$$
\begin{array}{r}
118.43 \\
\times 0.035 \\
\hline
59215 \\
3\,5529 \\
\hline
4.14505
\end{array}
$$

118.43 ← 2 decimal places
×0.035 ← 3 decimal places
4.14505 ← 5 decimal places

$4.14505 = n$

So, 3.5% of 118.43 is 4.14505.

---

practice ▷ **3.** 3.4% of 48.24 is what number? **4.** 21.5% of 217 is what number?

1.64016   46.655

Example 3

$6\frac{1}{4}$% of 175 is what number?

$6\frac{1}{4}$% of 175 is what number?

$\frac{1}{4} = 0.25;\ 6\frac{1}{4}\% = 6.25\%$ 6.25% of 175 $= n$

$0.06\,25 \times 175 = n$

Multiply.

$$
\begin{array}{r}
175 \leftarrow \text{0 decimal places} \\
\times 0.0625 \leftarrow \text{4 decimal places} \\
\hline
875 \\
350 \\
1050 \\
\hline
10.9375 \leftarrow \text{4 decimal places}
\end{array}
$$

$10.9375 = n$

So, $6\frac{1}{4}$% of 175 is 10.9375.

---

practice ▷ **5.** $3\frac{3}{4}$% of 185 is what number?   6.9375

6. $5\frac{1}{2}$% of 19.7 is what number?   1.0835

Example 4

A high school basketball team played 40 games. The team won 35% of the games they played. How many games did the team win?

The team won 35% of 40 games.
35% of 40 is what number?
$0.35 \times 40 = n$

Multiply.   It is easier to multiply by 40.

$$
\begin{array}{cc}
\begin{array}{r} 40 \\ \times 0.35 \end{array} & \text{or} \quad \begin{array}{r} 0.35 \leftarrow \text{2 decimal places} \\ \times 40 \leftarrow \text{0 decimal places} \\ \hline 14.00 \leftarrow \text{2 decimal places} \end{array}
\end{array}
$$

$14 = n$
So, the team won 14 games.

---

practice ▷ **7.** A team played 40 games. They won 25% of them. How many games did they win?   10

8. A test had 20 questions. Chris got 85% correct. How many were correct?   17

◇ ORAL EXERCISES ◇

To find each percent, by what decimal would you multiply?

1. 43% of 186   0.43
2. 32% of 45   0.32
3. 26% of 78   0.26
4. 49% of 16 0.49
5. 4% of 124   0.04
6. 6% of 89   0.06
7. 7% of 96   0.07
8. 3% of 88 0.03
9. 2.6% of 98   0.026
10. 3.1% of 59   0.031
11. 2.25% of 84   0.0225
12. 1.3% of 95 0.013
13. 5.3% of 88   0.053
14. 4.5% of 96   0.045
15. 6.2% of 112   0.062
16. 8.1% of 19 0.081

◇ EXERCISES ◇

Compute.

1. 27% of 135 is what number?   36.45
2. 43% of 192 is what number?   82.56
3. 54% of 87 is what number?   46.98
4. 24% of 164 is what number?   39.36
5. 32% of 24.4 is what number?   7.808
6. 26% of 32.8 is what number?   8.528
7. 43% of 24.7 is what number?   10.621
8. 28% of 19.3 is what number?   5.404
9. 4.3% of 185 is what number?   7.955
10. 2.8% of 19.6 is what number?   0.5488
11. 1.5% of 14.7 is what number?   0.2205
12. 4.7% of 79 is what number?   3.713
13. $4\frac{1}{2}$% of 145 is what number?   6.525
14. $3\frac{1}{4}$% of 243 is what number?   7.8975
15. $4\frac{3}{4}$% of 91 is what number?   4.3225
16. $6\frac{1}{2}$% of 78 is what number?   5.07
★ 17. $5\frac{1}{8}$% of 196.12 is what number?   10.05115
★ 18. $\frac{1}{16}$% of 19 is what number?   0.011875

Solve these problems.

19. A basketball team played 45 games. They won 60% of them. How many games did they win?   27
20. A test had 50 items. Ronnie got 70% correct. How many were correct?   35
21. A town has a population of 60,000. If 15% of the population is unemployed, how many people are unemployed?   9,000
22. The senior class consists of 260 students. If 90% of the students will graduate, how many students will graduate?   234

Calculator

Compute each percent to the nearest penny.

1. 65% of $189.45 $123.14
2. 45.3% of $2,478.32 $1,122.68
3. 49.5% of $246.85 $122.19
4. 62.3% of $11.48 $7.15

PERCENT OF A NUMBER

Finding Percents

◇ OBJECTIVE ◇

◇ OBJECTIVE ◇

To find the percent one number
is of another

◇ RECALL ◇

Solve. $n \cdot 6 = 24$

$n = 24 \div 6,\ \text{or } 4$ $n = \dfrac{24}{6},\ \text{or } 4$

Example 1

What % of 80 is 20?

What % of 80 is 20?

Let n = the percent.
Of means *multiply*.

$n \cdot 80 = 20$

Carry the division to hundredths.

$$n = 20 \div 80,\ \text{or } 80\overline{)20.00}$$

```
        0.25
 80)20.00
     16 0
      4 00
      4 00
         0
```

Write the percent for 0.25. $0.25 = 25\%$
So, 25% of 80 is 20.

Example 2

4 is what % of 7?

4 is what % of 7?

Let n = the percent.
Of means *multiply*. $4 = n \cdot 7$

Carry the division to hundredths. $4 \div 7 = n$, or

$$\begin{array}{r} 0.57 \text{ or } 0.57\tfrac{1}{7} \\ 7\overline{)4.00} \end{array}$$

Write the percent for $0.57\tfrac{1}{7}$. $0.57\tfrac{1}{7} = 57\tfrac{1}{7}\%$

So, 4 is $57\tfrac{1}{7}\%$ of 7.

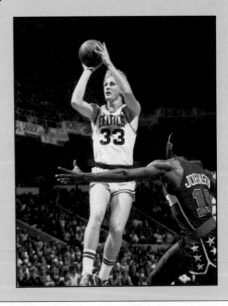

---

practice ▷   **1.** What % of 40 is 20?   50%   **2.** 5 is what % of 6? $83\tfrac{1}{3}\%$

Example 3

The Celtics played 20 basketball games. They lost one game. What % of the games played did they lose?

Think: Loss is what % of games played?

$$1 = n \cdot 20$$

Carry the division to hundredths.

$$
\begin{array}{r}
0.0 \\
20\overline{)1.00}
\end{array}
\qquad
\begin{array}{r}
0.05 \\
20\overline{)1.00} \\
\underline{1\,00} \\
0
\end{array}
$$

$$1 \div 20 = n, \text{ or } 20\overline{)1.00}$$

$$
\begin{array}{r}
0.05 \\
20\overline{)1.00} \\
\underline{1\,00} \\
0
\end{array}
$$

$0.05 = 5\%$

So, the Celtics lost 5% of the games played.

practice ▷ 3. A basketball team played 25 games. They lost 2 games. What % of the games played did they lose? 8%

◇ EXERCISES ◇

Compute.

1. What % of 80 is 40? 50%
2. What % of 10 is 8? 80%
3. What % of 40 is 30? 75%
4. What % of 70 is 14? 20%
5. What % of 20 is 18? 90%
6. What % of 30 is 12? 40%
7. 5 is what % of 7? $71\frac{3}{7}\%$
8. 5 is what % of 9? $55\frac{5}{9}\%$
9. 40 is what % of 60? $66\frac{2}{3}\%$
10. 20 is what % of 90? $22\frac{2}{9}\%$
11. 15 is what % of 21? $71\frac{3}{7}\%$
12. 21 is what % of 32? $65\frac{5}{8}\%$
13. What % of 62 is 45? $72\frac{18}{32}\%$
14. What % of 52 is 32? $61\frac{7}{13}\%$
★ 15. What % of 3.2 is 1.6? 50%
★ 16. What % of $24.75 is $4.95? 20%

Solve these problems.

17. The Blue Devils played 15 games. They lost 3 games. What % of the games played did they lose? 20%

18. John's salary was $50. He got a raise of $2. What % of his salary was his raise? 4%

19. Jane was at bat 80 times. She scored 6 hits. What % of the time did she score hits? $7\frac{1}{2}\%$

20. A bookkeeping test had 30 questions. Bill got 24 right. What % of the questions did he get right? 80%

21. Concha's salary is $140 a week. She saves $28 a week. What % of her salary does she save? 20%

22. Wendy weighed 60 kilograms. She lost 5 kilograms. What % of her weight did she lose? $8\frac{1}{3}\%$

23. Kazuo had $36. He spent $12 for a wallet. What % of his money did he spend? $33\frac{1}{3}\%$

24. In a class of 30 students, 7 students were absent. What % of the class was absent? $23\frac{1}{3}\%$

Finding the Number

 **OBJECTIVE**  ━━━━━━━━━━━ **RECALL** ━━━━━━━

To find a number given a
percent of the number

Round to the nearest whole number.

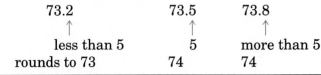

73.2	73.5	73.8
↑	↑	↑
less than 5	5	more than 5
rounds to 73	74	74

Example 1

15 is 30% of what number?

15 is 30% of what number?

Let n be the number. $15 = 30\%$ of n
30% of n means $0.30 \cdot n$. $15 = 0.30 \cdot n$
Divide both sides by 0.30.

$15 \div 0.30$ means $0.30\overline{)15}$. $15 \div 0.30 = n$ $0.30\overline{)15.00} \longrightarrow$

$$\begin{array}{r} 50 \\ 30\overline{)1500} \\ 150 \\ \hline 0 \\ 0 \\ \hline 0 \end{array}$$

Move the decimal point two places
to the right.

$50 = n$
So, 15 is 30% of 50.

---

practice ▷ **1.** 18 is 20% of what number? 90   **2.** 14 is 70% of what number? 20

Example 2

60% of what number is 40? Round the answer to
the nearest whole number.

Let n be the number.
60% of n means $0.60 \cdot n$.   60% of what number is 40?
Divide both sides by 0.60. $0.60 \cdot n = 40$

$40 \div 0.60$ means $0.60\overline{)40}$. $n = 40 \div 0.60$ $0.60\overline{)40.00} \longrightarrow$
Carry the division to tenths.

$$\begin{array}{r} 66.6 \\ 60\overline{)4000.0} \\ 360 \\ \hline 400 \\ 360 \\ \hline 400 \\ 360 \\ \hline 40 \end{array}$$

66.6
↑
more than 5
66.6 rounds to 67. $67 = n$
So, 40 is 60% of 67.

---

practice ▷ **Round the answer to the nearest whole number.**

3. 70% of what number is 40?   57 4.   7 is 3% of what number?   233

Example 3

Mr. Roth sold a TV and made a profit of $70. The profit was 20% of the cost. Find the cost.

$70 profit is 20% of cost.

Let c = cost. $70 = 20\%$ of c
20% of c means $0.20 \cdot c$. $70 = 0.20 \cdot c$

Divide both sides by 0.20. $70 \div 0.20 = c$
$70 \div 0.20$ means $0.20\overline{)70}$.

$$0.20\overline{)70.00} \rightarrow 20\overline{)7000}$$

$$\begin{array}{r} 350 \\ \hline 7000 \\ \underline{60} \\ 100 \\ \underline{100} \\ 0 \\ \underline{0} \\ 0 \end{array}$$

So, the cost of the TV was $350.

---

practice ▷ 5.   Ms. Giglio sold a stereo and made a profit of $70. The profit was 25% of the cost. Find the cost. $280

◇ EXERCISES ◇

Compute.

1. 16 is 40% of what number?   40
2. 45 is 30% of what number?   150
3. 18 is 90% of what number?   20
4. 14 is 50% of what number?   28
5. 6 is 8% of what number?   75
6. 32 is 40% of what number?   80

Compute. Round each answer to the nearest whole number.

7. 70% of what number is 50?   71
8. 60% of what number is 8?   13
9. 62% of what number is 30?   48
10. 41% of what number is 24?   59
11. 19 is 31% of what number?   61
12. 14 is 45% of what number?   31
★ 13. 4.2 is 12.5% of what number?   34
★ 14. 2.3 is 30.5% of what number?   8

Solve these problems.

15. Joe sold a car and made a profit of $850. The profit was 10% of the cost. What was the cost? $8,500

16. Mona got an $8.00 raise. The raise was 5% of her salary. What was her salary? $160

17. Luis won 18 chess games. The games won were 40% of the games played. How many games did he play?   45

18. Ms. Stein saves $40 a week. The amount she saves is 20% of her salary. What is her salary? $200

FINDING THE NUMBER

Commission

—◇ **OBJECTIVES** ◇—

To find commissions
To find total wages based on
   commission plus salary

—◇ **RECALL** ◇—

7% as a decimal is 0.07.

It is common practice to give salespeople extra
money, called *commission,* on the amount of sales.
The commission is computed using the formula
$c = r \times s$.

Example 1

John is paid a 7% commission on his camera sales.
This week he sold $275.45 worth of cameras. Find
his commission.

7% of means 0.07 times.

Let c = the commission.
   c = 7% of $275.45
   c = 0.07 × $275.45

   $275.45 ← sales
   ×   0.07 ← rate of commission

Round down to the lower cent. $19.2815 ← commission

   c = $19.2815 or $19.28

So, his commission is $19.28.

practice ▷   1.   Henrietta earns 6% commission on sales of $185.43.
Find her commission. $11.13

Example 2

Felicia is paid $125.00 a week plus 9% commission
on sales. Find her total weekly earnings if her sales
are $245.

Find her commission.
9% of $245 means 0.09 times $245.

c = 9% of $245
   $245 ← sales
   × 0.09 ← rate of commission
   $22.05 ← commission

Now find her total weekly earnings.

   $125.00 ← weekly salary
   + 22.05 ← commission
   $147.05 ← total weekly earnings

So, her total weekly earnings are $147.05.

practice ▷ 2. Bill is paid $215.00 a week plus 7% commission on sales. Find his total earnings if his sales are $1,275. $304.25

Example 3

Mona earns $115.45 a week plus 17% commission on sales. Find her total weekly earnings if she sold 3 TV's at $145.46 each, 2 freezers at $189.00 each, and 5 stereos at $139.85 each.

First, find the total sales.

Multiply to find the total sales for each item.

$145.46	$189	$139.85
× 3	× 2	× 5
$436.38	$378	$699.25

Find the total sales. Add. Keep the decimal points in line with each other.

$ 436.38
  378.00
+ 699.25
$1,513.63 ← total sales

Second, find her commission.

Find 17% of $1,513.63.
Multiply 0.17 × $1,513.63.

$ 1,513.63 ← total sales
×   0.17 ← rate of commission
1059541
151363
$257.3171 ← commission

Round down to the lower cent.

$257.3171 rounds down to $257.31.

Third, find her total earnings.
$ 257.31 ← commission
+ 115.45 ← regularly weekly salary
$ 372.76
So, her total weekly earnings are $372.76.

---

practice ▷ 3. Mr. Martinez is paid $165.65 a week plus 9% commission on sales. This week he sold 3 armchairs at $215 each, 4 end tables at $64.49 each, and 3 loveseats at $279.95 each. Find his total weekly earnings. $322.50

COMMISSION

227

◇ *EXERCISES* ◇

Solve these problems.

1. Gary earns 7% commission on sales of $149.93. Find his commission. $10.49

2. Marietta earns 9% commission on sales of $275.68. Find her commission. $24.81

3. Mr. Jenkins is paid a 15% commission on furniture sales. This week his sales were $546.53. Find his commission. $81.97

4. Ms. Olson is paid a 6% commission on each sewing machine sold. A sale was $189.63. Find the commission. $11.37

5. Laura is paid a 3% commission on reorders of tune-up kits. Find her commission on a reorder of $1,545. $46.35

6. Elliot is paid a 9% commission on each carpet sold. A sale was $415.49. Find the commission. $37.39

7. Wanda is paid $135 a week plus 8% commission on sales. Find her total earnings if her sales are $345. $162.60

8. José is paid $245 a week plus 9% commission on sales. Find his total earnings if his sales are $645. $303.05

9. Leona is paid $185 a week plus 7% commission on sales. Find her total earnings if her sales are $415. $214.05

10. Frank makes $275 a week plus 5% commission on sales. Find his total earnings if his sales are $775. $313.75

11. Mr. Langella earns $355 a week plus 15% commission on sales. Find his total earnings if his sales are $418.45. $417.76

★ 12. Mrs. Whitney earns $265 a week plus $8\frac{1}{2}$% commission on sales. Find her total earnings if her sales are $1,385.64. $382.77

13. Juan earns $179.95 a week plus 19% commission on sales. Find his total earnings if he sold 3 stereos at $114.95 each, 2 dishwashers at $314.65 each, and 4 TV's at $118.45 each. $445.06

14. Mrs. Hirsch earns $265.45 a week plus 16% commission on sales. Find her total earnings if she sold 3 chairs at $245 each, 4 end tables at $43.55 each, and 3 loveseats at $265.95 each. $538.57

Reading in Math

Match the expression in the left-hand column with the corresponding expression in the right-hand column.

d	53%	**a.**	53
a	53% of 100	**b.**	530
e	5.3 is what percent of 100?	**c.**	0.053
b	530 is what percent of 100?	**d.**	0.53
c	$\dfrac{53}{1,000}$	**e.**	5.3

Discounts

—◇ **OBJECTIVE** ◇—

To find the cost of an item if a
discount is given

————◇ **RECALL** ◇————

$\frac{1}{3}$ of 6 means	Shortcut
$\frac{1}{3} \times \frac{6}{1}$	$\frac{1}{3}$ of 6
$\frac{6}{3}$, or 2	$3\overline{)6}$ $\downarrow$ 2
	$\frac{1}{3}$ of 6 is 2.

Many stores try to increase sales by advertising a
discount, or a cut, in the regular price.

Example 1

Find the cost, after the discount is deducted, of a
$29.95 radio at a 45% discount.

$$\text{cost} = \text{regular price} - \text{discount}$$

Let c = cost. $c = \$29.95 - (45\% \times \$29.95)$

45% means 0.45.

$$\begin{array}{r} \$29.95 \leftarrow \text{regular price} \\ \times 0.45 \leftarrow \text{rate of discount} \\ \hline 14975 \\ 11980 \\ \hline \end{array}$$

Always round in favor of the
storekeeper. $\$13.4775 = \$13.47 \leftarrow \text{discount}$

$c = \$29.95 - \$13.47$

Subtract.

$$\begin{array}{r} \$29.95 \leftarrow \text{regular price} \\ -13.47 \leftarrow \text{discount} \\ \hline \$16.48 \leftarrow \text{cost after the discount is deducted} \end{array}$$

So, the cost after the discount is deducted is $16.48.

---

practice ▷   **Find the cost after the discount is deducted.**

1. $17.95 pocket radio
   15% discount $15.26

2. $129.95 coat
   20% discount $103.96

DISCOUNTS

229

Example 2

First, find the total cost.
Add.
Then, find 35% of $144.55 or (0.35)($144.55).

Find the cost of the following purchases at a 35% discount: suit, $79.95; jacket, $46.75; shirt, $17.85.

$ 79.95	$ 144.55 ← total cost
46.75	× 0.35 ← rate of discount
+17.85	7 2275
$144.55	43 365
	$50.5925 ← discount

$50.5925 rounds down to $50.59.

Subtract the discount from the total cost.

The discount is $50.59.

$144.55 ← total cost
−50.59 ← discount
$ 93.96

So, the cost is $93.96.

practice ▷ 3. Find the cost of the following purchases at a 15% discount: $4.97, record; $8.95, book; $37.65, hair dryer. $43.84

Some discounts are advertised as fractions.

Example 3

A drugstore advertised "$\frac{1}{3}$ off" on all prices.
Find the cost of a $4.94 bottle of cologne.

Find $\frac{1}{3}$ of $4.94.

$\frac{1}{3}$ × $4.94 means $\frac{\$4.94}{3}$.
Divide $4.94 by 3.

```
      $1.64
   3)$4.94
      3
      19
      18
      14
      12
       2
```

Round in favor of the store.

The discount rounds down to $1.64.

$4.94 ← original marked price
−1.64 ← discount
$3.30

Subtract.

So, the bottle of cologne cost $3.30.

practice ▷ 4. A music store advertised a "$\frac{1}{3}$ off" sale.

Find the cost of a $6.95 record. $4.64

◇ ORAL EXERCISES ◇

Find the cost after the discount is deducted.

1. $0.50 pen
$0.10 discount $0.40

2. $5.00 record
$1.00 discount $4.00

3. $250 TV
$100 discount $150

4. $1,200 used car
$300 discount $900

5. $75 bike
$20 discount $55

6. $135 suit
$45 discount $90

◇ EXERCISES ◇

Find the cost after the discount is deducted.

1. $19.75 shirt
15% discount $16.79

2. $5.95 book
45% discount $3.28

3. $49.95 tire
35% discount $32.47

4. $175.45 dishwasher
12% discount $154.40

5. $245.95 stereo
35% discount $159.87

6. $345.49 sofa
45% discount $190.02

7. $79.95 suit
25% discount $59.97

8. $749.95 used car
33% discount $502.47

9. $0.68 bottle of juice
10% discount $0.62

★ **10.** $10,275 car
16.5% discount $8,579.63

★ **11.** $4,995.95 boat
14.3% discount $4,281.53

12. Find the total cost of the following purchases at a 25% discount: $15.45, shirt; $7.50, tie; $34.65, pair of shoes. $43.20

13. Find the total cost of the following purchases at a 45% discount: $7.45, perfume; $3.95, razor; $49.68, hair dryer. $33.60

14. Find the total cost of the following items at a 35% discount: $39.75, dress; $16.49, blouse; $25.75, sweater. $53.30

15. Find the total cost of the following items at a 15% discount: $49.68, mixer; $19.63, coffee maker; $8.43, frying pan. $66.08

16. Find the cost of a $7.19 record at a "$\frac{1}{3}$ off" sale. $4.80

17. Find the cost of a $37.95 sports jacket at a "$\frac{1}{4}$ off" sale. $28.47

How many pennies can be placed next to each other with their edges touching for a distance of 1 kilometer? Give your answer to the nearest thousand. 56,000 pennies

DISCOUNTS

231

Problem Solving – Applications
Jobs for Teenagers

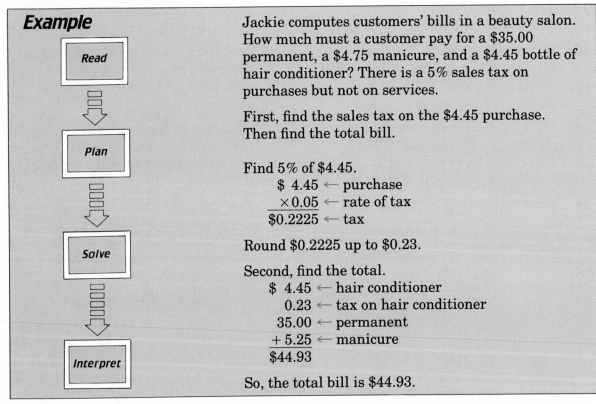

Example

Read

Plan

Solve

Interpret

Jackie computes customers' bills in a beauty salon. How much must a customer pay for a $35.00 permanent, a $4.75 manicure, and a $4.45 bottle of hair conditioner? There is a 5% sales tax on purchases but not on services.

First, find the sales tax on the $4.45 purchase. Then find the total bill.

Find 5% of $4.45.

$$\begin{array}{r} \$\ 4.45 \leftarrow \text{purchase} \\ \underline{\times 0.05} \leftarrow \text{rate of tax} \\ \$0.2225 \leftarrow \text{tax} \end{array}$$

Round $0.2225 up to $0.23.

Second, find the total.

$$\begin{array}{r} \$\ 4.45 \leftarrow \text{hair conditioner} \\ 0.23 \leftarrow \text{tax on hair conditioner} \\ 35.00 \leftarrow \text{permanent} \\ \underline{+\ 5.25} \leftarrow \text{manicure} \\ \$44.93 \end{array}$$

So, the total bill is $44.93.

practice ▷ Find the total cost of a $28.95 permanent, a $5.75 manicure, and a $6.75 bottle of skin conditioner. There is a 5% sales tax on purchases but not on services. $43.89

Solve these problems.

1. Find the total cost for a $49.95 permanent, a $15.75 manicure, and a $6.95 bottle of hair conditioner. There is a 5% sales tax on purchases only. $73.00

2. Find the total cost for a $19.95 haircut, a $15.25 manicure, and a $5.95 bottle of skin lotion. There is a 6% sales tax on purchases only. $41.51

3. Find the total cost for a $25.50 cut and blow dry, a $16.25 manicure, and a $3.75 bottle of hand lotion. There is a 5% sales tax on purchases only. $45.69

4. Find the total cost for a $53.50 permanent, a $10.25 facial massage, and a $6.25 bottle of hair conditioner. There is a 6% sales tax on purchases only. $70.38

5. Find the total cost of a $45.95 permanent and a $15.15 manicure. Find the change from four $20 bills. $61.10; $18.90

6. Find the total cost of a $52.50 permanent and a $10.50 manicure. Find the change from seven $10 bills. $63.00; $7.00

7. Find the total cost for a $44.75 permanent and a $6.25 bottle of lotion. There is a 5% sales tax on purchases only. Find the change from three $20 bills. $51.31; $8.69

8. Find the total for this bill and the change from a $100 bill. $24.95 change

CURLY'S	
Permanent	$54.25
Manicure	15.25
Lotion	5.30
Tax	.25
Total	$75.05

9. Janet works part time for a hairdresser. She is paid $5.15 per hour. Find her earnings for a 12-hour week. $61.80

10. Henry works part time for a beauty salon. He is paid $4.25 per hour. He worked 13 hours last week. How much did he earn? $55.25

11. Vincent's advertises a 20% discount on all permanents. Jane works at the register. How much should she charge a customer for a regularly priced $48 permanent? $38.40

12. This week, Cut 'n Curl is featuring a 15% discount on haircuts. Ivan is at the register this week. How much should he bill a customer for a haircut that regularly costs $15.00? $12.75

13. As a grand opening special, Hair Fair is featuring a 25% discount. Debbie is at the register. How much should she charge a customer for a cut and blow dry regularly priced at $25.60 and a frosting regularly-priced at $30.25? $41.89

14. On Wednesdays at Cerise's, all manicures are given at $\frac{1}{10}$ off. The regular price of a manicure is $15.00. How much is it on Wednesday? $13.50

15. On Tuesdays, senior citizens get a 15% discount at the Clip 'n Snip. The cost of a wash, cut, and blow dry is regularly $20.00. How much does a senior citizen pay on Tuesday? How much change will the senior citizen get from a $20 bill? $17.00; $3.00

Problem Solving – Careers
Sales Representatives

1. Mr. Skool is a yearbook sales representative. He figures that the cost per yearbook will be $4.75. What will be the total bill for a senior class of 275 students? $1,306.25

2. Julius is a traveling sales rep. He is allowed to claim $0.15 per kilometer for car expenses. Find his claim for a sales trip of 450 kilometers. $67.50

3. Maria sells encyclopedias. This month, the company is offering a 20% discount. How much should she charge a customer for a set that usually costs $485? $388

4. Goldie sells cosmetics. A customer bought a bottle of perfume for $24.75. The sales tax was 5%. The customer paid with three $10 bills. How much change should Goldie give? $4.01

5. Bill was told that there are 800 potential new car buyers in his sales district. He hopes to capture 25% of all sales. How many new cars does he expect to sell? 200

6. Mr. Martinson attended a sales convention. His expenses were $46 a night for 3 nights at a hotel, $81.72 for meals, and $65.49 for car expenses. Find his total expenses. $285.21

Non-Routine Problems

There is no simple method used to solve these problems. Use whatever methods you can. Do not expect to solve these easily.

See TE Answer Section for sample solutions to these problems.

1. Ms. Sobel traveled for 2 h at the rate of 50 mi/h. She made her return trip at the rate of 40 mi/h. What was the average speed for the entire trip? (*Hint:* It is *not* 45 mi/h.) $44\frac{4}{9}$ mi/h

2. A car travels 1 mile uphill at 20 mi/h. How fast should it travel 1 mile back downhill in order to end up with a 40 mi/h average speed for the entire 2-mile stretch?   Impossible, since the entire 3 min was used up in the first mile.

3. The Dearing family starts out on their vacation trip. They average 30 mi/h. The Osborne family starts out one hour later, and they average 40 mi/h. If they follow the same route, after how many hours of travel time are the Osbornes going to catch up with the Dearings?   3h

4. Two cars 400 mi apart start traveling toward each other. One car averages 45 mi/h and the other 35 mi/h. After how many hours will the two cars meet?   5h

5. A train covered a distance of 200 mi in 5 h. How fast must it travel another 200 mi in order to finish an entire trip with an average speed of 50 mi/h? $66\frac{2}{3}$ mi/h

6. Two trains started out traveling toward each other. They passed each other after 2 h of travel at a point 90 mi from the starting point of the first train. The second train's speed is 10 mi/h faster than the first train's speed. How many miles apart were the starting points of the two trains?   200 mi

7. You can bicycle at a speed of 15 mi/h. Your friend can bicycle at 12 mi/h. You are going to bicycle the distance of 3 mi to town, both of you starting from the same point. How much later should you start out bicycling in order for both of you to arrive there at the same time?   3 min

Mona deposits $1,000 in a bank that pays interest at a rate of 15% compounded annually (each year). This means that not only will the $1,000 draw interest, but the interest itself will, also. The following formula is used to compute the amount that the deposit will be worth at the end of any number of years:

$$A = D(1 + R)^Y \leftarrow \text{number of Years}$$

Amount   Deposit   Rate as a decimal

Example   Type in the following program for finding A in the compound interest formula $A = D(1 + R)^Y$. Then **RUN** the program to find the Amount a Deposit of $1,000 will be worth after 3 Years at a Rate of 15% (.15). First write the formula in BASIC.

$A = D \cdot (1 + R)^Y$ becomes $A = D*(1 + R) \wedge Y$.

```
10  INPUT "HOW MUCH IS DEPOSITED? ";D
20  INPUT "WHAT IS THE RATE AS A DECIMAL? ";R
30  INPUT "FOR HOW MANY YEARS? ";Y
40  LET A = D * (1 + R) ^ Y
50  PRINT "THE AMOUNT OF $"D" IS NOW WORTH $"A
60  END

]RUN
HOW MUCH IS DEPOSITED? 1000
WHAT IS THE RATE AS A DECIMAL? .15
FOR HOW MANY YEARS? 3
THE AMOUNT OF $1000 IS NOW WORTH $1520.875
```

You type 1000. No comma or $.
You type .15.
You type 3.
Round this to $1,520.88

See the Computer Section beginning on page 420 for more information.

Exercises

$7,049.37

1. **RUN** the program to find A if $4,000 is deposited at a rate of 12% for 5 years.
2. **RUN** the program to find A if $10,000 is deposited at a rate of 13% for 20 years. $115,230
3. Suppose the deposit in Exercise 2 above is doubled. Is the amount doubled for the same rate and years? Confirm by **RUN**ning the program.   yes, $230,461.70
4. Suppose the rate in Exercise 2 above is doubled to 26%. Is the amount doubled for the same deposit and years? Confirm your conclusion by **RUN**ning the program.   No, $1,017,210.67

Chapter Review

Change to decimals. [216]

1. 74% 0.74 **2.** 43.7% 0.437 **3.** $36\frac{1}{3}\%$ $0.36\frac{1}{3}$ **4.** 500% 5.00 **5.** 8% 0.08 **6.** 0.8% 0.008

Change to percents. [216]

7. 0.04 4% **8.** 0.733 73.3% **9.** $0.46\frac{1}{3}$ $46\frac{1}{3}\%$ **10.** 8 800% **11.** $\frac{2}{5}$ 40% **12.** $\frac{4}{7}$ $57\frac{1}{7}\%$

Compute. [219]

13. 45% of 156 is what number? 70.2   **14.** 3.4% of 52.23 is what number? 1.77582
15. $7\frac{1}{4}\%$ of 185 is what number? 13.4125

Compute. [222]

16. What % of 50 is 40? 80%   **17.** 4 is what percent of 9? $44\frac{4}{9}\%$

Compute. [224]

18. 12 is 60% of what number? 20   **19.** 40 is 5% of what number? 800

Round each answer to the nearest whole number. [224]

20. 3% of what number is 11? 367   **21.** 20 is 70% of what number? 29

Solve these problems.

22. [219] The Razors played 50 games. They won 30% of the games. How many games did they win? 15 games

23. [222] The Waves played 15 games. They lost 4 games. What % of the games played did they lose? $26\frac{2}{3}\%$

24. [224] Maria got a $6 raise. The raise was 4% of her salary. What was her salary? $150

25. [226] Henry earns 7% commission on all sales. This week his sales were $165.43. Find his commission. $11.58

26. [226] Jane earns $345 a week plus 7% commission on sales. Find her total earnings if her sales are $765. $398.55

27. [229] Find the cost, after the discount is taken, of a $29.95 radio at a 25% discount. $22.47

28. [229] Find the cost of the following purchases at a 25% discount: $14.95, shirt; $8.75, tie; $44.95, boots. $51.49

29. [229] Find the cost of a $19.95 electric razor at a "$\frac{1}{4}$ off" sale. $14.97

Change to decimals.

1. 29% 0.29   2. 62.3% 0.623   3. $47\frac{1}{3}$% $0.47\frac{1}{3}$ 4. 800% 8.00   5. 0.9% 0.009

Change to percents.

6. 0.09 9%   7. $0.49\frac{1}{3}$ $49\frac{1}{3}$%   8. 5 500%   9. $\frac{3}{5}$ 60%   10. $\frac{4}{9}$ $44\frac{4}{9}$%

Compute.

11. 27% of 519 is what number? 140.13

12. 4.2% of 46.83 is what number? 1.96686

13. $8\frac{1}{4}$% of 495 is what number? 40.8375

14. What % of 50 is 30? 60%

15. 5 is what % of 9? $55\frac{5}{9}$%

16. 18 is 60% of what number? 30

17. 8% of what number is 21? Round the answer to the nearest whole number. 263

Solve these problems.

18. A test had 40 questions. Martina got 60% of them right. How many did she get right? 24

19. The Jingles played 12 games. They lost 3 games. What % of the games played did they lose? 25%

20. Juan got a $15 raise. The raise was 5% of his salary. What was his salary? $300

21. Michele earns 7% commission on all sales. Last week her sales were $814.37. Find her commission. $57

22. Adam earns $385 a week plus 8% commission on sales. Find his total earnings if his sales are $735. $443.80

23. Find the cost, after the discount is taken, of a $49.95 broiler at a 35% discount. $32.47

24. Find the cost of the following purchases at a 15% discount: $19.95, slacks; $39.95, jacket; $14.35, shirt. $63.12

25. Find the cost of a $17.95 hair dryer at a "$\frac{1}{2}$ off" sale. $8.98

Statistics

Ohio's Great Serpent Mound, built by prehistoric Indians as a burial mound, is $\frac{1}{4}$ mile long.

Range, Mean, Median, Mode

To find the range, mean
(average), median, and
mode for a set of data

Scores: 8, 9, 7, 6, 10, 8, 9
Arrange in order: 10, 9, 9, 8, 8, 7, 6
Add: 10 + 9 + 9 + 8 + 8 + 7 + 6 = 57

The *range* is the difference between the highest and
lowest numbers.

Example 1

Find the range. Daily high temperatures:
24°, 25°, 20°, 18°, 21°, 23°, and 22°.

Arrange the numbers in order.   25°   24°   23°   22°   21°   20°   18°

To find the range, subtract the lowest $25° - 18° = 7°$
number from the highest number.

So, the range is 7°.

Example 2

Ivan scored 92, 97, 88, 93, and 100 on his
mathematics tests. What was his average?

Add the scores.   Add.   92   Divide.   94
Find out how many scores there are.   97   5)470
                                                   88   45
There are 5 scores.   93   20
Divide the sum by 5.   +100   20
                                                  470   0
94 is called the *mean*.   So, Ivan's average was 94.

The *mean* is the average. To find the mean, add the
numbers and divide by the number of terms.

---

practice ▷   **Find the range. Find the mean.**

   R = 12; M = 91.5   R = 6; M = 46.6   R = 8; M = 31.6
1. 91, 95, 96, 84   2. 48, 50, 46, 45, 44   3. 32, 30, 35, 34, 27

Example 3

What was Ivan's middle score?

Arrange the scores in order.   100   97   93   92   88

                                 2 above   middle score   2 below

93 is called the *median*.   So, 93 is the middle score.

The *median* is the middle number. To find the median, arrange the numbers in order and choose the middle one.

Example 4

When there are 2 middle scores, find their average.

$$\begin{array}{r} 7.5 \\ 2\overline{)15.0} \\ \underline{14} \\ 10 \\ \underline{10} \\ 0 \end{array}$$

Joe scored 7, 9, 9, 8, 6, and 5 in a dart game. What is his median score?

9 9	8 7	6 5
2 above	Middle scores	2 below

Average the middle scores: $\frac{8 + 7}{2} = \frac{15}{2}$

So, the median score is 7.5.

practice ▷ **Find the median.**

    4. 8, 7, 9, 6, 4, 8, 5   7   **5.** 4, 9, 8, 5, 10, 11   8.5 **6.** 99, 90, 88, 97, 86   90

Example 5

Group like scores. Find the score that occurs most often.

8 is called the *mode*.

Maxine scored 9, 8, 6, 7, 8, 10, 8, 9, and 10 on her English quizzes. What score did she achieve most often?

10 10	9 9	8 8 8	7 6
	twice	3 times	once

So, 8 is the score achieved most often.

The *mode* is the number that occurs most often. To find the mode, list each number as many times as it occurs and choose the one that occurs most frequently.

Example 6

Group like scores.

The high temperatures for a week were 31°, 29°, 30°, 31°, 30°, 28°, and 27°. The low temperatures were 17°, 19°, 18°, 20° 16°, 21°, and 15°. Find the mode(s) if they exist.

31° 31°	30° 30°	29° 28° 27°
twice	twice	once

So, the modes for the high temperatures are 30° and 31°. There is no mode for low temperatures.

practice ▷ **Find the mode(s) if they exist.**

    7. 8, 9, 9, 8, 7, 9   9   **8.** 3, 4, 6, 5, 2, 7   **9.** 10, 10, 9, 8, 7, 9, 9, 10
                                              none   10 and 9

◇ ORAL EXERCISES ◇

Find the range, mean (average), median, and mode(s).

1. 5, 3, 3, 3, 1   range = 4   mean = 3   median = 3; mode = 3

2. 5, 4, 3, 2, 1   range = 4   mean = 3   median = 3; mode = none

3. 10, 8, 6, 6, 3, 3   range = 7   mean = 6   median = 6; modes = 6,3

◇ EXERCISES ◇

Find the range. Find the mean.   See TE Answer Section.

1. 87, 84, 93, 92
2. 10, 12, 8, 14, 16
3. 18, 20, 22, 16, 18, 14
4. 99, 87, 99, 100, 95
5. 14, 12, 10, 14, 15
6. 16, 19, 18, 22, 19, 20
7. 26, 33, 32, 41, 58
8. 89, 78, 73, 72, 83, 91
9. 68, 72, 69, 76, 70, 77

Find the median.

10. 97, 86, 93, 84, 94   93
11. 89, 91, 78, 47, 96   89
12. 68, 71, 78, 48, 63   68
13. 8, 14, 17, 16, 19, 22   16.5
14. 84, 78, 63, 94, 96, 85   84.5
15. 14, 16, 12, 10, 9, 18   13
16. 3, 7, 14, 6, 5   6
17. 2, 4, 6, 3, 5   4
18. 4, 8, 13, 14, 87, 46   13.5

Find the mode(s) if they exist.

19. 96, 94, 96, 93, 100, 97   96
20. 98, 87, 86, 98, 87   98 and 87
21. 9, 8, 8, 9, 8, 7, 6   8
22. 24, 26, 23, 19, 18   none
23. 10, 12, 10, 9, 8, 11   10
24. 97, 96, 96, 97, 98, 100   97 and 96
25. 86, 84, 100, 84, 83   84
26. 46, 48, 53, 47, 49   none
27. 97, 96, 100, 98, 99   none

Find the range, mean, median, and mode(s).   See TE Answer Section.

28. 29, 36, 45, 28, 36, 54
29. 9, 8, 9, 8, 9, 7, 9
30. 48, 47, 52, 56, 52, 58
31. 94, 91, 88, 91, 94, 96
32. 94, 90, 90, 87, 87
33. 8, 8, 9, 9, 8, 8, 7, 6, 9
34. 8, 10, 10, 9, 9, 8, 10
35. 28, 30, 30, 27, 26
36. 99, 98, 86, 100, 98, 100

★ **37.** A basketball team averaged 76 points for 8 games. The scores for 7 games were 82, 78, 72, 76, 81, 80, and 73. What was the team's score in the eighth game?   66

★ **38.** A worker averaged $102 per week for 10 weeks. The earnings for 9 of the weeks were $98, $105, $100, $95, $115, $101, $99, $100, and $106. What was the pay for the tenth week? $10

$$9(9) = 81$$
$$99(99) = 9801$$
$$999(999) = 998001$$
$$9999(9999) = 99980001$$

Look for a pattern; then use it to find these products.

1. 99999(99999)   9999800001

2. 999999(999999)   999998000001

Mean of Grouped Data

 **OBJECTIVES**

To make a frequency table
To find a mean of grouped data

 **RECALL**

Find the mean.   10, 9, 8, 9, 7, 5
$10 + 9 + 8 + 9 + 7 + 5 = 48$
$48 \div 6 = 8$
So, the mean is 8.

Example 1

Make a frequency table.

Scores for 18 Holes of Golf

| 3 | 5 | 5 | 4 | 4 | 7 | 5 | 6 | 3 |
| 3 | 4 | 4 | 5 | 4 | 6 | 3 | 7 | 6 |

Label the columns.
Tally.
Count the tallies to get the frequency.

Score (s)	Tally	Frequency (f)	Sum ($f \cdot s$)
3	////	4	12
4	//// /	5	20
5	////	4	20
6	///	3	18
7	//	2	14

Sum = frequency × score.

Total		18	84

Example 2

Make a frequency table. Then find the mean.

Class Test Scores

95	85	100	95	75	90	100	75	100	90
90	90	85	90	85	95	90	70	95	70
100	95	80	75	80	100	75	95	90	95

Label the columns.
Tally.
Count the tallies to get the frequency.
Sum = frequency × score.
The mean is the sum divided by the frequency.

$$\begin{array}{r} 88.3 \\ 30\overline{)2,650.0} \\ \underline{2\ 40} \\ 250 \\ \underline{240} \\ 100 \\ \underline{90} \\ 10 \end{array}$$

Score (s)	Tally	Frequency (f)	Sum ($f \cdot s$)
100	////	5	500
95	//// //	7	665
90	//// //	7	630
85	///	3	255
80	//	2	160
75	////	4	300
70	//	2	140
Total		30	2,650

So, the mean $= \dfrac{2,650}{30}$, or 88.3.

practice ▷ 1. **Make a frequency table, and then find the mean.**   8.65

7 8 9 10 10 8 8 9 8 8 9 10 9 10 9 9 7 8
10 7   Check students' tables.

Sometimes it is desirable to group the data in intervals.

Example 3

Group the scores in intervals of 5. Make a frequency table. Find the mean.

Scores

48	24	26	30	34	45	28	13	21	21
6	22	31	44	36	37	12	10	27	34
35	41	48	7	14	29	31	36	42	43
9	15	23	26	29	32	36	43	10	18
20	43	30	29	27	21	27	33	28	27

Label the columns.

Choose a convenient interval by which to group the scores. For these scores, use 5.
The interval midpoint is the middle number of each interval.
Use the interval midpoint for *s*.
Sum = frequency × interval midpoint.

Score(*s*)	Tally	Frequency(*f*)	Interval(*s*) Midpoint	Sum(*f · s*)
45–49	///	3	47	141
40–44	＃＃＃ /	6	42	252
35–39	＃＃＃	5	37	185
30–34	＃＃＃ ///	8	32	256
25–29	＃＃＃ ＃＃＃ /	11	27	297
20–24	＃＃＃ //	7	22	154
15–19	//	2	17	34
10–14	＃＃＃	5	12	60
5–9	///	3	7	21
0–4		0	2	0
Total		50		1,400

Mean = $\dfrac{\text{sum}}{\text{total frequency}} = \dfrac{1,400}{50}$, or 28.

So, the mean is 28.

practice ▷ 2. **Group the scores. Choose a convenient interval. Make a frequency table. Find the mean.**   Check students' tables. 12.56

22	22	22	21	21	20	19	19	19	19	18	18
17	16	16	16	16	16	15	14	14	13	13	13
12	12	11	11	11	11	10	9	8	8	8	8
8	7	7	7	6	6	6	5	4	4	3	2

1. Copy and complete the frequency table. Then find the mean.

Scores (s)	Tally	Frequency (f)	Sum (f · s)
90	///	3	270
80	++++	5	400
70	++++ //	7	490
60	////	4	240
50	/	1	50
Total		20	1,450

Use these test scores for Exercises 2 and 3.

    95   90   85   90   95   100   80   85
    75   85   90   100   70   100   85   90
    70   75   95   75   75   95   70   100

2. Make a frequency table.

    Check students' tables.

3. Find the mean.

    86.25

Use these scores for Exercises 4 and 5.

    44   43   43   43   42   42   41   41   41   41   40   40
    40   39   39   38   37   37   36   36   36   35   34   34
    34   34   34   33   33   32   32   31   31   31   31   30
    30   30   29   29   28   28   27   27   27   26   26

4. Make a frequency table. Group the scores in intervals of three.

    Check students' tables.

5. Find the mean.   34.79

6. Collect a set of data about the heights in inches of 20 students in your class. Make a frequency table and then find the mean of the data.   Answers may vary.

7. Collect a set of data, for example, the scores you made on several tests. Make a frequency table. Find the mean.   Answers may vary.

8. Keep a record of the number of hours your mathematics class spent doing homework during one week. Choose a convenient interval and make a frequency table for the set of data. How do the mean, median, and mode compare?   Answers may vary.

Analyzing Data

─── ◇ **OBJECTIVES** ◇ ─── ─────── ◇ **RECALL** ◇ ───────

To construct a bar graph
To calculate mean
To collect data

The mean is found by adding the numbers
and dividing the sum by the number of terms.

$$70 + 80 + 90 + 100 + 110 + 120 = 570$$

6 terms

$$\frac{570}{6} = 95 \longleftarrow \text{average}$$

Example

Wildcats **44**

Visitors **30**

The Wildcats football team defeated all of their
opponents. Scores were: 24–8, 42–6, 26–18, 35–10,
44–30, and 21–18. Construct a bar graph to
compare the scores. Find the average score for the
Wildcats and for their opponents.

SCORES OF FOOTBALL GAMES

Teams—Wildcats: ▨
Opponent: ▨

To find the mean, add the scores.

Wildcats: $24 + 42 + 26 + 35 + 44 + 21 = 192$
Opponent: $8 + \ \ 6 + 18 + 10 + 30 + 18 = \ \ 90$

Divide the sum by the number of scores,
which is 6.

Wildcats	Opponent
$\frac{192}{6} = 32$	$\frac{90}{6} = 15$

So, the Wildcats averaged 32 points per game, and
their opponents averaged 15 points per game.

◇ EXERCISES ◇

**Construct a bar graph to compare scores. Calculate the
averages.**   Check students' graphs.

1. Bears vs opponents. Scores:
   18–6, 37–32, 14–21, 12–8   20.25; 16.75

2. Panthers vs opponents. Scores:
   6–8, 12–12, 24–6, 42–21   21; 11.75

3. Collect data from 5 team games in your school. Construct bar
   graphs and calculate the averages.   Check students' graphs.

Problem Solving – Applications
Jobs for Teenagers

1. Each scarf in a school store is to be sold at a discount of 10%. The marked price of each scarf is $9.00. The sales tax is 6%. What is the total sale price? $8.59

2. The marked price of a box of pencils is $1.20. The pencils are sold at a discount of 20%. The sales tax is 4%. Find the total sale price. $1.00

3. School sweaters are sold at a discount of 15%. The marked price on a school sweater is $12. The sales tax is 7%. What is the total sale price? $10.91

4. The marked price on a pair of sneakers in the school store is $24. There is a 25% discount on sneakers. The sales tax is 5%. What is the total sale price? $18.90

5. To cover the cost of operations, a school store marks up the selling price at 20% over the cost. A pen costs $4. What is the selling price? $4.80

6. A banner costs the school store $2.25. The store marks up the selling price at 10% over the cost. What is the selling price of the banner? $2.48

7. A spiral notebook costs the school store $1.10. The store marks up the selling price at 30% over the cost. Find the selling price. $1.43

8. A school store marks up the selling price at 10% over the cost. A gym suit costs the school store $8. What do they sell the gym suit for? $8.80

9. A package of construction paper costs the school store $1.50. The store marks up the selling price at 20% over the cost. What is the selling price of the construction paper? $1.80

10. A ruler costs the school store $1.30. The store marks up the selling price at 20% over the cost. The store gives a 10% discount on rulers. The sales tax is 4%. Find the total sale price. $1.47

Simple Events

To find the probability of a
  simple event
To determine the probability of
  an event that is certain to
  occur and one that cannot
  happen

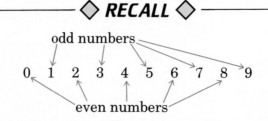

odd numbers

0 1 2 3 4 5 6 7 8 9

even numbers

Example 1

Each part of the circle is the same size.	What is the chance that the spinner will stop on red?
List all possible outcomes.	There are four possible outcomes:
Find the number of favorable outcomes—the number of times red can occur.	blue, red, green, red. There are two favorable outcomes; red occurs twice.

Red | Blue
Green | Red

Write a fraction. $\dfrac{\text{Number of favorable outcomes}}{\text{Total number of all possible outcomes}} = \dfrac{2}{4}$

2 out of 4 is the same as 1 out of 2, or $\dfrac{1}{2}$. So, the chance that the spinner will stop on red is $\dfrac{2}{4}$, or 2 out of 4.

---

practice

1. What is the chance that the spinner will stop on green? $\dfrac{1}{4}$

2. What is the chance that the spinner will stop on blue? $\dfrac{1}{4}$

Probability of an event = $\dfrac{\text{Number of favorable outcomes}}{\text{Total number of all possible outcomes}}$

Example 2

What is the probability that the spinner will stop on 1? on 2? on an even number?

Each part of the circle is the same size.
Assume that the spinner will not stop on a line.

There are 8 possible outcomes.
The number 1 occurs twice.
The number 2 occurs 3 times.
An even number occurs 4 times.

2 | 3
1 | 1
5 | 4
2 | 2

$P(1)$ means the probability that the spinner will stop on 1.

So, $P(1) = \dfrac{2}{8}$, $P(2) = \dfrac{3}{8}$, $P(\text{even number}) = \dfrac{4}{8}$.

The probability of an event that is certain to occur is 1.
The probability of an event that cannot occur is 0.

Example 3

$$P(\text{event}) = \frac{\text{Favorable outcomes}}{\text{All possible outcomes}}$$

Find. $P(\text{Red})$
Red occurs twice.

Find. $P(\text{Blue})$
Blue does not occur.

So, $P(\text{Red}) = \frac{2}{2}$, or 1.

So, $P(\text{Blue}) = \frac{0}{2}$, or 0.

Example 4

$P(1) = \frac{1}{2}$, or 1 out of 2.

About $\frac{1}{2}$ of the 50 spins should stop on 1.

$\frac{1}{2} \cdot 50 = 25$

Spin the arrow 50 times. About how many times should it stop on 1?
There are 8 possible outcomes.
The number 1 occurs 4 times.

$P(1) = \frac{4}{8}$, or $\frac{1}{2}$

So, for 50 spins, the arrow should stop on 1 about $\frac{1}{2} \cdot 50$, or 25 times.

---

practice ▷   **Use the spinner in Example 4. Spin the arrow 80 times.**

3. About how many times should it stop on 1?   40 times

4. About how many times should it stop on 2?   30 times

◇ ORAL EXERCISES ◇

1. How many possible outcomes?   4
2. How many blue outcomes?   0
3. How many green outcomes?   3
4. How many red outcomes?   1
5. What is the chance that the spinner will stop on green? $\frac{3}{4}$
6. What is the chance that the spinner will stop on red? $\frac{1}{4}$

◇ EXERCISES ◇

Find the probability for each event.

1. blue $\frac{1}{4}$
2. green $\frac{1}{4}$
3. red $\frac{2}{4}$ or $\frac{1}{2}$
4. white   0

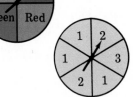

Compute.

5. $P(1)$ $\frac{3}{6}$ or $\frac{1}{2}$
6. $P(2)$ $\frac{2}{6}$ or $\frac{1}{3}$
7. $P(3)$ $\frac{1}{6}$
8. $P(\text{even number})$ $\frac{2}{6}$ or $\frac{1}{3}$
9. $P(\text{odd number})$ $\frac{4}{6}$ or $\frac{2}{3}$

10. Spin the arrow 100 times. About how many times should it stop on 1?   50

11. Spin the arrow 75 times. About how many times should it stop on 2?   25

Independent and Dependent Events

— ◇ **OBJECTIVES** ◇ —

To determine the probability of
  independent events
To determine the probability of
  dependent events

—————— ◇ **RECALL** ◇ ——————

$$P(\text{event}) = \frac{\text{number of favorable outcomes}}{\text{Total number of all possible outcomes}}$$

Find the probability that
the spinner will stop on
an odd number.

$$P(\text{odd number}) = \frac{4}{8} \text{ or } \frac{1}{2}$$

Tree diagrams can be used to determine all possible
outcomes of independent events.

Example 1

Toss one coin, and roll one die. What is the
probability of getting a head on the coin and a 4 on
the die? Use tree diagrams to find all possible
outcomes.

Toss a coin.
Roll a die.
All possible outcomes.

H						T					
1	2	3	4	5	6	1	2	3	4	5	6
H, 1	H, 2	H, 3	H, 4	H, 5	H, 6	T, 1	T, 2	T, 3	T, 4	T, 5	T, 6

There is 1 favorable outcome, H, 4.
There are 12 possible outcomes.

$P(H, 4)$ means $P(H \text{ and } 4)$.

$$P(\text{event}) = \frac{\text{Number of favorable outcomes}}{\text{Total number of all possible outcomes}}$$

$$P(H, 4) = \frac{1}{12}$$

So, the probability of getting a head on the coin and
a 4 on the die is $\frac{1}{12}$.

---

practice ▷ **Toss a coin and roll a die. Use a tree diagram to find the
probability.**   Check students' diagrams.

1. $P(H, 6)$ $\frac{1}{12}$

2. $P(T, 1)$ $\frac{1}{12}$

3. $P(H, 3)$ $\frac{1}{12}$

Tossing a coin and rolling a die are *independent
events*. One event does not depend on the other.
If A and B are independent events,
$P(A \text{ and } B) = P(A) \cdot P(B)$

practice ▷   **4.** What is the probability of tossing a tail on a coin and rolling a 5 on a die? Use the formula $P(A \text{ and } B) = P(A) \cdot P(B)$. $\frac{1}{2} \cdot \frac{1}{6} = \frac{1}{12}$

Example 3

4 green marbles and 6 red marbles are in a bag. What is the probability of picking a green marble? If you do not put it back, what is the probability of picking a green marble on the second draw?

There are 4 green marbles. $P(\text{Green Marble}) = \frac{4}{10} \text{ or } \frac{2}{5}$ First draw
10 marbles altogether.

Now, there are 3 green marbles, $P(\text{Second Green Marble}) = \frac{3}{9} \text{ or } \frac{1}{3}$ Second draw
9 marbles altogether.

So, the probability of picking a green marble on the second draw if the first is not replaced is $\frac{1}{3}$.

Example 3 is an example of *dependent events*. The second event depends on the first event.

◇ EXERCISES ◇

One box contains color cards: a white, a green, and a yellow. Another box contains number cards: 1, 2, 3, and 4. Pick one card from each.

1. Draw a tree diagram.   **2.** Find $P(W, 4)$. $\frac{1}{12}$ **3.** Find $P(G, 3)$. $\frac{1}{12}$ **4.** Find $P(Y, 1)$. $\frac{1}{12}$

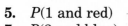 Check students' diagrams.

Look at the spinners on the right. Use the formula to find each probability.

5. $P(1 \text{ and red})$ $\frac{3}{16}$ **6.** $P(3 \text{ and white})$ $\frac{1}{16}$

7. $P(2 \text{ and blue})$ $\frac{1}{8}$ **8.** $P(4 \text{ and red})$ $\frac{1}{16}$

A bag contains 9 white marbles and 6 blue marbles. A white marble is drawn and it is not put back. On the second draw find the probability

9. of picking a white marble. $\frac{4}{7}$ **10.** of picking a blue marble. $\frac{3}{7}$

Counting Principle

To apply the counting principle in finding the number of possible arrangements

List all possible two-digit numbers using the digits 3, 5, or 7.

33	35	37
53	55	57
73	75	77

There are nine possible numbers.

Example

All possible arrangements:

Toni	Toni	Toni
Mary	Pat	Ed
Rory	Rory	Rory
Mary	Pat	Ed
Cal	Cal	Cal
Mary	Pat	Ed
Juanita	Juanita	Juanita
Mary	Pat	Ed

There are 4 candidates for president:
Toni, Cal, Rory, and Juanita.
There are 3 candidates for vice-president:
Mary, Pat, and Ed.
How many different ways can the two offices be filled?
One way: Make a tree diagram to find all possible arrangements.

Toni ⟨ Mary / Pat / Ed   3 ways   Rory ⟨ Mary / Pat / Ed   3 ways

Cal ⟨ Mary / Pat / Ed   3 ways   Juanita ⟨ Mary / Pat / Ed   3 ways

So, the offices can be filled in 12 different ways.
Another way: There are 4 choices for president and 3 choices for vice-president.
So, there are 4 · 3, or 12, ways that the offices can be filled.

counting principle
One event can happen in m ways. Another event can happen in n ways. The total number of ways both events can happen is $m \cdot n$ ways.

practice ▷

1. There are 5 candidates for president and 4 candidates for vice-president. How many different ways can the offices be filled?   20

2. Lee has 4 watches and 4 watchbands. How many different watch arrangements can be made?   16

1. There are 3 candidates for president of the student council and 4 candidates for vice-president. How many different ways can the two offices be filled?   12

2. In a restaurant, there are 5 choices of soup and 3 choices of salad. How many different soup and salad orders are there?   15

3. There are 2 positions and 5 candidates. How many different ways can the positions be filled?   10

4. A closet contains 8 blouses and 3 skirts. How many different outfits are possible?   24

★ **6.** At a restaurant, there are 4 main courses, 5 beverage choices, and 6 dessert choices. How many different dinners can be formed?   120

5. A member of a baseball team has 3 jerseys and 3 pairs of pants. How many different outfits may she wear?   9

Find the number of different license plates that can be formed.

★ **7.** The license plate must use 6 spaces. Letters and numerals can be used in any order and can be repeated. Assume that 0 cannot be used but the letter O can be used.   1,838,265,625

★ **8.** The license plate must have 3 numerals followed by 3 letters. Letters and numerals can be repeated. Assume that 0 cannot be used but the letter O can be used. 12,812,904

Algebra Maintenance

1. Find the area: $A = l \cdot w$.

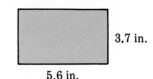

3.7 in.

5.6 in.

$A = 20.72$ in.^{2}

2. Find the perimeter: $P = 2(l + w)$

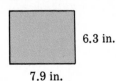

6.3 in.

7.9 in.

$P = 28.4$ in.

3. Find the area: $A = \dfrac{b \cdot h}{2}$

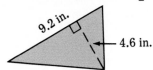

9.2 in.

4.6 in.

$A = 21.16$ in.^{2}

4. Find the perimeter: $P = a + b + c$.

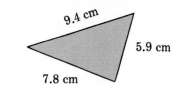

9.4 cm

5.9 cm

7.8 cm

$P = 23.1$ cm

Solve and check.

5. $\dfrac{3}{4}x = 12$ 16

6. $0.02y = 3.2$ 160

7. $r + 3\dfrac{1}{2} = 7\dfrac{1}{4}$ $3\dfrac{3}{4}$

Problem Solving – Careers
Manufacturing Inspectors

1. George Rey, an inspector at the Keystone Shirt Manufacturing Company, checks about 425 shirts in one day. About 12% of the shirts have minor defects. How many shirts have minor defects? 51 shirts

2. A shirt with no defects sells for $24. A shirt with a minor defect is discounted 20%. How much does a shirt with a minor defect sell for? $19.20

3. What is the total amount of money made on 250 shirts with minor defects? $4,800

4. A shirt with a major defect is discounted 60%. What is the total amount of money made on 125 shirts with major defects? $1,800

5. In one day, Agnes inspected 425 shirts and found that 34 of them had major defects. What percent of the shirts had major defects? 8%

6. There are 5 inspectors at the Sunshine Records and Tapes Company. Each inspector checks about 275 cassettes a day. How many cassettes are inspected in a 5-day week? $6,875

7. In a batch of 650 records, 6% were defective. How many records were defective? 39

8. In a batch of 275 cassettes, 22 were defective. What percent of the batch of cassettes was defective? 8%

Mathematics Aptitude Test

This test will prepare you for taking standardized tests.
Choose the best answer for each question.

1. The class average of 22 students was 82. If the two highest and two lowest scores were eliminated, the average of the remaining scores was 80. What was the average of the eliminated scores?
   - a. 68
   - b. 72
   - c. 88
   - **d. 91**

2. How many positive whole numbers are factors of 48?
   - a. 5
   - b. 7
   - **c. 10**
   - d. 12

3. If $A = \frac{3}{9} + \frac{3}{9} + \frac{3}{9} + \ldots$ and $B = \frac{4}{9} + \frac{4}{9} + \frac{4}{9} \ldots$, then which of the following is true?
   - **a. $A < B$**
   - b. $B < A$
   - c. $A = B$
   - d. cannot be determined

4. Twenty minutes per pound is needed to cook a 4 lb 12 oz roast. The roast was placed in the oven at 6:00 P.M. What time should the roast be removed?
   - a. 7:05 P.M.
   - b. 7:30 P.M.
   - **c. 7:35 P.M.**
   - d. 7:36 P.M.

5. $\dfrac{5(0.2) + 4(0.25) - 2(0.5)}{8(0.125)} = ?$
   - a. $\frac{1}{10}$
   - b. $\frac{7}{8}$
   - **c. 1**
   - d. 2

6. A salesperson has a weekly salary of $300. She takes a 10% reduction in pay. She then receives a 10% increase in pay. What is the correct salary?
   - a. $300
   - **b. $297**
   - c. $299.97
   - d. $330

7. The cost of m ounces of nuts at n cents a pound is?
   - **a. $\frac{m(n)}{16}$**
   - b. $\frac{16m}{n}$
   - c. $\frac{16n}{m}$
   - d. $16m(n)$

8. A student does $\frac{1}{4}$ of his homework before lunch. After lunch he completes $\frac{2}{3}$ of the remainder of his homework and then goes to a football game. What part of his homework must he still complete?
   - a. $\frac{1}{6}$
   - b. $\frac{1}{2}$
   - **c. $\frac{3}{4}$**
   - d. $\frac{11}{12}$

9. Swimming 10 laps in a swimming pool is equivalent to swimming 2 km. How many laps are equivalent to 3.2 km?
   - a. 6.25
   - b. 6.40
   - **c. 16**
   - d. 32

If a die is tossed, any 1 of 6 different numbers can come up at *random*. The probability of getting a 4 is 1 out of 6, or $P(4) = \frac{1}{6}$. So, if you were to toss the die 240 times, a 4 could be expected $\frac{1}{6} \cdot 240 = 40$ times. This is a mathematical probability. Physically tossing a die 240 times will not necessarily produce exactly 40 4's.

The program below *simulates* (shows a model of) the tossing of a die 240 times. The number of times a 4 turns up is counted. It should be close to 40.

Exercise Write a program to produce 240 random numbers from 1 through 6. Use a counter to count the number of times a 4 occurs.

C is the counter. Start with C = 0.	10	`LET C = 0`
	20	`FOR K = 1 TO 240`
	30	`LET R = INT (6 * RND (1) + 1)`
Print toss number and the result.	40	`PRINT "TOSS "K,R" COMES UP."`
If a 4 occurs, count it. Increase the number of 4's by 1 each time.	50	`IF R = 4 THEN C = C + 1`
After 240 tosses, **END**.	60	`NEXT K`
Print the number of 4's after 240 tosses.	70	`PRINT "4 OCCURS "C" TIMES."`
	80	`END`

See the Computer Section beginning on page 420 for more information.

Exercises

1. Type and **RUN** the program above 5 times. Are the results close to the predicted 40 occurrences of 4's out of 240 tosses?   yes
2. If a die is tossed 480 times, what is $P(4)$?   80
3. Modify the program above to allow for 480 tosses. **RUN** the program 5 times. Does increasing the number of tosses result in a closer degree of accuracy to the predicted $P(4)$?   Change line 20 to read **FOR K = 1 to 480**. yes
★ 4. Write a program to simulate the tossing of a coin. Allow for 200 tosses of the coin. Allow for counting the number of times the coin comes up heads. **RUN** the program 5 times.   See TE Answer Section for program and results.

Find the range. Find the mean. [240]

1. 12, 9, 13, 14, 17   R = 8; M = 13

2. 93, 97, 86, 76   R = 21; M = 88

Find the median. [240]

3. 78, 96, 70, 63, 58   70

4. 14, 18, 20, 12, 16, 23   17

Find the mode(s). [240]

5. 79, 63, 80, 79, 80, 79   79

6. 10, 9, 9, 8, 10, 7, 9, 8, 7, 8   9 and 8

Use these test scores for Exercises 7 and 8. [243]

| 100 | 95 | 100 | 85 | 90 | 95 | 100 | 95 | 80 | 90 |
| 80 | 90 | 90 | 100 | 90 | 85 | 90 | 100 | 85 | 95 |

7. Make a frequency table.
   Check students' tables.

8. Find the mean.   91.75

Use these scores for Exercises 9 and 10. [243]

99	99	98	97	97	97	69	95	95	95	95	95	94	93	93
93	92	91	91	91	91	90	90	89	89	89	88	87	87	87
86	86	86	85	85	85	85	85	84	83	83	82	81	80	80
78	78	77	75	74	74	74	73	72	72	72	71	71	71	70

9. Make a frequency table. Group the scores in intervals of 5.   Check students' tables.

10. Find the mean.   85.2

11. Construct a bar graph to compare the scores. Tigers vs opponents: 58−46, 73−56, 67−72, 46−44. Calculate the averages.   Check students' graphs.
[246]   61; 54.5

Answer these questions about the spinner. [248]

12. How many possible outcomes?   8

13. How many red outcomes?   4   **14.** blue outcomes?   3

15. P(Red)? $\frac{1}{2}$ **16.** P(Blue)? $\frac{3}{8}$ **17.** P(Green)? $\frac{1}{8}$

18. Spin the arrow 100 times. About how many times should it stop on red?   50

19. P(Red and Blue) $\frac{3}{16}$ **20.** P(Green and Blue) $\frac{3}{64}$
[250]

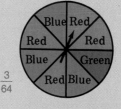

Solve these problems. [252]

21. A bag contains 10 green marbles and 6 red marbles. A red marble is drawn and not put back. Find the probability of picking another red marble. $\frac{1}{3}$

22. A closet contains 9 shirts and 3 pairs of pants. How many different outfits are possible?   27

Find the range. Find the mean.

1. 12, 9, 10, 17, 22   R = 13; M = 14

2. 83, 93, 86, 90   R = 10; M = 88

Find the median.

3. 26, 17, 83, 22, 71   26

4. 12, 16, 14, 18, 22, 10   15

Find the mode(s).

5. 8, 7, 8, 6, 7, 5, 7, 8   8 and 7

6. 27, 96, 27, 96, 96, 25   96

Use these quiz scores for Exercises 7 and 8.

10	7	6	8	9	10	10	8	6	7
8	9	8	8	7	10	9	8	6	7

7. Make a frequency table.   Check students' tables.

8. Find the mean.   8.05

Use these scores for Exercises 9 and 10.

25	25	25	24	24	23	22	22	21	21	21	21	19
19	18	17	17	17	17	16	15	15	15	15	14	14
13	13	12	11	11	10	10	10	9	8	8	7	

9. Make a frequency table. Group the scores in intervals of 3.   Check students' tables.

10. Find the mean.   16.42

11. Construct a bar graph to compare the scores. Owls vs opponents: 60–50, 48–42, 52–56, 70–68. Calculate the averages.
Check students' graphs.   57.5; 54

Answer these questions about the spinner.

12. How many possible outcomes?   8

13. $P(1)$? $\frac{1}{2}$ 14. $P(2)$? $\frac{1}{4}$ 15. $P(3)$? $\frac{1}{8}$

16. Spin the arrow 500 times. About how many times should it stop on 1?   250

17. $P(1$ and $4)$ $\frac{5}{8}$ 18. $P(2$ and $3)$ $\frac{3}{8}$

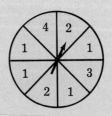

Solve these problems.

19. A bag contains 4 white marbles and 3 blue marbles. A white marble is drawn and not put back. Find the probability of picking another white marble. $\frac{1}{2}$

20. A closet contains 12 blouses and 5 skirts. How many different outfits are possible?   60

Ratio and Proportion

11

Ten million bricks were used to build New York's famous Empire State Building.

Ratios

—◇ **OBJECTIVES** ◇—

To find ratios
To determine the fraction of a
   job completed
To write ratios in two forms

—◇ **RECALL** ◇—

$\frac{3}{5} = 3 \div 5$

A comparison of 3 parts to 5 parts

Example 1

All parts are the same size.

Find the number of shaded parts and
the total number of parts. Compare the
number of shaded parts to the total
number of parts.

Write the ratio in two ways. Find the number of
shaded parts to the total number of parts.

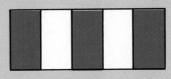

3 of 5 parts are shaded means

$\frac{3}{5}$ or 3:5

Read: *three fifths* or *the ratio 3 to 5*

So, the ratio of the number of shaded parts to the
total number of parts is $\frac{3}{5}$, or 3:5.

A *ratio* is a comparison of two numbers by division.
A fraction is a ratio.

Example 2

won 5, lost 4, total 9

A soccer team won 5 games and lost 4 games.

Find the ratio: $\frac{\text{wins}}{\text{total games}}$; $\frac{\text{losses}}{\text{wins}}$.

wins   to   total   losses   to   wins
 ↓   ↓   ↓   ↓   ↓   ↓
 5   :   9   4   :   5

So, the ratio of $\frac{\text{wins}}{\text{total games}}$ is $\frac{5}{9}$ or 5:9; the ratio of

$\frac{\text{losses}}{\text{wins}}$ is $\frac{4}{5}$ or 4:5.

---

practice ▷ **A volleyball team won 13 games and lost 3 games. Write each
            ratio in two ways.**

$\frac{13}{3}$ or 13:3 **1.** wins to losses

$\frac{3}{13}$ or 3:13
2. losses to wins

$\frac{13}{15}$ or 13:16
3. wins to total

Example 3

Jennifer can cut the lawn in 5 hours. What part can she cut in 2 hours?

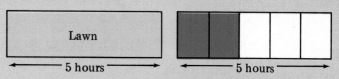

Lawn

5 hours   5 hours

Write the ratio of the number of hours worked to the total number of hours.

$$\frac{\text{Number of hours worked}}{\text{Number of hours to do the job}} = \frac{2}{5}$$

So, Jennifer can cut $\frac{2}{5}$ of the lawn in 2 hours.

Example 4

George can wash the windows in 5 h. What part can he wash in 1 h? in 3 h? in n h?

Write the ratio:
$$\frac{\text{Number of hours worked}}{\text{Number of hours to do the job}}$$

1 h worked: $\frac{1}{5}$ of job   3 h worked: $\frac{3}{5}$ of job

n h worked: $\frac{n}{5}$ of job

So, George can wash $\frac{1}{5}$ of the windows in 1 hour, $\frac{3}{5}$ in 3 hours, and $\frac{n}{5}$ in n hours.

practice ▷ Lucille can clean the house in 7 hours. What part can she clean in the given number of hours?

4. 1 $\frac{1}{7}$ **5.** 3 $\frac{3}{7}$ **6.** 4 $\frac{4}{7}$ **7.** 7 $\frac{7}{7}$ or 1   **8.** n $\frac{n}{7}$

Ratios can also be used to compare different types of rates, such as miles per hour or quantities such as number of items per cost.

Example 5

Write the ratio in two ways. Simplify when possible.

150 mi in 3 h   4 slices of pizza for $3.20

miles → $\frac{150}{3} = \frac{3 \cdot 50}{3 \cdot 1}$ $\frac{4}{3.20}$ or $\frac{4}{320}$ ← slices
hours →   ← cents

$= \frac{50}{1}$ or 50:1 $\frac{\overset{1}{\cancel{2}} \cdot \overset{1}{\cancel{2}} \cdot 1}{\underset{1}{\cancel{2}} \cdot \underset{1}{\cancel{2}} \cdot 80} = \frac{1}{80}$ or 1:80

50 mi in 1 h   1 slice of pizza for 80 cents

practice ▷ Write each ratio in two ways. Simplify when possible. $\frac{1}{35}$ or 1:35

9. 300 miles in 6 hours $\frac{50}{1}$ or 50:1   **10.** 3 pounds of apples for $1.05

RATIOS   **261**

◇ ORAL EXERCISES ◇

A fruit basket contains 7 apples, 20 oranges, 5 bananas, and 3 peaches.
Find each ratio.

1. apples to oranges   7:20
2. bananas to oranges   5:20 or 1:4
3. peaches to oranges   3:20
4. bananas to apples   5:7
5. apples to bananas   7:5
6. peaches to apples   3:7

◇ EXERCISES ◇

A hockey team won 12 games and lost 5 games. Find each ratio.

1. wins to losses   12:5
2. losses to wins   5:12
3. wins to total   12:17
4. total to losses   17:5
5. total to wins   17:12
6. losses to total   5:17

A farmer can milk the cows in 4 hours. What part of the milking can be completed in the given number of hours?

Alberta can paint a house in 13 days. What part of the painting can be completed in the given number of days?

7. 1 $\frac{1}{4}$
8. 2 $\frac{2}{4}$ or $\frac{1}{2}$
9. 3 $\frac{3}{4}$
10. x $\frac{x}{4}$
11. 4 $\frac{4}{13}$
12. 7 $\frac{7}{13}$
13. 10 $\frac{10}{13}$
14. n $\frac{n}{13}$

Write each ratio in two ways. Simplify when possible.

15. 4 cans for 84 cents $\frac{1}{21}$ or 1:21
16. 330 mi in 6 h $\frac{55}{1}$ or 55:1
17. 16 km in 60 min $\frac{4}{15}$ or 4:15
18. $66 for 2 shirts $\frac{1}{33}$ or 1:33
19. 3 pounds for $6.45 $\frac{1}{215}$ or 1:215
20. 1,024 revolutions in 4 min $\frac{256}{1}$ or 256:1
★ 21. 3.2 mi in $\frac{1}{3}$ h $\frac{9.6}{1}$ or 9.6:1
★ 22. $2\frac{1}{2}$ in. per 2 ft. $\frac{5}{4}$ or 5:4

Algebra Maintenance

Evaluate if $x = 5$, $y = 2\frac{3}{4}$, $a = 0.3$, and $b = 2.7$.

1. $2x + 3y$ $18\frac{1}{4}$
2. $5y - 2x$ $3\frac{3}{4}$
3. $2b - 4a$ 4.2
4. $2(x - y)$ $4\frac{1}{2}$
5. $\frac{x - y}{6}$ $1\frac{7}{11}$
6. $\frac{b - 3a}{b}$ 0.6

Solve. Check.

7. $x + 9 = 26$ 17
8. $6 = 12 - \frac{2}{3}m$ 9
9. $0.36k = 7.2$ 20

Solve.

10. After spending twice as much as you intended to spend, you were left with $0.46 from a $20 bill. How much did you intend to spend? $9.77

11. Your friend gave you this puzzle. Start with a number, divide it by 2, multiply the quotient by 24. The result will be 6. What number should you start with? $\frac{1}{2}$

Proportions

—◇ **OBJECTIVES** ◇—

To identify a proportion
To find the product of the means
    and of the extremes
To solve proportions

————————◇ **RECALL** ◇————————

Simplify. $\frac{12}{30}$

$$\frac{\overset{1}{\cancel{2}} \cdot 2 \cdot \overset{1}{\cancel{3}}}{\cancel{2} \cdot \cancel{3} \cdot 5}$$ Factor into primes.
        Divide out like factors.

$$\frac{1 \cdot 2 \cdot 1}{1 \cdot 1 \cdot 5}$$

$$\frac{2}{5}$$

Example 1

Are the ratios equal?

$\frac{18}{24}$ and $\frac{3}{4}$ $\frac{2}{7}$ and $\frac{6}{20}$

$\frac{18}{24}$	$\frac{3}{4}$		$\frac{2}{7}$	$\frac{6}{20}$

Factor into primes.

| $\frac{2 \cdot 3 \cdot 3}{2 \cdot 2 \cdot 2 \cdot 3}$ | $\frac{3}{4}$ | | $\frac{2}{7}$ | $\frac{2 \cdot 3}{2 \cdot 2 \cdot 5}$ |

$$\frac{\overset{1}{\cancel{2}} \cdot \overset{1}{\cancel{3}} \cdot 3}{\cancel{2} \cdot 2 \cdot 2 \cdot \cancel{3}} \qquad\qquad\qquad \frac{\overset{1}{\cancel{2}} \cdot 3}{\cancel{2} \cdot 2 \cdot 5}$$

$$\frac{1 \cdot 1 \cdot 3}{1 \cdot 2 \cdot 2 \cdot 1} \qquad\qquad\qquad \frac{1 \cdot 3}{1 \cdot 2 \cdot 5}$$

$$\frac{3}{4} \qquad\qquad\qquad\qquad \frac{3}{10}$$

So, $\frac{18}{24} = \frac{3}{4}$, but $\frac{2}{7}$ does not equal $\frac{6}{20}$.

A *proportion* is an equation which states that two
ratios are equal. $\frac{2}{3} = \frac{4}{6}$ or $2:3 = 4:6$ is a proportion.

Each is read: *two is to three as four is to six.*

practice ▷ **Which equations are proportions? Read each proportion.**

1. $\frac{2}{3} = \frac{12}{18}$ yes   **2.** $\frac{16}{20} = \frac{4}{5}$ yes   **3.** $\frac{1}{2} = \frac{7}{13}$ no   **4.** $\frac{4}{3} = \frac{20}{15}$ yes

Example 2

Identify the means and the extremes of the proportion $\frac{2}{3} = \frac{4}{6}$.

$2:3 = 4:6$
means
extremes

$\frac{2}{3} = \frac{4}{6}$ $\frac{2}{3} = \frac{4}{6}$

means   extremes

So, the means are 3 and 4; the extremes are 2 and 6.

Example 3

For the proportion $\frac{2}{5} = \frac{8}{20}$, find the product of the means. Then find the product of the extremes.

Means: 5, 8
Extremes: 2, 20

$\frac{2}{5} = \frac{8}{20}$ $\frac{2}{5} = \frac{8}{20}$

$5 \cdot 8 = 40$ $2 \cdot 20 = 40$

So, the product of the means is 40; the product of the extremes is 40.

In a *true proportion,* the product of the means equals the product of the extremes.

practice ▷ **For each proportion, find the product of the means. Then find the product of the extremes.**

5. $\frac{1}{5} = \frac{4}{20}$ 20   **6.** $\frac{2}{3} = \frac{10}{15}$ 30   **7.** $\frac{18}{30} = \frac{3}{5}$ 90   **8.** $\frac{5}{2} = \frac{10}{4}$ ^{20}

Example 4

Solve the proportions.

$\frac{2}{3} = \frac{n}{6}$ $x:6 = 3:2$

Think: What value of n makes the ratios equal?
The product of the means equals the product of the extremes.
Divide 12 by 3.

$\frac{2}{3} = \frac{n}{6}$ $\frac{x}{6} = \frac{3}{2}$

$3n = 12$ $2x = 18$

$n = \frac{12}{3}$, or 4 $x = \frac{18}{2}$, or 9

So, the solution is 4.   So, the solution is 9.

practice ▷ **Solve each proportion.**

9. $\frac{2}{5} = \frac{n}{10}$ 4   **10.** $\frac{3}{7} = \frac{n}{21}$ 9   **11.** $\frac{2}{3} = \frac{6}{x}$ 9   **12.** $x:6 = 4:3$ ^{8}

Read each proportion.

1. $\frac{2}{3} = \frac{10}{15}$
   $2:3 = 10:15$

2. $\frac{6}{5} = \frac{12}{10}$
   $6:5 = 12:10$

3. $\frac{1}{5} = \frac{6}{30}$
   $1:5 = 6:30$

4. $\frac{6}{15} = \frac{2}{5}$
   $6:15 = 2:5$

5. $\frac{2}{9} = \frac{6}{27}$
   $2:9 = 6:27$

◇ *EXERCISES* ◇

Which equations are proportions?

1. $\frac{1}{2} = \frac{4}{8}$ yes

2. $\frac{9}{12} = \frac{3}{4}$ yes

3. $\frac{1}{3} = \frac{4}{11}$ no

4. $\frac{5}{6} = \frac{15}{18}$ yes

5. $\frac{3}{7} = \frac{12}{28}$ yes

For each proportion find the product of the means. Then find the product of the extremes.

6. $\frac{1}{4} = \frac{4}{16}$ m = 16, e = 16

7. $\frac{5}{3} = \frac{15}{9}$ m = 45, e = 45

8. $\frac{2}{5} = \frac{12}{30}$ m = 60, e = 60

9. $\frac{3}{4} = \frac{18}{24}$ m = 72, e = 72

10. $\frac{2}{9} = \frac{8}{36}$ m = 72, e = 72

11. $\frac{5}{10} = \frac{3}{6}$ m = 30, e = 30

12. $\frac{4}{16} = \frac{5}{20}$ m = 80, e = 80

13. $\frac{4}{12} = \frac{6}{18}$ m = 72, e = 72

14. $\frac{2}{18} = \frac{3}{27}$ m = 54, e = 54

15. $\frac{4}{20} = \frac{2}{10}$ m = 40, e = 40

Solve each proportion.

16. $\frac{2}{5} = \frac{n}{15}$ 6

17. $\frac{2}{7} = \frac{n}{28}$ 8

18. $\frac{3}{5} = \frac{x}{15}$ 9

19. $\frac{2}{3} = \frac{x}{12}$ 8

20. $\frac{4}{5} = \frac{x}{25}$ 20

21. $\frac{1}{3} = \frac{5}{n}$ 15

22. $\frac{4}{5} = \frac{8}{n}$ 10

23. $\frac{3}{4} = \frac{15}{n}$ 20

24. $\frac{5}{6} = \frac{20}{x}$ 24

25. $\frac{1}{8} = \frac{3}{n}$ 24

26. $\frac{n}{18} = \frac{4}{9}$ 8

27. $\frac{n}{24} = \frac{7}{8}$ 21

28. $\frac{x}{20} = \frac{2}{5}$ 8

29. $\frac{x}{15} = \frac{1}{3}$ 5

30. $\frac{x}{16} = \frac{3}{4}$ 12

31. $x:12 = 5:6$ 10

32. $x:15 = 2:3$ 10

33. $x:21 = 3:7$ 9

Calculator

Solve. $\frac{2}{1.78} = \frac{n}{49.31}$ Round to the nearest hundredth. A calculator can be used, as shown below, to solve the proportion.

$$\frac{2}{1.78} = \frac{n}{49.31}$$
$$1.78\,n = 2 \cdot 49.31$$
$$n = \frac{2 \cdot 49.31}{1.78}$$

Press 2 ⊗ 49.31 ⊘ 1.78 ⊜
Display 55.404494
Round to the nearest hundredth, 55.40.

Solve. Round to the nearest hundredth.

1. $\frac{3.2}{2.56} = \frac{n}{38.37}$ 47.96

2. $\frac{1.6}{5.8} = \frac{4.3}{n}$ 15.59

3. $\frac{n}{49.3} = \frac{6.2}{8.75}$ 34.93

Applying Proportions

To solve problems by using
  proportions
To determine if a table of data
  represents direct variation

A team won 7 games and lost 4 games.
Find the ratio of losses to wins.

$$\frac{\text{losses}}{\text{wins}} = \frac{4}{7}$$

Example 1

Greta can buy a book for $6. Find the cost of 4
books. Then find the cost of 6 books.

1 book costs $6
$4 \cdot \$6 = \$24$ 4 books cost $24
$6 \cdot \$6 = \$36$ 6 books cost $36

Display this data in a table.

NUMBER OF BOOKS	COST
1	$6
4	$24
6	$36

Notice that the ratio of the number of books to the
cost remains the same.

$$\frac{1}{6} = \frac{4}{24} = \frac{6}{36}$$

As the number of books increases, the cost also
increases. This is an example of *direct variation*.

Example 2

From the table, determine if y varies directly as x.

x	y
3	9
2	6
1	2

Determine if the ratio $\frac{x}{y}$ remains
the same.

$$\frac{3}{9} = \frac{1}{3} \qquad \frac{2}{6} = \frac{1}{3} \qquad \frac{1}{2} \neq \frac{1}{3}$$

So, y does not vary directly as x.

Determine if _y_ varies directly as _x_.

1.

_x_	_y_
3	15
2	10
4	20

yes

2.

_x_	_y_
7	14
4	8
2	4

yes

3.

_x_	_y_
6	18
9	27
8	32

no

4.

_x_	_y_
2	8
3	12
5	25

no

Proportions can be used to solve problems involving direct variation.

Example 3

Lou can buy 2 book covers for 25¢. How many can he buy for 75¢?

Write a proportion. $\dfrac{2}{25} = \dfrac{n}{75}$ ← Let _n_ be the number of covers he can buy for 75¢.

The product of the means = the product of the extremes. $25n = 2 \cdot 75$
$25n = 150$

Divide.
$6 \cdot 25 = 150$ $n = \dfrac{150}{25}$, or $25\overline{)150}^{\,6}$

So, Lou can buy 6 book covers for 75¢.

practice ▷ **5.** Jorge can buy 2 pairs of trousers for $36. How much will 3 pairs cost? $54

Example 4

Clark received $15 for 6 hours of babysitting. How many hours did he sit if he earned $20?

Write a proportion.
Compare dollars to hours.

pay →
hours → $\dfrac{15}{6} = \dfrac{20}{n}$ ← Let _n_ be the number of hours he babysat.

$6 \cdot 20 = 120$ $15n = 120$

Divide. $n = \dfrac{120}{15}$, or $15\overline{)120}^{\,8}$

So, Clark babysat for 8 hours.

practice ▷ **6.** Anne got $6 every 3 days for feeding a neighbor's dog. How many days did she feed the dog if she earned $14? 7 days

State each proportion.

1. 10 boxes for $0.70 $\frac{10}{70} = \frac{n}{120}$
   n boxes for $1.20

2. 17 cans for $3.90 $\frac{17}{39} = \frac{43}{n}$
   43 cans for n

3. 8 hits for every 20 times at bat
   n hits for 465 times at bat $\frac{8}{20} = \frac{n}{465}$

4. 3 shirts for $20 $\frac{3}{20} = \frac{5}{n}$
   5 shirts for n

◇ *EXERCISES* ◇

Determine if *y* varies directly as *x*.

1.

x	y
1	3
8	24
2	8

no

2.

x	y
4	6
8	12
10	15

yes

3.

x	y
4	5
12	15
2	5

no

4.

x	y
2	7
6	21
4	7

no

Solve each problem. Use a proportion.

5. Ben can buy 2 pairs of shoes for $60. How many pairs of shoes can he buy for $90? 3 pairs

6. Marge used 2 skeins of yarn to make a scarf. How many skeins of yarn will she need to make 3 scarves? 6 skeins

7. James was given $5 for 2 hours of babysitting. How many hours must he babysit in order to earn $7.50? 3 h

8. It takes 3 meters of cloth to make 2 skirts. How many meters of cloth are needed to make 3 skirts? $2\frac{1}{2}$ m

9. Walter types term papers for $1.25 per page. How many pages must he type in order to earn $13.75? 11 pages

10. Arlene made $6.50 for 2 hours of cutting grass. How much did she earn for 3 hours of grass cutting? $9.75

11. Chipo saves $6 out of every $20 he earns. He earned $70 one month. How much did he save? $21

12. It cost Bernie $2.52 to bake 6 loaves of bread. How much did it cost him to bake 8 loaves? $3.36

13. Ms. Rich drove 200 km in 4 h. How long will it take her to drive 300 km? 6 h

14. Akiko drove 400 km in 6 h. How far can she drive in 9 h? 600 km

15. Cynthia bought 5 donuts for 75¢. How many donuts can she buy for $1.65? 11 donuts

16. A Chinese restaurant charges $2.70 for 6 egg rolls. How much should they charge for 7 egg rolls? $3.15

★ 17. A close-out sale features ties at $3.98 each, 6 for $19.99, and $44.99 a dozen. Which is the best buy? 6 for $19.99

★ 18. A supermarket sells apples at $1.56 a dozen and oranges at $1.44 a dozen. What is the total cost of 3 apples and 7 oranges? $1.23

Inverse Variation

—◇ **OBJECTIVE** ◇—

To determine if a table of data represents inverse variation

—————◇ **RECALL** ◇—————

Area of rectangle = length · width

$$A = l \cdot w$$
$$36 = 36 \cdot 1$$

Example 1

Find the area of each rectangle.
Display this data in a table.

l	w	$A = l \cdot w$
36	1	36
18	2	36
12	3	36

Notice that the product of the length and width remains the same for all three rectangles. As the width increases, the length decreases. This is an example of *inverse variation*.

Example 2

From the table, determine if y varies inversely as x.

x	y
3	16
6	8
24	2

Determine if the product $x \cdot y$ remains the same.
$$3 \cdot 16 = 48$$
$$6 \cdot 8 = 48$$
$$24 \cdot 2 = 48$$

The product $y \cdot x$ is always 48. So, y varies inversely as x.

◇ **EXERCISES** ◇

Determine if y varies inversely as x.

1.

x	y
7	8
14	4
2	28

yes

2.

x	y
4	25
10	10
30	3

no

3.

x	y
4	18
9	8
24	3

yes

4.

x	y
7	12
21	4
6	12

no

Circle Graphs

◇ OBJECTIVES ◇

To read circle graphs
To make circle graphs

◇ RECALL ◇

The sum of the measures of the central angles in a circle is 360°.

Example 1

The sum of the percents in a circle is 100%.

$30\% + 20\% + 40\% + 10\% = 100\%$

The circle graph shows how a senior class spent $4,000. How much was spent for each event?

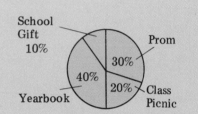

$30\% = 0.30; 20\% = 0.20;$
$40\% = 0.40; 10\% = 0.10$

	Prom	Class Picnic	Yearbook	School Gift
	$4,000	$4,000	$4,000	$4,000
	×0.30	×0.20	×0.40	×0.10
	$1,200.00	$800.00	$1,600.00	$400.00

So, $1,200 was spent on the prom, $800 on the class picnic, $1,600 on the yearbook, and $400 on the school gift.

Example 2

$10\% + 10\% + 5\% + 15.2\% + 59.8\% = 100\%$

One year, a family spent $1,500 on the use of energy. How much was spent for each use?

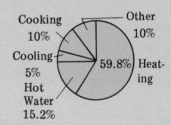

$5\% = 0.05; 10\% = 0.10;$
$59.8\% = 0.598; 15.2\% = 0.152$

$1,500
×0.598
12 000
135 00
750 0
$897.000

Cooling	Cooling	Other
$1,500	$1,500	$1,500
×0.05	×0.10	×0.10
$75.00	$150.00	$150.00

Heating	Hot Water
$1,500	$1,500
0.598	0.152
$897.00	$228.00

So, the family spent $75 on cooling, $150 on cooking, $150 on other, $897 on heating, and $228 on hot water.

Transportation, $174; Food, $366; Savings, $120; Housing, $300;

practice ▷ 1. Total monthly budget: $1,200. How much budgeted for each expense?

Clothing, $180; Other, $60

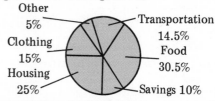

Other 5%

Transportation 14.5%

Clothing 15%

Food 30.5%

Housing 25%

Savings 10%

Example 3

Carlos earns about $60 per month. Make a circle graph to show how he spends his money.

$$\frac{1}{6} = 6\overline{)1.0000}^{\,0.1666}, \text{ or } 16.7\%$$

6
40
36
40
36
40
36
4

	Amount Spent	Fractional Part	Measure of Central Angle	Percent Spent
School supplies	$10	$\frac{10}{60}$, or $\frac{1}{6}$	$\frac{1}{6}(360°) = 60°$	16.7%
Clothing	$15	$\frac{15}{60}$, or $\frac{1}{4}$	$\frac{1}{4}(360°) = 90°$	25%
Miscellaneous	$5	$\frac{5}{60}$, or $\frac{1}{12}$	$\frac{1}{12}(360°) = 30°$	8.3%
Entertainment	$20	$\frac{20}{60}$, or $\frac{1}{3}$	$\frac{1}{3}(360°) = 120°$	33.3%
Savings	$10	$\frac{10}{60}$, or $\frac{1}{6}$	$\frac{1}{6}(360°) = 60°$	16.7%
Total	$60	1	360°	100%

From the table, find the measure of each central angle.
Use a protractor and make each central angle in the circle.
From the table, find the corresponding percent for each item. Label the graph.

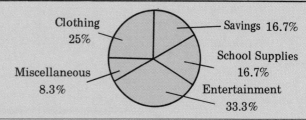

Clothing 25%

Savings 16.7%

School Supplies 16.7%

Miscellaneous 8.3%

Entertainment 33.3%

---

practice ▷ 2. Copy and complete the table. Then make a circle graph.

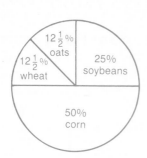

$12\frac{1}{2}\%$ oats
$12\frac{1}{2}\%$ wheat
25% soybeans
50% corn

Crop	Hectares Planted	Fractional Part	Measure of Central Angle	Percent Planted
Soybean	50	$\frac{50}{200}$ or $\frac{1}{4}$	$\frac{1}{4}(360°) = 90°$	25%
Wheat	25	$\frac{25}{200}$ or $\frac{1}{8}$	$\frac{1}{8}(360°) = 45°$	$12\frac{1}{2}\%$
Corn	100	$\frac{100}{200}$ or $\frac{1}{2}$	$\frac{1}{2}(360°) = 180°$	50%
Oats	25	$\frac{25}{200}$ or $\frac{1}{8}$	$\frac{1}{8}(360°) = 45°$	$12\frac{1}{2}\%$
Total	200	1	360°	100%

CIRCLE GRAPHS

◇ ORAL EXERCISES ◇

The circle graph shows an advertising
budget. What percent was budgeted for
each item?

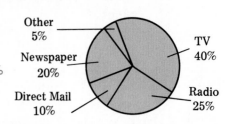

1. TV 40% 2. Radio 25% 3. Direct mail 10%
4. Newspaper 20%   5. Other 5%

◇ EXERCISES ◇

How much was spent on each item?

1. Automobile expenses for a year:
    $3,000.

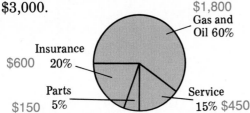

2. Summer living expenses: $4,400.

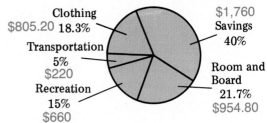

Make a circle graph.

3. Summer earnings: $300
    Mowing lawns, $150
    Washing cars, $30
    Babysitting, $90
    Walking dogs, $30

4. Audio equipment sales: $200,000
    TV's, $100,000
    CB's, $40,000
    Tape recorders, $25,000
    Radios, $25,000
    Other, $10,000

Challenge

$$\frac{1}{5} < \frac{1}{4} \text{ and } \frac{1}{6} < \frac{1}{5} \rightarrow \frac{1}{5} + \frac{1}{6} < \frac{1}{4} + \frac{1}{5}$$

$$\frac{1}{3} > \frac{1}{4} \text{ and } \frac{1}{2} > \frac{1}{3} \rightarrow \frac{1}{3} + \frac{1}{2} > \frac{1}{4} + \frac{1}{3}$$

**Replace __?__ with <, >, or = to make the sentence true.
Explain why the sentence is true.**

1. $\frac{1}{6} + \frac{1}{5} + \frac{1}{4} + \frac{1}{3}$ __?__ $\frac{1}{5} + \frac{1}{4} + \frac{1}{3} + \frac{1}{2}$ <

2. $\frac{1}{9} + \frac{1}{7} + \frac{1}{5} + \frac{1}{3}$ __?__ $\frac{1}{10} + \frac{1}{8} + \frac{1}{6} + \frac{1}{4}$ >

3. $\frac{1}{10} + \frac{1}{20} + \frac{1}{5} + \frac{1}{4}$ __?__ $0.10 + 0.05 + 0.20 + 0.25$ =

See TE Answer Section.

Problem Solving – Careers
Electric Sign Service

Hanssen's Electric Signs charges $25 for a service call and $18 per hour for services or labor, plus parts. Hanssen's charges 15¢/km after the first 15 km. There is no charge for the first 15 km.

1. It took a servicer 3 hours to repair a thermostat. The parts cost $24. What was the total bill? $103

2. A servicer traveled 105 kilometers. Find the travel charge. $13.50

3. Joanne Hanssen repaired an electric sign. She started at 9:20 A.M. and finished at 10:50 A.M. She traveled 58 kilometers, and the parts cost $18.25. What was the total bill? $76.70

4. Hanssen's sent two servicers to repair the Burger Time sign. They each worked 6 hours, the parts cost $23.45, and they traveled 126 kilometers. What was the total bill? $281.10

5. The odometer on the Hanssen's van read 29,977 when the servicer left Hanssen's and 30,019 when the servicer returned. What was the travel charge? $4.05

6. Two servicers installed a new sign. They each worked 9 hours. The odometer read 16,905 when they left Hanssen's and 16,978 when they returned. What was the total bill? $357.70

7. Monte Hanssen worked $4\frac{1}{2}$ hours repairing a sign. The odometer read 53,429 when he left and 53,492 when he returned. The parts cost $73. What was the total bill? $186.20

Scale Drawing

To find distances by using a scale

$$\begin{array}{r} 30 \\ \times 1.5 \\ \hline 15\ 0 \\ 30 \\ \hline 45.0 \end{array} \qquad \begin{array}{r} 16 \\ \times 3.25 \\ \hline 80 \\ 3\ 2 \\ 48 \\ \hline 52.00 \end{array} \qquad \begin{array}{r} 2.5 \\ \times 2.5 \\ \hline 1\ 25 \\ 5\ 0 \\ \hline 6.25 \end{array}$$

Example 1

The distance between two towns on a map is 2.9 cm. The scale on the map is 1 cm → 15 km. Find the distance between the towns.

Write a proportion.
Compare cm to km.
The product of the means equals the product of the extremes.

$$\frac{1 \text{ cm}}{15 \text{ km}} = \frac{2.9 \text{ cm}}{n \text{ km}}$$

$$n = 15(2.9)$$
$$n = 43.5$$

$$15 \times 2.9 = 4.35$$

So, the distance between the towns is 43.5 km.

Example 2

Use a centimeter ruler. Measure the distance between the towns in centimeters.

Find the distance: from Fort Morris to Darby; from Darby to Westbury.

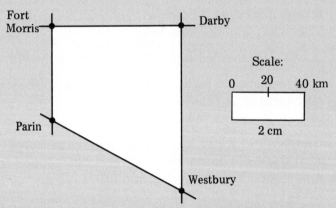

From the scale on the map,
2 cm → 40 km, or
1 cm → 20 km.

Multiply.

$$\begin{array}{r} 20 \\ \times 3.5 \\ \hline 10\ 0 \\ 60 \\ \hline 70.0 \end{array} \qquad \begin{array}{r} 20 \\ \times 4.4 \\ \hline 8\ 0 \\ 80 \\ \hline 88.0 \end{array}$$

Between Fort Morris and Darby is about 3.5 cm.

$$\frac{1 \text{ cm}}{20 \text{ km}} = \frac{3.5 \text{ cm}}{n \text{ km}}$$
$$1\, n = 20(3.5)$$
$$n = 70$$

Between Darby and Westbury is about 4.4 cm.

$$\frac{1 \text{ cm}}{20 \text{ km}} = \frac{4.4 \text{ cm}}{n \text{ km}}$$
$$1\, n = 20(4.4)$$
$$n = 88$$

So, it is about 70 km from Fort Morris to Darby and 88 km from Darby to Westbury.

practice ▷ **Find the distance. Use the map in Example 2.**

    1. from Westbury to Parin 60 km **2.** from Parin to Fort Morris 37.5 km

---

Example 3

Find the distance.
from A to B; from B to C

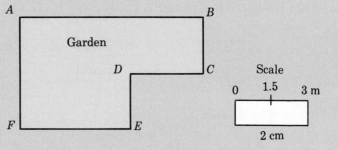

Scale

0   1.5   3 m

2 cm

Between A and B is 5 cm.

$$\begin{array}{r} 1.5 \\ \times 5 \\ \hline 7.5 \end{array}$$

$$\dfrac{1 \text{ cm}}{1.5 \text{ m}} = \dfrac{5 \text{ cm}}{n \text{ m}}$$

$$1n = 1.5(5) = 7.5$$

Between B and C is 1.5 cm.

$$\begin{array}{r} 1.5 \\ \times 1.5 \\ \hline 7\,5 \\ 1\,5 \\ \hline 2.2\,5 \end{array}$$

$$\dfrac{1 \text{ cm}}{1.5 \text{ m}} = \dfrac{1.5 \text{ cm}}{n \text{ m}}$$

$$1\,n = 1.5(1.5) = 2.25$$

So, the distance between A and B is 7.5 m and the distance between B and C is 2.25 m.

---

practice ▷ **Find the distance. Use the drawing in Example 3.**

    3. from D to E 2.25 m **4.** from E to F 4.5 m **5.** from F to A 4.5 m

---

◇ ORAL EXERCISES ◇

How far is it between towns with the given map distance? Use the scale 1 cm → 30 km.

1. 2 cm 60 km **2.** 3 cm 90 km **3.** 4 cm 120 km **4.** 5 cm 150 km **5.** 10 cm 300 km

---

◇ EXERCISES ◇

How far is it between towns with the given map distance? Use the scale 1 cm → 25 km.

1. 3 cm 75 km **2.** 7 cm 175 km **3.** 4.1 cm 102.5 km **4.** 6.3 km 157.5 km **5.** 10.5 cm 262.5 km

How far is it between towns with the given map distance? Use the scale 2 cm → 60 km.

6. 6 cm 180 km **7.** 9 cm 270 km **8.** 2.7 cm 81 km **9.** 5.4 cm 162 km **10.** 11.1 cm 333 km

Find the distance.

11. from Bradford to Eaton   58.5 km
12. from Eaton to Cramer   17.25 km
13. from Cramer to Randall   70.5 km
14. from Randall to Sealy   63 km
15. from Sealy to Bradford   75 km
16. from Bradford to Cramer   75 km

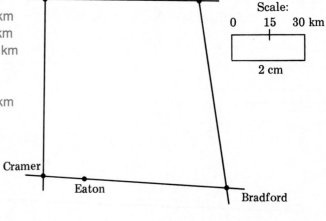

Find the distance from:

17. Landing to Filton   26 km
18. Landing to Battle   21 km
19. Battle to Kaldor   33 km
20. Kaldor to Grandville   40 km
21. Grandville to Filton   66 km
22. Landing to Grandville   57 km

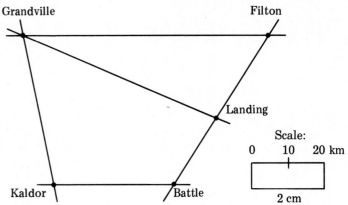

Find the distance.

23. from *A* to *B*   3.75 m
24. from *B* to *C*   2.25 m
25. from *C* to *D*   5.25 m
26. from *D* to *E*   2.25 m
27. from *E* to *F*   9 m
28. from *F* to *A*   4.5 m
29. Find the perimeter of
    the room.   27 m

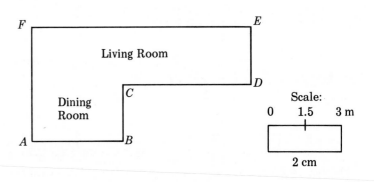

Make scale drawings.   Check students' drawings.

	Item	Dimensions	Scale
★ 30.	soccer field	50 m by 100 m	1 cm → 10 m
★ 31.	table	1 m by 2 m	1 cm → 0.5 m
★ 32.	room	3.5 m by 4.5 m	1 cm → 1 m

Similar Triangles

To find the lengths of the sides
of similar triangles
To solve word problems using
similar triangles

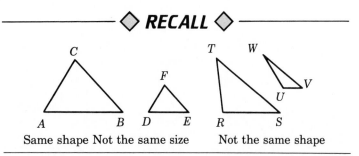

Same shape Not the same size   Not the same shape

Triangles *ABC* and *DEF* have the same shape. They
are similar triangles.

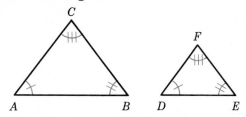

Corresponding sides:
$\overline{AB}$ and $\overline{DE}$,
$\overline{BC}$ and $\overline{EF}$,
$\overline{AC}$ and $\overline{DF}$

Corresponding angles:
$\angle A$ and $\angle D$,
$\angle B$ and $\angle E$,
$\angle C$ and $\angle F$

Read: Triangle *ABC* is similar to triangle *DEF*.

Write: $\triangle ABC$ ~ $\triangle DEF$.

Example 1

$\angle R$ means angle R.
$m\angle R$ means measure of $\angle R$.

Corresponding angles:
$\angle R$ and $\angle U$,
$\angle S$ and $\angle V$,
$\angle T$ and $\angle W$

A right angle measures 90°.
$m\angle R = 90° = m\angle U$
$m\angle S = 37° = m\angle V$
$m\angle T = 53° = m\angle W$

Corresponding sides:
$\overline{RS}$ and $\overline{UV}$,
$\overline{ST}$ and $\overline{VW}$,
$\overline{TR}$ and $\overline{WU}$

$\triangle RST$ is similar to $\triangle UVW$. Show that the
measures of the corresponding angles are equal.
Show that the lengths of the corresponding sides
have the same ratio.

$m\angle R = m\angle U$
$m\angle S = m\angle V$
$m\angle T = m\angle W$

$\dfrac{RS}{UV} = \dfrac{8}{4} = \dfrac{2}{1}$

$\dfrac{ST}{VW} = \dfrac{10}{5} = \dfrac{2}{1}$ same ratio

$\dfrac{TR}{WU} = \dfrac{6}{3} = \dfrac{2}{1}$

So, the measures of the corresponding angles are
equal and the lengths of the corresponding sides
have the same ratio.

If two triangles are *similar*, then the measures of the corresponding angles are equal and the lengths of the corresponding sides have the same ratio.

Example 2

$\triangle ABC \sim \triangle DEF$. Find x and y.

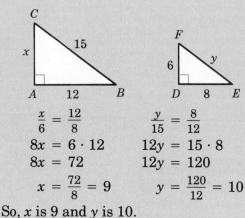

$\triangle ABC \sim \triangle DEF$.
The lengths of the corresponding sides have the same ratio.
Write and solve proportions.

$$\frac{x}{6} = \frac{12}{8} \qquad \frac{y}{15} = \frac{8}{12}$$

The product of the means equals the product of the extremes.

$$8x = 6 \cdot 12 \qquad 12y = 15 \cdot 8$$

$$8x = 72 \qquad 12y = 120$$

Divide.

$$x = \frac{72}{8} = 9 \qquad y = \frac{120}{12} = 10$$

So, x is 9 and y is 10.

practice ▷   1. $\triangle ABC \sim \triangle DEF$. Find x and y.
$x = 9 \qquad y = 4$

Example 3

A vertical meterstick casts a shadow 4 m long. At the same time, a pole casts a shadow 28 m long. How tall is the pole?

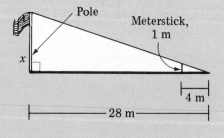

Two similar triangles are formed.
Let x = the height of the pole.
Write and solve a proportion.

$$\frac{x}{1} = \frac{28}{4}$$

$$4x = 28 \cdot 1$$

$$x = \frac{28}{4}, \text{ or } 7$$

So, the pole is 7 m tall.

practice ▷   2.   A man 2 m tall casts a shadow 4 m long. At the same time, a pole casts a shadow 16 m long. How tall is the pole?   8m

◇ ORAL EXERCISES ◇

△ ABC ~ △ DEF. Which side corresponds to the given side?

1. $\overline{AC}$ DF 2. $\overline{AB}$ DE 3. $\overline{BC}$ EF
4. $\overline{DE}$ AB 5. $\overline{DF}$ AC 6. $\overline{EF}$ BC

Which angle corresponds to the given angle?

7. $\angle A$ ∠D 8. $\angle C$ ∠F 9. $\angle B$ ∠E
10. $\angle E$ ∠B 11. $\angle D$ ∠A 12. $\angle F$ ∠C

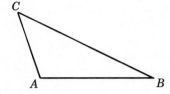
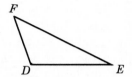

◇ EXERCISES ◇

△ ABC ~ △ DEF. Find x and y.

1.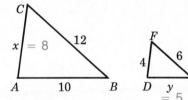

2.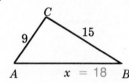

3.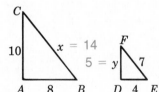

4.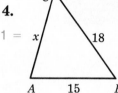

5.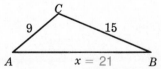

6.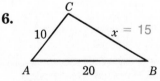

7. A vertical 2-meter stick casts a shadow 10 m long. At the same time, a flagpole casts a shadow 15 m long. How tall is the flagpole? 3 m

8. A child 1 m tall casts a shadow 2 m long. At the same time, a telephone pole casts a shadow 20 m long. How tall is the telephone pole? 10 m

★ 9. A scout used similar triangles to find the width of a river. The scout constructed similar triangles and made measurements as shown. Find the width of the river. 48 m

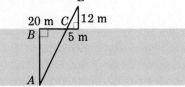

SIMILAR TRIANGLES

Trigonometric Ratios

◇ OBJECTIVES ◇

To find ratios of the sides of a
  right triangle
To compute tangent, sine, and
  cosine of an acute angle of a
  right triangle

◇ RECALL ◇

$\triangle ABC \sim \triangle DEF$

$$\frac{a}{b} = \frac{d}{e}; \frac{a}{c} = \frac{d}{f}; \frac{b}{c} = \frac{e}{f}$$

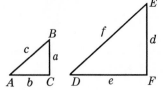

Trigonometry means *three-angle measure* or *triangle measure*. Only right triangles will be used. The side opposite the right angle in a right triangle is called the *hypotenuse*.

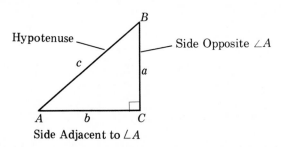

Example 1

What is the length of
the side opposite $\angle R$?
of the side adjacent to
$\angle R$? of the hypotenuse?

r is the side *opposite* $\angle R$
t is the side *adjacent* to $\angle R$
s is the *hypotenuse*

So, the length of the side opposite $\angle R$ is 12, the length of the side adjacent to $\angle R$ is 5, and the length of the hypotenuse is 13.

practice ▷ **Find the indicated lengths.**

1. side opposite $\angle B$ 4
2. side adjacent to $\angle B$ 3
3. the hypotenuse   5

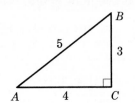

280   CHAPTER ELEVEN

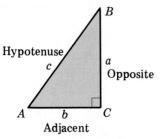

trigonometric ratios

In right $\triangle ABC$, the ratios of the sides have special names:

$$\text{tangent of } \angle A = \frac{\text{length of side opposite } \angle A}{\text{length of side adjacent to } \angle A}$$

$$\text{sine of } \angle A = \frac{\text{length of side opposite } \angle A}{\text{length of hypotenuse}}$$

$$\text{cosine of } \angle A = \frac{\text{length of side adjacent to } \angle A}{\text{length of hypotenuse}}$$

tan stands for *tangent*.
sin stands for *sine*.
cos stands for *cosine*.

$$\tan A = \frac{a}{b}, \sin A = \frac{a}{c}, \cos A = \frac{b}{c}$$

Example 2

Find each ratio. tan A, sin A, and cos A

$$\tan A = \frac{\text{opp. } (\angle A)}{\text{adj. (to } \angle A)} \qquad \tan A = \frac{a}{b} \text{ or } \frac{5}{12}$$

$$\sin A = \frac{\text{opp. } (\angle A)}{\text{hyp.}} \qquad \sin A = \frac{a}{c} \text{ or } \frac{5}{13}$$

$$\cos A = \frac{\text{adj. (to } \angle A)}{\text{hyp.}} \qquad \cos A = \frac{b}{c} \text{ or } \frac{12}{13}$$

So, $\tan A = \frac{5}{12}$, $\sin A = \frac{5}{13}$, and $\cos A = \frac{12}{13}$.

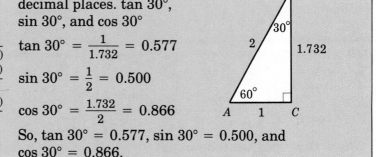

---

practice ▷ **Find each ratio. Use $\triangle ABC$ in Example 2.**

4. $\tan B$ $\frac{12}{5}$ 5. $\sin B$ $\frac{12}{13}$ 6. $\cos B$ $\frac{5}{13}$

Example 3

Find each ratio to three decimal places. tan $30°$, sin $30°$, and cos $30°$

$$\tan B = \frac{\text{opp. } (\angle B)}{\text{adj. (to } \angle B)} \qquad \tan 30° = \frac{1}{1.732} = 0.577$$

$$\sin B = \frac{\text{opp. } (\angle B)}{\text{hyp.}} \qquad \sin 30° = \frac{1}{2} = 0.500$$

$$\cos B = \frac{\text{adj. (to } \angle B)}{\text{hyp.}} \qquad \cos 30° = \frac{1.732}{2} = 0.866$$

So, $\tan 30° = 0.577$, $\sin 30° = 0.500$, and $\cos 30° = 0.866$.

---

practice ▷ **Find each ratio to three decimal places. Use $\triangle ABC$ in Example 3.**

7. $\tan 60°$ 1.732   8. $\sin 60°$ 0.866   9. $\cos 60°$ 0.500

---

1. What is the length of the side opposite ∠D? 8
2. What is the length of the side adjacent to ∠D? 6
3. What is the length of the side opposite ∠E? 6
4. What is the length of the side adjacent to ∠E? 8
5. What is the length of the hypotenuse? 10

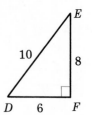

◇ *EXERCISES* ◇

Find each ratio. Leave answers in fractional form.

1. $\tan A$ $\frac{4}{3}$
2. $\sin A$ $\frac{4}{5}$
3. $\cos A$ $\frac{3}{5}$
4. $\tan B$ $\frac{3}{4}$
5. $\sin B$ $\frac{3}{5}$
6. $\cos B$ $\frac{4}{5}$

Find each ratio. Leave answers in fractional form.

7. $\tan D$ $\frac{12}{15}$
8. $\sin D$ $\frac{12}{13}$
9. $\cos D$ $\frac{5}{13}$
10. $\tan E$ $\frac{5}{12}$
11. $\sin E$ $\frac{5}{13}$
12. $\cos E$ $\frac{12}{13}$

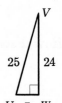

Find each ratio. Leave answers in fractional form.

13. $\tan R$ $\frac{8}{15}$
14. $\sin R$ $\frac{8}{17}$
15. $\cos R$ $\frac{15}{17}$
16. $\tan S$ $\frac{15}{8}$
17. $\sin S$ $\frac{15}{17}$
18. $\cos S$ $\frac{8}{17}$

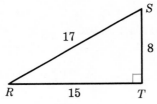

Find each ratio. Leave answers in fractional form.

19. $\tan U$ $\frac{24}{7}$
20. $\sin U$ $\frac{24}{25}$
21. $\cos U$ $\frac{7}{25}$
22. $\tan V$ $\frac{7}{24}$
23. $\sin V$ $\frac{7}{25}$
24. $\cos V$ $\frac{24}{25}$

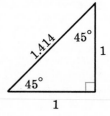

Find each ratio to three decimal places.

25. $\tan A$ 0.500
26. $\sin A$ 0.447
27. $\cos A$ 0.894
28. $\tan B$ 2.000
29. $\sin B$ 0.894
30. $\cos B$ 0.447

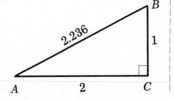

Find each ratio to three decimal places.

31. $\tan 45°$ 1.000
32. $\sin 45°$ 0.707
33. $\cos 45°$ 0.707

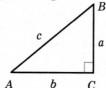

Show that each statement is true. Use the given figure.

★ 34. $\sin A = \cos B$ $\frac{a}{c} = \frac{a}{c}$
★ 35. $\sin B = \cos A$ $\frac{b}{c} = \frac{b}{c}$
★ 36. $\tan A = \dfrac{1}{\tan B}$ $\frac{a}{b} = \frac{1}{\frac{b}{a}}$ or $\frac{a}{b} = \frac{a}{b}$

Non-Routine Problems

There is no simple method used to solve these problems. Use whatever methods you can. Do not expect to solve these easily.

1. You have one of each of the following coins: 1¢, 5¢, 10¢, 25¢, and 50¢. How many different amounts of money can you obtain by forming all possible 2-coin combinations? 10

2. You have a choice of 2 different routes from Niceville to Warsaw and 3 different routes from Warsaw to Farmville. How many different routes can you take from Niceville to Farmville? 6

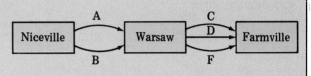

3. You have the five digits shown on the slips at the right.
   a. Form the smallest possible five-digit number using each of the digits once. 12,359
   b. Form the largest possible five-digit number using each of the digits once. 95,321
   c. How many different five-digit numbers can you form using each digit once? 120
   d. How many different four-digit numbers? 120
   e. Three-digit numbers? 60
   f. Two-digit numbers? 20

4. You have 5 different colors available with which to form 2-color flags. How many different flags can you form? (*Hint:* Number the colors 1, 2, 3, 4, 5; how many colors are available for the top portion? After that choice is made, how many colors are available for the lower portion?) 20; 5; 4

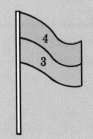

5. How many different automobile plates can be made using one letter followed by two digits? You can use the same digit twice. (*Hint:* How many choices do you have for the letter spot? After you choose the letter, how many choices do you have for the first digit? For the second digit?) 2,600

The program below determines whether an equation is a proportion. Writing such a program depends upon the ability of the programmer to state the general problem. Discovery of the generalization can be aided by first looking at several specific examples.

Example Determine if the equations are proportions.

$$\frac{3}{4} = \frac{6}{8} \qquad\qquad \frac{5}{10} = \frac{1}{2}$$

Check: $4 \cdot 6 = 3 \cdot 8$ Check: $10 \cdot 1 = 5 \cdot 2$

$24 = 24$ $10 = 10$

Now in *general* $\frac{A}{B} = \frac{X}{Y}$ if $B \cdot X = A \cdot Y$

Write and **RUN** a program to determine if $\frac{5}{17} = \frac{10}{32}$ is a proportion.

Computer checks if product of means = product of extremes.

```
10   INPUT "WHAT IS A? ";A
20   INPUT "WHAT IS B? ";B
30   INPUT "WHAT IS X? ";X
40   INPUT "WHAT IS Y? ";Y
50   IF B * X = A * Y THEN 80
60   PRINT "THIS IS NOT A PROPORTION."
70   END
80   PRINT "THE EQUATION IS A PROPORTION."
```

See the Computer Section beginning on page 420 for more information.

Exercises

Use the program above to determine if each of the following is a proportion. In each case first make the determination on your own by computation.

1. $\frac{7}{14} = \frac{35}{70}$ yes   2. $\frac{14}{19} = \frac{98}{133}$ yes   3. $\frac{43}{11} = \frac{215}{55}$ yes   4. $\frac{17}{37} = \frac{68}{168}$ no

5. Modify the program of the lesson so that it is not necessary to type **RUN** each time. Allow for checking three equations. Add line 5 **For K = 1 to 3** and line 65 **NEXT K.**

6. Use the results of **Exercise 5** above to determine if the following are proportions.

$$\frac{37}{65} = \frac{105}{195}; \frac{119}{207} = \frac{238}{415}; \frac{39}{117} = \frac{156}{468}$$

    no   no   yes

Chapter Review

A football team won 8 games and lost 3 games. Find each ratio. [260]

3:11

1. wins to losses   8:3
2. wins to total   8:11
3. losses to total

4. It takes 12 hours to plow a field. [260] What part can be plowed in 5 hours? $\frac{5}{12}$

5. Write the ratio in two ways. [260] Simplify. 8 pencils for 96 cents.

$\frac{8}{96}$ or $\frac{1}{12}$; 8:96 or 1:12

Solve each proportion. [263]

6. $\frac{2}{3} = \frac{n}{9}$ 6
7. $\frac{3}{4} = \frac{6}{x}$ 8
8. $\frac{n}{10} = \frac{2}{5}$ 4
9. $x:15 = 1:5$

3

Solve. [266]

10. A baker used 3 cups of flour to make 60 cookies. How many cups of flour are needed to make 90 cookies? $4\frac{1}{2}$ c

11. Mr. Flax drove 160 kilometers in 2 hours. How far can he drive in 6 hours? 480 km

12. The circle graph shows how a scout [270] troop raised $3,000. How much did they earn from each event?

13. Janet earned $2,000 by working part [270] time. Make a circle graph to show how she spent her earnings.

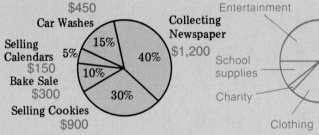

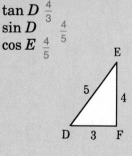

Clothing, $250
Savings, $1,000
Entertainment, $500
School supplies, $200
Charity, $50

14. How far is it between Sandland and [274] Auburn?   150 km

15. Find each ratio. Leave answers in [280] in fractional form.
tan D $\frac{4}{3}$
sin D $\frac{4}{5}$
cos E $\frac{4}{5}$

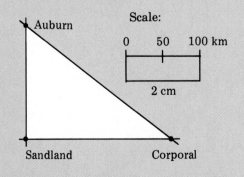

A baseball team won 20 games and lost 7 games. Find each ratio.

1. losses to total   7:27

2. wins to losses   20:7

Solve.

3. It takes 5 hours to fill the pool. What part can be filled in 3 hours? $\frac{3}{5}$

4. Write the ratio two ways. Simplify. 110 miles in 2 hours.
   $\frac{110}{2} = \frac{55}{1}$ or 110:2 = 55:1

Solve each proportion.

5. $\frac{3}{7} = \frac{n}{21}$ 9

6. $\frac{6}{x} = \frac{2}{5}$ 15

7. $n{:}10 = 3{:}5$
   6

Solve.

8. A dressmaker used 5 meters of silk to make 2 dresses. How many meters of silk are needed to make 8 dresses?   20 m

9. The circle graph shows how a band raised $4,000 for a trip. How much did they earn from each event?

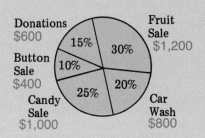

Donations $600
Fruit Sale $1,200
Button Sale $400
Candy Sale $1,000
Car Wash $800
15%   30%   10%   25%   20%

10. How far is it between Concord and Rochester?   38 km

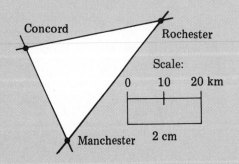

Concord
Rochester
Manchester
Scale:
0   10   20 km
2 cm

11. Find each ratio. Leave answers in fractional form.
    $\sin A\ \frac{12}{13}$
    $\cos B\ \frac{12}{13}$
    $\tan A\ \frac{12}{5}$

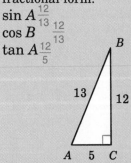

B
13   12
A   5   C

Adding and
Subtracting Integers

The main section of the Golden Gate Bridge spans 4,200 feet across San Francisco Bay.

Integers on a Number Line

To tell what integer corresponds
to a given point

To tell what integer comes just
after or just before a given
integer

To compare integers

To think of integers in terms of
"trips" on a number line

Whole numbers can be shown on a number line.

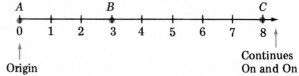

0 corresponds to point A, 3 corresponds
to point B, and 8 corresponds to point C.

A number line can go on forever in both directions.

Integers on a number line

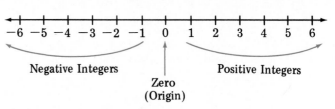

Negative Integers   Positive Integers

Zero
(Origin)

Positive integers   1   2   3

Read.   positive 1   positive 2   positive 3

Zero is halfway between -1 and 1.   Zero is neither positive nor negative.

Negative integers -1 -2 -3

Read.   negative 1   negative 2   negative 3

Example 1

What integer corresponds to point A? to point M?
to point R?

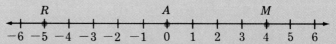

Read on the number line above.   So, 0 corresponds to point A, 4 corresponds to point
M, and -5 corresponds to point R.

---

practice ▷   **What integer corresponds to the point?**

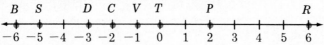

1. P 2. S 3. T 4. V 5. C 6. R 7. B 8. D
   2   -5   0   -1   -2   6   -6   -3

CHAPTER TWELVE

Example 2

On a number line, 4 is to the right of 3; 0 is to the right of −1; 1 is to the left of 2; and −5 is to the left of −4.

What integer comes just after 3? just after −1? just before 2? just before −4?

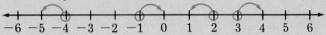

So, 4 comes just after 3, 0 comes just after −1, 1 comes just before 2, and −5 comes just before −4.

practice ▷ **What integer comes just before the given integer? just after the given integer?**

9. 4   3 before   5 after
10. 1   0 before   2 after
11. −3   −4 before   −2 after
12. −5   −6 before   −4 after
13. 0   −1 before   1 after

Example 3

> means is *greater than*.
< means is *less than*.

4 comes after 3. 4 > 3
−2 comes before 1. −2 < 1

Read 4 > 3 as 4 *is greater than* 3.
Read 3 < 4 as 3 *is less than* 4.

Compare. Use > or <.

4 _?_ 3, 3 _?_ 4   0 _?_ −1, −1 _?_ 0
−2 _?_ 1, 1 _?_ −2   −4 _?_ −6, −6 _?_ −4

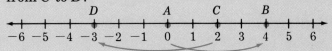

So, 4 > 3 and 3 < 4, 0 > −1 and −1 < 0, −2 < 1 and 1 > −2, and −4 > −6 and −6 < −4.

practice ▷ **Compare. Use > or <.**

14. 2 _?_ 3   **15.** −2 _?_ −3   **16.** 0 _?_ 1   **17.** −4 _?_ 0
       <   >   <   <

Example 4

What integer corresponds to the "trip" from A to B? from C to D?

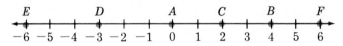

The trip from A to B is in the positive direction. The trip from A to B is 4. The trip from C to D is in the negative direction. The trip from C to D is −5. So, 4 corresponds to the trip from A to B and −5 corresponds to the trip from C to D.

practice ▷ **What integer corresponds to the trip?**

E   D   A   C   B   F
−6 −5 −4 −3 −2 −1 0 1 2 3 4 5 6

18. from A to F   6   **19.** from D to B   7   **20.** from C to D   −5

INTEGERS ON A NUMBER LINE

◇ ORAL EXERCISES ◇

Read.

1. 4   positive 4   **2.** −5   negative 5   **3.** 3   positive 3   **4.** −1   negative 1

5. −2   negative 2   **6.** 10   positive 10   **7.** 8   positive 8   **8.** −7   negative 7

◇ EXERCISES ◇

What integer corresponds to the point?

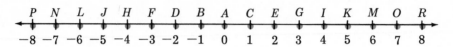

1. M 6   **2.** A 0   **3.** J −5   **4.** D −2

5. P −8   **6.** B −1   **7.** R 8   **8.** E 2

9. K 5   **10.** N −7   **11.** F −3   **12.** L −6

What integer comes just after the given integer?

13. 6   7   **14.** 0   1   **15.** −2   −1   **16.** −8   −7

17. −1   0   **18.** 7   8   **19.** −5   −4   **20.** 1   2

What integer comes just before the given integer?

21. 7   6   **22.** 0   −1   **23.** −2   −3   **24.** 5   4

25. 4   3   **26.** −6   −7   **27.** −1   −2   **28.** 1   0

Compare. Use > or <.

29. 5 $\underline{\ ?\ }$ 6   <   **30.** −3 $\underline{\ ?\ }$ −2   <   **31.** 0 $\underline{\ ?\ }$ 1   <   **32.** −5 $\underline{\ ?\ }$ −1   <

33. 3 $\underline{\ ?\ }$ 10   <   **34.** 4 $\underline{\ ?\ }$ −4   >   **35.** 0 $\underline{\ ?\ }$ −5   >   **36.** −3 $\underline{\ ?\ }$ 2   <

What integer corresponds to the trip?

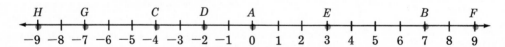

37. from A to B 7   **38.** from B to A −7   **39.** from A to C −4

40. from D to G −5   **41.** from E to F 6   **42.** from F to E −6

43. from H to F 18   **44.** from F to H −18   **45.** from B to G −14

Graph the numbers on a number line.   Check students' graphs.

★ **46.** The whole numbers less than 10.   0, 1, 2, 3, 4, 5, 6, 7, 8, 9

★ **47.** The odd integers between −10 and 10.   −9, −7, −5, −3, −1, 1, 3, 5, 7, 9

★ **48.** The odd integers between −12 and −1.   −11, −9, −7, −5, −3

★ **49.** The even integers between −5 and 10.   −4, −2, 0, 2, 4, 6, 8

290   *CHAPTER TWELVE*

Adding Integers

◇ OBJECTIVE ◇

To add integers

◇ RECALL ◇

$9 + 8 = 17$ $7 + 6 = 13$ $5 + 9 = 14$

$$\begin{array}{r} 28 \\ +39 \\ \hline 67 \end{array} \qquad \begin{array}{r} 56 \\ +97 \\ \hline 153 \end{array} \qquad \begin{array}{r} 139 \\ +485 \\ \hline 624 \end{array}$$

We can use trips on a number line to add integers.

Example 1

$2 + 6$ is read *positive two add positive six.*
A number line can be vertical.
A trip of 2 means move 2 units in the positive direction.
Start at 0.
The second trip starts where the first trip ended.

Add. $2 + 6$
$2 + 6$ means a trip of 2 followed by a trip of 6.

Think: Start at 0; move 2 in the positive direction, then 6 in the positive direction; finish at 8.

So, $2 + 6 = 8$.

$$\begin{array}{c} 9 \\ 8 \\ 7 \\ 6 \\ 5 \\ 4 \\ 3 \\ 2 \\ 1 \\ 0 \\ -1 \end{array}$$

The sum of two positive integers is a positive integer.

Example 2

$-3 + (-4)$ is read *negative three add negative four.*
A trip of -3 means move 3 units in the negative direction.
Start at 0.
The second trip starts where the first trip ended.

Add. $-3 + (-4)$
$-3 + (-4)$ means a trip of -3 followed by a trip of -4.

$$-9 \quad -8 \quad -7 \quad -6 \quad -5 \quad -4 \quad -3 \quad -2 \quad -1 \quad 0 \quad 1 \quad 2 \quad 3$$

Think: Start at 0; move 3 in the negative direction, then 4 in the negative direction; finish at -7.
So, $-3 + (-4) = -7$.

The sum of two negative integers is a negative integer.

practice ▷   **Add. Draw trips on a number line.**   Check students' drawings.

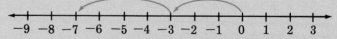

1. $3 + 5$ 8   **2.** $4 + 3$ 7   **3.** $-2 + (-1)$ -3 **4.** $-4 + (-2)$ -6

Example 3

-5 + 2 is read *negative five add positive two.*
A trip of -5 means move 5 units in the negative direction.
A trip of 2 means move 2 units in the positive direction.
Start at 0.
The second trip starts where the first trip ended.

Add. -5 + 2
-5 + 2 means a trip of -5 followed by a trip of 2.

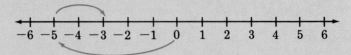

Think: Start at 0; move 5 units in the negative direction, then 2 units in the positive direction; finish at -3.
So, -5 + 2 = -3.

Example 4

6 + (-3) is read *positive six add negative three.*
A trip of 6 means move 6 units in the positive direction.
A trip of -3 means move 3 units in the negative direction.
Start at 0.
The second trip starts where the first trip ended.

Add. 6 + (-3)
6 + (-3) means a trip of 6 followed by a trip of -3.

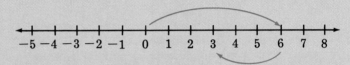

Think: Start at 0; move 6 units in the positive direction, then 3 units in the negative direction; finish at 3.
So, 6 + (-3) = 3.

practice ▷ **Add. Draw trips on a number line.** Check students' drawings.

5. -6 + 4   -2 **6.** -2 + 5   3   **7.** 8 + (-7)   1   **8.** 2 + (-6)   -4

Example 5

-4 + 4 is read *negative four add positive four.*
A trip of -4 means move 4 units in the negative direction.
A trip of 4 means move 4 units in the positive direction.
Start at 0.
The second trip starts where the first trip ended.

Add. -4 + 4
-4 + 4 means a trip of -4 followed by a trip of 4.

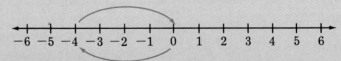

Think: Start at 0; move 4 units in the negative direction, then 4 units in the positive direction; finish at 0.
So, -4 + 4 = 0.

practice ▷ **Add. Draw trips on a number line.** Check students' drawings.

9. 2 + (-2)   0   **10.** -1 + 1   0   **11.** -3 + 3   0   **12.** 5 + (-5)   0

Read.   See TE Answer Section.

1. $3 + 9$	**2.** $6 + 8$	**3.** $9 + 4$	**4.** $-6 + (-2)$	**5.** $-7 + (-5)$
6. $3 + (-7)$	**7.** $1 + (-9)$	**8.** $-2 + 8$	**9.** $-7 + 3$	**10.** $8 + (-3)$
11. $-2 + 2$	**12.** $7 + (-7)$	**13.** $9 + 0$	**14.** $0 + (-2)$	**15.** $-9 + 0$

Add.

16. $2 + 3$ 5	**17.** $3 + 7$ 10	**18.** $4 + 9$ 13	**19.** $7 + 8$ 15
20. $9 + 9$ 18	**21.** $-1 + (-2)$ -3	**22.** $-2 + (-6)$ -8	**23.** $-5 + (-7)$ -12
24. $-9 + (-2)$ -11	**25.** $-7 + (-8)$ -15	**26.** $-2 + 3$ 1	**27.** $-4 + 6$ 2
28. $-7 + 2$ -5	**29.** $-9 + 12$ 3	**30.** $-8 + 1$ -7	**31.** $5 + (-4)$ 1
32. $7 + (-2)$ 5	**33.** $9 + (-4)$ 5	**34.** $2 + (-6)$ -4	**35.** $3 + (-7)$ -4
36. $7 + (-7)$ 0	**37.** $-9 + 9$ 0	**38.** $8 + (-8)$ 0	**39.** $10 + (-10)$ 0

Add. Draw trips on a number line.   Check students' drawings.

1. $4 + 5$ 9	**2.** $6 + 8$ 14	**3.** $9 + 0$ 9	**4.** $8 + 5$ 13
5. $7 + 7$ 14	**6.** $8 + 11$ 19	**7.** $13 + 4$ 17	**8.** $14 + 6$ 20
9. $5 + 20$ 25	**10.** $8 + 19$ 27	**11.** $-3 + (-1)$ -4	**12.** $-2 + (-7)$ -9
13. $-6 + (-2)$ -8	**14.** $0 + (-4)$ -4	**15.** $-5 + (-5)$ -10	**16.** $-9 + (-4)$ -13
17. $-7 + (-3)$ -10	**18.** $-9 + (-9)$ -18	**19.** $-6 + (-11)$ -17	**20.** $-8 + (-12)$ -20
21. $-4 + 2$ -2	**22.** $-7 + 0$ -7	**23.** $-5 + 9$ 4	**24.** $-8 + 3$ -5
25. $-1 + 6$ 5	**26.** $-8 + 11$ 3	**27.** $-10 + 3$ -7	**28.** $-12 + 1$ -11
29. $-5 + 14$ 9	**30.** $-6 + 15$ 9	**31.** $6 + (-2)$ 4	**32.** $1 + (-8)$ -7
33. $7 + (-6)$ 1	**34.** $8 + (-9)$ -1	**35.** $4 + (-7)$ -3	**36.** $10 + (-3)$ 7
37. $5 + (-12)$ -7	**38.** $18 + (-9)$ 9	**39.** $6 + (-19)$ -13	**40.** $11 + (-15)$ -4
41. $-6 + 6$ 0	**42.** $11 + (-11)$ 0	**43.** $-12 + 12$ 0	**44.** $13 + (-13)$ 0

Step 1   Write down the year of your birth.
This answer will vary step by step.

Step 2   Add it to the year of some important event in your life.

Step 3   Add the age you will be this year.

Step 4   Add the number of years since the important event took place.

Step 5   Multiply the current year by 2.
Steps 4 and 5 are the same.

Step 6   What do you discover? Try to explain why the answers in Steps 4 and 5 are the same.
Steps 2, 3, and 4: Add twice the difference between the current year and the year of birth to twice the year of birth.

ADDING INTEGERS

Problem Solving – Applications
Jobs for Teenagers

<table>
<tr><td>Example</td><td colspan="4">Percent can be used in describing a rate of increase or decrease.</td></tr>
</table>

NUMBER OF PAPERS SOLD		
	Lou	Maria
last week	28	96
this week	42	72

Find the amount of change.
Divide this amount by the original.
Write the percent.

$42 - 28 = 14$ $96 - 72 = 24$
$n = 14 \div 28$ $n = 24 \div 96$
$n = 0.50 = 50\%$ $n = 0.25 = 25\%$

So, the number of papers sold by Lou increased by 50% while those sold by Maria decreased by 25%.
To find the % increase or decrease:
1. Find the amount of increase or decrease.
2. Find what % this amount is of the original.

Solve these problems.

1. Glenda delivered 45 papers yesterday. She delivered 54 papers today. Find the percent increase. 20%

2. Joshua delivered 125 papers yesterday. He delivered 105 today. Find the percent decrease. 16%

3. Lora's earnings increased from $23.50 to $29.61. Find the percent increase. 26%

4. Paulo rode 12 km on his paper route yesterday. He rode only 8 km today. Find the percent decrease. $33\frac{1}{3}\%$

5. Sele earned $42 last week. He earned $48.30 this week. Find the percent increase. 15%

6. Issac's earnings increased from $43.20 to $47.52. Find the percent increase. 10%

Subtracting Integers

—◇ **OBJECTIVE** ◇— ——————— ◇ **RECALL** ◇ ———————

To subtract integers

Add. $3 + (-7)$

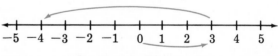

So, $3 + (-7) = -4$.

Example 1

$6 - 2$ is read *positive six subtract positive two.*

Show the trips on a number line.

The second trip starts where the first trip ended.

Subtract. $6 - 2$

Subtract a positive integer means move in the negative direction. $6 - 2$ means a trip of 6 followed by a trip of 2 in the negative direction.

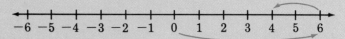

Think: Start at 0; move 6 units in the positive direction, then 2 units in the negative direction; finish at 4.

So, $6 - 2 = 4$.

Example 2

$3 - 7$ is read *positive three subtract positive seven.*

Show the trips on a number line.

Subtract. $3 - 7$

$3 - 7$ means a trip of 3 followed by a trip of 7 in the negative direction.

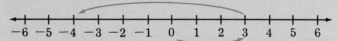

Think: Start at 0; move 3 units in the positive direction, then 7 units in the negative direction; finish at -4.

So, $3 - 7 = -4$.

Example 3

$-5 - 3$ is read *negative five subtract positive three.*

Show the trips on a number line.

Subtract. $-5 - 3$

$-5 - 3$ means a trip of -5 followed by a trip of 3 in the negative direction.

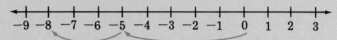

Think: Start at 0; move 5 units in the negative direction, then 3 units in the negative direction; finish at -8.

So, $-5 - 3 = -8$.

---

1. $5 - 1$ 4   **2.** $2 - 7$ −5   **3.** $-2 - 1$ −3   **4.** $-4 - 2$ −6

When we subtracted a positive integer, we moved in the negative direction on a number line. We shall agree to move in the positive (opposite) direction on a number line when we subtract a negative integer.

Example 4

$-5 - (-3)$ is read *negative five subtract negative three.*

Think: To subtract a negative integer means to move in the positive direction.

Subtract. $-5 - (-3)$
$-5 - (-3)$ means a trip of -5 followed by a trip of 3 in the positive direction.

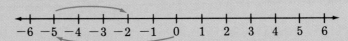

Think: Start at 0; move 5 units in the negative direction, then 3 units in the positive direction; finish at -2.
So, $-5 - (-3) = -2$.

Example 5

$2 - (-4)$ is read *positive two subtract negative four.*

Think: To subtract a negative integer means to move in the positive direction.

Subtract. $2 - (-4)$
$2 - (-4)$ means a trip of 2 followed by a trip of 4 in the positive direction.

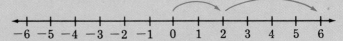

Think: Start at 0; move 2 units in the positive direction, then 4 units in the positive direction; finish at 6.
So, $2 - (-4) = 6$.

Example 6

Subtract. $0 - (-3)$.
$0 - (-3)$ means a trip of 0 followed by a trip of 3 in the positive direction.

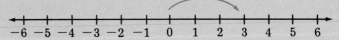

Think: Start at 0; finish at 3.
So, $0 - (-3) = 3$.

practice ▷ **Subtract. Draw trips on a number line.**   Check students' drawings.

5. $-4 - (-2)$ −2   **6.** $3 - (-5)$ 8   **7.** $0 - (-5)$ 5   **8.** $0 - (-7)$ 7

Read.   See TE Answer Section.

1. $4 - 2$	**2.** $7 - 5$	**3.** $9 - 4$	**4.** $5 - 2$
5. $3 - 5$	**6.** $2 - 8$	**7.** $4 - 6$	**8.** $5 - 7$
9. $-2 - 3$	**10.** $-1 - 4$	**11.** $-3 - 7$	**12.** $-4 - 6$
13. $-4 - (-2)$	**14.** $-1 - (-1)$	**15.** $-2 - (-3)$	**16.** $-7 - (-2)$
17. $5 - (-1)$	**18.** $3 - (-3)$	**19.** $7 - (-2)$	**20.** $4 - (-6)$
21. $0 - (-2)$	**22.** $0 - 5$	**23.** $0 - (-4)$	**24.** $0 - 1$

◇ *EXERCISES* ◇

Subtract. Draw trips on a number line.   Check students' drawings.

1. $6 - 1$ 5	**2.** $5 - 3$ 2	**3.** $4 - 2$ 2	**4.** $6 - 4$ 2
5. $2 - 4$ -2	**6.** $3 - 4$ -1	**7.** $2 - 6$ -4	**8.** $5 - 8$ -3
9. $-2 - 4$ -6	**10.** $-1 - 1$ -2	**11.** $-3 - 2$ -5	**12.** $-5 - (-1)$ -4
13. $-6 - (-2)$ -4	**14.** $-4 - (-3)$ -1	**15.** $-5 - (-1)$ -4	**16.** $-2 - (-1)$ -1
17. $-1 - (-7)$ 6	**18.** $-4 - (-10)$ 6	**19.** $-3 - (-6)$ 3	**20.** $-5 - (-10)$ 5
21. $2 - (-2)$ 4	**22.** $4 - (-1)$ 5	**23.** $7 - (-3)$ 10	**24.** $8 - (-5)$ 13
25. $3 - (-6)$ 9	**26.** $1 - (-5)$ 6	**27.** $4 - (-9)$ 13	**28.** $2 - (-8)$ 10
29. $5 - (-6)$ 11	**30.** $0 - 3$ -3	**31.** $0 - 8$ -8	**32.** $0 - (-1)$ 1
33. $0 - (-4)$ 4	**34.** $0 - (-7)$ 7	**35.** $5 - 0$ 5	**36.** $-3 - 0$ -3
37. $-4 - (-4)$ 0	**38.** $6 - 6$ 0	**39.** $-5 - 5$ -10	**40.** $9 - (-6)$ 15
41. $-8 - 2$ -10	**42.** $-7 - (-6)$ -1	**43.** $0 - (-8)$ 8	**44.** $4 - (-3)$ 7

Algebra Maintenance

Solve each proportion.

1. $\frac{4}{7} = \frac{k}{21}$ 12

2. $\frac{12}{m} = \frac{3}{5}$ 20

3. $x:12 = 7:24$

$3\frac{1}{2}$

Solve. Check.

4. $y - 2 = 12$ 14

5. $6 = 62 - 4r$ 14

6. $0.02g = 5.1$

255

7. What is the perimeter of a rectangle whose length is r and width is u? $2r + 2u$

8. What is the area of a circle whose radius is w cm long? $A = \pi w^2$

Problem Solving – Careers
Farm Equipment Mechanics

1. Ms. Contreras repairs an average of 15 tractors per week. How many tractors does she repair in a year? 780 tractors/yr

2. Last year, Mr. Robinson repaired 625 diesel-powered tractors. About how many did he repair each week? 12 tractors/wk

3. During August, Sato Repair received $3,000 for repairs of machinery. In September, Sato received $600 more than in August. In October, Sato received $800 less than in September. How much did Sato receive in October? $2,800

4. In May, Mr. Fitch spent 150% of what he spent in April to repair his crop dryer. In April he spent $30. How much did he spend in May? $45

5. In August, Ms. Royce spent 95% of what she spent in July to repair her combine. In July she spent $120. How much did she spend in August? $114

6. Ms. Bonk spends $80 per month on the maintenance of her machinery. How much does she spend in a year? $960/yr

7. It takes George $1\frac{1}{2}$ hours to repair a corn picker. How many hours will it take him to repair 36 corn pickers? 54 h

8. Melinda can repair 3 hay balers in 2 hours. How many hay balers can she repair in 8 hours? 12 hay balers

Opposites

─── ◇ *OBJECTIVES* ◇ ───

To identify the opposite of an integer

To find the missing integer given a sentence like
3 + ___?___ = 0

To express real-life situations with integers

─────── ◇ *RECALL* ◇ ───────

Add. −3 + 3

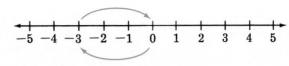

So, −3 + 3 = 0

Example 1

Find two integers that are each 4 units from 0.

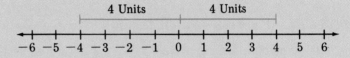

So, 4 and −4 are each 4 units from 0.

4 and −4 are opposites.   *Opposites* are the same distance from 0 on a number line. 0 is its own opposite.

Example 2

Give the opposite of 2 and of −5.

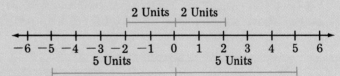

So, the opposite of 2 is −2 and the opposite of −5 is 5.

---

practice ▷   **Give the opposite.**

    1. 3   −3   **2.** −8   8   **3.** 7   −7   **4.** −10   10   **5.** 21   −21   **6.** 0   0

Example 3

Find the sum. −5 + 5

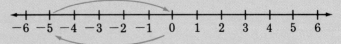

So, −5 + 5 = 0.

The sum of an integer and its opposite is 0.

practice ▷ **Copy and complete.**

7. $-9 + 9 =$ __?__   0   8. __?__ $+ (-25) = 0$ 25   9. $12 +$ __?__ $= 0$
   $^{-12}$

Example 4

Express as integers: a gain of $10; a loss of $10
Gain of $10: 10
Loss of $10: -10
So, 10 shows a gain of $10 and -10 shows a loss of $10.

---

practice ▷ **Express as an integer.**

10. a gain of $25   25   11. a loss of 17 points -17 12. 33 m up
                                                                    33

◇ *ORAL EXERCISES* ◇

What is the opposite of the given integer?

1. 6 -6 2. -5 5   3. 1 -1 4. -2 2   5. 0   0   6. -11
                                                                       11

Express as an integer.

7. a gain of $5   5   8. a loss of $3 -3 9. a gain of 20 points
10. 100 m above sea level   100   11. 5 km up   5   12. 64 m down -64
                                                                20

◇ *EXERCISES* ◇

What is the opposite of the given integer?

1. 15 -15 2. -12 12   3. 25 -25 4. 78 -78 5. -56 56   6. -30
7. 100 -100 8. -29 29   9. -32 32   10. 19 -19 11. -20 20   12. -365
                                                                                  30
                                                                                  365

Copy and complete.

13. $-21 + 21 =$ __?__   0   14. $-60 +$ __?__ $= 0$ 60   15. $43 +$ __?__ $= 0$
16. __?__ $+ (-27) = 0$ 27   17. __?__ $+ 91 = 0$ -91 18. $82 + (-82) =$ __?__
                                                                          $^{-43}$
                                                                          0

Express as an integer.

19. a gain of $950   950   20. a loss of 56 points -56 21. 1,200 m up
22. 3,000 m above sea level   3,000 23. 4,520 m below sea level -4520 24. 38 km down -38
25. a loss of $75 -75 26. a gain of 10 kg   10   27. 6 m up   6
28. a gain of 8 cm³   8   29. a loss of 5 g -5 30. 742 m up   742
                                                                          1,200

Subtracting by Adding

To subtract an integer by adding its opposite

−5 is the opposite of 5.
9 is the opposite of −9.
0 is the opposite of 0.
(0 is its own opposite.)

The number line on the left shows $7 - 4 = 3$.
The number line on the right shows $7 + (-4) = 3$.

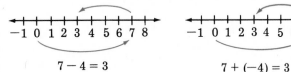

$7 - 4 = 3$ $7 + (-4) = 3$

This suggests that instead of subtracting, we can add.

$$7 - 4 = 7 + (-4)$$
opposites

To subtract an integer, add its opposite.

Example 1

Subtract. $3 - 9$

−9 is the opposite of 9.

$3 - 9$
$3 + (-9)$
-6

Check on a number line. So, $3 - 9 = -6$

---

practice ▷ **Subtract by adding the opposite.**

1. $10 - 7$ 3
2. $12 - 3$ 9
3. $4 - 9$ −5
4. $9 - 15$ −6

Example 2

Subtract. $-8 - 7$

−7 is the opposite of 7.

$-8 - 7$
$-8 + (-7)$
-15

So, $-8 - 7 = -15$.

---

practice ▷ **Subtract by adding the opposite.**

5. $-6 - 7$ −13
6. $-4 - 9$ −13
7. $-13 - 8$ −21
8. $-15 - 12$ −27

Example 3

3 is the opposite of -3.

Subtract. $-12 - (-3)$
$-12 - (-3)$
$-12 + 3$
-9
So, $-12 - (-3) = -9$.

Example 4

14 is the opposite of -14.

Subtract. $-5 - (-14)$
$-5 - (-14)$
$-5 + 14$
9
So, $-5 - (-14) = 9$.

practice ▷ **Subtract by adding the opposite.**

9. $-9 - (-12)$ **10.** $-13 - (-4)$ **11.** $-10 - (-12)$ **12.** $-8 - (-13)$
   3   -9 2   5

Example 5

9 is the opposite of -9.

Subtract. $3 - (-9)$
$3 - (-9)$
$3 + 9$
12
So, $3 - (-9) = 12$.

practice ▷ **Subtract by adding the opposite.**

13. $8 - (-2)$ 10 **14.** $10 - (-5)$ 15 **15.** $4 - (-4)$ 8 **16.** $2 - (-7)$ 9

◇ EXERCISES ◇

Subtract by adding the opposite.

1. $12 - 5$ 7
2. $16 - 9$ 7
3. $15 - 7$ 8
4. $18 - 9$ 9
5. $4 - 10$ -6
6. $7 - 27$ -20
7. $3 - 24$ -21
8. $12 - 26$ -14
9. $-12 - 10$ -22
10. $-30 - 15$ -45
11. $-50 - 20$ -70
12. $-42 - 31$ -73
13. $-25 - 30$ -55
14. $-46 - 70$ -116
15. $-60 - 100$ 160
16. $-20 - 130$ -150
17. $-10 - (-5)$ -5
18. $-32 - (-14)$ -18
19. $-15 - (-12)$ -3
20. $-50 - (-30)$ -2
21. $-5 - (-20)$ 15
22. $-10 - (-70)$ 60
23. $-12 - (-26)$ 14
24. $-31 - (-47)$ 16
25. $4 - (-5)$ 9
26. $11 - (-8)$ 19
27. $14 - (-1)$ 15
28. $7 - (-7)$ 14
29. $21 - (-8)$ 29
30. $38 - (-15)$ 53
31. $42 - (-26)$ 68
32. $19 - (-28)$ 47
33. $29 - 0$ 29
34. $0 - 37$ -37
35. $-23 - 0$ -23
36. $0 - (-80)$ 80

★ **37.** Subtract around the track. What is the final result? 59

Start at 0. Subtract -5, and
continue subtracting.

Using Integers

◇ *OBJECTIVE* ◇ ──── ──── ◇ *RECALL* ◇ ────

To solve problems using integers

$4 + 5 = 9$	$5 + (-2) = 3$
$5 - 2 = 3$	$5 - 8 = -3$
$3 + 0 = 3$	$0 + (-5) = -5$
$5 + (-8) = -3$	$-2 + (-3) = -5$
$-2 - (-5) = 3$	$-2 - (-1) = -1$
$3 - 0 = 3$	$0 - (-2) = 2$

Example 1

A football team gained 3 yards, lost 5 yards, and gained 12 yards. What was the net result?

Think of a gain as positive and a loss as negative. 3, −5, 12
Add the three numbers.
Positive 10

$3 + (-5) + 12$
$-2 + 12$
10

So, the net result was a gain of 10 yards.

practice ▷
1. A football team lost 6 yards, gained 18 yards, and lost 7 yards. What was the net result?   a gain of 5 yd

2. A football team gained 17 yards, lost 9 yards, and lost 8 yards. What was the net result?   no gain or loss

Example 2

Above sea level: positive
Below sea level: negative

A mountain climber is located at 4,000 m above sea level. A submarine is located at 1,000 m below sea level. What is the difference between the levels?

Subtract to find the difference.

$4,000 - (-1,000)$
$4,000 + 1,000$
$5,000$

So, the difference between 4,000 m above sea level and 1,000 m below sea level is 5,000 m.

practice ▷
3. A mountain climber is located at 900 m above sea level. A submarine is located at 800 m below sea level. What is the difference between the levels?
1,700 m

4. A mountain climber is located at 1,200 m above sea level. A submarine is located at 1,200 m below sea level. What is the difference between the levels?
2,400 m

1. $-4 + (-9) - 3$ -16 **2.** $6 - 3 + (-5)$ -2 **3.** $5 + 18 - (-15)$ 38

4. $-17 - (-13) + 8$ 4   **5.** $-3 - 7 + (-9)$ -19 **6.** $4 + 0 - (-18)$ 22

7. $-9 - (-4) + 15$ 10   **8.** $10 + (-11) - 8$ -9 **9.** $-13 - (-20) + (-7)$ 0

10. $4 + 9 - (-12)$ 25   **11.** $-11 - 7 - (-3)$ -15 **12.** $1 - (-1) - 1$ 1

13. $28 + (-8) - 17$ 3   **14.** $-15 + (-9) - 12$ -36 **15.** $22 + (-13) - 9$ 0

16. $-9 + (-4) - (-13)$ 0   **17.** $-30 + 19 - (-11)$ 0   **18.** $23 - (-7) - (-32)$ 62

19. $-2 + 21 - (-7)$ 26   **20.** $-17 - (-9) + 56$ 48

Solve these problems.

21. A football team gained 9 yards, lost 6 yards, and gained 11 yards. What was the net result?   a gain of 14 yd

22. A football team lost 15 yards, gained 9 yards, and lost 7 yards. What was the net result?   loss of 13 yd

23. A mountain climber is located at 800 m above sea level. A submarine is located at 700 m below sea level. What is the difference between the levels?   1,500 m

24. A mountain climber is located at 3,000 m above sea level. A submarine is located at 1,500 m below sea level. What is the difference between the levels?   4,500 m

25. Brian started with 19 points. Then he won 9 points, lost 6 points, and lost 7 points. How many points does he now have?   15

26. Yoko started with 23 points. Then she won 12 points, lost 7 points, and won 2 points. How many points does she now have?   30

27. Jason had $400 in a checking account. He made deposits of $150 and $230. He has to write checks for $200, $250, and $300. Does he have enough money?   yes

28. Muna had $250 in a checking account. She made deposits of $110 and $85. She has to write checks for $165, $140, and $145. Does she have enough money?   no

29. At noon, the temperature was 14°C. The temperature rose 4°C between noon and 1:00 P.M. and rose 5°C during the next hour. What was the temperature at 2:00 P.M.?   23°C

30. At 6:00 P.M., the temperature was 5°C. The temperature dropped 3°C between 6:00 P.M. and 7:00 P.M. and dropped 4°C during the next hour. What was the temperature at 8:00 P.M.? -2°C

★ **31.** Five years ago, the population of a city was 120,000. During the last five years these changes took place: gained 5,000, lost 3,000, gained 2,000, gained 1,000, and lost 7,000. What is the population now?   118,000

★ **32.** Four years ago, the population of a town was 8,650. During the last four years these changes took place: lost 150, gained 225, gained 175, and lost 100. What is the population now?   8,800

★ **33.** Lucy's kite was flying at a height of 800 m. She lowered it 200 m, raised it 500 m, lowered it 300 m, and raised it 700 m. At what height is the kite flying now?   1,500 m

★ **34.** Adam's kite was flying at a height of 650 m. He raised it 200 m, lowered it 350 m, raised it 125 m, and lowered it 175 m. At what height is the kite flying now?   450 m

Equations

◇ OBJECTIVE ◇

To solve equations like
$x - 5 = -2, x + 4 = -7,$
and $x + a = b$

◇ RECALL ◇

Solve.

$$x + 9 = 12$$
$$x + 9 - 9 = 12 - 9$$
$$x = 3$$

$$x - 6 = 3$$
$$x - 6 + 6 = 3 + 6$$
$$x = 9$$

Undo addition by subtraction.

Undo subtraction by addition.

Example 1

Undo addition.
Subtract 5.

Undo subtraction.
Add 8.

Solve and check.

$$x + 5 = -7$$
$$x + 5 - 5 = -7 - 5$$
$$x = -12$$

$$x - 8 = -15$$
$$x - 8 + 8 = -15 + 8$$
$$x = -7$$

Check: $\dfrac{x + 5 = -7}{\begin{array}{c|c} -12 + 5 & -7 \\ -7 & \end{array}}$

$$-7 = -7$$
true

Check: $\dfrac{x - 8 = -15}{\begin{array}{c|c} -7 - 8 & -15 \\ -15 & \end{array}}$

$$-15 = -15$$
true

So, -12 is the solution
of $x + 5 = -7$.

So, -7 is the solution
of $x - 8 = -15$.

practice ▷ Solve and check.

1. $x + 3 = -21$ -24
2. $x - 9 = -14$ -5
3. $x + 12 = 4$ -8

Example 2

Solve. $x - (-4) = -3$

$$x - (-4) = -3$$
$$x + 4 = -3$$

Undo addition by subtraction. $x + 4 - 4 = -3 - 4$
$$x = -7$$

Check on your own. So, -7 is the solution of $x - (-4) = -3$.

practice ▷ Solve and check.

4. $x - (-7) = 10$ 3
5. $x + (-9) = -11$ -2
6. $x - (-5) = -3$ -8

Summary

To solve an equation that shows addition, undo the addition by subtraction.
For example: $x + a = b$
$$x = b - a$$
So, $b - a$ solves the equation.

To solve an equation that shows subtraction, undo the subtraction by addition.
For example: $x - a = b$
$$x = b + a$$
So, $b + a$ solves the equation.

Opposites can also be used to solve equations involving integers.

Example 3

Add the opposite of -4 to each side of the equation.
$(-4) + 4 = 0$

Solve and check. $-4 + x = 11$
$$-4 + x = 11$$
$$-4 + 4 + x = 11 + 4$$
$$x = 15$$

Check:

$-4 + x = 11$	
$-4 + 15$	11
11	
$11 = 11$ true	

So, 15 is the solution of $-4 + x = 11$.

practice ▷   **7.** $-8 + x = 10$ 18   **8.** $-9 + x = -15$ -6 **9.** $3 + x = -5$ -8

◇ EXERCISES ◇

Solve and check.

1. $x + 2 = -4$ -6
2. $x + (-9) = -11$ -2
3. $5 + x = -12$ -17

4. $x + (-12) = -14$ -2
5. $x + (-9) = -2$ 7
6. $x + (-8) = -1$ 7

7. $x + (-11) = -4$ 7
8. $x + (-15) = -5$ 10
9. $x + 4 = 3$ -1

10. $x + 7 = -1$ -8
11. $x + 5 = -9$ -14
12. $x + 15 = 0$ -15

13. $-5 + x = -2$ 3
14. $-9 + x = -8$ 1
15. $-10 + x = -1$ 9

16. $-12 + x = 4$ 16
17. $-9 + x = 4$ 13
18. $-12 + x = 1$ 13

19. $-20 + x = 10$ 30
20. $-17 + x = 0$ 17
21. $x - 5 = -2$ 3

22. $x - 12 = -5$ 7
23. $x - 13 = 15$ 28
24. $x - 20 = 30$ 50

25. $x - (-6) = 2$ -4
26. $x - (-10) = -3$ -13
27. $x - (-15) = 20$ 5

Solve for x.

★ **28.** $x + r = s$ $x = s - r$
★ **29.** $x - p = q$ $x = q + p$
★ **30.** $r = x + s$ $x = r - s$

★ **31.** $-a + x = b$ $x = b + a$
★ **32.** $c = x - d$ $x = c + d$
★ **33.** $p = q + x$ $x = p - q$

The **BASIC** language has a very useful device for adding or accumulating numbers. The **ACCUMULATOR** makes it unnecessary to repeat writing the "+" symbol when adding a long series of numbers. The use of the **ACCUMULATOR** is illustrated in the program below.

Example   Write a program that will allow you to enter any number of numbers and then find their sum. Then **RUN** the program to find the sum $-58 + 119 - 73 - 98$.

Computer allows you to choose number of numbers to be added	`10  INPUT "NUMBER  OF NUMBERS TO ADD? ";N`
Clears all previous numbers from memory so they will not be added to the new numbers	`20  LET A = 0`
Allows you to enter N numbers	`30  FOR J = 1 TO N`
You type in the values of X to be added.	`40  INPUT "ENTER A NUMBER ";X`
Replaces each value of X with the new X value added to it; when J = 2, −58 will be replaced with −58 + 119	`50  LET A = A + X`
	`60  NEXT J`
Prints the sum of all numbers entered	`70  PRINT : PRINT "SUM IS "A`
	`80  END`
	`]RUN`
Type 4. (There are 4 numbers to add.)	`NUMBER  OF NUMBERS TO ADD? 4`
Enter the first number; −58.	`ENTER A NUMBER -58`
119 is second number to be added.	`ENTER A NUMBER 119`
Type third number; −73.	`ENTER A NUMBER -73`
Enter the last number; −98.	`ENTER A NUMBER -98`
PRINT: leaves a blank line for readability.	`SUM IS -110`

See the Computer Section beginning on page 420 for more information.

Exercises

Find each sum. Then type and **RUN** the program above to check your answers.

1. $-67 - 89 + 278 - 45$
   77

2. $98 - 121 - 453 + 89 - 76$
   -463

3. $-114 + 139 - 229$
   -204

Compare. Use > or <. [288]

1. 8 __?__ 9   <
2. −6 __?__ −1   <
3. −2 __?__ 4   <
4. 0 __?__ −6   >

Add. [291]

5. 8 + 3   11
6. −5 + (−7)   −12
7. 12 + (−3)   9
8. −6 + 8   2
9. 9 + (−9)   0
10. −7 + 7   0
11. 0 + 2   2
12. −3 + 0 −3

Copy and complete. [299]

13. −8 + 8 = __?__   0
14. 17 + __?__ = 0   −17
15. 10 + (−10) = __?__   0

Subtract. [295, 301]

16. 13 − 5   8
17. 12 − 26   −14
18. −27 − 52   −79
19. −36 − (−54)   18
20. 16 − (−10)   26

Solve these problems. [303]

21. A kite was flying at a height of 700 m. Then it was lowered 300 m, raised 100 m, lowered 200 m, and raised 500 m. At what height was the kite flying then?   800 m

22. At 6:00 A.M., the temperature was 9°C. The temperature rose 9°C between 6:00 A.M. and noon and rose 6°C between noon and 4:00 P.M. What was the final temperature? 24°C

23. A football team gained 11 yards, lost 13 yards, and lost 7 yards. What was the net result?   loss of 9 yd

24. Louise started with 30 points. Then she won 12 points, lost 8 points, and won 16 points. How many points does she have now?   50

Solve for x. [305]

25. $x + (−5) = −9$ −4
26. $4 + x = −5$ −9
27. $x − (−6) = −12$ −18
28. $x − (−12) = 34$ 22

Calculator

4! means $1 \cdot 2 \cdot 3 \cdot 4$

└ read *four factorial*

$4! = 1 \cdot 2 \cdot 3 \cdot 4$
$5! = 1 \cdot 2 \cdot 3 \cdot 4 \cdot 5$

Generally, $n! = 1 \cdot 2 \cdot 3 \cdot \ldots \cdot (n − 2)(n − 1)n$

Compute. Use a calculator.

1. 2!   2
2. 3!   6
3. 4!   24
4. 5!   120
5. 6!   720
6. 7!   5,040
7. 8!   40,320
8. 9!   362,880
9. 10!   3,628,800
10. 11!   39,916,800

Compare. Use > or <.

1. 8 __?__ -10 >
2. -5 __?__ -3 <
3. -12 __?__ 0 <
4. -6 __?__ -11
   >

Add.

5. $-2 + 13$ 11
6. $-8 + (-6)$ -14
7. $-11 + 4$ -7

8. $4 + (-8) + 7$ 3
9. $-15 + (-4) + 8$ -11
10. $9 + (-3) + (-6)$
    0

Copy and complete.

11. $5 + (-5) =$ __?__ 0
12. __?__ $+ (-12) = 0$ 12
13. $-6 +$ __?__ $= 0$
    6

Subtract.

14. $36 - 42$ -6
15. $-12 - (-15)$ 3
16. $27 - (-14)$ 41

17. $-16 - 23$ -39
18. $6 - (-6)$ 12
19. $-8 - (-7)$ -1

Solve these problems.

20. A kite was flying at a height of 500 m. Then it was lowered 200 m, raised 300 m, lowered 100 m, and raised 400 m. At what height was the kite flying then? 900 m

21. At 7:00 A.M. the temperature was 3°C. The temperature rose 6°C between 7:00 A.M. and 11:00 A.M., and rose 5°C between 11:00 A.M. and 3:00 P.M. What was the temperature at 3:00 P.M.? 14°C

Solve for x.

22. $x + (-6) = -8$ -2
23. $-9 + x = 12$ 21

24. $x - (-7) = 7$ 0
25. $x - (-5) = -5$ -10

Cumulative Review

Evaluate if $a = -4$, $b = \frac{1}{5}$, and $c = 9$.

1. $\frac{1}{2}c - b$ $\quad 4\frac{3}{10}$

2. $a - 3b + \frac{2}{3}c$ $\quad 1\frac{2}{5}$

3. $\frac{a + 2c}{4b}$ $\quad 17\frac{1}{2}$

Solve.

4. $x + 9 = 4$ $\quad -5$

5. $5y - 3 = 17$ $\quad 4$

6. $-4 + z = -13$ $\quad -9$

Compute.

7. 32% of 362 is what number? $\quad$ 115.84

8. What % of 75 is 15? $\quad$ 20

9. What % of 50 is 250? $\quad$ 500

10. 4 is what percent of 5? $\quad$ 80

11. Find the range, mean, median, and mode for these scores.
72, 89, 94, 99, 70, 68, 92, 94, 96 $\quad$ Range: 68–99; Mean: 86; Median: 92; Mode: 94

Answer these exercises about the spinner.

12. How many possible outcomes? $\quad$ 4

13. $P(4)$ $\quad \frac{1}{8}$

14. $P(2)$ $\quad \frac{1}{2}$

15. $P(3)$ $\quad \frac{1}{4}$

16. $P(2 \text{ or } 3)$ $\quad \frac{3}{4}$

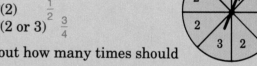

17. If you spin the spinner 200 times, about how many times should it stop on 2? $\quad$ 100

Solve each proportion.

18. $\frac{3}{7} = \frac{n}{28}$ $\quad$ 12

19. $\frac{4}{y} = \frac{2}{9}$ $\quad$ 18

20. $n:12 = 9:4$ $\quad$ 27

Add or subtract.

21. $7 + (-4) + (-9)$ $\quad -6$

22. $12 - 19$ $\quad -7$

23. $-7 - 4 - (-22)$ $\quad$ 11

Solve these problems.

24. A test had 50 questions. Teresa got 94% right. How many questions did she miss? $\quad$ 3

25. Ms. Todd is earning $31,000. Next year she will get a raise of 7%. What will be her salary next year? $\quad$ $33,170

26. The circle graph shows how Gregg spent $60. How much did he spend on each item?

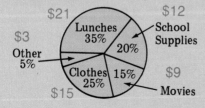

27. There are 5 candidates for president and 4 for vice-president. How many different ways can these offices be filled? $\quad$ 20

Multiplying and Dividing Integers

13

The Okefenokee Swamp in Georgia is a wildlife refuge covering 293,826 acres.

Multiplying Integers

To multiply two positive
  integers
To multiply a positive integer
  and a negative integer

$$3 \cdot 4 = 12 \qquad 5 \cdot 9 = 45 \qquad 9 \cdot 8 = 72$$

$$
\begin{array}{r} 26 \\ \times 14 \\ \hline 104 \\ 26 \\ \hline 364 \end{array}
\qquad
\begin{array}{r} 38 \\ \times 23 \\ \hline 114 \\ 76 \\ \hline 874 \end{array}
$$

Example 1

$4 - 1 = 3; 20 - 5 = 15$

$3 - 1 = 2; 15 - 5 = 10$
$2 - 1 = 1; 10 - 5 = 5$

Continue the pattern for two more steps.
$5 \cdot 4 = 20$
$5 \cdot 3 = 15$
  Subtract 5 each time.
  Subtract 1 each time.
So, $5 \cdot 2 = 10,$
    $5 \cdot 1 = 5.$

The product of two positive integers is a positive
integer.

Example 2

Continue the pattern for three more steps.
$5 \cdot 1 = 5$
$5 \cdot 0 = 0$
  Subtract 5 each time.
  Subtract 1 each time.

$0 - 1 = -1; 0 - 5 = -5$
$-1 - 1 = -2; -5 - 5 = -10$
$-2 - 1 = -3; -10 - 5 = -15$

So, $5 \cdot (-1) = -5,$
    $5 \cdot (-2) = -10,$
    $5 \cdot (-3) = -15.$

Example 3

Continue the pattern for three more steps.
$2 \cdot 3 = 6$
$1 \cdot 3 = 3$
$0 \cdot 3 = 0$
  Subtract 3 each time.
  Subtract 1 each time.

$0 - 1 = -1; 0 - 3 = -3$
$-1 - 1 = -2; -3 - 3 = -6$
$-2 - 1 = -3; -6 - 3 = -9$

So, $-1 \cdot 3 = -3,$
    $-2 \cdot 3 = -6,$
    $-3 \cdot 3 = -9.$

The product of a positive integer and a negative
integer is a negative integer.

Multiply.

    1. $6 \cdot 9$ 54   **2.** $9 \cdot (-7)$ -63 **3.** $-8 \cdot 6$ -48 **4.** $-12 \cdot 3$ -36

Example 4

Multiply. $29 \cdot (-48)$
First, multiply as with whole numbers.

$$\begin{array}{r} 48 \\ \times 29 \\ \hline 432 \\ 96 \\ \hline 1{,}392 \end{array}$$

Determine the sign.

(positive) $\cdot$ (negative) = negative   So, $29 \cdot (-48) = -1{,}392$.

---

practice ▷ **Multiply.**

    5. $12 \cdot (-42)$ **6.** $29 \cdot (-57)$ **7.** $-67 \cdot 98$ **8.** $-19 \cdot 46$
        -504 $-1{,}653$ $-6{,}566$ -874

Example 5

Multiply. $27 \cdot 0;\ -13 \cdot 0$

$$\begin{array}{c|c} 27 \cdot 0 & -13 \cdot 0 \\ 0 & 0 \end{array}$$

So, $27 \cdot 0 = 0$ and $-13 \cdot 0 = 0$.

The product of any integer and 0 is 0.

practice ▷ **Multiply.**

    9. $19 \cdot 0$ 0   **10.** $0 \cdot 62$ 0   **11.** $-28 \cdot 0$ 0   **12.** $0 \cdot (-57)$ 0

◇ ORAL EXERCISES ◇

Multiply.

                                                           -36
1. $12 \cdot 2$ 24   **2.** $13 \cdot 3$ 39   **3.** $3 \cdot 15$ 45   **4.** $5 \cdot (-7)$ -35 **5.** $-9 \cdot 4$
6. $9 \cdot 0$ 0   **7.** $11 \cdot 7$ 77   **8.** $9 \cdot (-3)$ -27 **9.** $0 \cdot (-97)$ 0   **10.** $-6 \cdot 7$
                                                               -42

◇ EXERCISES ◇

Multiply.

  1. $32 \cdot 4$ 128   **2.** $5 \cdot 24$ 120   **3.** $6 \cdot 13$ 78   **4.** $14 \cdot 8$ 112
  5. $26 \cdot 71$ 1,846   **6.** $19 \cdot 82$ 1,558   **7.** $29 \cdot 34$ 986   **8.** $76 \cdot 41$ 3, 116
  9. $26 \cdot (-31)$ -806 **10.** $-45 \cdot 45$ $-2{,}025$ **11.** $93 \cdot (-91)$ $-8{,}463$ **12.** $26 \cdot (-25)$ -650
13. $42 \cdot (-98)$ $-4{,}116$ **14.** $-11 \cdot 78$ -858 **15.** $62 \cdot (-12)$ -744 **16.** $14 \cdot (-17)$ -238
17. $0 \cdot 35$ 0   **18.** $-48 \cdot 0$ 0   **19.** $0 \cdot 88$ 0   ★ **20.** $77 \cdot 0 \cdot (-10)$ 0

MULTIPLYING INTEGERS

Multiplying Negative Integers

To multiply negative integers

The product of any integer and 0 is 0.
$$-8 \cdot 0 = 0 \qquad 0 \cdot 7 = 0$$

The product of a positive integer and a negative integer is a negative integer.
$$-3 \cdot 2 = -6 \qquad 4 \cdot (-5) = -20$$

Example 1

Continue the pattern for three more steps.

$$-3 \cdot 3 = -9$$
$$-3 \cdot 2 = -6$$
$$-3 \cdot 1 = -3$$
$$-3 \cdot 0 = 0$$

⌐ Add 3 each time.

└ Subtract 1 each time.

$0 - 1 = -1; 0 + 3 = 3$
$-1 - 1 = -2; 3 + 3 = 6$
$-2 - 1 = -3; 6 + 3 = 9$

So, $-3 \cdot (-1) = 3$,
$\quad -3 \cdot (-2) = 6$,
$\quad -3 \cdot (-3) = 9$.

Example 2

Continue the pattern for three more steps.

$$2 \cdot (-7) = -14$$
$$1 \cdot (-7) = -7$$
$$0 \cdot (-7) = 0$$

⌐ Add 7 each time.

└ Subtract 1 each time.

$0 - 1 = -1; 0 + 7 = 7$
$-1 - 1 = -2; 7 + 7 = 14$
$-2 - 1 = -3; 14 + 7 = 21$

So, $-1 \cdot (-7) = 7$,
$\quad -2 \cdot (-7) = 14$,
$\quad -3 \cdot (-7) = 21$.

The product of two negative integers is a positive integer.

practice ▷ **Multiply.**

1. $-5 \cdot (-3)$ ₁₅   **2.** $-7 \cdot (-4)$ ₂₈   **3.** $-3 \cdot (-9)$ ₂₇   **4.** $-9 \cdot (-9)$ ₈₁

Example 3

Multiply. $-37 \cdot (-86)$

First multiply as with whole numbers.

$$\begin{array}{r} 86 \\ \times 37 \end{array} \qquad 86 \times 37 = 3{,}182$$

Determine the sign.

(negative) $\cdot$ (negative) = positive

So, $-37 \cdot (-86) = 3{,}182$.

practice ▷ **Multiply.**

5. $-47 \cdot (-81)$ 3,807 **6.** $-63 \cdot (-46)$ 2,898 **7.** $-45 \cdot (-92)$ 4,140 **8.** $-83 \cdot (-200)$ 16,600

◇ *ORAL EXERCISES* ◇

Multiply.

1. $-1 \cdot (-2)$ 2   **2.** $-2 \cdot (-3)$ 6   **3.** $-5 \cdot (-1)$ 5   **4.** $-6 \cdot (-2)$ 12
5. $-3 \cdot (-3)$ 9   **6.** $-4 \cdot (-2)$ 8   **7.** $-9 \cdot (-7)$ 63   **8.** $-4 \cdot (-8)$ 32

◇ *EXERCISES* ◇

Multiply.

1. $-15 \cdot (-3)$ 45   **2.** $-25 \cdot (-4)$ 100   **3.** $-35 \cdot (-2)$ 70
4. $-6 \cdot (-15)$ 90   **5.** $-8 \cdot (-25)$ 200   **6.** $-5 \cdot (-22)$ 110
7. $-17 \cdot (-15)$ 255   **8.** $-36 \cdot (-12)$ 432   **9.** $-30 \cdot (-11)$ 330
10. $-78 \cdot (-19)$ 1,482   **11.** $-29 \cdot (-76)$ 2,204   **12.** $-55 \cdot (-55)$ 3,025
13. $-20 \cdot (-55)$ 1,100   **14.** $-16 \cdot (-16)$ 256   **15.** $-100 \cdot (-14)$ 1,400
16. $-300 \cdot (-11)$ 3,300 **17.** $-21 \cdot (-400)$ 8,400 **18.** $-15 \cdot (-300)$ 4,500
★ **19.** $-300 \cdot (-10)^2$ $-30{,}000$ ★ **20.** $-400 \cdot (-5)^2$ $-10{,}000$ ★ **21.** $(-4)^2 \cdot (-5)^2$ 400

Challenge

1	2	3	4	5	6	7	8	9	10	11	12	13
A	B	C	D	E	F	G	H	I	J	K	L	M

14	15	16	17	18	19	20	21	22	23	24	25	26
N	O	P	Q	R	S	T	U	V	W	X	Y	Z

Decode the message.

First word: $-13 \cdot (-1)$; $-1 \cdot (-1)$; $-5 \cdot (-5)$
Second word: $-25 + 50$; $30 - 15$; $-7 \cdot (-3)$; $-3 \cdot (-6)$
Third word: $-2 - (-6)$; $-12 - (-13)$; $100 - 75$
Fourth word: $-1 \cdot (-2)$; $-1 \cdot (-5)$
Fifth word: $-20 - (-21)$
Sixth word: $-2 \cdot (-11)$; $2 - (-3)$; $-9 \cdot (-2)$; $-50 - (-75)$
Seventh word: $-2 + 10$; $-4 - (-5)$; $-2 \cdot (-8)$; $20 - 4$;
$-26 - (-51)$
Eighth word: $-3 \cdot (-5)$; $-2 \cdot (-7)$; $-10 - (-15)$

May your day be a very happy one

Properties of Multiplication

—◇ *OBJECTIVE* ◇—   —◇ *RECALL* ◇—

To use properties of multiplication of integers

The product of any whole number and 1 is that whole number.
$$26 \cdot 1 = 26 \qquad 1 \cdot 95 = 95$$

Example 1

Multiply integers as you would whole numbers.

Multiply. $5 \cdot 1$; $1 \cdot 367$; $-270 \cdot 1$

$5 \cdot 1$	$1 \cdot 367$	$-270 \cdot 1$
5	367	-270

So, $5 \cdot 1 = 5$, $1 \cdot 367 = 367$, and $-270 \cdot 1 = -270$.

---

property of 1 for multiplication
The product of any integer and 1 is that integer.
$$1 \cdot n = n \text{ and } n \cdot 1 = n$$

Example 2

Multiply. $7 \cdot (-1)$; $-1 \cdot 150$; $-300 \cdot (-1)$

$7 \cdot (-1)$	$-1 \cdot 150$	$-300 \cdot (-1)$
-7	-150	300

So, $7 \cdot (-1) = -7$, $-1 \cdot 150 = -150$, and $-300 \cdot (-1) = 300$.

---

property of −1 for multiplication
The product of any integer and -1 is the opposite of that integer.
$$-1 \cdot n = -n \text{ and } n \cdot (-1) = -n$$

practice ▷ **Multiply.**

    1. $36 \cdot 1$ 36   **2.** $1 \cdot (-525)$ –525   **3.** $18 \cdot (-1)$ -18 **4.** $-1 \cdot 98$ -98

Example 3

Show that $-5 \cdot (-2) = -2 \cdot (-5)$.

$-5 \cdot (-2)$	$-2 \cdot (-5)$
10	10

(negative) · (negative) = positive   So, $-5 \cdot (-2) = -2 \cdot (-5)$.

---

commutative property of multiplication
When multiplying two integers, it does not matter in which order they are multiplied. For all integers a and b, $a \cdot b = b \cdot a$.

Rewrite by using the commutative property. Then multiply.

5. $-5 \cdot (-12)$
$-12 \cdot (-5) = 60$

6. $-6 \cdot 8$
$8 \cdot (-6) = -48$

7. $9 \cdot (-7)$
$-7 \cdot 9 = -63$

8. $-11 \cdot 10$
$10 \cdot (-11) = -110$

Example 4

Show that
$-5 \cdot [-2 \cdot (-11)] = [-5 \cdot (-2)] \cdot (-11)$.

$-5 \cdot [-2 \cdot (-11)]$	$[-5 \cdot (-2)] \cdot (-11)$
$-5 \cdot (22)$	$10 \cdot (-11)$
-110	-110

So, $-5 \cdot [-2 \cdot (-11)] = [-5 \cdot (-2)] \cdot (-11)$.

---

associative property of multiplication
For all integers a, b, and c, $a \cdot (b \cdot c) = (a \cdot b) \cdot c$.

practice ▷ **Multiply. Use the associative property.**

9. $-6 \cdot 5 \cdot (-2)$ 60

10. $-9 \cdot 2 \cdot (-10)$
180

11. $-1 \cdot (-20) \cdot (-30)$
-600

Example 5

Show that
$-5 \cdot [-7 + (-3)] = -5 \cdot (-7) + (-5) \cdot (-3)$.

$-7 + (-3) = -10$

$-5 \cdot [-7 + (-3)]$	$-5 \cdot (-7) + (-5) \cdot (-3)$
$-5 \cdot (-10)$	$35 + 15$
50	50

So, $-5 \cdot [-7 + (-3)] = -5 \cdot (-7) + (-5) \cdot (-3)$.

---

distributive property of multiplication over addition
For all integers a, b, and c, $a \cdot (b + c) = a \cdot b + a \cdot c$.

practice ▷ **Multiply. Use the distributive property.**

12. $-8 \cdot (-23 + 3)$ 160

13. $-8 \cdot (-23) + (-8) \cdot 3$
160

Example 6

Show that
$-8 \cdot [-5 - (-7)] = -8 \cdot (-5) - (-8) \cdot (-7)$.

$-5 - (-7) = -5 + 7 = 2$

$-8 \cdot [-5 - (-7)]$	$-8 \cdot (-5) - (-8) \cdot (-7)$
$-8 \cdot 2$	$40 - 56$
-16	-16

So, $-8 \cdot [-5 - (-7)] = -8 \cdot (-5) - (-8) \cdot (-7)$.

---

practice ▷ **Multiply. Use the distributive property.**

14. $-6 \cdot (7 - 2)$
-30

15. $-6 \cdot 7 - (-6) \cdot 2$
-30

16. $-4 \cdot [-7 - (-9)]$
-8

Multiply.

1. $1 \cdot 5$ 5 2. $-6 \cdot 1$ −6 3. $-8 \cdot (-1)$ 8 4. $-1 \cdot (-93)$ 93 5. $-63 \cdot 1$ −63
6. $-1 \cdot 73$ 7. $73 \cdot (-1)$ 8. $1 \cdot (-963)$ 9. $-963 \cdot 1$ −963 10. $-107 \cdot (-1)$
   −73   −73   −963   107

◇ *EXERCISES* ◇

Multiply.

1. $412 \cdot 1$ 412 2. $1 \cdot 329$ 329 3. $-157 \cdot 1$ −157 4. $1 \cdot (-222)$ −222 5. $1 \cdot 0$ 0
6. $807 \cdot (-1)$ 7. $-1 \cdot 584$ 8. $-1 \cdot (-396)$ 9. $-920 \cdot (-1)$ 10. $-1 \cdot 0$
   −807   −584   396   920   0

Rewrite by using the commutative property. Then multiply.

                                                                                126
11. $-6 \cdot (-4)$ 24 12. $9 \cdot (-9)$ −81 13. $-15 \cdot 6$ −90 14. $-18 \cdot (-7)$
15. $22 \cdot (-15)$ −330 16. $-31 \cdot 31$ −961 17. $-12 \cdot (-48)$ 576 18. $-52 \cdot 11$
                                                                            −572

Multiply. Use the associative property.

19. $-17 \cdot (-5) \cdot 2$ 170   20. $26 \cdot 25 \cdot (-4)$ −2,600
21. $-83 \cdot (-34) \cdot 0$ 0   22. $-36 \cdot 27 \cdot (-1)$ 972
23. $-63 \cdot 2 \cdot (-5)$ 630   24. $-20 \cdot 5 \cdot 13$ −1,300
25. $40 \cdot (-25) \cdot 3$ −3,000   26. $-12 \cdot (-2) \cdot (-50)$ −1,200

Multiply. Use the distributive property.

27. $-9 \cdot (-17 + 7)$ 90   28. $-9 \cdot (-17) + (-9) \cdot 7$ 90
29. $13 \cdot 37 + 13 \cdot (-17)$ 260   30. $13 \cdot [37 + (-17)]$ 260
31. $-21 \cdot (5 - 25)$ 420   32. $-21 \cdot 5 - (-21) \cdot 25$ 420
33. $-50 \cdot 51 - (-50) \cdot 1$ −2,500 34. $-50 \cdot (51 - 1)$ −2,500

Calculator

1. Compute. Use a calculator.
   $1^3 + 2^3$ 9 $(1 + 2)^2$ 9
   $1^3 + 2^3 + 3^3$ 36 $(1 + 2 + 3)^2$ 36
   $1^3 + 2^3 + 3^3 + 4^3$ 100 $(1 + 2 + 3 + 4)^2$ 100
   $1^3 + 2^3 + 3^3 + 4^3 + 5^3$ 225 $(1 + 2 + 3 + 4 + 5)^2$ 225

2. Do you see a pattern? Use the pattern to compute these.
   $1^3 + 2^3 + 3^3 + 4^3 + 5^3 + 6^3$ 441
   $1^3 + 2^3 + 3^3 + 4^3 + 5^3 + 6^3 + 7^3$ 784
   $1^3 + 2^3 + 3^3 + 4^3 + 5^3 + 6^3 + 7^3 + 8^3$ 1,296
   $1^3 + 2^3 + 3^3 + 4^3 + 5^3 + 6^3 + 7^3 + 8^3 + 9^3$ 2,025
   $1^3 + 2^3 + 3^3 + 4^3 + 5^3 + 6^3 + 7^3 + 8^3 + 9^3 + 10^3$ 3,025

Problem Solving – Applications
Jobs for Teenagers

1. As a camp counselor, Tina earned $240 over $2\frac{1}{2}$ pay periods. How much did she make per pay period? $96

2. A group of campers went on a 20-kilometer hike. The counselor let them rest after every $3\frac{1}{3}$ kilometers. How many times did they rest? 6

3. Night patrol at the Green Hills Camp starts at 9:45 P.M. It ends at 12:30 A.M. How long is night patrol? $2\frac{3}{4}$ h

4. One summer, a camp had 176 campers and 22 counselors. How many campers were there per counselor? 8

5. A counselor earned $210 in July and $1\frac{1}{7}$ times as much in August. How much did the counselor earn for the two months? $450

6. Last summer, Kaoni worked in a camp 27 days. He expects to work $1\frac{1}{3}$ as many days next summer. How many days will he work next summer? 36

7. A group of campers on a canoe trip rowed upstream for $1\frac{1}{2}$ hours. Coming back downstream, they took $\frac{2}{3}$ as long. How long did it take them to come back? 1 h

8. A counselor had waterfront duty for $\frac{3}{4}$ hour in the afternoon. Waterfront duty in the morning was $\frac{2}{3}$ as long. How long was waterfront duty in the morning? $\frac{1}{2}$ h

Dividing Integers

To divide integers

Multiplication and division are related.

$$5 \cdot 2 = 10 \overbrace{}^{\begin{array}{l}10 \div 2 = 5\\10 \div 5 = 2\end{array}}$$

Example 1

Write related division and
multiplication sentences.

$7 \cdot 3 = 21$ and $21 \div 3 = 7$.

7 is the *quotient*.

Divide. $21 \div 3$
$21 \div 3 = n$, so $n \cdot 3 = 21$.
Think: What number times 3 is 21?
Answer: $7 \cdot 3 = 21$.
So, $21 \div 3 = 7$.

Example 2

Write related division and
multiplication sentences.

$9 \cdot (-2) = -18$ and
$-18 \div (-2) = 9$.

Divide. $-18 \div (-2)$
$-18 \div (-2) = n$, so $n \cdot (-2) = -18$.
Think: What number times -2 is -18?
Answer: $9 \cdot (-2) = -18$.
So, $-18 \div (-2) = 9$.

The quotient of two positive integers or two negative
integers is a positive integer.

practice ▷   **Divide.**

1. $35 \div 5$ 7 **2.** $42 \div 6$ 7 **3.** $-72 \div (-9)$ 8 **4.** $-63 \div (-7)$ 9

Example 3

Write related division and
multiplication sentences.

$-6 \cdot 3 = -18$ and $-18 \div 3 = -6$.

Divide. $-18 \div 3$
$-18 \div 3 = n$, so $n \cdot 3 = -18$.
Think: What number times 3 is -18?
Answer: $-6 \cdot 3 = -18$.
So, $-18 \div 3 = -6$.

Example 4

Write related division and
multiplication sentences.

$-5 \cdot (-4) = 20$ and
$20 \div (-4) = -5$.

Divide. $20 \div -4$
$20 \div (-4) = n$, so $n \cdot (-4) = 20$.
Think: What number times -4 is 20?
Answer: $-5 \cdot (-4) = 20$.
So, $20 \div (-4) = -5$.

The quotient of a positive integer and a negative
integer is a negative integer.

Divide.

     5. $-25 \div 5$ -5 **6.** $-30 \div 6$ -5 **7.** $48 \div (-6)$ -8 **8.** $64 \div (-8)$ -8

Example 5

Write related division and multiplication sentences.

Divide. $-17 \div 1$
$-17 \div 1 = n,$ so $n \cdot 1 = -17.$
$-17 \cdot 1 = -17$
So, $-17 \div 1 = -17.$

Example 6

25 is the opposite of -25.

Divide. $-25 \div (-1)$
$-25 \div (-1) = n,$ so $n \cdot (-1) = -25.$
$25 \cdot (-1) = -25$
So, $-25 \div (-1) = 25.$

The quotient of an integer and 1 is that integer. The quotient of an integer and -1 is the opposite of that integer.

practice ▷ **Divide.**

    9. $44 \div 1$ 44 **10.** $-36 \div 1$ **11.** $-85 \div (-1)$ 85 **12.** $99 \div (-1)$ -99
                        -36

Example 7

Write related division and multiplication sentences.

Divide. $0 \div (-8)$
$0 \div (-8) = n,$ so $n \cdot (-8) = 0.$
Think: What number times -8 is 0?
Answer: $0 \cdot (-8) = 0.$
So, $0 \div (-8) = 0.$

Example 8

Write related division and multiplication sentences.

Divide. $-5 \div 0$
$-5 \div 0 = n,$ so $n \cdot 0 = -5.$
Think: What number times 0 is -5?
There is no such number, since any number times 0 is 0.
So, dividing -5 by 0 has no answer.

The quotient of 0 and an integer is 0. Division by 0 has no answer.

practice ▷ **Divide, if possible.**

    13. $0 \div 5$ 0 **14.** $0 \div (-25)$ 0 **15.** $-2 \div 0$ **16.** $98 \div 0$
                                        not possible   not possible

DIVIDING INTEGERS

Divide, if possible.

1. $36 \div 9$ 4	**2.** $48 \div 8$ 6	**3.** $35 \div 7$ 5	**4.** $56 \div 7$ 8
5. $63 \div 9$ 7	**6.** $64 \div 8$ 8	**7.** $30 \div 5$ 6	**8.** $60 \div 5$ 12
9. $48 \div 4$ 12	**10.** $51 \div 3$ 17	**11.** $-27 \div (-3)$ 9	**12.** $-32 \div (-8)$ 4
13. $-40 \div (-5)$ 8	**14.** $-81 \div (-9)$ 9	**15.** $-42 \div (-7)$ 6	**16.** $-35 \div (-5)$ 7
17. $-45 \div (-9)$ 5	**18.** $-33 \div (-3)$ 11	**19.** $-36 \div (-2)$ 18	**20.** $-80 \div (-5)$ 16
21. $-28 \div 7$ −4	**22.** $-64 \div 8$ −8	**23.** $-18 \div 6$ −3	**24.** $-24 \div 4$ −6
25. $-36 \div 6$ −6	**26.** $-15 \div 5$ −3	**27.** $-54 \div 6$ −9	**28.** $-57 \div 3$ −19
29. $-60 \div 6$ −10	**30.** $-75 \div 5$ −15	**31.** $28 \div (-4)$ −7	**32.** $18 \div (-9)$ −2
33. $49 \div (-7)$ −7	**34.** $63 \div (-7)$ −9	**35.** $56 \div (-8)$ −7	**36.** $20 \div (-5)$ −4
37. $48 \div (-8)$ −6	**38.** $93 \div (-3)$ −31	**39.** $84 \div (-4)$ −21	**40.** $72 \div (-6)$ −12
41. $-8 \div 1$ −8	**42.** $27 \div 1$ 27	**43.** $-32 \div 1$ −32	**44.** $-11 \div 1$ −11
45. $59 \div 1$ 59	**46.** $40 \div (-1)$ −40	**47.** $-21 \div (-1)$ 21	**48.** $-19 \div (-1)$ 19
49. $59 \div (-1)$ −59	**50.** $-78 \div (-1)$ 78	**51.** $0 \div 3$ 0	**52.** $0 \div (-7)$ 0
53. $0 \div 15$ 0	**54.** $0 \div (-24)$ 0	**55.** $0 \div (-35)$ 0	**56.** $4 \div 0$ not possible
57. $-8 \div 0$ not possible	★ **58.** $12^2 \div (-9)$ −16	★ **59.** $(-8)^2 \div (-16)$ −4	★ **60.** $(-15)^2 \div (-25)$ −9

Reading in Math

For each group of four mathematical terms, identify the term that does not belong with the other three.

    Example: integer
                     fraction
                     line
                     percent

Line does not belong; the other three terms refer to numbers.

1.	ratio	**2.**	mean
	triangle		mode
	proportion		score
	extremes		median
3.	temperature	**4.**	kilometer
	probability		gram
	experiment		centimeter
	event		millimeter

From Integers to Rationals

To locate rational numbers on a number line

To divide rational numbers

$7 \div 3 = n,$ so $n \cdot 3 = 7.$
There is no whole number which when multiplied by 3 gives 7.
The quotient of two numbers is not always a whole number.

$$\frac{2}{3} \div \frac{4}{5} = \frac{2}{3} \cdot \frac{5}{4}$$

In general, $\dfrac{a}{b} \div \dfrac{c}{d} = \dfrac{a}{b} \cdot \dfrac{d}{c}.$

Example 1

What number corresponds to point A? to point B?

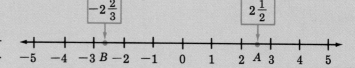

A is halfway between 2 and 3. B is two-thirds of the way from -2 to -3.

So, $2\frac{1}{2}$ corresponds to point A and $-2\frac{2}{3}$ corresponds to point B.

---

practice ▷   **What number corresponds to the point?**

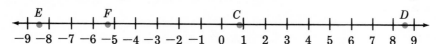

1. C $\frac{3}{4}$ 2. D $8\frac{1}{2}$ 3. E $-8\frac{1}{2}$ 4. F $-5\frac{1}{3}$

Positive fractions, negative fractions, positive mixed numbers, and negative mixed numbers are rational numbers:

$$\frac{1}{2}, \quad \frac{3}{7}, \quad \frac{17}{5}, \quad 4\frac{5}{8}, \quad -\frac{2}{3}, \quad -\frac{4}{9}, \quad -\frac{12}{5}, \quad -5\frac{2}{3}.$$

Decimals that terminate or repeat are also rational numbers: $1.6, -0.5, -2.65, 0.007, 0.\overline{6}.$
All integers are also rational numbers: $4, 21, -3, -15.$

Dividing rational numbers is like dividing fractions.

Example 2 Divide. $\dfrac{2}{3} \div \left(-\dfrac{5}{7}\right)$ $\qquad$ $-3\dfrac{3}{4} \div \left(-2\dfrac{1}{2}\right)$

$$\dfrac{2}{3} \div \left(-\dfrac{5}{7}\right) \qquad\qquad -3\dfrac{3}{4} \div \left(-2\dfrac{1}{2}\right)$$

$$\dfrac{2}{3} \cdot \left(-\dfrac{7}{5}\right) \qquad\qquad -\dfrac{15}{4} \div \left(-\dfrac{5}{2}\right)$$

$$\dfrac{2(-7)}{3 \cdot 5} \qquad\qquad -\dfrac{15}{4} \cdot \left(-\dfrac{2}{5}\right)$$

(positive) · (negative) = negative
(negative) · (negative) = positive

$$-\dfrac{14}{15} \qquad\qquad \dfrac{\overset{3}{-\cancel{15}} \cdot (-\cancel{2})}{\underset{2}{\cancel{4}} \cdot \underset{1}{\cancel{5}}} = \dfrac{3}{2}, \text{ or } 1\dfrac{1}{2}$$

So, $\dfrac{2}{3} \div \left(-\dfrac{5}{7}\right) = -\dfrac{14}{15}$ and $-3\dfrac{3}{4} \div \left(-2\dfrac{1}{2}\right) = 1\dfrac{1}{2}$.

practice ▷ Divide.

5. $\dfrac{4}{5} \div \left(-\dfrac{1}{3}\right)$ $\;-2\dfrac{2}{5}$ **6.** $-\dfrac{2}{3} \div \left(-\dfrac{4}{9}\right)$ $\;1\dfrac{1}{2}$ **7.** $-3\dfrac{1}{2} \div \left(-5\dfrac{1}{4}\right)$ $\;\dfrac{2}{3}$ **8.** $-4 \div 2\dfrac{1}{3}$ $\;-1\dfrac{5}{7}$

◇ ORAL EXERCISES ◇

Is the quotient an integer?

1. $13 \div (-1)$ yes **2.** $-21 \div (-7)$ yes **3.** $-15 \div (-6)$ no **4.** $-17 \div 2$ no
5. $14 \div (-3)$ no **6.** $27 \div (-3)$ yes **7.** $-42 \div 6$ yes **8.** $45 \div (-9)$ yes

---

◇ EXERCISES ◇

What number corresponds to the point?

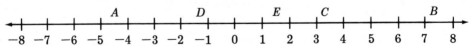

1. A $\;-4\dfrac{1}{2}$ **2.** B $\;7\dfrac{1}{3}$ **3.** C $\;3\dfrac{1}{3}$ **4.** D $\;-1\dfrac{1}{4}$ **5.** E $\;1$

Divide.

6. $\dfrac{4}{5} \div \dfrac{3}{7}$ $\;1\dfrac{13}{15}$ **7.** $\dfrac{9}{4} \div \dfrac{3}{10}$ $\;7\dfrac{1}{2}$ **8.** $\dfrac{2}{9} \div \dfrac{3}{5}$ $\;\dfrac{10}{27}$ **9.** $\dfrac{4}{9} \div \dfrac{8}{13}$ $\;\dfrac{13}{18}$

10. $-\dfrac{2}{5} \div \left(-\dfrac{4}{3}\right)$ $\;\dfrac{3}{10}$ **11.** $-\dfrac{5}{8} \div \left(-\dfrac{9}{10}\right)$ $\;\dfrac{25}{36}$ **12.** $-\dfrac{3}{7} \div \dfrac{5}{6}$ $\;-\dfrac{18}{35}$ **13.** $-\dfrac{2}{3} \div \dfrac{9}{4}$ $\;-\dfrac{8}{27}$

14. $-2\dfrac{1}{3} \div \dfrac{3}{7}$ $\;-5\dfrac{4}{9}$ **15.** $-5\dfrac{2}{5} \div 1\dfrac{1}{2}$ $\;-3\dfrac{3}{5}$ **16.** $-3 \div \left(-2\dfrac{1}{4}\right)$ $\;1\dfrac{1}{3}$ **17.** $4\dfrac{1}{4} \div \left(-5\dfrac{1}{5}\right)$ $\;-\dfrac{8}{1}$

$\qquad\qquad\qquad$ *CHAPTER THIRTEEN*

Problem Solving – Careers
Industrial Machinists

1. Ms. Barkley repairs industrial sewing machines. One week, she repaired 12 machines. The following week, she repaired $2\frac{1}{3}$ times as many. How many did she repair then?   28

2. Mr. Bing works in a printing plant. One week, 24 pieces of machinery broke down. On the average, he can repair $1\frac{1}{2}$ pieces per day. How many days will it take him to repair the 24 pieces?   16

3. Ms. Garcia's factory spent \$1,500 for new parts in one month. The following month, it spent $1\frac{1}{4}$ times as much. How much was that?   \$1,875

4. Mr. Ferguson runs the repair shop in a food products plant. One month, he ordered 1,200 parts for the shop. The following month he ordered 3 times as many parts. How many parts did he order in the two months?   4,800

5. Lucille works in a textile mill. She repaired 78 pieces of machinery during January. During February, she repaired $1\frac{1}{2}$ times as many. How many pieces did she repair in the two months?   195

6. Raymond's Packaging Plant spent \$2,100 for parts one month. The next month the plant spent $\frac{2}{3}$ as much for parts. How much did the plant spend on parts in the two months?   \$3,500

PROBLEM SOLVING

Simplifying Expressions

— ◇ **OBJECTIVES** ◇ —

To simplify expressions like
$3x + (-5x)$, $4 - (-6x)$,
and $-4 \cdot (-3y)$

To simplify expressions like
$3(x - 5)$, $-2(3m - 1)$,
and $-(-3 - 2x)$

— ◇ **RECALL** ◇ —

$5 + (-9) = -4$ $3[5 + (-1)] = 3(5) + 3(-1)$
$-3 - (-4) = 1$
$-30 \div 5 = -6$ $2(4) + (-3)(4) = [2 + (-3)]4$
$-3 \cdot (-4) = 12$
$5 \cdot (-7) = -35$ $-2 \cdot [-3 \cdot 4] = [-2 \cdot (-3)]4$

Example 1

Use the distributive property.
$5x$ means $5 \cdot x$; $-9x$ means $-9 \cdot x$;
$5 + (-9) = -4$.

Simplify. $5x + (-9x)$
$5x + (-9x)$
$[5 + (-9)]x$
$-4x$
So, $5x + (-9x) = -4x$.

---

practice ▷ **Simplify.**

1. $3x + (-7x)$ $_{-4x}$ 2. $-5x + 3x$ $_{-2x}$ 3. $-6x + (-2x)$ $_{-8x}$ 4. $8x + (-8x)$ $_{0}$

Example 2

Use the distributive property.
$-3 - (-4) = -3 + 4 = 1$
$1x$ means $1 \cdot x$, or x.

Simplify. $-3x - (-4x)$
$-3x - (-4x)$
$[-3 - (-4)]x$
$1x$, or x
So, $-3x - (-4x) = x$.

---

practice ▷ **Simplify.**

5. $-2x - (-6x)$ $_{4x}$ 6. $5x - 9x$ $_{-4x}$ 7. $8x - (-3x)$ $_{11x}$ 8. $-4x - x$ $_{-5x}$

Example 3

$-6y$ means $-6 \cdot y$.
Use the associative property.
$-5 \cdot (-6) = 30$

Simplify. $-5 \cdot (-6y)$
$-5 \cdot (-6y)$
$-5 \cdot (-6) \cdot y$
$[-5 \cdot (-6)]y$
$30y$
So, $-5 \cdot (-6y) = 30y$.

---

practice ▷ **Simplify.**

9. $3 \cdot 7x$ $_{21x}$ 10. $-4 \cdot 5y$ $_{-20y}$ 11. $-2 \cdot (-6y)$ $_{12y}$ 12. $6 \cdot (-3x)$ $_{-18x}$

Example 4

Use the distributive property.

$6x + (-8)$ means $6x - 8$.

Simplify. $2(3x - 4)$

$$2(3x - 4) = 2(3x) + 2(-4)$$
$$6x + (-8)$$
$$6x - 8$$

So, $2(3x - 4) = 6x - 8$.

Example 5

Use the distributive property.
$-3x + (-15)$ means $-3x - 15$.

Simplify. $-3(x + 5)$

$$-3(x + 5) = -3(x) + (-3)(5)$$
$$-3x + (-15)$$
$$-3x - 15$$

So, $-3(x + 5) = -3x - 15$.

Example 6

$-(6 - 5y)$ means $-1(6 - 5y)$.
Use the distributive property.

Simplify. $-(6 - 5y)$

$$-(6 - 5y) = -1(6 - 5y)$$
$$-1(6) + (-1)(-5y)$$
$$-6 + 5y$$

So, $-(6 - 5y) = -6 + 5y$.

practice ▷ **Simplify.**

13. $3(2x - 1)$ **14.** $-2(y - 4)$ **15.** $-(3 + 4x)$ **16.** $-(5 - 3y)$
$6x - 3$ $-2y + 8$ $-3 - 4x$ $-5 + 3y$

◇ EXERCISES ◇

Simplify.   See TE Answer Section.

1. $3x + (-7x)$ **2.** $-6y + (-7y)$ **3.** $4z + 9z$ **4.** $-10a + 4a$
5. $-6c + (-6c)$ **6.** $10m + (-14m)$ **7.** $-3n + (-7n)$ **8.** $4c + (-2c)$
9. $17t + (-20t)$ **10.** $-19p + 7p$ **11.** $-2x - (-3x)$ **12.** $-5x - 2x$
13. $7u - (-1u)$ **14.** $-6v - 6v$ **15.** $-8a - (-8a)$ **16.** $4y - (-10y)$
17. $-12h - (-30h)$ **18.** $14k - (-5k)$ **19.** $r - (-r)$ **20.** $-d - 9d$
21. $-3 \cdot (-7m)$ **22.** $5 \cdot 9d$ **23.** $-7 \cdot 2b$ **24.** $8 \cdot (-9n)$
25. $-6 \cdot (-6t)$ **26.** $3(2x + 3)$ **27.** $2(5y - 1)$ **28.** $-2(4x + 1)$
29. $-4(3z - 2)$ **30.** $6(2h + 1)$ **31.** $4(n - 2)$ **32.** $3(p + 7)$
33. $-5(x - 3)$ **34.** $-2(y + 2)$ **35.** $7(a - 1)$ **36.** $-(r - 3)$
37. $-(8 - z)$ **38.** $-(4 - 2b)$ **39.** $-(4 + 3w)$ **40.** $-(5y - 3)$
★ **41.** $-8(2m - p) + 5$ ★ **42.** $6(-m - n) - 10$

SIMPLIFYING EXPRESSIONS

Solving Equations

◇ **OBJECTIVE** ◇

To solve equations like

$-4x = -20$, $\frac{x}{-3} = -2$, and

$-3x + 4 = -8$

◇ **RECALL** ◇

Solve. $3x = 12$

$x = \frac{12}{3}$, or 4

So, 4 is the solution.

Solve. $\frac{x}{4} = 5$

$x = 5 \cdot 4$, or 20

So, 20 is the solution.

Example 1

Undo multiplication by division.

Solve and check.

$$5x = -30 \qquad\qquad -3y = -39$$

$$\frac{5x}{5} = \frac{-30}{5} \qquad\qquad \frac{-3y}{-3} = \frac{-39}{-3}$$

$$x = -6 \qquad\qquad y = 13$$

Check: Check:

$$\begin{array}{c|c} 5x = -30 & -3y = -39 \\ \hline 5 \cdot (-6) \mid -30 & -3 \cdot 13 \mid -39 \\ -30 = -30 & -39 = -39 \end{array}$$

So, -6 is the solution of $5x = -30$, and 13 is the solution of $-3y = -39$.

---

practice ▷   **Solve and check.**

1. $2x = -18$ -9 **2.** $-4x = 28$ -7 **3.** $-6y = -42$ 7

Example 2

Undo division by multiplication.

Solve.

$$\frac{x}{-3} = 4 \qquad\qquad\qquad \frac{a}{-7} = -5$$

$$-3 \cdot \left(\frac{x}{-3}\right) = 4 \cdot (-3) \qquad -7 \cdot \left(\frac{a}{-7}\right) = -5 \cdot (-7)$$

$$x = -12 \qquad\qquad\qquad a = 35$$

Check on your own.   So, -12 is the solution of $\frac{x}{-3} = 4$, and 35 is the

solution of $\frac{a}{-7} = -5$.

---

practice ▷   **Solve and check.**

4. $\frac{x}{-5} = 4$ -20 **5.** $\frac{n}{-4} = -3$ 12   **6.** $\frac{w}{6} = -8$ -48

Example 3

Solve and check. $-4x + 7 = -1$

Undo addition by subtraction.	$-4x + 7 = -1$
Subtract 7 from each side.	$-4x + 7 - 7 = -1 - 7$
Undo multiplication by division.	$-4x = -8$
Divide each side by -4.	$\dfrac{-4x}{-4} = \dfrac{-8}{-4}$
	$x = 2$

Check: $\underline{ -4x + 7 = -1}$

$$-4 \cdot 2 + 7$$
$$-8 + 7$$
$$-1 = -1$$

So, 2 is the solution of $-4x + 7 = -1$.

---

practice ▷ **Solve and check.**

7. $2x - 8 = -14$ -3 **8.** $-4x + 7 = 11$ -1 **9.** $\dfrac{x}{-3} - 6 = 2$ -24

◇ EXERCISES ◇

Solve and check.

1. $3x = -6$ -2 **2.** $-7y = 28$ -4 **3.** $-4z = -36$ 9 **4.** $9a = 81$ 9

5. $6t = -42$ -7 **6.** $-8m = -64$ 8 **7.** $\dfrac{x}{-4} = -1$ 4 **8.** $\dfrac{y}{-2} = 6$ -12

9. $\dfrac{m}{-6} = -9$ 54 **10.** $\dfrac{r}{9} = -7$ -63 **11.** $\dfrac{c}{-8} = 7$ -56 **12.** $\dfrac{a}{-4} = -12$ 48

13. $3x - 5 = 1$ 2 **14.** $-2y + 7 = -9$ 8 **15.** $-4a - 2 = -10$ 2**16.** $5x + 8 = -7$ -3

17. $\dfrac{x}{-2} + 8 = -1$ 8 **18.** $\dfrac{y}{3} - 7 = 4$ 33 **19.** $\dfrac{z}{-5} - 3 = 2$ -25 **20.** $\dfrac{a}{4} + 6 = 1$ -20

1. A farmer has some chickens and some goats. Altogether there are 43 heads and 108 legs. How many chickens does the farmer have? How many goats does the farmer have?

2. You need 5 liters of water. Three containers are available to you. One holds 3 liters; the second, 4 liters; and the third, 7 liters. How can you get 5 liters of water?

See TE Answer Section.

SOLVING EQUATIONS

Scientific Notation and Small Numbers

── ◇ **OBJECTIVES** ◇ ──

To multiply decimals by
  negative powers of ten
To write small numbers in
  scientific notation

─────── ◇ **RECALL** ◇ ───────

Powers of 10

$$100 = 10^2 \qquad 1,000 = 10^3 \qquad 10,000 = 10^4$$

Example 1

Divide by 10: Move the decimal point 1 place left.

Divide. $57.9 \div 10$
$57.9 \div 10$
$57.9 \div 10 = 57.9$
     1 zero   1 place

Multiply. 57.9×0.1
57.9×0.1 57.9
              $\times 0.1$
              5.79

So, $57.9 \div 10 = 5.79$ and $57.9 \times 0.1 = 5.79$.

practice ▷ Compute.

1. 3.62×0.1
   0.362

2. $3.62 \div 10$
   0.362

3. 4.5×0.01
   0.045

4. $4.5 \div 100$
   0.045

Example 2

$10^{-2} = 0.01$

Divide by 100: Move the decimal point 2 places left.

Multiply. 0.79×10^{-2}
0.79×10^{-2}
0.79×0.01
$0.79 \div 100 = 00.79$
   2 zeros   2 places

Multiply. 81.3×10^{-3}
81.3×10^{-3} $10^{-3} = 0.001$
81.3×0.001
$81.3 \div 1,000 = 081.3$
   3 zeros   3 places

So, $0.79 \times 10^{-2} = 0.0079$ and $81.3 \times 10^{-3} = 0.0813$.

practice ▷ Multiply.

5. 6.9×10^{-1}
   0.69

6. 0.45×10^{-2}
   0.0045

7. 50.9×10^{-3}
   0.0509

8. 8.73×10^{-4}
   0.000873

You can write small numbers using scientific notation. Complete the pattern.

Number		Scientific Notation
$45 = 4.5 \times 10$	$= 4.5 \times 10^1$	
$4.5 = 4.5 \times 1$	$= 4.5 \times 10^0$	
$0.45 = 4.5 \times 0.1$	$= 4.5 \times 10^{-1}$	
$0.045 = 4.5 \times 0.01$	$= 4.5 \times 10^{-2}$	
$0.0045 = 4.5 \times \square$	$= \underline{\ ?\ }$	
0.001	4.5×10^{-3}	

A number is in *scientific notation* if it is written as a product of a number from 1 to 10 and a power of 10.

Example 3

To make 0.513 a number between 1 and 10, move the decimal point and multiply by a power of 10.

Write 0.513 and 0.0079 in scientific notation.

$$0.513 = 0.513 \times 10^n$$
1 place

$$n = -1$$

$$0.513 = 5.13 \times 10^{-1}$$

So, $0.513 = 5.13 \times 10^{-1}$ in scientific notation.

$$0.0079 = 0.0079 \times 10^n$$
3 places

$$n = -3$$

So, $0.0079 = 7.9 \times 10^{-3}$ in scientific notation.

---

practice ▷ **Write in scientific notation.**

9. 0.48
4.8×10^{-1}

10. 0.036
3.6×10^{-2}

11. 0.05
5×10^{-2}

12. 0.0004
4×10^{-4}

◇ EXERCISES ◇

Multiply.

1. 4.13×0.1 0.413
2. 7.2×0.01 0.072
3. 576×0.001 0.576
4. 9.185×0.0001 0.0009185
5. 3.2×10^{-1} 0.32
6. 0.68×10^{-2} 0.0068
7. 49.98×10^{-2} 0.4998
8. 5.01×10^{-4} 0.000501
9. 9.0634×10^{-3} 0.0090634
10. 23×10^{-4} 0.0023
11. 0.09×10^{-5} 0.0000009
12. 1.6×10^{-6} 0.0000016

Write in scientific notation.

13. 0.72 7.2×10^{-1}
14. 0.659 6.59×10^{-1}
15. 0.5 5×10^{-1}
16. 0.05 5×10^{-2}
17. 0.083 8.3×10^{-2}
18. 0.0072 7.2×10^{-3}
19. 0.00091 9.1×10^{-4}
20. 0.0006 6×10^{-4}
21. 23.0024 2.30024×10^1
22. 318.007 3.18007×10^2
23. 0.00006 6×10^{-5}
24. 0.0000074 7.4×10^{-6}

SCIENTIFIC NOTATION AND SMALL NUMBERS

Chapter Review

Multiply. [312, 314]

1. $11 \cdot 23$ 253
2. $24 \cdot (-18)$ −432
3. $-29 \cdot 71$ −2,059
4. $48 \cdot 0$ 0
5. $-49 \cdot (-62)$ 3,038
6. $-68 \cdot (-91)$ 6,188
7. $-100 \cdot (-12)$ 1,200
8. $-17 \cdot (-200)$ 3,400

Multiply. [316]

9. $1 \cdot 275$ 275
10. $-312 \cdot (-1)$ 312
11. $42 \cdot 25 \cdot (-4)$ −4,200
12. $-61 \cdot (-22) \cdot 0$ 0

Multiply. Use the distributive property. [316]

13. $-7 \cdot (-19 + 9)$ 70
14. $14 \cdot 26 + 14 \cdot (-6)$ 280
15. $-12 \cdot (7 - 27)$ 240
16. $-30 \cdot 41 - (-30) \cdot 1$ −1,200

Divide, if possible. [320]

17. $-38 \div 1$ −38
18. $0 \div (-5)$ 0
19. $88 \div (-1)$ −88
20. $-42 \div 0$ not possible
21. $48 \div 6$ 8
22. $-22 \div 2$ −11
23. $-55 \div (-5)$ 11
24. $57 \div (-3)$ −19
25. $150 \div (-5)$ −30
26. $-200 \div 4$ −50
27. $-500 \div 50$ −10
28. $-600 \div (-30)$ 20

Divide. [323]

29. $\frac{3}{7} \div \frac{2}{9}$ $1\frac{13}{14}$
30. $-\frac{3}{5} \div \left(-\frac{2}{3}\right)$ $\frac{9}{10}$
31. $-\frac{4}{5} \div \frac{1}{2}$ $-1\frac{3}{5}$
32. $-3\frac{1}{4} \div \left(-4\frac{1}{5}\right)$ $\frac{65}{84}$

Simplify. [326]

33. $3y + (-16y)$ −13y
34. $-5t - (-11t)$ 6t
35. $-4 \cdot 5x$ −20x
36. $-7 \cdot (-8m)$ 56m
37. $3(2y - 5)$ 6y − 15
38. $5(a + 6)$ 5a + 30
39. $-(6 + 2x)$ −6 − 2x
40. $-4(4n - 3)$ −16n + 12

Solve. [328]

41. $-9x = 54$ −6
42. $-7y = -63$ 9
43. $\frac{t}{-6} = 2$ −12
44. $2k - 7 = -3$ 2

Multiply.

1. $12 \cdot 47$ 564

2. $54 \cdot (-23)$ $-1{,}242$

3. $-64 \cdot (-93)$ 5,952

4. $32 \cdot 0$ 0

5. $-1 \cdot 518$ -518

6. $20 \cdot (-5) \cdot 37$ $-3{,}700$

Multiply. Use the distributive property.

7. $-22 \cdot (-9 + 15)$ -132

8. $9 \cdot (-11) - 9 \cdot (-24)$ 117

Divide, if possible.

9. $16 \div (-1)$ -16

10. $-800 \div (-40)$ 20

11. $12 \div 0$ not possible

12. $0 \div (-31)$ 0

Divide.

13. $-\dfrac{5}{7} \div \dfrac{2}{3}$ $-1\dfrac{1}{14}$

14. $-\dfrac{2}{5} \div \left(-\dfrac{3}{8}\right)$ $1\dfrac{1}{15}$

15. $-5\dfrac{1}{2} \div \left(-6\dfrac{1}{3}\right)$ $\dfrac{33}{38}$

Simplify.

16. $-5t + 16t$ $11t$

17. $-9 \cdot (-4m)$ $36m$

18. $-2(2x - 1)$ $-4x + 2$

Solve.

19. $-7x = 63$ -9

20. $\dfrac{y}{-4} = -8$ 32

21. $-2x + 5 = -7$ 6

Challenge

Replace each letter by a digit so that correct additions will result. Within the same problem, the same letters should be replaced by the same digit, different letters by different digits.

1.   85,771   and three
   + 4,862   other solutions
   90,633

1.
```
  MERRY
+ XMAS
 TO ALL
```

2.   9,567
   +1,085
   10,652

2.
```
  SEND
+MORE
MONEY
```

3.   92,836
   + 12,836
   105,672

3.
```
 HOCUS
+POCUS
PRESTO
```

4.   47,474
   + 5,272
   52,746

4.
```
AHAHA
+ TEHE
TEHAW
```

Equations and Inequalities

The mirror in the Mayall telescope at the Kitt Peak National Observatory weighs 14 t.

Solving Equations

◇ OBJECTIVE ◇

To solve an equation like
$$4x - 9 = 2x + 5$$

◇ RECALL ◇

Solve. $6x - 5 = 19$

Add 5 to each side. $6x - 5 + 5 = 19 + 5$

$$6x = 24$$

Divide each side by 6. $\dfrac{6x}{6} = \dfrac{24}{6}$

So, 4 is the solution. $x = 4$

Equation properties can be used to solve equations in which the variable is on each side of the equation.

Example 1

Solve and check. $5x + 4 = x + 16$

	$5x + 4 = x + 16$
Subtract 4 from each side.	$5x + 4 - 4 = x + 16 - 4$
	$5x = x + 12$
Subtract x from each side.	$5x - x = x + 12 - x$
Combine like terms.	$4x = 12$
Divide each side by 4.	$\dfrac{4x}{4} = \dfrac{12}{4}$
	$x = 3$

Check.

$$5x + 4 = x + 16$$

$5 \cdot 3 + 4$	$3 + 16$
$15 + 4$	19
19	
$19 = 19$	
true	

So, 3 is the solution of $5x + 4 = x + 16$.

practice ▷ Solve.

1. $3x + 8 = x + 2$ -3

2. $6x + 9 = x + 4$ -1

Example 2

Solve and check. $6x - 7 = 4x + 5$

	$6x - 7 = 4x + 5$
Add 7 to each side.	$6x - 7 + 7 = 4x + 5 + 7$
	$6x = 4x + 12$
Subtract $4x$ from each side.	$6x - 4x = 4x + 12 - 4x$
Combine like terms.	$2x = 12$
Divide each side by 2.	$\dfrac{2x}{2} = \dfrac{12}{2}$
	$x = 6$

Check.

$$6x - 7 = 4x + 5$$

$6 \cdot 6 - 7$	$4 \cdot 6 + 5$
$36 - 7$	$24 + 5$
29	29
$29 = 29$	
true	

So, 6 is the solution of $6x - 7 = 4x + 5$.

practice ▷ Solve.

3. $3x + 7 = 5x - 3$ 5

4. $9x - 2 = 2x + 12$ 2

SOLVING EQUATIONS

335

Solve.

1. $2x = 12$ 6
4. $12x = 6$ $\frac{1}{2}$
7. $x + 3 = 5$ 2

2. $3x = 24$ 8
5. $15x = 5$ $\frac{1}{3}$
8. $x - 7 = 2$ 9

3. $5x = 30$ 6
6. $12x = 3$ $\frac{1}{4}$
9. $x - 5 = 20$ 25

◇ EXERCISES ◇

Solve.

1. $7x + 3 = 17$ 2

2. $9x + 4 = 4$ 0

3. $12x + 5 = 17$ 1

4. $2x - 5 = 7$ 6

5. $5x - 4 = 11$ 3

6. $4x - 8 = 12$ 5

7. $4x + 9 = x + 6$ -1

8. $2x + 7 = x + 1$ -6

9. $7x + 8 = x + 2$ -1

10. $3x - 6 = x - 2$ 2

11. $7x - 2 = x - 14$ -2

12. $8x - 1 = x - 8$ -1

13. $4x - 5 = 2x + 1$ 3

14. $7x - 5 = 5x + 5$ 5

15. $12x + 3 = 5x - 18$ -3

16. $3x + 2 = 6x - 13$ 5

17. $4x - 5 = 9x + 15$ -4

18. $9x - 4 = 3x + 2$ 1

★ 19. $2x - 3 = 5x - 4$ $\frac{1}{3}$

★ 20. $12 - 3x = 13 - 5x$ $\frac{1}{2}$

★ 21. $5x - 7 = x - 8$ $-\frac{1}{4}$

★ 22. $-7x + 4 = 7 - x$ $-\frac{1}{2}$

★ 23. $-5x - 3 = -2x + 2$ $-1\frac{2}{3}$

★ 24. $4 - 2x = 5 + 7x$ $-\frac{1}{9}$

Calculator

You can use a calculator to check whether a given number is a solution of an equation.

Which of the numbers 5 or 7 is the solution of the equation $37x + 9 = 54x - 76$?
Replace x with 5 in $37x + 9$ first.
Press 37 ⊗ 5 ⊕ 9 ⊜
Display 194
Now replace x with 5 in $54x - 76$.
Press 54 ⊗ 5 ⊖ 76 ⊜
Display 194   So, 5 is the solution of $37x + 9 = 54x - 76$.

Which of the given numbers is the solution of the equation?

1. $45x + 27 = 74x - 234$ 8, 9, 10 9
2. $79x - 56 = 120x - 384$ 6, 7, 8 8
3. $94x + 63 = 136x - 441$ 9, 12, 16 12

Equations with Parentheses

—◇ **OBJECTIVE** ◇—   ——————◇ **RECALL** ◇——————

To solve an equation like
$2(x - 1) = -(8x - 18)$

$$4(x + 5) = 4(x) + 4(5)$$
$$= 4x + 20$$
$$-3(y - 6) = -3(y) + (-3)(-6)$$
$$= -3y + 18$$
$$-(4 + 2x) = -1(4 + 2x)$$
$$= -1(4) + (-1)(2x)$$
$$= -4 - 2x$$

If an equation contains parentheses, first rewrite the equation without parentheses. This can be done by using the distributive property as shown in the recall. Then solve the resulting equation.

Example 1

Solve $5(x - 2) = 20$. Then check your solution.

$5(x - 2) = 5(x) + 5(-2)$

$$5(x - 2) = 20$$
$$5(x) + 5(-2) = 20$$
$$5x - 10 = 20$$

Add 10 to each side. $5x - 10 + 10 = 20 + 10$

Combine like terms. $5x = 30$

Divide each side by 5. $\dfrac{5x}{5} = \dfrac{30}{5}$

$x = 6$ So, 6 is the solution.

Check.

$$5(x - 2) = 20$$
$$\overline{5(6 - 2)} \,\big|\, 20$$
$$5(4)$$
$$20 \quad = 20$$

Example 2

Solve $2(n - 6) = -4(n - 3)$.

$$2(n - 6) = -4(n - 3)$$
$$2(n) + 2(-6) = -4(n) + (-4)(-3)$$
$$2n - 12 = -4n + 12$$

Add $4n$ to each side. $2n - 12 + 4n = -4n + 12 + 4n$

Combine like terms. $6n - 12 = 12$

Add 12 to each side. $6n - 12 + 12 = 12 + 12$

$$6n = 24$$

Divide each side by 6. $\dfrac{6n}{6} = \dfrac{24}{6}$

$$n = 4$$

Check on your own.   So, 4 is the solution.

practice ▷   **Solve.**

      1. $4(n + 2) = 24$ 4   **2.** $6(z - 3) = 3(z + 4)$ 10

Example 3   Solve $4x - (-6 - 3x) = -8$.

$$4x - (-6 - 3x) = -8$$

$-(-6 - 3x) = -1(-6 - 3x)$ $4x - 1(-6 - 3x) = -8$

$$4x + (-1)(-6) + (-1)(-3x) = -8$$

Combine like terms. $4x + 6 + 3x = -8$

$4x + 3x = 7x$ $7x + 6 = -8$

Subtract 6 from each side. $7x + 6 - 6 = -8 - 6$

$$7x = -14$$

Divide each side by 7. $\dfrac{7x}{7} = \dfrac{-14}{7}$

$$x = -2$$

So, -2 is the solution.

---

practice ▷   **Solve.**

      3. $-(c + 8) = 5$ -13 **4.** $2x - (4 - 3x) = 11$ 3

◇ *EXERCISES* ◇

Solve.

1. $3(x + 1) = 15$ 4	**2.** $-2(x + 3) = -12$ 3
3. $4(x - 5) = 16$ 9	**4.** $-7(y - 1) = 21$ -2
5. $6(4 - z) = 18$ 1	**6.** $-20 = -5(a + 7)$ -3
7. $4(x - 3) = 8(x + 1)$ -5	**8.** $-6(y + 4) = 3(y - 2)$ -2
9. $2(z - 6) = -4(z - 9)$ 8	**10.** $-5(c + 3) = -2(c - 6)$ -9
11. $-3(x + 7) = -4(x - 1)$ 25	**12.** $7(x + 1) = -4(x + 12)$ -5
13. $-(c + 6) = 9$ -15	**14.** $-(y - 4) = 17$ -13
15. $-(7 - z) = 13$ 20	**16.** $-(6 + a) = 10$ -16
17. $17 = -(y + 4)$ -21	**18.** $-20 = -(x - 9)$ 29
19. $4z - (7 + 3z) = 5$ 12	**20.** $-6y - (3 - 4y) = 11$ -7
21. $-(-5a + 8) + 3a = 32$ 5	**22.** $41 = 3n - (7 - 3n)$ 8
23. $3(x - 4) = -(7 - 2x)$ 5	**24.** $-(6 + 4x) = -2(x + 9)$ 6
25. $2(y - 1) = -(8y - 18)$ 2	**26.** $-(12 - 3z) = 6(z + 2)$ -8
★ **27.** $2(3x + 1) - 5x = -6$ -8	★ **28.** $4(-2x - 3) + 6x = 18$ -15
★ **29.** $4(2y - 9) - (5y + 6) = 0$ 14	★ **30.** $-3(5 - 2c) - (4c + 7) = 0$ 11
★ **31.** $\frac{2}{3}(3y + 6) = -10$ -7	★ **32.** $-\frac{1}{2}(6 + 4x) = \frac{1}{3}(9 - 3x)$ -6
★ **33.** $5z - \frac{3}{4}(8z - 4) = -1$ 4	★ **34.** $\frac{3}{5}(10a - 5) - \frac{3}{2}(8a + 10) = 0$ -3

Number Problems

To write an equation for a word
problem
To solve a word problem

5 increased by 8

$5 + 8$

7 less than twice a number

$2n - 7$

Example 1

Let n = the number.
Six more than n

$n + 6$

Write an equation for the sentence.
Six more than a number is 9.

Six more than a number is 9.

$$n + 6 = 9$$

So, the equation is $n + 6 = 9$.

Example 2

Let n = the number.

n decreased by 4

$n - 4$

Write an equation for the sentence.
A number decreased by 4 is equal to 12.

A number decreased by 4 is equal to 12.

$$n - 4 \quad = \quad 12$$

So, the equation is $n - 4 = 12$.

Example 3

Let n = the number.
2 less than   3 times n

$3n - 2$

Write an equation for the sentence.
Two less than 3 times a number is 16.

Two less than 3 times a number is 16.

$$3n - 2 = 16$$

So, the equation is $3n - 2 = 16$.

---

practice ▷   **Write an equation for the sentence.**

1. A number increased by 7 is
equal to 15. $n + 7 = 15$

2. Three less than 5 times a
number is 7. $5x - 3 = 7$

Example 4

A number increased by 5 is equal to 19. Find the number.

Let n = the number.
n increased by 5 is equal to 19.

Write an equation.
Subtract 5 from each side.

$$n + 5 = 19$$
$$n + 5 - 5 = 19 - 5$$
$$n = 14$$

Now check 14 in the problem.

Check: n increased by 5 is 19.

$$14 + 5 \mid 19$$
$$19 \mid \quad \text{true}$$

So, the number is 14.

Example 5

Eight less than a number is 17. Find the number.

Let n = the number.
8 less than n is 17.

Write an equation.
8 less than n
$n \leftarrow - \rightarrow 8$
Add 8 to each side.
Check 25 in the problem.
$25 - 8 = 17$

$$n - 8 = 17$$
$$n - 8 + 8 = 17 + 8$$
$$n = 25$$

So, the number is 25.

Example 6

Three more than twice a number is equal to -11.
Find the number.
Let n = the number.

Write an equation.
3 more than twice n is -11.

$$2n + 3 = -11$$

Subtract 3 from each side.
Divide each side by 2.
Check -7 in the problem.

$$2n + 3 - 3 = -11 - 3$$
$$2n = -14$$
$$n = -7$$

So, the number is -7.

---

practice ▷ **Write an equation for the sentence.**

3. Six dollars less than what you paid for a blouse is $11. How much did you pay for the blouse? $n - 6 = 11$; $17

4. Five more than 3 times Paul's score is 23 points. What is Paul's score? $3n + 5 = 23$; 6

◇ EXERCISES ◇

Write an equation for each sentence.

1. $x + 4 = 15$ 5. $x + 12 = 23$ 2. $x - 6 = 9$ 4. $9 + x = 17$ 6. $x + 3 = 18$

1. Four more than a number is 15.
2. Six less than a number is 9.
3. A number decreased by 8 is 7. $x - 8 = 7$
4. Nine increased by a number is 17.
5. A number increased by 12 is 23.
6. Three more than a number is 18.
7. Seven less than twice a number is equal to 19. $2x - 7 = 19$
8. Eight, decreased by three times a number, is equal to 2. $8 - 3x = 2$
9. Five more than four times a number is equal to 21. $4x + 5 = 21$
10. Twice a number, increased by 11, is equal to 25. $2x + 11 = 25$
11. Four times a number, decreased by 5, is 11. $4x - 5 = 11$
12. Eight less than 3 times a number is equal to 7. $3x - 8 = 7$

Solve each problem.

13. Eight less than a number is 15. Find the number. 23
14. Three more than a number is 18. Find the number. 15
15. A number increased by 12 is 21. Find the number. 9
16. Sixteen decreased by a number is 7. Find the number. 9
17. Twenty-four increased by a number is equal to 31. Find the number. 7
18. A number decreased by 13 is equal to 7. Find the number. 20
19. Five more than a number is equal to 7. Find the number. 2
20. Ten less than a number is equal to 12. Find the number. 22
21. Four increased by twice the number of tomato plants you planted this year is 30. How many tomato plants did you plant this year? 13
22. Six less than ten times the number of absences you had this year is 14. How many absences did you have this year? 2
23. Five times the number of books Maria carries to school each day increased by 11 is 31. How many books does Maria carry to school each day? 4
24. Four more than 9 times the width of a sidewalk is 49 feet. How wide is the sidewalk? 5 feet
25. Ten times the number of students on a committee decreased by 7 is 43. What is the number of students on the committee? 5
26. Seven less than 6 times the number of students that boarded the bus is equal to 35. How many students boarded the bus? 7
★ 27. Twice the number of good friends you have increased by 3 times the number is equal to 45. How many good friends do you have? 9
★ 28. Four times the number of days you went to school during the last two weeks decreased by 3 is equal to 3 times that number, increased by 7. How many days did you go to school during the last two weeks? 10

Problem Solving – Applications
Jobs for Teenagers

Example	Jane weighed 32 lb in March. That was an increase of 5 lb in a month. How much did she weigh a month ago?
	You can write an equation for this problem. Let w = weight a month ago. Weight in March $\longrightarrow w + 5$ Equation $\qquad w + 5 = 32$
	Solve. $\qquad w + 5 - 5 = 32 - 5$ $\qquad\qquad\qquad w = 27$
	Check 27 lb in the problem. So, Jane weighed 27 lb a month ago.

practice ▷ In June, Mrs. Sadowski paid $27.50 for diapers. That was an increase of $5.75 over the cost in May. How much was the cost in May? $21.75

CHAPTER FOURTEEN

Solve these problems.

1. Bob weighed 29 lb in May. This was an increase of 6 lb in 3 months. What was Bob's weight 3 months ago?   23 lb

2. In Exercise 1, assume that Bob's weight increased the same amount each month. What were Bob's weights in February, March, and April?
23 lb, 25 lb, 27 lb

3. Sharon's age is 2 more than 3 times Tod's age. Tod is now 3 years old. How old is Sharon?   11 years

4. Lori charges $1.00 an hour for babysitting. She worked from 8:00 P.M. to 11:00 P.M. How much did she earn? $3

5. For babysitting, Jerry charges $1.00 per hour before midnight and $5.00 to stay overnight. Saturday night, he arrived at 6:30 P.M. and stayed overnight. How much did he earn? $10.50

6. Jackie works during the summer as a mother's helper. She is paid $8 per day and works 5 days a week. How much will she earn in 11 weeks? $440

7. Julie charges $0.75 per hour before midnight and $1.25 per hour after midnight for babysitting. She babysat from 7:00 P.M. to 2:00 A.M. How much did she earn? $6.25

8. Karen took two children to a museum. Bus fare was $0.50 per person each way. Admission to the museum was $0.75 per person. How much did the trip cost for all three? $5.25

9. Bill works at a day-care center 5 days a week. He is paid $25 per day. How much will he earn in 2 weeks? $250

10. When a new baby came along, the Katsaris family had to increase the purchase of baby foods by 25%. They are now spending $44 per week. How much were they spending previously?
$35.20

11. The Happy Hour Nursery had 36 babies to take care of during the month of May. This was 20% more than were there during April. How many babies were there in April?   30

12. Mr. Willis pays his babysitter $1.25 per hour and gives a $0.50 tip. How much does he pay a babysitter for 5 hours of babysitting? $6.75

13. Mrs. Reed pays her babysitter $1.25 per hour and gives a $0.50 tip. She uses the formula $1.25h + 0.50$, where h is the number of hours. Use the formula to find a babysitter's pay for sitting from 7:00 P.M. to midnight. $6.75

14. Pablo gets babysitting jobs through an agency. He is paid $2.00 per hour, but he must pay the agency $3.00 for each job. Thursday he babysat from 8:00 P.M. until 12:30 A.M. How much did he earn after paying the agency?
$6

15. Pablo uses the formula $2.00h - 3.00$ to help figure his earnings. Use the formula to find his earnings for sitting from 7:30 P.M. until 1:00 A.M.
$8

16. Rosemarie earns $1.75 per hour babysitting, but pays an agency $2.50 for each job. How much does she earn for babysitting for 6 hours? $8

Properties for Inequalities

—— ◇ **OBJECTIVE** ◇ ——   —————— ◇ **RECALL** ◇ ——————

To write a true inequality after
performing a given operation
on each side of an inequality
like $-2 < 5$

$6 < 4$ false	$-3 \geq 2$ false
$6 > 4$ true	$-3 \leq 2$ true
$2 \geq 2$ true	$3 \leq 3$ true
$2 > 2$ false	$3 < 3$ false

Example 1

Add 3 to each side of the inequality $2 < 5$. Then use
$<$ or $>$ to write a true inequality.

Different numbers
Add the same numbers.
Different numbers, same order

$$2 < 5 \qquad \text{same}$$
$$2 + 3 \; ? \; 5 + 3 \qquad \text{order}$$
So, $\qquad 5 < 8.$

Example 2

Subtract 2 from each side of $7 > -1$. Then write a
true inequality.

$$7 > -1 \qquad \text{same}$$
$$7 - 2 \; ? \; -1 - 2 \qquad \text{order}$$
So, $\;\; 5 > -3.$

$6 > -1$ ← true inequality	
$\underline{+2 \quad +2}$ ← Add 2 to each side.	
$8 > \quad 1$ ← same order	

addition/subtraction properties for inequalities
Adding or subtracting the same number to or from
each side of a true inequality does not change the
order.

practice   1. Subtract 5 from each side of
$-3 \leq 4$. Then write a true
inequality. $-8 \leq -1$

2. Add 4 to each side of
$-2 \geq -5$. Then write a
true inequality. $2 \geq -1$

Example 3

Multiply each side of $3 < 6$ by 2 and write a true
inequality. Then multiply each side of $3 < 6$ by -2
and write a true inequality.

Multiplying by a negative number
reverses the order.

$$3 < 6 \qquad \text{same} \qquad\qquad 3 < 6 \qquad\qquad \text{reverse}$$
$$3 \cdot 2 \; ? \; 6 \cdot 2 \quad \text{order} \qquad 3 \cdot (-2) \; ? \; 6 \cdot (-2) \quad \text{order}$$
So, $6 < 12.$ $\qquad\qquad\qquad$ So, $\;\; -6 > -12.$

Example 4

Multiply each side of $2 \geq -4$ by 3 and write a true inequality. Then multiply each side of $2 \geq -4$ by -3 and write a true inequality.

Multiplying by a negative number reverses the order.

$2 \geq -4$	same		$2 \geq -4$	reverse
$2 \cdot 3$? $-4 \cdot 3$	order		$2 \cdot (-3)$? $-4 \cdot (-3)$	order
So, $6 \geq -12$.			So, $-6 \leq 12$.	

$-5 < -2 \leftarrow$ true inequality
$-5(-3)$? $-2(-3) \leftarrow$ Multiply by -3.
$15 > 6 \leftarrow$ reverse order

multiplication property for inequalities
Multiplying each side of a true inequality by the same positive number does not change the order. Multiplying each side of a true inequality by the same negative number reverses the order.

practice ▷ 3. Multiply each side of $-1 < 5$ by 4. Then write a true inequality. $-4 < 20$

4. Multiply each side of $-1 < 5$ by -4. Then write a true inequality. $4 > -20$

Example 5

Divide each side of $-6 < 8$ by 2 and write a true inequality. Then divide each side of $-6 < 8$ by -2 and write a true inequality.

$-6 < 8$			$-6 < 8$	
$\frac{-6}{2}$? $\frac{8}{2}$	same order		$\frac{-6}{-2}$? $\frac{8}{-2}$	reverse order
So, $-3 < 4$.			So, $3 > -4$.	

Reverse the order for a true inequality when multiplying or dividing by a negative number.

division property for inequalities
Dividing each side of a true inequality by the same positive number does not change the order. Dividing each side of a true inequality by the same negative number reverses the order.

practice ▷ 5. Divide each side of $-6 \geq -9$ by 3. Then write a true inequality.
$-2 \geq -3$

6. Divide each side of $-6 \geq -9$ by -3. Then write a true inequality.
$2 \leq 3$

PROPERTIES FOR INEQUALITIES

**If the indicated operation is performed on an inequality, will
the order of the inequality change?**

1. Add 6. no	**2.** Multiply by 4. no	**3.** Divide by 2. no
4. Multiply by -2. yes	**5.** Subtract -4. no	**6.** Multiply by 6. no
7. Divide by -4. yes	**8.** Multiply by -5. yes	**9.** Divide by 6. no
10. Subtract 5. no	**11.** Add 1. no	**12.** Multiply by 3. no
13. Divide by -3. yes	**14.** Divide by 4. no	**15.** Subtract -2. no
16. Add 0. no	**17.** Multiply by -1. yes	**18.** Divide by -1. yes

◇ *EXERCISES* ◇

**Perform the operation on each side of the inequality. Then write
a true inequality.**

14. $-12 < -8$ **22.** $15 \geq -30$ **28.** $-50 < -20$

1. $2 < 5$	Add 2. $4 < 7$	**2.** $3 > 1$	Add -1. $2 > 0$
3. $3 < 4$	Multiply by 3. $9 < 12$	**4.** $5 > 2$	Divide by -1. $-5 < -$
5. $6 > -1$	Subtract 6. $0 > -7$	**6.** $-4 < 6$	Divide by 2. $-2 < 3$
7. $1 > -2$	Multiply by -4. $-4 < 8$	**8.** $-1 < 4$	Multiply by 5. $-5 < 2$
9. $6 > -3$	Divide by 3. $2 > -1$	**10.** $-2 > -4$	Add. 0. $-2 > -4$
11. $-1 < 3$	Add 6. $5 < 9$	**12.** $4 > -3$	Multiply by -2. $-8 <$
13. $-9 < 3$	Divide by -3. $3 > -1$	**14.** $-3 < -2$	Multiply by 4.
15. $-2 > -6$	Subtract 5. $-7 > -11$	**16.** $-5 > -30$	Divide by -5. $1 < 6$
17. $6 \geq -12$	Divide by 6. $1 \geq -2$	**18.** $4 \geq 4$	Add 5. $9 \geq 9$
19. $-2 \leq 6$	Multiply by 1. $-2 \leq 6$	**20.** $-3 \leq -1$	Add 10. $7 \leq 9$
21. $12 \geq -8$	Divide by -4. $-3 \leq 2$	**22.** $-3 \leq 6$	Multiply by -5.
23. $-6 < 7$	Subtract -8. $2 < 15$	**24.** $4 > -6$	Divide by -2. $-2 < 3$
25. $3 \geq -1$	Multiply by -1. $-3 \leq 1$	**26.** $2 \leq 5$	Add. -7. $-5 \leq -2$
27. $14 > -7$	Divide by 7. $2 > -1$	**28.** $-5 < -2$	Multiply by 10.
★ **29.** $2 \leq 5$	Multiply by 0. $0 \leq 0$	★ **30.** $-4 \geq -1$	Multiply by 0. $0 \geq 0$

Challenge

Using each of the numbers 1 through 9 exactly once, place a
number in each circle so that the sum of the three numbers
along every straight line is 15.

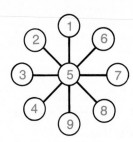

Solving Inequalities

---◇ **OBJECTIVE** ◇--- ---◇ **RECALL** ◇---

To solve inequalities like
$x + 1 > -4$, $x - 5 \leq 3$,
$-3x \geq 9$, and $\frac{1}{2}x < -5$
using integers

Solve. $x - 4 = -13$
$$x - 4 = -13$$
$$x - 4 + 4 = -13 + 4$$
$$x = -9$$
So, the solution is -9.

Solve. $-4y = 24$
$$\frac{-4y}{-4} = \frac{24}{-4}$$
$$y = -6$$
So, the solution is -6.

The sum of a number and 3 is greater than -1.
$$x \quad + \ 3 \quad\quad > \quad\quad -1$$

Example 1

Find the integers that make $x + 3 > -2$ true.

Subtract 3 from each side.
Subtracting the same number from each side of a true inequality does not change the order.

$$x + 3 > -2$$
$$x + 3 - 3 > -2 - 3 \quad \text{same}$$
$$x + 0 > -5 \quad\quad \text{order}$$
$$x > -5$$

{ subtraction property for inequalities

Integers greater than -5:
$-4, -3, -2, -1, 0, \ldots$
Try -4 and 1.

Check.
$x + 3 > -2$		$x + 3 > -2$	
$-4 + 3$	-2	$1 + 3$	-2
-1		4	
$-1 > -2$ true		$4 > -2$ true	

All integers greater than -5 make $x + 3 > -2$ true.

So, $-4, -3, -2, \ldots$ are solutions of $x + 3 > -2$.

Example 2

Solve. $y - 2 \leq -1$

Add 2 to each side.

$$y - 2 \leq -1$$
$$y - 2 + 2 \leq -1 + 2 \quad \text{same}$$
$$y + 0 \leq \quad 1 \quad\quad \text{order}$$
$$y \leq \quad 1$$

{ addition property for inequalities

Integers less than or equal to 1:1, 0,
$-1, -2, -3, \ldots$ Try 1 and -4.

Check.
$y - 2 \leq -1$		$y - 2 \leq -1$	
$1 - 2$	-1	$-4 - 2$	-1
-1		-6	
$-1 \leq -1$ true		$-6 \leq -1$ true	

The solutions are $1, 0, -1, -2, \ldots$

So the solutions are integers less than or equal to 1.

$n \geq -4$

1. $x + 2 > -5$ $x > -7$ **2.** $y + 1 \leq -7$ $y \leq -8$ **3.** $n - 5 \geq -9$

Let n be the number.   A number multiplied by 4 is greater than -12.

$n \cdot 4 > -12$ may also be written   ↓   ↓   ↓   ⌒   ↓

$4n > -12$. n ·   4   > -12

Example 3

Find the integers that make $4n > -12$ true.

$$4n > -12$$

Divide each side by 4.

$$\frac{4n}{4} > \frac{-12}{4} \quad \text{same} \quad \left\{\begin{array}{l}\text{division property} \\ \text{for inequalities}\end{array}\right.$$

Dividing each side of a true inequality $1n > -3 \quad \text{order}$
by the same positive number does not
change the order. Any integer greater $n > -3$
than -3 is a solution.

So, the integers that make $4n > -12$ true are -2, $-1, 0, \ldots$

Example 4

Solve. $-5x \leq -15$

$$-5x \leq -15$$

Divide each side by -5.

$$\frac{-5x}{-5} \geq \frac{-15}{-5} \quad \begin{array}{l}\text{reverse} \\ \text{order}\end{array} \quad \left\{\begin{array}{l}\text{division property} \\ \text{for inequalities}\end{array}\right.$$

$$1x \geq 3$$

$$x \geq 3$$

Dividing each side of a true inequality
by the same negative number reverses   So, any integer greater than or equal to 3 solves the
the order. The solutions are   inequality.
$3, 4, 5, \ldots$

Example 5

Solve. $\frac{2}{3}x < -6$ Solve. $-\frac{1}{2}n \geq -3$

$$\frac{2}{3}x < -6$$ $$-\frac{1}{2}n \geq -3$$

$\frac{3}{2}$ is the reciprocal of $\frac{2}{3}$. $\frac{3}{2} \cdot \frac{2}{3}x < -6 \cdot \frac{3}{2}$ $-2 \cdot \left(-\frac{1}{2}n\right) \leq -3 \cdot (-2)$

Use the multiplication property for $1x < \frac{-18}{2}$ $1n \leq 6$
inequalities.
                                          $x < -9$ $n \leq 6$

So, $-10, -11, -12, \ldots$ So, $6, 5, 4, \ldots$ are
are solutions.   solutions.

---

4. $2x < -8$ $x < -4$ **5.** $-3y > 15$ $y < -5$ **6.** $-\frac{3}{4}y \leq -6$

$y \geq 8$

Solve.   See TE Answer Section.

1. $x + 1 > 3$ 2. $y + 6 \leq 7$ 3. $z + 2 \leq 6$ 4. $a + 5 > 8$

5. $c + 6 \geq 2$ 6. $w + 5 < 3$ 7. $d + 7 \geq 2$ 8. $x + 1 < 0$

9. $m + 1 < -6$ 10. $e + 3 \leq -2$ 11. $y + 7 > -1$ 12. $n + 2 \leq -4$

13. $r - 3 > 5$ 14. $k - 5 \geq 0$ 15. $z - 10 \geq 3$ 16. $b - 7 < 1$

17. $s - 9 < -5$ 18. $n - 6 < -3$ 19. $p - 4 \geq -1$ 20. $a - 8 \geq -7$

21. $v - 3 \geq -12$ 22. $x - 7 \leq -9$ 23. $t - 6 < -8$ 24. $z - 1 > -5$

25. $2n \geq -6$ 26. $3x < 6$ 27. $4y \leq -12$ 28. $3a > -18$

29. $8k < -16$ 30. $3n \geq 3$ 31. $5d \leq 15$ 32. $7z > -14$

33. $-2m \leq 10$ 34. $-4x > -4$ 35. $-2c < 8$ 36. $-5s \leq 5$

37. $-7t > -14$ 38. $-3s \geq -12$ 39. $-8v < 16$ 40. $-12n \leq -12$

41. $\frac{1}{2}x > -1$ 42. $\frac{1}{3}y > 2$ 43. $\frac{1}{5}a \geq -2$ 44. $\frac{1}{4}m < -7$

45. $\frac{3}{4}a \leq -9$ 46. $\frac{2}{5}c \geq 4$ 47. $\frac{7}{10}d < -21$ 48. $\frac{5}{6}w > 10$

49. $-\frac{1}{3}c < -4$ 50. $-\frac{1}{5}s \leq 1$ 51. $-\frac{2}{3}y \geq -4$ 52. $-\frac{3}{8}z < 15$

1. Find a fraction equal to $\frac{2}{3}$ such that the sum of the numerator and denominator is 100. $\frac{40}{60}$

2. Write a ten-digit numeral so that the first digit on the left tells how many 0's there are in the entire numeral, the second digit tells how many 1's, the third digit how many 2's, and so on.   6,210,001,000

3. A car traveled for 2 hours and averaged 80 km/h. On the return trip it averaged 70 km/h. What was the average speed for the entire trip?   74.7 km/h

Problem Solving – Careers
Railroad Brakers

1. Mr. Scorsese is a railroad braker. He is stationed at the rear of the train to display warning lights and signals. Last year he earned $24,640. What were his average earnings per month? $2,053.33

2. Ms. Nicholas is a passenger braker. She often helps the conductor by collecting or selling tickets. A passenger bought four one-way tickets at $3.40 each. He gave Ms. Nicholas a $20 bill. How much change should she give him? $6.40

3. John bought a one-way ticket for $2.55. He paid with a $10 bill. How much change did he get? $7.45

4. A round-trip ticket cost $7.60. How much did Shelly pay for three round-trip tickets? $22.80

5. As a yard braker, Mr. Martinez earns about $2,250 per month. As a passenger braker, Ms. Rosen earns about $2,517 per month. How much more does Ms. Rosen earn in a year than Mr. Martinez? $3,204

6. Sometimes passenger brakers regulate car lighting and temperature. The braker saw that the car thermostat read 32°C. Was the car too hot or too cold? too hot

7. In 1980, there were nearly 72,000 brakers in the U.S.A. By 1990 that number may decrease by about 5%. If so, how many brakers will there be in the U.S.A. in 1990? 68,400

Exponents

— ◇ **OBJECTIVES** ◇ —

To evaluate expressions like
$-3xy^2$ if x is 4 and y is -1
To evaluate expressions like
$(-2a)^2$ if a is 3

— ◇ **RECALL** ◇ —

Evaluate $4x^3$ if x is 2.
$$4x^3$$
$$\downarrow\downarrow$$
$$4 \cdot 2^3$$
$$4 \cdot 2 \cdot 2 \cdot 2$$
$$4 \cdot 8$$
$$32$$

Example 1

Evaluate x^3 if x is -3.
$$x^3$$
$$\downarrow$$

Substitute -3 for x.

$$(-3)^3$$
$$\underbrace{(-3)(-3)(-3)}$$

$(-3)(-3) = 9$
$(9)(-3) = -27$

$$(9)(-3)$$
$$-27$$

So, the value of x^3 is -27 if $x = -3$.

Example 2

Evaluate $-4y^2$ if $y = -5$.
$$-4y^2$$
$$\downarrow\downarrow$$

Substitute -5 for y.

$$-4(-5)^2$$
$$\underbrace{-4(-5)(-5)}$$

$(-5)(-5) = 25$

$$-4 \cdot 25$$
$$-100$$

So, the value of $-4y^2$ is -100 if $y = -5$.

Example 3

Evaluate $-a^2b^3$ if $a = 6$ and $b = -1$.
$$-a^2b^3$$
$$(-1)a^2b^3$$

Substitute 6 for a and -1 for b.

$$(-1)(6)^2(-1)^3$$
$$(-1)(6)(6)(-1)(-1)(-1)$$
$$\downarrow \qquad \downarrow$$

$(-1)(-1)(-1) = -1$

$$(-1)\,(36)(-1)$$
$$36$$

So, the value of $-a^2b^3$ is 36 if $a = 6$ and $b = -1$.

EXPONENTS

351

Evaluate.

1. y^2 if $y = -3$ 9   2. $-3x^3$ if $x = -2$ 24   3. $-mn^3$ if $m = 4$ 32
                                                              and $n = -2$

Example 4

Evaluate $(3x)^2$ if $x = -4$.
$(3x)^2$

Substitute -4 for x. $[(3)(-4)]^2$
$(3)(-4) = -12$ $(-12)^2$
$(-12)^2 = (-12)(-12)$ $(-12)(-12)$
$(-12)(-12) = 144$ 144

So, the value of $(3x)^2$ is 144 if $x = -4$.

Evaluate.

4. $(4y)^2$ if $y = -2$ 64   5. $(-2x)^3$ if $x = 2$ -64 6. $(-6a)^4$ if $a = -1$ 1,296

◇ EXERCISES ◇

Evaluate.

1. x^2 if $x = 7$ 49   2. y^3 if $y = -1$ -1 3. z^3 if $z = -4$ -64
4. a^5 if $a = 2$ 32   5. b^4 if $b = -3$ 81   6. c^5 if $c = -1$ -1
7. $2x^2$ if $x = 5$ 50   8. $-3n^2$ if $n = -2$ -12 9. $5a^3$ if $a = -1$ -5
10. $-4c^4$ if $c = 1$ -4 11. $-2x^4$ if $x = -3$ -162 12. $3y^5$ if $y = -1$ -3
13. $-n^2$ if $n = 7$ -49 14. $-z^2$ if $z = -8$ -64 15. $-x^3$ if $x = 3$ -27
16. $-a^3$ if $a = -4$ 64   17. $-b^4$ if $b = 2$ -16 18. $-c^5$ if $c = -1$ 1
19. $2xy^2$ if $x = 3, y = 2$ 24   20. $3a^2b$ if $a = -1, b = 8$ 24   21. $-4m^2n^2$ if $m = 2, n = 1$ -16
22. $-3ab^4$ if $a = -7, b = 1$ 21   23. $-xy^3$ if $x = 5, y = 2$ -40 24. $-r^2s^3$ if $r = 9, s = -1$ 81
25. $(6x)^2$ if $x = -1$ 36   26. $(3a)^2$ if $a = 2$ 36   27. $(4n)^2$ if $n = -5$ 400
28. $(-2c)^2$ if $c = 5$ 100   29. $(-4n)^2$ if $n = -2$ 64   30. $(3x)^3$ if $x = -2$ -216
31. $(-y)^3$ if $y = -7$ 343   32. $(-4d)^3$ if $d = 1$ 64   33. $(-5z)^3$ if $z = -1$ 125
★ 34. x^3y^2z if $x = 2, y = 3, z = 1$ 72   ★ 35. ab^3c if $a = 3, b = -2, c = -1$ 24
★ 36. d^2ef^3 if $d = 1, e = 4, f = -2$ -32 ★ 37. $-xy^3z$ if $x = 3, y = 2, z = -1$ 24
★ 38. $-r^2st^2$ if $r = 5, s = -1, t = 2$ 100   ★ 39. $-a^3b^2c$ if $a = 1, b = 3, c = -4$ 36
★ 40. $(xy)^2$ if $x = 4, y = 2$ 64   ★ 41. $(ab)^3$ if $a = 2, b = -1$ -8
★ 42. $(x^2y)^3$ if $x = 3, y = -1$ -729 ★ 43. $(rs)^3$ $r = 4, s = -2$ -512
★ 44. $(p^2q^2)^2$ if $p = -1, q = 3$ 81   ★ 45. $(-x^2y^3)^2$ if $x = 2, y = -1$ 16
★ 46. $-(pq)^2$ if $p = -3, q = 1$ -9 ★ 47. $-(3r^2s)^2$ if $r = -1, s = 2$ -36

Properties of Exponents

To simplify expressions like
$(-2x^3)(5x^2)$, $(x^4)^3$, and $(-3a)^4$

$x^3 = x \cdot x \cdot x \leftarrow 3$ factors | $x^1 = x$
x^3 is the third power of x

In this lesson, you will learn some shortcuts for multiplying expressions with exponents.

Example 1

Show that $y^2 \cdot y^3 = y^5$.

| $y^2 \cdot y^3$ | y^5 |
5 factors, y | $y \cdot y \cdot y \cdot y \cdot y$ | y^5 |

$$y^5$$

So, $y^2 \cdot y^3 = y^5$.

In Example 1, notice that the factors y^2 and y^3 have y as a base; the product y^5 also has y as a base; the exponent of the product is the sum of the exponents of the factors. This suggests the following property.

When multiplying with the same base, add the exponents.

product of powers
$$x^m \cdot x^n = x^{m+n}$$

Example 2

Simplify. $a^3 \cdot a^4$
$$a^3 \cdot a^4$$
$$a^{3+4}$$
$$a^7$$
So, $a^3 \cdot a^4 = a^7$.

Simplify. $x^3 \cdot x$
$$x^3 \cdot x$$
$$x^3 \cdot x^1$$
$$x^{3+1}$$
$$x^4$$
So, $x^3 \cdot x = x^4$.

Example 3

Simplify. $(7c^4)(-3c^2)$
$$(7c^4)(-3c^2)$$

Group numbers together. $7 \cdot (-3) \cdot c^4 \cdot c^2$
Add the exponents. $-21c^{4+2}$
$$-21c^6$$

So, $(7c^4)(-3c^2) = -21c^6$.

---

practice ▷   **Simplify.**

1. $b^2 \cdot b^2$ b^4

2. $(-2y^2)(6y^5)$ $-12y^7$

3. $(-5r)(-7r^8)$ $35r^9$

Example 4

Show that $(x^2)^3 = x^{2 \cdot 3}$.

3 factors, x^2

$$\begin{array}{c|c} (x^2)^3 & x^{2 \cdot 3} \\ x^2 \cdot x^2 \cdot x^2 & x^6 \\ x^{2+2+2} & \\ x^6 & \text{So, } (x^2)^3 = x^{2 \cdot 3}. \end{array}$$

Example 4 suggests that a power is raised to a power by multiplying the exponents.

Multiply the exponents.

power of a power
$$(x^m)^n = x^{m \cdot n}$$

Example 5

Simplify $(a^4)^2$.

Multiply the exponents.

$$(a^4)^2$$
$$a^{4 \cdot 2}$$
$$a^8$$
So, $(a^4)^2 = a^8$.

Simplify $(y^3)^3$.

$$(y^3)^3$$
$$y^{3 \cdot 3}$$
$$y^9$$
So, $(y^3)^3 = y^9$.

practice ▷ **Simplify.**

4. $(z^5)^2$ z^{10}

5. $(x^4)^3$ x^{12}

6. $(c^2)^4$ c^8

Example 6

Show that $(x \cdot y)^3 = x^3 \cdot y^3$.

3 factors, $(x \cdot y)$

Group x's and y's together.

$$\begin{array}{c|c} (x \cdot y)^3 & x^3 \cdot y^3 \\ (x \cdot y)(x \cdot y)(x \cdot y) & x^3 \cdot y^3 \\ x \cdot x \cdot x \cdot y \cdot y \cdot y & \\ x^3 \cdot y^3 & \end{array}$$

$x^3 \cdot y^3$ can be written x^3y^3.   So, $(x \cdot y)^3 = x^3 \cdot y^3$.

Example 6 suggests that a product can be raised to a power by distributing the exponent.

Distribute the exponent.

power of a product
$$(x \cdot y)^n = x^n \cdot y^n$$

Example 7

Simplify $(2a)^3$

$(2a)^3 = (2^1 \cdot a^1)^3 = 2^{1 \cdot 3} \cdot a^{1 \cdot 3}$

$$(2a)^3$$
$$2^3 \cdot a^3$$
$$8a^3$$
So, $(2a)^3 = 8a^3$.

Simplify $(-3c)^2$.

$$(-3c)^2$$
$$(-3)^2 \cdot c^2$$
$$9c^2$$
So, $(-3c)^2 = 9c^2$.

practice ▷ **Simplify.**

7. $(ab)^4$ a^4b^4

8. $(4y)^2$ $16y^2$

9. $(-2s)^3$ $-8s^3$

◇ ORAL EXERCISES ◇

Should you add, multiply, or distribute exponents?

1. $a^3 \cdot a^5$ add
2. $(x^2)^5$ multiply
3. $(3a)^2$ distrib.
4. $(x^3)^2$ multiply
5. $c^4 \cdot c$ add
6. $(x \cdot y)^3$ distrib.
7. $(4x)^3$ distrib.
8. $y \cdot y^5$ add
9. $(-3y)^3$ distrib.
10. $(z^5)^3$ multiply
11. $(a \cdot b)^5$ distrib.
12. $x^4 \cdot x^3$ add
13. $(-2c)^6$ distrib.
14. $(b^2)^4$ multiply
15. $x \cdot x^3$ add
16. $(-4y)^4$ distrib.
17. $(3c)^4$ distrib.
18. $c^2 \cdot c^6$ add
19. $(x^5)^2$ multiply
20. $y^6 \cdot y$ add

◇ EXERCISES ◇

Simplify.

16. $-15r^5$ 20. $-14y^6$

1. $x^2 \cdot x^2$ x^4
2. $y^3 \cdot y$ y^4
3. $z^4 \cdot z^3$ z^7
4. $c^5 \cdot c$ c^6
5. $r^4 \cdot r^5$ r^9
6. $b \cdot b^6$ b^7
7. $x^7 \cdot x$ x^8
8. $y \cdot y^4$ y^5
9. $(x^2)(2x^3)$ $2x^5$
10. $(3y^2)(2y^2)$ $6y^4$
11. $(-4c^3)(c^4)$ $-4c^7$
12. $(7x^2)(x^5)$ $7x^7$
13. $(5y)(3y^2)$ $15y^3$
14. $(-4x)(2x^3)$ $-8x^4$
15. $(5d^2)(-2d)$ $-10d^3$
16. $(3r^4)(-5r)$
17. $(-7x^2)(-2x^4)$ $14x^6$
18. $(-3y)(-7y^4)$ $21y^5$
19. $(-4a)(-3a^5)$ $12a^6$
20. $(2y^5)(-7y)$
21. $(r^5)^2$ r^{10}
22. $(x^3)^2$ x^6
23. $(y^4)^3$ y^{12}
24. $(x^4)^2$ x^8
25. $(z^4)^3$ z^{12}
26. $(m^6)^2$ m^{12}
27. $(c^2)^5$ c^{10}
28. $(r^3)^4$ r^{12}
29. $(x^2)^2$ x^4
30. $(y^3)^3$ y^9
31. $(a^2)^4$ a^8
32. $(c^4)^4$ c^{16}
33. $(xy)^3$ x^3y^3
34. $(2a)^2$ $4a^2$
35. $(ab)^4$ a^4b^4
36. $(-3x)^3$ $-27x^3$
37. $(4y)^2$ $16y^2$
38. $(-2z)^3$ $-8z^3$
39. $(-3y)^2$ $9y^2$
40. $(rs)^5$ r^5s^5
41. $(-2x)^4$ $16x^4$
42. $(cd)^2$ c^2d^2
43. $(5p)^2$ $25p^2$
44. $(-2y)^5$ $-32y^5$
★ 45. $(5x^3)^2$ $25x^6$
★ 46. $(-3c^2)^2$ $9c^4$
★ 47. $(2x^3)^4$ $16x^{12}$
★ 48. $(-2r^3)^3$ $-8r^9$

Calculator

A calculator can be used to evaluate $3xy^2$ if x is 9 and y is 7.

$$3 \quad x \quad y^2$$

Press 3 ⊗ 9 ⊗ 7 ⊗ 7 ⊜
Display 1323
So, the value of $3xy^2$ is 1,323 when x is replaced with 9 and y with 7.

Evaluate.

1. z^5 if z is 7   16,807
2. $4x^3$ if x is 8   2,048
3. $5a^2b$ if a is 4, b is 11   880
4. $7c^3d^4$ if c is 5, d is 3
5. $(6y)^4$ if y is 5   810,000
6. $(rs)^3$ if r is 6, s is 7

4. 70,875   6. 74,088

Polynomials

◇ OBJECTIVES ◇

To classify polynomials
To simplify polynomials
  by combining like terms

◇ RECALL ◇

Simplify. $4x + 2x$

like terms

So, $4x + 2x = 6x$.

Thus far, you have been working with polynomials of one term. These are called *monomials*. The prefix *mono* means *one*.

Following are examples of polynomials.

$$3x^2 + 2x - 1 \qquad\qquad 7y + 4 \qquad\qquad 4a^2$$

    3 terms   2 terms   1 term
   Trinomial   Binomial   Monomial

Tri means *three* and *bi* means *two*.

Example 1

Classify each polynomial.

$$4x^2 - 3 \qquad\qquad 9y \qquad\qquad 6a^2 - 5a + 1$$

  2 terms   1 term   3 terms
 Binomial   Monomial   Trinomial

---

practice ▷   **Classify each polynomial.**

   1. $2y^2 - 3y + 1$ trinomial   **2.** $5x^2$ monomial   **3.** $3c - 2$ binomial

---

Example 2

Simplify. $6x^2 - 4x + 3 - 4x^2 - 3x - 5$

                                 $6x^2 - 4x + 3 - 4x^2 - 3x - 5$

Group like terms together. $(6x^2 - 4x^2) + (-4x - 3x) + (3 - 5)$
Combine like terms. $2x^2 - 7x - 2$

So, $6x^2 - 4x + 3 - 4x^2 - 3x - 5 = 2x^2 - 7x - 2$.

Example 3

Simplify and write the result in descending order of exponents. $3x + 6x^2 - 4 - 8x^2 + 7x + 3$

Descending means in order from highest $3x + 6x^2 - 4 - 8x^2 + 7x + 3$
to lowest. $(6x^2 - 8x^2) + (3x + 7x) + (-4 + 3)$
Group like terms together, with highest $-2x^2 + 10x - 1$
exponents first.
Combine like terms. $3x + 6x^2 - 4 - 8x^2 + 7x + 3 = -2x^2 + 10x - 1$.

practice ▷ **Simplify.**

    4. $7 + 6y^2 - 3y + 8y - 9 - 2y^2$ $4y^2 + 5y - 2$

Example 4   Simplify. $x^2 + 3x^3 - 4 + 6x - 2x^2 - 3x$

                            $x^2 = 1x^2$ $1x^2 + 3x^3 - 4 + 6x - 2x^2 - 3x$

Group like terms in descending order. $3x^3 + (1x^2 - 2x^2) + (6x - 3x) - 4$

            Combine like terms. $3x^3 - 1x^2 + 3x - 4$

             $-1x^2 = -x^2$ So, $x^2 + 3x^3 - 4 + 6x - 2x^2 - 3x =$
                                 $3x^3 - x^2 + 3x - 4$.

---

practice ▷ **Simplify.**

    5. $5a^2 - 3a^4 + 6 - 2a + a^2 - 7$ $-3a^4 + 6a^2 - 2a - 1$

◇ ORAL EXERCISES ◇

Arrange each polynomial in descending order of exponents.

                                                        $-3a^2 + 2a + 7$

1. $6x + 3x^2 - 1$ $3x^2 + 6x - 1$ **2.** $1 - 4y$ $-4y + 1$ **3.** $7 + 2a - 3a^2$

4. $2 - x^2 + 3x$ $-x^2 + 3x + 2$ **5.** $6x^2 - 4 + 2x$ $6x^2 + 2x - 4$ **6.** $3 + 5z^2$ $5z^2 + 3$

7. $3x^2 - 2x + 1 + 3x^3$ **8.** $6y - 5y^3 + 4 - 3y^2$ **9.** $7a - 3 - a^3 + 4a^2$
    $3x^3 + 3x^2 - 2x + 1$ $-5y^3 - 3y^2 + 6y + 4$ $-a^3 + 4a^2 + 7a - 3$

◇ EXERCISES ◇

Classify each polynomial. See TE Answer Section.

1. $4y^2 - 1$ **2.** $3x^2 - 2x + 1$ **3.** $3x$ **4.** $-7 + a$

5. $-2x^3 + 3x - 2$ **6.** 6 **7.** $-5x^2$ **8.** $4y^2 + 3y - 1$

9. $3c^2 + 4c$ **10.** $a^2 + a + 2$ **11.** $6x - 1$ **12.** -2

Simplify. See TE Answer Section.

13. $5x^2 - 3x + 2 - 2x^2 - 4x + 1$ **14.** $6y^2 + 2y - 1 + y^2 - 3y + 5$

15. $4a^2 - 2a - 1 - 3a^2 + 5a - 4$ **16.** $n^2 + 5n + 8 + 4n^2 - 2n - 9$

17. $-y^2 + 3y + 5 - 2y^2 - 3y + 2$ **18.** $3x^2 + 4x - 1 - 3x^2 - 2x + 8$

19. $5y + 4y^2 - 3 - 8y^2 + 6y + 1$ **20.** $3 + 2a - 5a^2 + a^2 - 4 - 6a$

21. $7x^2 - 3 + x - 4x^2 - 6x + 2$ **22.** $3c + 2c^2 - 4c + 1 + c^2 - 5$

23. $9 + 8z^2 - 6z - 5 + 4z - 3z^2$ **24.** $7x^2 - 8 + 4x - 1 - 5x^2 - 4x$

25. $2y - 2y^2 - 2 + 5 - 3y - 4y^2$ **26.** $-6n^2 + 2 - 4n + 3 + 6n^2 - 5n$

27. $y^2 + 4y^3 - 3 + 7y - 3y^2 - 2y$ **28.** $6r^2 - 2r^4 + 5 - 3r + r^2 - 8$

29. $3x - 2x^2 + 5x^3 + 2 - x - 4x^2$ **30.** $3x^4 - 2x^2 - 5x + 2x^2 - 5 + 9$

31. $7c^3 - 5 + 6c^2 - 6c + 4c - 6c^2$ **32.** $5a - 2a^4 + 4a^2 - 7 + 4a^2 - 2$

33. $6z - 8z^3 + 3 - 2z^2 + z^3 - 3$ **34.** $6a^3 - 5a^4 + 2a - 1 + 5a^4 - a^3$

★ **35.** $x^2 + x^4 - 3x^3 - 5x + 2 + 3x^4 - 5x^3 + 7x - 5 + x^2$

★ **36.** $5 - 8a^2 + 3a^4 - 2a^3 + 7 - 4a - 3a^4 + 6a - a^2 + 5a^3$

POLYNOMIALS   **357**

Simplifying Polynomials

── ◇ **OBJECTIVES** ◇ ── ── ◇ **RECALL** ◇ ──

To multiply a monomial by a
  polynomial
To simplify polynomials

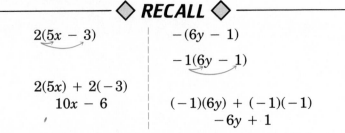

The recall shows that the distributive property can
be used to multiply a monomial by a polynomial
when the monomial is an integer. The distributive
property can also be used when the monomial
contains a variable.

Example 1

Multiply. $x(4x^2 + x)$
$x(4x^2 + x)$

By commutative property of
multiplication, $x(4x^2) = (4x^2)x$.

$x(4x^2) + x(x)$
$4 \cdot x^2 \cdot x + x \cdot x$
$4x^3 + x^2$
So, $x(4x^2 + x) = 4x^3 + x^2$.

Example 2

Distribute $3y^2$.

Multiply. $3y^2(2y^2 - 4y + 1)$
$3y^2(2y^2 - 4y + 1)$

$3y^2(2y^2) + 3y^2(-4y) + 3y^2(1)$
$3 \cdot 2 \cdot y^2 \cdot y^2 + 3 \cdot (-4) \cdot y^2 \cdot y + 3 \cdot 1 \cdot y^2$
$6y^4 - 12y^3 + 3y^2$
So, $3y^2(2y^2 - 4y + 1) = 6y^4 - 12y^3 + 3y^2$.

---

practice ▷ **Multiply.**

1. $a(a^3 - 5a)$ $\quad a^4 - 5a^2$

$-12c^4 - 18c^3 + 6c^2$
2. $-3c^2(4c^2 + 6c - 2)$

Example 3

$a^2 = 1a^2; -a = -1 \cdot a$
$3(1a^2 - 1a + 2)$
$= 3(1a^2) + 3(-1a) + 3(2)$
$= 3a^2 - 3a + 6$

Simplify. $a^2 + 5a - 4 + 3(a^2 - a + 2)$
$a^2 + 5a - 4 + 3(a^2 - a + 2)$
$1a^2 + 5a - 4 + 3(1a^2 - 1a + 2)$
$1a^2 + 5a - 4 + 3a^2 - 3a + 6$
$(1a^2 + 3a^2) + (5a - 3a) + (-4 + 6)$
$\quad 4a^2 \quad + \quad 2a \quad + \quad 2$
So, $a^2 + 5a - 4 + 3(a^2 - a + 2) = 4a^2 + 2a + 2$.

358

CHAPTER FOURTEEN

Simplify.

    3. $x^2 - 2x + 6 - 2(3x^2 - x + 5)$ $-5x^2 - 4$

Example 4

Simplify. $-(y^2 - 5y + 4)$

$$-y = -1 \cdot y \qquad -(y^2 - 5y + 4)$$
$$\text{Distribute } -1. \qquad -1(y^2 - 5y + 4)$$
$$-1(y^2) + (-1)(-5y) + (-1)(4)$$
$$-y^2 + 5y - 4$$

So, $-(y^2 - 5y + 4) = -y^2 + 5y - 4$.

Example 5

Simplify. $4x^2 + 6x - 3 - (x^2 - 5x + 2)$

$$-1(x^2 - 5x + 2) = \qquad 4x^2 + 6x - 3 - (x^2 - 5x + 2)$$
$$(-1)(x^2) + (-1)(-5x) + (-1)(2) \qquad 4x^2 + 6x - 3 - 1(x^2 - 5x + 2)$$
$$4x^2 + 6x - 3 - 1x^2 + 5x - 2$$
$$\text{Group like terms.} \quad (4x^2 - 1x^2) + (6x + 5x) + (-3 - 2)$$
$$3x^2 + 11x - 5$$

So, $4x^2 + 6x - 3 - (x^2 - 5x + 2) = 3x^2 + 11x - 5$.

---

practice ▷ **Simplify.**

                                                         $2c^2 - 3c - 3$

    4. $-(a^2 + 7a - 4)$ $-a^2 - 7a + 4$ **5.** $3c^2 + 2c - 1 - (c^2 + 5c + 2)$

◇ EXERCISES ◇

Multiply.   See TE Answer Section.

  1. $x(3x^2 + 2)$ **2.** $y(4y - 1)$ **3.** $a(-5a^2 + a)$
  4. $2x(x^2 + 1)$ **5.** $3c(2c - 2)$ **6.** $-4y(2y^2 - 3y)$
  7. $3c(5c^2 + 2c - 3)$ **8.** $-4x(x^2 + 6x + 2)$ **9.** $5a(a^2 + 5a - 2)$
10. $2z^2(z^2 - 4z + 2)$ **11.** $y^2(3y^2 + 4y - 5)$ **12.** $-3b^2(-b^2 + 2b - 1)$

Simplify.   See TE Answer Section.

13. $-(x^2 + 3x)$ **14.** $-(r^2 - 6r + 2)$ **15.** $-(3c^2 + 2c + 5)$
16. $-(-2a^2 + 4a - 3)$ **17.** $-(6x^2 - 5x - 4)$ **18.** $-(-4y^2 - y + 1)$
19. $x^2 + 2x - 5 + 4(x^2 - 3x - 2)$ **20.** $y^2 - 3y - 2 + 3(y^2 - 2y + 1)$
21. $2z^2 - 6z - 3 + 2(z^2 - 5z + 4)$ **22.** $3c^2 + 4c - 1 - 3(c^2 - 6c - 7)$
23. $5x^2 - 3x + 4 - 4(2x^2 + 6x - 1)$ **24.** $-4y^2 + 6y - 2 - 5(-y^2 + 2y + 3)$
25. $3a^2 + 2a - 1 - (a^2 - 4a + 2)$ **26.** $c^2 - 5c + 6 - (2c^2 + 5c - 4)$
27. $z^2 - 3z + 4 - (4z^2 - 2z + 1)$ **28.** $-y^2 + 3y - 1 - (4y^2 + 3y - 2)$
29. $-6x^2 + 2x - 1 - (-x^2 + 4x - 3)$ **30.** $-z^2 - 8z - 1 - (-3z^2 + 2z - 2)$
★ **31.** $2x^2(3x^2 - 6x + 4) - x^2(x^2 + 4x - 2)$ ★ **32.** $7y^2(3y^2 - y + 1) - 2y^2(4y^2 + 3y - 1)$
★ **33.** $5x(2x^3 - 3x^2 + 1) - 4x^2(2x + 5)$ ★ **34.** $6a^2(5a^2 + 7a - 1) - a^2(a^2 + 4a - 2)$

SIMPLIFYING POLYNOMIALS

Common Factors

◆ OBJECTIVE ◆

To factor a common monomial
from a polynomial

◆ RECALL ◆

$3(x^2 + 1)$ ← distributive property

$3(x^2) + 3(1)$

$3x^2 + 3$

We can rewrite $2x + 6$ as $2(x + 3)$ by using
the distributive property in reverse.
$2x + 6 = 2(x + 3)$
   ↑2 is a common monomial factor.

Example 1

Factor a common monomial factor from $3y + 9$.

$3y + 9$

$3(y) + 3(3)$
$3(y + 3)$

3 is a common factor of $3y$ and 9.
Distributive property in reverse.

So, $3y + 9 = 3(y + 3)$.

---

practice ▷ **Factor a common monomial factor from each.** $6(2 - a)$

1. $5c - 10$ $5(c - 2)$ 2. $21 + 7z$ $7(3 + z)$ 3. $12 - 6a$

Example 2

Factor the greatest common factor (GCF) from
$8c^2 - 12c + 4$.

The GCF is the greatest number that is
a common factor of all three terms.

$8c^2 - 12c + 4$

$4(2c^2) + 4(-3c) + 4(1)$

$4(2c^2 - 3c + 1)$

4 is the GCF.
Factor out 4.

So, $8c^2 - 12c + 4 = 4(2c^2 - 3c + 1)$.

---

practice ▷ **Factor the GCF from each.**

4. $6x^2 - 18$
$6(x^2 - 3)$

5. $2y^2 - 6y + 10$
$2(y^2 - 3y + 5)$

6. $4n^2 + 6n - 12$
$2(2n^2 + 3n - 6)$

Example 3

Factor the GCF from $x^2 + 5x$.

$$x^2 + 5x$$

$$x \cdot x + x \cdot 5$$

$$x(x + 5)$$

x is the GCF of x^2 and $5x$.
Factor out x.

So, $x^2 + 5x = x(x + 5)$.

practice ▷ **Factor the GCF from each.**

7. $y^2 + 4y$ $y(y + 4)$

8. $c^3 - 2c^2$ $c^2(c - 2)$

9. $n^4 + 3n^2$ $n^2(n^2 + 3)$

Example 4

To factor $4a^2 + 6a$ means to factor the GCF from $4a^2 + 6a$.

Factor $4a^2 + 6a$.

First, look for the greatest common whole number factor.

2 is the greatest common whole number factor.
Factor out 2.

$$4a^2 + 6a$$

$$\underline{2}(2a^2) + \underline{2}(3a)$$

$$2(2a^2 + 3a)$$

Now look for the greatest common variable factor.

a is the greatest common variable factor.
Factor out a.

$$2(2a^2 + 3a)$$

$$2(\underline{a} \cdot 2a + \underline{a} \cdot 3)$$

$$2 \cdot a(2a + 3)$$

So, $4a^2 + 6a = 2a(2a + 3)$.

Example 5

Factor $6x^3 - 8x^2 + 2x$.

$$6x^3 - 8x^2 + 2x$$

2 is the greatest common whole number factor.

$$\underline{2}(3x^3) + \underline{2}(-4x^2) + \underline{2}(x)$$

$$2(3x^3 - 4x^2 + x)$$

x is the greatest common variable factor.

$$2(\underline{x} \cdot 3x^2 + \underline{x} \cdot -4x + \underline{x} \cdot 1)$$

$$2 \cdot x(3x^2 - 4x + 1)$$

So, $6x^3 - 8x^2 + 2x = 2x(3x^2 - 4x + 1)$.

practice ▷ **Factor.**

10. $3y^2 - 12y$
$3y(y - 4)$

11. $10y^3 + 15y^2 - 5y$
$5y(2y^2 + 3y - 1)$

12. $3y^4 - 6y^2 + 9y$
$3y(y^3 - 2y + 3)$

COMMON FACTORS

What is the GCF of each?

1. $2x + 4$ 2
2. $3y - 9$ 3
3. $4a + 16$ 4
4. $10c^2 - 5$ 5
5. $3z^2 + 9z - 6$ 3
6. $4x^2 - 2x + 10$ 2
7. $5y^2 + 15y - 20$ 5
8. $8a^2 - 4a + 12$ 4
9. $x^2 + 4x$ x
10. $y^2 - 7y$ y
11. $2a^3 + 6a^2 + a$ a
12. $5c^3 - 3c^2 + 2c$ c
13. $3y^2 + 6y$ 3y
14. $8b^2 - 4b$ 4b
15. $2x^3 - 4x^2 + 6x$ 2x
16. $10a^3 - 15a^2 + 5a$ 5a

◇ EXERCISES ◇

Factor.   See TE Answer Section.

1. $4n + 8$
2. $6x - 9$
3. $3c + 12$
4. $8y - 2$
5. $2d - 10$
6. $15y + 5$
7. $7x^2 - 14$
8. $16c^2 + 8$
9. $2x^2 + 4x - 8$
10. $3y^2 - 9y + 6$
11. $5x^2 - 10x + 20$
12. $4a^2 - 12a - 20$
13. $6z^2 + 12z + 3$
14. $8x^2 - 6x + 2$
15. $10n^2 + 5n - 15$
16. $12y^2 - 8y - 4$
17. $y^2 + 2y$
18. $x^2 - 3x$
19. $c^3 + 6c^2$
20. $n^3 - 5n$
21. $3z^3 - 2z^2 + 4z$
22. $7y^3 + 2y^2 - 3y$
23. $6n^3 - 2n^2 + n$
24. $5x^3 - 4x^2 - x$
25. $2x^2 + 4x$
26. $3y^2 + 6y$
27. $8a^2 - 2a$
28. $10d^2 - 20d$
29. $14n^2 + 7n$
30. $8y^3 + 4y^2$
31. $6n^3 - 12n^2$
32. $15b^3 - 5b^2$
33. $4x^3 + 2x^2 - 6x$
34. $3c^3 - 6c^2 + 12c$
35. $5r^3 + 15r^2 - 10r$
36. $4y^3 - 8y^2 - 12y$
37. $10c^3 - 4c^2 + 2c$
38. $20n^3 - 30n^2 + 10n$
39. $6a^3 + 18a^2 - 6a$
★ 40. $4x^4 - 8x^3 + 6x^2 - 2x$
★ 41. $3y^4 - 9y^3 + 6y^2 + 3y$
★ 42. $10c^4 - 20c^3 + 30c^2 - 5c$
★ 43. $12n^4 - 6n^3 - 18n^2 + 6n$

Algebra Maintenance

Solve.

1. $x - 3 = -10$ -7
2. $-12 + x = -17$ -5
3. $x + 9 = -4$ -13
4. $-8x = 40$ -5
5. $\frac{u}{-6} = -5$ 30
6. $-4k + 6 = -14$ 5

Solve for x.

7. $x + u = t$ $x = t - u$
8. $t - x = r$ $x = t - r$
9. $e + x = k$ $x = k - e$

Solve.

10. A number added to twice the number is -36. What is the number? -12
11. A number diminished by 3 times the number is -8. What is the number?   4

Mathematics Aptitude Test

This test will prepare you for taking standardized tests.
Choose the best answer for each question.

1. The average of 4 numbers is 6. If two of the numbers are 2 and 6,
    which of the following cannot be one of the other two numbers?
    a. 8   **b.** 10   **c.** 16   **(d.)** 18

2. What is the value of $1^n + 1^{2n} + 1^{3n}$ for any whole number n?
    a. 1   **b.** 1^{6n} **c.** 3^{6n} **(d.)** 3

3. A student has grades of 90, 78, and 82. What grade must she get on
    her fourth test to get an average of 85?
    a. 85   **(b.)** 90   **c.** 95   **d.** 100

4. Which of the following is a list of numbers in decreasing order?
    (a.) 8, 4/5, -0.2, -2 **b.** -2, -0.2, 4/5, 8
    c. 8, 4/5, -2, -0.2 **d.** -0.2, -2, 4/5, 8

5. If $n = 3^2$, then $n^2 =$ _____?
    a. 12   **b.** 27   **c.** 36   **(d.)** 81

6. If the sum of two numbers is divisible by 5, then which of the following is true?
    a. The difference of the two is divisible by 5.
    b. At least one of the numbers is divisible by 5.
    c. None of the two numbers is divisible by 5.
    (d.) none of the above

7. A micromillimeter is defined as one millionth of a millimeter.
    What is the length of 12 micromillimeters?
    a. 12,000,000 mm   **b.** 0.00000012 mm   **c.** 0.0000012 mm   **(d.)** 0.000012 mm

8. The original price of an item is \$6.80. It is reduced by 20%. What is the new price?
    a. \$1.36   **b.** \$8.16   **(c.)** \$5.44   **d.** \$4.80

9. $m\overline{)n}\,^{q}$ indicates division of n by m.
    $\dfrac{r}{s}$ Which of the following is true?
    (a.) $n = m \cdot q + s$ **b.** $m \cdot q \cdot n = s$ **c.** $m \cdot s + q = n$ **d.** $q = m \cdot n - s$

You have learned to solve equations of the type $-2x + 10 = 4x - 8$. You can write a program to solve any equation of this type. First, it is necessary to write a general equation.

Specific equation: $-2x + 10 = 4x + (-8)$
General equation: $Ax + B = Cx + D$

Second, you must write a formula for solving the general equation $Ax + B = Cx + D$ for x. The solution is shown below.

$$Ax + B = Cx + D$$

Subtract Cx from each side.
$$Ax - Cx + B = Cx - Cx + D$$
$$Ax - Cx + B = D$$

Subtract B from each side.
$$Ax - Cx + B - B = D - B$$
$$Ax - Cx = D - B$$

Factor out common factor x.
$$(A - C)x, = D - B$$

Divide each side by $(A - C)$.
$$x = \frac{D - B}{A - C}$$

Example Write a program to solve any equation of the form
$AX + B = CX + D$.

```
10  INPUT "TYPE IN A VALUE FOR A ";A
20  INPUT "TYPE IN A VALUE FOR B ";B
30  INPUT "TYPE IN A VALUE FOR C ";C
40  INPUT "TYPE IN A VALUE FOR D ";D
50  LET X = (D - B) / (A - C)
60  PRINT "THE SOLUTION IS X = "X
70  END
```

See the Computer Section beginning on page 420 for more information.

Exercises

1. **RUN** the program above to solve $-2x + 10 = 4x - 8$. (*Hint:* A $= -2$, B $= 10$, C $= 4$, D $= -8$) 3
2. Modify the program above with a loop to allow you to solve any 3 equations without having to retype. Add line 5, **FOR K = 1 to 3** and line 65 **NEXT K.**

Use the program from Exercise 2 to solve each of these equations.
3. $3x - 2 = 7x + 10$ -3 4. $-5x - 7 = 10x + 8$ -1 5. $6x - 4 = 14$ 3
(*Hint:* What is the value of D?)

Solve. [335]

1. $4x + 9 = 2x + 3$ -3 2. $8x - 5 = 2x - 17$ -2 3. $7x + 2 = 2x - 23$ -5

Solve. [337]

4. $2(x - 3) = 18$ 12 5. $-4(z + 7) = 24$ -13 6. $-(2 + 2y) = 4(y - 2)$ 1

Write in mathematical terms. [339]

7. 5 increased by 3 $5 + 3$ 8. 8 less than x $x - 8$ 9. 4 more than 7 times a number
$7n + 4$

Solve. [339]

10. Three more than twice a number is equal to 36. Find the number. $16\frac{1}{2}$

Perform the operation on each side of the inequality.
Then write a true inequality. [344]

11. $5 < 6$ Subtract -3. $8 < 9$ 12. $-8 > -12$ Divide by -4. $2 < 3$

Solve. [347]

13. $5n \leq -25$ $n \leq -5$ 14. $-\frac{1}{7}t > -5$ $t < 35$

Evaluate. [351]

15. $4xy^2$ if $x = 8$ and $y = -1$ 32 16. $(-3b)^2$ if $b = -4$ 144

Simplify. [353]

17. $y^4 \cdot y$ y^5 18. $(3c^2)(-8c^5)$ $-24c^7$ 19. $(-3z)^3$ $-27z^3$

Classify each polynomial. [356]

20. $3 - 2x$ binomial   21. $-6y$ monomial   22. $3x^2 - 5 + 9x$ trinomial

Simplify. [356]

23. $2c - 3c^2 + 6 - 8 - 3c + c^2$ 24. $3d - 4d^4 + 5d^2 - 6 + 3d^2 + 5$
$-2c^2 - c - 2$ $-4d^4 + 8d^2 + 3d - 1$

Multiply. [358]

25. $m(-3m^2 + m)$ $-3m^3 + m^2$ 26. $-2y(y^2 + 3y - 1)$ $-2y^3 - 6y^2 + 2y$

Simplify. [358]

27. $-(a^2 + 3a)$ $-a^2 - 3a$ 28. $-5c^2 + 3c - 2 - (4c^2 + 2c - 1)$
$-9c^2 + c - 1$

Factor. [360]

29. $3y^2 - 6y + 24$ $3(y^2 - 2y + 8)$ 30. $5a^3 + 10a^2 - 25a$ $5a(a^2 + 2a - 5)$

Solve.

1. $3x + 5 = 2x + 1$ -4
2. $9x - 8 = 4x - 23$ -3
3. $3x + 7 = 5x - 7$ 7
4. $3(z - 8) = 9$ 11
5. $3(r - 1) = -(6 - 2r)$ -3

Write in mathematical terms.

6. 4 increased by x $4 + x$
7. 6 less than 5 times a number $5n - 6$

Solve.

8. Seven less than 4 times a number is equal to -3. Find the number. 1

Perform the operation on each side of the inequality.
Then write a true inequality.

9. $-4 < 3$ Add 6. $2 < 9$
10. $6 \geq -6$ Divide by -3. $-2 \leq 2$

Solve.

11. $x - 6 > -10$ $x > -4$
12. $-9x \geq 18$ $x \leq -2$

Evaluate.

13. x^5 if $x = -2$ -32
14. $-4y^2$ if $y = -3$ -36
15. $3a^2b$ if $a = -1, b = 9$ 27
16. $(-5n)^2$ if $n = -2$ 100

Simplify.

17. $x^5 \cdot x$ x^6
18. $(4d^3)(-7d^4)$ $-28d^7$
19. $(m^4)^2$ m^8
20. $(-2y)^3$ $-8y^3$

Classify each polynomial.

21. $-5d^3$ monomial
22. $4x^2 - 2x + 1$ trinomial

Simplify.

23. $2x^2 + 3x - 1 - 2x^2 - 2x + 4$ $x + 3$
24. $6r^3 - 3 + 4r^2 - 5r + 3r - 2r^2$
    $6r^3 + 2r^2 - 2r - 3$

Multiply.

25. $z(3z^2 - 1)$ $3z^3 - z$
26. $-2x(-x^2 + 5x - 4)$ $2x^3 - 10x^2 + 8x$

Simplify.

27. $-(3x^2 - 4x - 1)$ $-3x^2 + 4x + 1$
28. $d^2 + 3d - 2 - (5d^2 - 6d + 1)$
    $-4d^2 + 9d - 3$

Factor.

29. $9x - 18$ $9(x - 2)$
30. $t^3 + 6t^2$ $t^2(t + 6)$
31. $4b^3 - 8b^2 + 20b$ $4b(b^2 - 2b + 5)$

Some Ideas from Geometry

George Washington's head on Mt. Rushmore is as tall as a five-story building.

Points, Lines, and Planes

—— ◇ **OBJECTIVES** ◇ ——

To show that two points
determine a unique line
To show that three noncollinear
points determine a plane

———— ◇ **RECALL** ◇ ————

$A \bullet$ point A

   line containing points X and Y

In mathematics some key words are not defined.
They are used to define other words. Three key
geometric words that are not defined are *point, line,*
and *plane.*
Think of a point as being a location in space.

Example 1

Use a dot on paper to represent a point.

Choose any two points. Name them.
Draw a line through the points.

Choose any two points. Name them.
Draw a line segment between the
points.

Draw and name a point, a line, and a line segment.

A
• point A

We say two points determine a *unique line.*

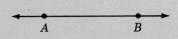   Line AB, or $\overleftrightarrow{AB}$

A line continues without end in both directions.
The arrowheads are used to suggest that the line
has no end.

A part of a line made up of two points and all the
points between them is called a *segment.*

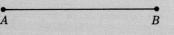   Line segment AB, or $\overline{AB}$

Points A and B are called *endpoints.* A line
segment has a definite length.

Points that are on the same line are called *collinear
points.*

Example 2

K, M, and T are on the same line.

Name the collinear points.

So, K, M, and T are collinear points.

---

practice ▷ **Draw a picture of**

1. a line 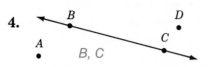 2. a point • 3. a segment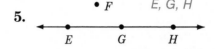

Name the collinear points.

4. 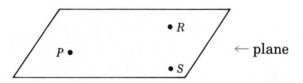 B, C

5. • F   E, G, H

Three points can be noncollinear.
Noncollinear means *not on the same line.* Three
noncollinear points determine a *unique plane.*

• R

P •   ← plane

• S

◇ EXERCISES ◇

1. Draw a picture of a line. Mark on it four points. Name them A, B, C, and D. Why are the four points collinear?   A B C D   They lie on the same line.

2. Choose a point *not* on the line in Exercise 1. Name it E. How many planes are determined by points A, B, and E?   one

3. How many points are there in space?   an infinite number

★ 4. Can four points in space be coplanar (in one plane)? Can four points be noncoplanar?   yes   yes

Algebra Maintenance

Write in mathematical terms.

1. -8 increased by t $-8 + t$ 2. 7 less than 8 times a number $8x - 7$

Solve.

$1\frac{1}{2}$

3. $5x + 3 = 2x + 15$ 4   4. $4(y - 3) = 12$ 6   5. $4(m - 2) = -(5 - 2m)$
6. $x - 4 > -6$ 7. $3x \le 12$ 8. $-5x \ge 15$
   all integers greater than -2 all integers less than or   all integers less than or equal to
                                      equal to 4 -3

Evaluate.

9. x^4 if $x = -2$ 16   10. $-5z^2$ if $z = -4$ -80
11. $4c^2g^2$ if $c = -2, g = 3$ 144   12. $(-2m)^2$ if $m = -3$ 36

POINTS, LINES, AND PLANES

Angles

◇ OBJECTIVES ◇

To classify angles as acute,
  right, or obtuse
To define complementary and
  supplementary angles
To solve problems involving
  complementary and
  supplementary angles

◇ RECALL ◇

$\overline{AB}$ has two
endpoints and has
a definite length.

A *ray* has one endpoint and continues indefinitely in
one direction. A ray is part of a line.

ray AB, or $\overrightarrow{AB}$

An *angle* is made up of two rays
that have the same endpoint.

vertex ⟶

angle BAC, or $\angle BAC$

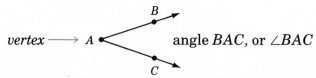

Example 1

Use a protractor to measure angle COB.

Place the center of the protractor at the
vertex O. Line up the bottom with $\overrightarrow{OB}$.
Read the number of degrees where
$\overrightarrow{OC}$ falls on your protractor.

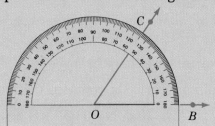

m∠ $COB = 55°$ So angle COB measures 55 degrees, or 55°.

Angles are classified according to their measures.

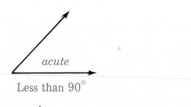

acute

Less than 90°

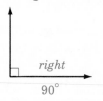

right

90°

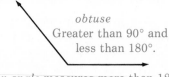

obtuse
Greater than 90° and
less than 180°.

A *reflex angle* measures more than 180°.

practice ▷ **Measure each angle and classify it as acute, right, or obtuse.**

1.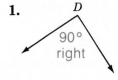
   D
   90°
   right

2.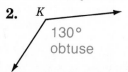
   K
   130°
   obtuse

3.
   acute
   75°
   T

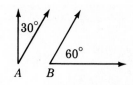

$\angle A$ is the complement of $\angle B$.

complementary angles
Two angles are said to be *complementary* if the sum of their measures is 90°.

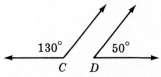

$\angle C$ is the supplement of $\angle D$.

supplementary angles
Two angles are said to be *supplementary* if the sum of their measures is 180°.

Example 2

Tell whether the angles are complementary or supplementary.

$120° + 60° = 180°$ 120°, 60°   50°, 40° $50° + 40° = 90°$

supplementary   complementary

---

practice ▷ **Complementary or supplementary?**

4. 25°, 65°   **5.** 98°, 82°   **6.** 1°, 179°   **7.** 5°, 85°
complementary   supplementary   supplementary   complementary

---

Example 3

One of two complementary angles is twice the other. What are the measures of the two angles?
Let $x =$ measure of one angle
$2x =$ measure of the other angle

Write an equation. $x + 2x = 90$
The two angles are complementary.
Solve. $x = 30$
So, one angle measures 30° and the other 60°.

◇ *EXERCISES* ◇

Measure each angle and classify it as acute, right, or obtuse.

1.   *A*   2.   *B*   3. *C*   acute   4.   obtuse 160°
90°   140°   30°
right   obtuse   *D*

Complementary or supplementary?

supplementary

5. 17°, 163°   supplementary **6.** 45°, 45° complementary **7.** 90°, 90° supplementary **8.** 101°, 79°
9. One of two supplementary angles is 5 times the size of the other.
What are the measures of the two angles?   30°, 150°

Problem Solving – Applications
Jobs for Teenagers

1. Jim works as a stockperson at the Josco Supermarket. He works 7 hours a day Monday through Friday and 5 hours on Saturday. He is paid $4.50 per hour. How much does he earn in a week? $180

2. In the stockroom at Josco's, Jim counted 10 crates of avocados. There are 48 avocados in each crate, and avocados sell for 49¢ apiece. What is the value of the 10 crates of avocados? $235.20

3. Josco's usually sells 3 cases of canned peaches during the week and 2 cases over the weekend. How many cases should Jim recommend that Josco's buy to last for 4 weeks? 20

4. During one week, Josco's sold $68.40 worth of potatoes. They made 20% profit on potatoes. How much profit does Jim figure that Josco's made on potatoes that week? $13.68

5. There is an 8¢ profit on each can of tomatoes. What is the profit on 10 cases of tomatoes if there are 12 cans in a case? $9.60

6. Maria has worked at the Josco Supermarket for a year. She works 40 hours a week and is paid $4.50 per hour. When her pay is raised to $4.75 per hour, how much more will she earn each week? $10

7. Maria unpacked 8 cases of peaches. There are 24 cans in a case. How many cans did she unpack? 192

8. The store makes a profit of 9¢ on each can of corn sold. There are 24 cans in a case of corn. How much profit does Maria figure the store will make on 8 cases of corn? $17.28

9. Josco's paid $5.88 for a crate of 12 melons. At what price should Maria mark each melon so that Josco's will make a profit of $4.80 on a crate of melons? $0.89

10. Maria allows 40% of the canned vegetable shelf to be stock by Pixie brand products. The shelf is 15 meters long. How many meters of the shelf will she fill with Pixie products? 6 meters

CHAPTER FIFTEEN

Parallel Lines and Angles

To identify pairs of alternate
   interior angles, corresponding
   angles, and vertical angles
To determine measures of angles
   on the basis of congruence of
   special pairs of angles

is parallel to

A line continues without end in both directions.

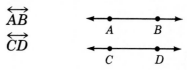

Two lines in the same plane which do not intersect
are called *parallel lines*.

When lines intersect, angles are formed.

Vertical angles are opposite angles formed by two
intersecting lines.

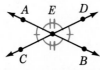

Vertical angles are congruent. They
have the same measure. The symbol
for congruence is ≅.
$\angle AEC \cong \angle DEB$ and $\angle AED \cong \angle CEB$

Example 1

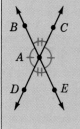

← Note the convenient
way of marking
congruent angles.

$\overleftrightarrow{BE}$ and $\overleftrightarrow{CD}$ intersect at point A.
Name the vertical angles.

$\angle BAC$ and $\angle DAE$ are opposite angles.
$\angle BAD$ and $\angle CAE$ are opposite angles.

So, $\angle BAC$ and $\angle DAE$ are vertical angles and
$\angle BAD$ and $\angle CAE$ are vertical angles.

Line a ∥ Line b

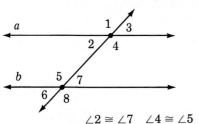

$\angle 2 \cong \angle 7$ $\angle 4 \cong \angle 5$
$\angle 1 \cong \angle 5$ $\angle 2 \cong \angle 6$ $\angle 3 \cong \angle 7$
$\angle 4 \cong \angle 8$

A *transversal* is a line that intersects two or more
lines.

When a transversal intersects parallel lines, pairs of
congruent angles are formed.

$\angle 2$ and $\angle 7$, $\angle 4$ and $\angle 5$ are *alternate interior angles*.
$\angle 1$ and $\angle 5$, $\angle 2$ and $\angle 6$, $\angle 3$ and $\angle 7$, $\angle 4$ and $\angle 8$
are *corresponding angles*.

practice ▷ **In the figure at the right, identify all pairs of vertical angles, alternate interior angles, and corresponding angles.** See TE Answer Section.

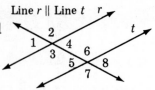

Line r ‖ Line t

If the sides of an angle lie on a straight line, then that angle is called a *straight angle*. A straight angle measures 180°.

Example 2

The 65° angle and angle 1 form a straight angle.
65° + 115° = 180°
So, m∠1 = 115°.

Determine the measures of the angles labeled 1 through 7.

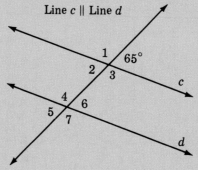

Line c ‖ Line d

angle 1: 115°

angle 2: 65°

angle 3: 115°

angle 4: 115°

angle 5: 65°

angle 6: 65°

angle 7: 115°

◇ EXERCISES ◇

1. In the figure at the right, name two pairs of **congruent** Answers may vary.
   a. vertical angles ∠1 and ∠4, ∠5 and ∠7
   b. alternate interior angles ∠4 and ∠6   ∠3 and ∠5
   c. corresponding angles. ∠1 and ∠6
      Answers may vary.   ∠3 and ∠7

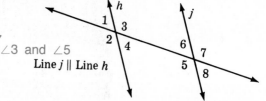

Line j ‖ Line h

2. Determine the measures of angles labeled 1 through 7.
   m∠1 = 143°   m∠5 = 37°
   m∠2 = 143°   m∠6 = 143°
   m∠3 = 37°   m∠7 = 37°
   m∠4 = 143°

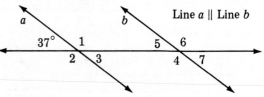

Line a ‖ Line b

★ 3. Line d ‖ line r. The measure of the marked angle is x. Express the measure of angle 1 in terms of x. m∠1 = 180° − x

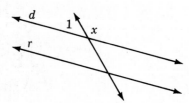

Perpendicular Lines

To construct a line perpendicular to a given line and passing through a point not on the given line

A right angle is an angle that has a measure of 90°.

_____

If two lines intersect in such a way that they form right angles, then they are *perpendicular lines*.

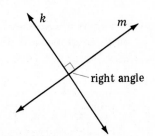

right angle

$k \perp m$

⌐ is perpendicular to

Example

Construct a line perpendicular to line *t* that passes through point *P*.

Step 1   Draw line *t* and mark the point *P*.

Step 2   Place the tip of your compass at point *P* and make two arcs that cut line *t* at points *A* and *B*.

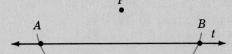

Step 3   Place the tip of the compass at *A* and make an arc below the line. Do the same with the same opening of the compass but now with the tip of the compass on the point *B*.

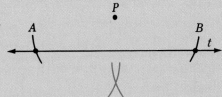

Step 4   Name the point of intersection of the two arcs below line *t*, *R*. Draw a line through *R* and *P*.

$\overleftrightarrow{RP}$ is perpendicular to line *t* and passes through point *P* not on line *t*.

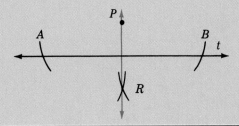

◇ *ORAL EXERCISES* ◇

1. What is the measure of a right angle? 90°

2. Define perpendicular lines. lines that intersect to form right angles

---

◇ *EXERCISES* ◇   Check students' constructions.

1. Construct a line perpendicular to line *k* that passes through point *D*. First duplicate the picture on your paper.

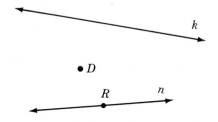

★ 2. Construct a line perpendicular to line *n* that passes through point *R* on the line. First duplicate the picture on your paper.

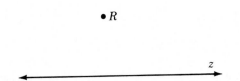

★ 3. Construct a line that is parallel to line *z* and passes through point *R*. First duplicate the picture on your paper. (*Hint:* First construct a line perpendicular to line *z* that passes through point *R;* then construct a line through point *R* that is perpendicular to the line you constructed.)

Calculator

Solve each equation. Round the answer to the nearest hundredth.

$$2.6x + 1.9 = 0.3$$
$$2.6x + 1.9 = 8.3$$
$$2.6x + 1.9 - 1.9 = 8.3 - 1.9$$
$$2.6x = 6.4$$
$$x = \frac{6.4}{2.6}$$
$$= 2.462$$

The solution to the nearest hundredth is 2.46.

1. $3.1x + 2.7 = 2.3x + 1.6$ -1.38

2. $4.8(x - 2.9) = 1.4$ 3.19

3. $9.3(x - 0.4) = -2.7(3.3 - 5.6x)$ 0.89

4. $2.5(5.4 - 2.5x) = 1.3(0.5 - 10.7x)$ -1.68

376 *CHAPTER FIFTEEN*

Triangles

To classify triangles according to measures of their angles
To classify triangles according to measures of their sides
To solve problems involving angle measures in triangles

An acute angle measures less than 90°.
A right angle measures 90°.
An obtuse angle measures more than 90° and less than 180°.

Triangles can be classified according to the measures of their angles.

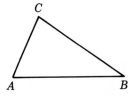
△ *ABC* is *acute*, because each of its three angles is acute.

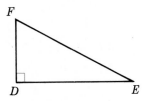
△ *DEF* is a *right* triangle, because it has one right angle.

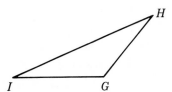
△ *GHI* is *obtuse*, because it has one obtuse angle.

practice ▷ **Classify these triangles according to the measures of their angles.**

1.
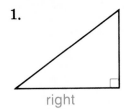
right

2.
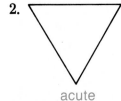
acute

3.
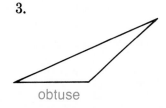
obtuse

The sum of the measures of the three angles in a triangle is 180°.

Example

Find the measure of the third angle, given the measures of two angles of a triangle. Let x = the measure of the third angle.

$$90°, 12°$$
$$90 + 12 + x = 180$$
$$102 \quad + x = 180$$
$$x = 78°$$

$$130°, 11°$$
$$130 + 11 + x = 180$$
$$141 \quad + x = 180$$
$$x = 39°$$

TRIANGLES

Triangles can be classified according to the relationships among their sides.

No two sides are congruent.	Two sides are congruent.	All three sides are congruent.
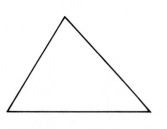	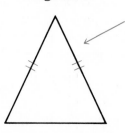 Sides are marked with the same symbol to indicate that they are congruent.	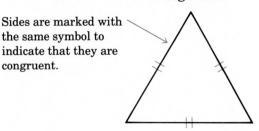
Scalene triangle	Isosceles triangle	Equilateral triangle

◇ EXERCISES ◇

Classify the triangles according to the measures of their angles.

1.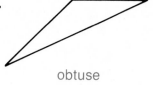
obtuse

2.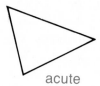
acute

3.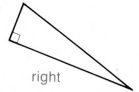
right

4. 35°, 37°, 108°   obtuse

5. 40°, 90°, 50°   right

6. 70°, 70°, 40°   acute

Find the measure of the third angle, given the measures of two angles of a triangle.

7. 47°, 92°   41°

8. 50°, 65°   65°

★ **9.** 179°, $\frac{7}{8}$° $\frac{1}{8}$°

★ **10.** 178°, $1\frac{1}{2}$° $\frac{1}{2}$°

★ **11.** 90°, $89\frac{3}{8}$° $\frac{5}{8}$°

★ **12.** 89°, 90.6°   0.4°

Classify the triangles according to the measures of their sides.

13.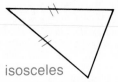
isosceles

14.
equilateral

15.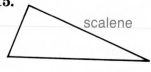
scalene

isosceles

16. 5 in., 3 in., 4 in.   scalene

17. 4 in., 4 in., 4 in.   equilateral

18. 7 in., 7 in., $8\frac{1}{2}$ in.

19. 7 ft, 7ft, 12 ft   isosceles

★ **20.** 3 ft, 30 in., 4 ft   scalene

★ **21.** 13 in., 15 in., 2 ft   scalene

Polygons

To classify polygons according
to the number of their sides
To classify quadrilaterals
To determine measures of sides
and angles in special polygons

Segments are parts of lines; they have
definite lengths.

A *polygon* is made up of three or more segments that
intersect only at their endpoints, and each endpoint
is shared by exactly two segments.

Polygons are named according to the number of sides they have.

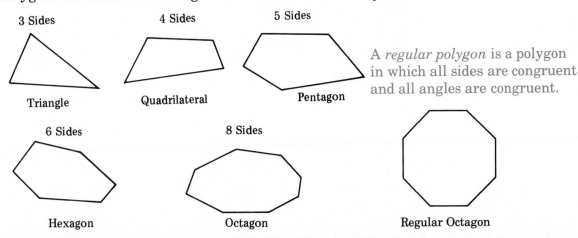

3 Sides

4 Sides

5 Sides

Triangle

Quadrilateral

Pentagon

A *regular polygon* is a polygon
in which all sides are congruent
and all angles are congruent.

6 Sides

8 Sides

Hexagon

Octagon

Regular Octagon

Example 1

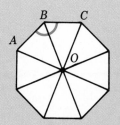

Find the measure of each angle and central angle
in a regular octagon.

∠*ABC* is one angle of the regular octagon.
Angles like ∠*AOB* are called *central angles.*

The octagon is divided into 8 triangles.

Sum of the measures of angles in the
8 triangles: $8 \times 180°$ or $1{,}440°$.
Sum of the measures of all central
angles: $360°$.
To find the sum of the measures of the
angles of the octagon, subtract:
$1{,}440° - 360° = 1{,}080°.$

Measure of each angle: $\frac{1080°}{8} = 135°$

So, each angle in a regular octagon measures $135°$.

Measure of each central angle: $\frac{360°}{8} = 45°$.

So, each central angle in a regular octagon
measures $45.°$

Quadrilaterals can be classified according to their special properties.

Parallelogram

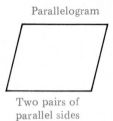

Two pairs of
parallel sides

Trapezoid

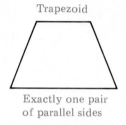

Exactly one pair
of parallel sides

Rectangle

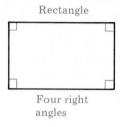

Four right
angles

Rhombus

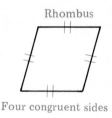

Four congruent sides

Square

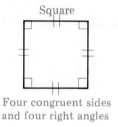

Four congruent sides
and four right angles

◇ EXERCISES ◇

1. How many sides does each one of the following polygons have?

   a. pentagon 5   b. octagon 8   c. quadrilateral 4
   d. triangle 3   e. hexagon 6   ★ f. decagon 10

2. Find the measure of each angle of a regular pentagon. 108°

Name each of the special quadrilaterals.

3.

rhombus

4.
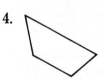
trapezoid

5.

rectangle

6.

square

7. In what sense is a square a special rhombus? It is a rhombus with right angles.

8. *ABCD* is a square. The length of $\overline{CB}$ is 5 cm.
   What is the length of $\overline{CD}$? Why is this true?
   5 cm; a square has 4 congruent sides.

9. Find the measure of each angle in a regular hexagon. 120°

Problem Solving – Careers
Tile Setters

1. Mr. Rodriguez is a marble setter. He has been hired to set a marble wall. The wall is 12 m long and 3 m high. How many marble slabs each measuring 1 m by 1.5 m will he need to cover the wall? 24 slabs

2. Ms. Ostfeld is a tile setter who will set a tile floor in a kitchen that measures 4 m by 6 m. Each tile is 20 cm by 20 cm. How many tiles will she need? 600 tiles

3. Mrs. Aaron is a terrazzo worker. Terrazzo is tinted concrete with marble chips, and it is used for decorative floors. Mrs. Aaron has been hired to set a terrazzo star in the floor of a new building. She will use metal strips 100 cm long and 173 cm long to separate the parts of the design. How many strips of each length will she need? 6 strips of 173 cm lengths; 18 strips of 100 cm lengths

100 cm

173 cm

4. As a marble setter, Mr. Alexander earns $8.30 per hour. Mr. Blake works as a marble setter in a different city and earns $8.75 per hour. How much more does Mr. Blake earn in a 40-hour week? $18 more

5. A room measures 3 m by 4 m. How many tiles are needed to cover the floor if 36 tiles cover 1 m^2? 432 tiles

Square Roots

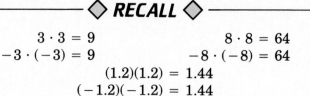

—◇ *OBJECTIVES* ◇—   ——————◇ *RECALL* ◇——————

To identify the perfect squares from 1 to 100

To find approximations of square roots from a table

To solve area problems involving square roots

$$3 \cdot 3 = 9 \qquad\qquad 8 \cdot 8 = 64$$
$$-3 \cdot (-3) = 9 \qquad\qquad -8 \cdot (-8) = 64$$
$$(1.2)(1.2) = 1.44$$
$$(-1.2)(-1.2) = 1.44$$

Definition of square root If $x \cdot x = n$, then x is a *square root* of n.

Every positive number has two square roots, one positive and one negative.

Number	Positive Square Root	Negative Square Root
9	3	-3
64	8	-8
1.44	1.2	-1.2

The two square roots of 49 are 7 and -7 because $7 \cdot 7 = 49$ and $-7 \cdot (-7) = 49$. The two square roots of 100 are 10 and -10 because $10 \cdot 10 = 100$ and $-10 \cdot (-10) = 100$. In most instances, especially in problems dealing with lengths of sides, only the positive square root is used. This has a special name, as shown in the definition below.

principal square root

The positive square root of a number is called the *principal square root*. It is symbolized by $\sqrt{}$.

The principal square root of 9 is 3. $\sqrt{9} = 3$ $\sqrt{64} = 8$ $\sqrt{1.44} = 1.2$

perfect square

Definition of perfect square
9 is a perfect square.

A *perfect square* is a number whose principal square root is a whole number.

Example 1

$6 \cdot 6 = 36$
$7 \cdot 7 = 49$ and $8 \cdot 8 = 64$
52 is between 49 and 64.

Are 36 and 52 perfect squares?
$\sqrt{36} = 6$, and 6 is a whole number.
$\sqrt{52}$ is not a whole number.
So, 36 is a perfect square, but 52 is not.

practice ▷ **Which are perfect squares?** 25, 81

1. 48 **2.** 25 **3.** 81 **4.** 66

We can use the table on page 473 to find an approximation for $\sqrt{52}$ correct to three decimal places.

$\doteq$ means *is approximately equal to.*

$$\sqrt{52} \doteq 7.211$$
$$7.211 \times 7.211 = 51.998521$$

Note that $(7.211)^2$ is not exactly equal to 52. We are not able to find an exact decimal which when squared will equal 52. For this reason, $\sqrt{52}$ is called an *irrational number.*

Example 2

Approximate $\sqrt{48}$ and $\sqrt{82}$ to the nearest tenth. Use the table on page 473.

Round to the nearest tenth.

$$\sqrt{48} \doteq 6.9\underline{2}8 \qquad\qquad \sqrt{82} \doteq 9.0\underline{5}5$$
$$\doteq 6.9 \nwarrow \qquad\qquad\qquad \doteq 9.1 \uparrow$$
$$\text{less than 5} \qquad\qquad \text{5 or greater}$$
$$\text{So, } \sqrt{48} \doteq 6.9 \text{ and } \sqrt{82} \doteq 9.1$$

practice ▷ **Approximate to the nearest tenth. Use the table on page 473.**

5. $\sqrt{8}$ 2.8 **6.** $\sqrt{67}$ 8.2 **7.** $\sqrt{33}$ 5.7 **8.** $\sqrt{75}$ 8.7

Example 3

The area of a square is 62 cm². Find the length of a side, to the nearest tenth of a cm.

Formula for area of a square. $s^2 = A$
Substitute 62 for A. $s^2 = 62$
$s = \sqrt{62}$
From the table, $\sqrt{62} \doteq 7.874$. $s \doteq 7.874$
Round to the nearest tenth. $s \doteq 7.9$

s

Area:
62 cm²

s

So, the length of a side is 7.9 cm.

practice ▷ **9.** The area of a square is 54 m². Find the length of a side, to the nearest tenth of a m. 7.3 m $\doteq s$

SQUARE ROOTS

383

Give two square roots of each number.

9, −9

1. 25 5, −5	**2.** 9 3, −3	**3.** 64 8, −8	**4.** 16 4, −4	**5.** 81
6. 36 6, −6	**7.** 1 1, −1	**8.** 4 2, −2	**9.** 144 12, −12	**10.** 121

11, −11

◇ EXERCISES ◇

Which are perfect squares?

1. 25 yes	**2.** 42 no	**3.** 1 yes	**4.** 9 yes	**5.** 87 no
6. 8 no	**7.** 64 yes	**8.** 18 no	**9.** 79 no	**10.** 45 no
11. 36 yes	**12.** 28 no	**13.** 56 no	**14.** 81 yes	**15.** 125 no

Approximate to the nearest tenth. Use the table on page 473.

16. $\sqrt{80}$ 8.9	**17.** $\sqrt{3}$ 1.7	**18.** $\sqrt{90}$ 9.5	**19.** $\sqrt{35}$ 5.9	**20.** $\sqrt{53}$ 7
21. $\sqrt{12}$ 3.5	**22.** $\sqrt{62}$ 7.9	**23.** $\sqrt{14}$ 3.7	**24.** $\sqrt{79}$ 8.9	**25.** $\sqrt{37}$ 6
26. $\sqrt{44}$ 6.6	**27.** $\sqrt{58}$ 7.6	**28.** $\sqrt{46}$ 6.8	**29.** $\sqrt{54}$ 7.3	**30.** $\sqrt{2}$ 1.4

Find *s* to the nearest tenth of a unit. Use the table on page 473.

31.
s
4.9 cm
Area:
24 cm^2
s

32.
s
3.7 m
Area:
14 m^2
s

33.
s
3.2 cm
Area:
10 cm^2
s

34.
s
9.6 m
Area:
92 m^2
s

Give answers to the nearest tenth of a unit.

35. The area of a square is 42 cm^2.
Find the length of a side.   6.5 cm

36. The area of a square room is 38 m^2.
Find the length of a wall.   6.2 m

★ **37.** The area of a rectangle is 56 m^2.
The length is twice the width.   L = 10.6 m
Find the length and the width.   W = 5.3 m

★ **38.** The area of a rectangle is 68 cm^2.
The length is 4 times the width.
Find the length and the width.
L = 16.4 cm   W = 4.1 cm

Challenge

There is a pattern when adding the cubes of consecutive numbers. Observe this pattern in the following:

$$1^3 + 2^3 = 9 \quad \text{also } (1 + 2)^2 = 9$$
$$1^3 + 2^3 + 3^3 = 36 \quad \text{also } (1 + 2 + 3)^2 = 36$$

$1^3 + 2^3 + 3^3 + 4^3 = 100 \qquad 1^3 + 2^3 + 3^3 + 4^3 + 5^3 = 225$

Give the next 3 lines in this pattern. Using the pattern,

$1^3 + 2^3 + 3^3 + 4^3 + 5^3 + 6^3 = 441$

1. give quickly the sum of the cubes of the first 10 counting numbers.   3,025

2. give the formula for the sum of the cubes of the first k natural numbers. $(1 + \ldots + k)^2$

The Pythagorean Theorem

To find the length of a side of a right triangle by applying the Pythagorean theorem

Right Triangle ABC

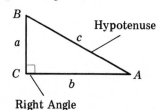

Hypotenuse

Right Angle

Square $DEFG$

$A = s \cdot s$, or s^2

Example 1

Triangle RST is a right triangle. Figures I, II, and III are squares. Find the area of each square. Add the areas of Squares I and II. Then compare that sum with the area of Square III.

Formula for area of a square

$A = s \cdot s$
Area of I $= 6 \cdot 6 = 36$
Area of II $= 8 \cdot 8 = 64$
Area of III $= 10 \cdot 10 = 100$

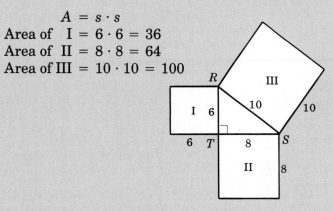

Add the areas of Squares I and II.

$$36 \quad + \quad 64 \quad = \quad 100$$
$$\downarrow \qquad\qquad \downarrow \qquad\qquad \downarrow$$

This is true for all right triangles.

So, $\dfrac{\text{Area of}}{\text{Square I}} + \dfrac{\text{Area of}}{\text{Square II}} = \dfrac{\text{Area of}}{\text{Square III}}$.

Pythagorean theorem
In any right triangle, the square of the length of the hypotenuse equals the sum of the squares of the lengths of the other two sides.

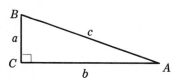

c is the length of the hypotenuse. a and b are the lengths of the other two sides.

If $\triangle ABC$ is a right triangle, then $a^2 + b^2 = c^2$.

Example 2

The lengths of two sides of a right triangle are 4 cm and 6 cm. Find the length of the hypotenuse, to the nearest tenth of a cm.

<table>
<tr><td>Pythagorean theorem.</td><td>$a^2 + b^2 = c^2$</td></tr>
<tr><td>Replace a with 4 and b with 6.</td><td>$4^2 + 6^2 = c^2$</td></tr>
<tr><td></td><td>$16 + 36 = c^2$</td></tr>
<tr><td></td><td>$52 = c^2$</td></tr>
<tr><td>Use the table on page 473.</td><td>$\sqrt{52} = c$</td></tr>
<tr><td>$\sqrt{52} \doteq 7.211 \doteq 7.2$</td><td>$7.2 \doteq c$</td></tr>
</table>

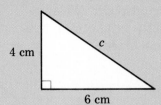

So, the length of the hypotenuse is about 7.2 cm.

practice ▷ **Find c to the nearest tenth of a unit.**

1.

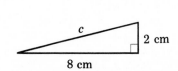

8.2 cm

2.

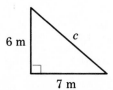

9.2 m

Example 3

The length of the hypotenuse of a right triangle is 12 meters, and the length of one side is 7 meters. Find the length of the other side, to the nearest tenth of a meter.

<table>
<tr><td>Pythagorean theorem.</td><td>$a^2 + b^2 = c^2$</td></tr>
<tr><td>Replace b with 7 and c with 12.</td><td>$a^2 + 7^2 = 12^2$</td></tr>
<tr><td></td><td>$a^2 + 49 = 144$</td></tr>
<tr><td>Subtract 49 from each side.</td><td>$a^2 + 49 - 49 = 144 - 49$</td></tr>
<tr><td></td><td>$a^2 = 95$</td></tr>
<tr><td>Use the table on page 473.</td><td>$a = \sqrt{95}$</td></tr>
<tr><td>$\sqrt{95} \doteq 9.747 \doteq 9.7$</td><td>$a \doteq 9.7$</td></tr>
</table>

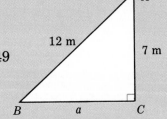

So, the length of the other side is about 9.7 m.

practice ▷ **Find a or b to the nearest tenth of a unit.**

3.

7.1 cm

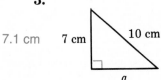

4.

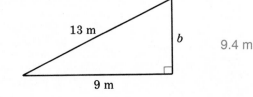

9.4 m

◇ EXERCISES ◇

For each right triangle, find the missing length, to the nearest tenth.

1. $a = 2, b = 3 \quad c \doteq 3.6$
2. $a = 3, b = 4 \quad c = 5$
3. $a = 5, b = 5 \quad c \doteq 7.1$
4. $a = 3, b = 6 \quad c \doteq 6.7$
5. $a = 4, b = 5 \quad c \doteq 6.4$
6. $a = 1, b = 9 \quad c \doteq 9.1$
7. $a = 8, c = 10 \quad b = 6$
8. $b = 12, c = 15 \quad a = 9$
9. $a = 4, c = 8 \quad b \doteq 6.9$
10. $b = 2, c = 6 \quad a \doteq 5.7$
11. $b = 10, c = 14 \quad a \doteq 9.8$
12. $a = 21, c = 25 \quad b \doteq 13.6$

13.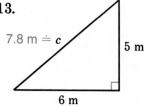

14.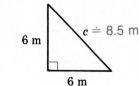

15.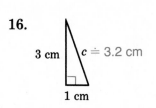

16.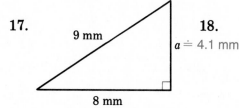

17.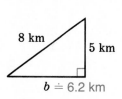

18.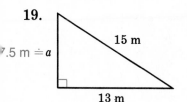

19.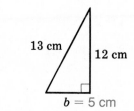

20.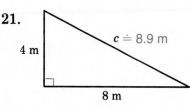

21.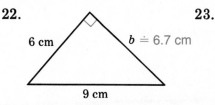

22.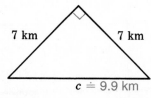

23.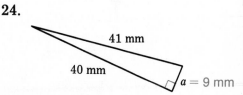

24.

25. The lengths of two sides of a right triangle are 4 m and 7 m. Find the length of the hypotenuse. $h \doteq 8.1$ m

26. The length of the hypotenuse of a right triangle is 10 cm, and the length of one side is 5 cm. Find the length of the other side. 8.7 cm $\doteq$ b

★ 27. Bill's room measures 3 m by 5 m. Find the length of the diagonal connecting two opposite corners. $d \doteq 5.8$ m

★ 28. Jessie walked 2 km south and 5 km west. How far was she from her starting point? nearly 5.4 km

★ 29. A 4-m ladder is 1 m from the base of a house. At what height does the top of the ladder touch the house? at about 3.9 m up

★ 30. The diagonal of a square measures 8 cm. Find the length of a side of the square. about 5.7 cm

THE PYTHAGOREAN THEOREM

Non-Routine Problems

Finish drawing the figure on the right so that it is congruent to the figure on the left.

1.
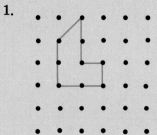

2.
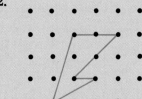

3.
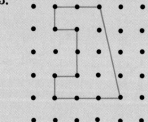
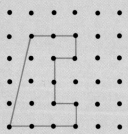

4.
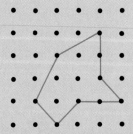
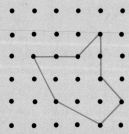

COMPUTER ACTIVITIES

Recall that for a right triangle, the formula for finding the hypotenuse if you know the lengths of the legs is $c = \sqrt{a^2 + b^2}$. You can write a program for finding c if you are given a and b.

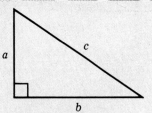

First, you need to know the **BASIC** symbol for square root. The square root of 25 is symbolized by

$$\text{SQR}(25) \qquad \text{no space between SQR and (}$$

Example 1 Compute SQR($3 \wedge 2 + 4 \wedge 2$)

$3 \wedge 2$ means $3^2 = 9$ SQR($3 \wedge 2 + 4 \wedge 2$)
$4 \wedge 2$ means $4^2 = 16$ SQR($9 + 16$)
                        SQR(25)
SQR(25) $= \sqrt{25} = 5$ 5

Example 2 Write a program to find C in the formula $C = \sqrt{A^2 + B^2}$. Allow for 2 sets of values for A and B.

First write the formula in **BASIC**: $C = \sqrt{A^2 + B^2}$
$$C = \text{SQR}(A \wedge 2 + B \wedge 2)$$

| Two sets of data. | ```
10   FOR M = 1 TO 2
20   INPUT "TYPE IN LENGTH OF A LEG ";A
30   INPUT "TYPE IN LENGTH OF OTHER LEG ";B
``` |
| Formula in **BASIC** for finding the hypotenuse of a right triangle. | ```
40   LET C =  SQR (A ^ 2 + B ^ 2)
50   PRINT "C = "C
60   PRINT : PRINT
70   NEXT M
80   END
``` |

See the Computer Section beginning on page 420 for more information.

Exercises

RUN the program above to find C for each of the following values of A and B.

1. A = 6, B = 8   10   2. A = 5, B = 12   13   3. A = 6, B = 6
                                                                 8.485281374

★ 4. Write a program to find the leg of a right triangle given the hypotenuse and one of the legs. **RUN** the program to find B if C = 26 and A = 10.
     See TE Answer Section for program and results.

1. You are given two points. How many lines can contain these two points? [368] exactly one

2. You are given four points. Describe all possible relationships that can exist among these four points. [368] See TE Answer Section.

3. Measure each angle and classify as acute, right, or obtuse. [370]

A
90°
right

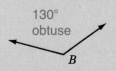
130°
obtuse
B

C
47°
acute

4. Complementary or supplementary? [370]
   a. 24°, 156°   b. 89°, 1°   c. 120°, 60°   d. 45°, 45°
      supplementary   complementary   supplementary   complementary

5. a. In the figure at the right, name two pairs of congruent vertical angles, alternate interior angles, and corresponding angles. See TE Answer Section.
   [373]
   b. If m∠1 = 37°, determine the measures of all labeled angles.

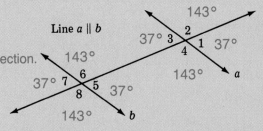

Line a ∥ b
143°
37° 3
2
143°
4
1 37°
37° 7
6
143° a
37°
8
5
143° b

6. Construct a line perpendicular to line t and passing through point A. [375]
   Check students' constructions.

A•   t

7. Name three special parallelograms. Describe their properties. [380]
   See TE Answer Section.

8. Classify each of the triangles in the picture according to its angles and according to its sides. Tell why you classified each triangle as you did. [377]
   See TE Answer Section.

   right
   scalene

   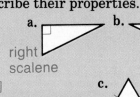
   a.

   b.
   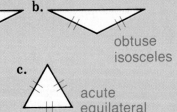
   obtuse
   isosceles

   c.
   acute
   equilateral

9. Which are perfect squares? [382] (a.) 81 b. 93 c. 105 (d.) 169

10. Approximate to the nearest tenth. Use the table on page 473. [382]
    a. $\sqrt{67}$ 8.2   b. $\sqrt{19}$ 4.4   c. $\sqrt{83}$ 9.1   d. $\sqrt{5}$ 2.2

11. For each right triangle, find the missing length, to the nearest tenth. a and b are legs. c is the hypotenuse. [385] $b = 15.0$
    a. $a = 11, b = 15$ $c \doteq 18.6$ b. $a = 8, c = 17$

12. Find the missing lengths. [385] a.
    $c \doteq 9.8$ in.
    4 in.
    9 in.

    b.
    13 ft
    10 ft
    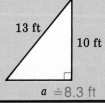
    $a \doteq 8.3$ ft

Chapter Test

1. What does it mean to say that two points determine a unique line?   There is exactly one line that contains these two points.

2. There are two ways in which three points can be arranged. Describe these two ways.   (1) all are collinear   (2) two collinear, one off the line

3. Measure each angle and classify as acute, right, or obtuse.

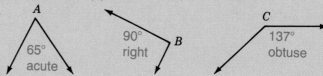

   A   65° acute   90° right B   C 137° obtuse

4. Complementary or supplementary?
   a. 36°, 144°   b. 72°, 18°   c. 150°, 30°   d. 90°, 90°
      supplementary   complementary   supplementary   supplementary

5. a. In the figure at the right, name two pairs of congruent vertical angles, alternate interior angles, and corresponding angles.   See TE Answer Section.
   b. If m∠1 = 83°, determine the measures of all labeled angles.

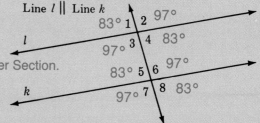

   Line l ∥ Line k
   83° 1   2 97°
   97° 3   4 83°
   83° 5   6 97°
   97° 7   8 83°

6. Construct a line perpendicular to line m and passing through point X.   Check students' constructions.

   •X   m

7. Name a special rhombus. Describe its properties.   square; all right angles

8. Classify each of the triangles in the picture according to its angles and according to its sides. Tell why you classified each triangle as you did.   See TE Answer Section.

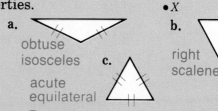

   a. obtuse isosceles   b. right scalene   c. acute equilateral

9. Which are perfect squares?   (a.) 49   b. 74   (c.) 196   d. 110

10. Approximate to the nearest tenth. Use the table on page 473.
    a. $\sqrt{59}$ 7.7   b. $\sqrt{27}$ 5.2   c. $\sqrt{91}$ 9.5   d. $\sqrt{8}$ 2.8

11. For each right triangle, find the missing length, to the nearest tenth. a and b are legs. c is the hypotenuse.
    a. $a = 14, b = 19$ $c \doteq 23.6$ b. $a = 5, c = 21$ $b \doteq 20.4$

12. Find the missing lengths.   a.

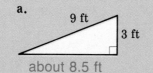

   9 ft   3 ft   about 8.5 ft

   b.

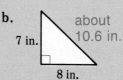

   7 in.   about 10.6 in.   8 in.

Coordinate Geometry

16

Dams built by the TVA help control flood waters along the Tennessee River.

Graphing on a Number Line

◇ **OBJECTIVES** ◇ — — ◇ **RECALL** ◇ —

To graph the integers between
two given integers

To graph equations like
$3x - 2 = 4$

To graph inequalities like
$x < 2$ and $-1 \le y$

Horizontal Number Line

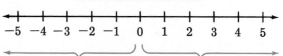

Negative numbers
go on forever in the
negative direction.

Positive numbers
go on forever in the
positive direction.

The integers between 2 and 6 are 3, 4, 5. 2 and 6 are
not included. There are no integers between 1 and 2.
1 and 2 are not included. The integers greater than
3 are 4, 5, 6, 7, ...

. . . means *goes on forever.*

Example 1

−2 and 4 are not included. Draw a
horizontal or vertical number line.
Use a heavy dot to graph each number.

Graph the integers between −2 and 4.
The integers between −2 and 4 are −1, 0, 1, 2, 3.

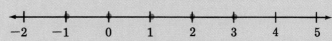

Example 2

2 and 3 are not included.

The graph is blank.

Graph the integers between 2 and 3.
There are no integers between 2 and 3.

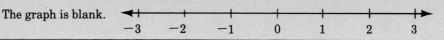

---

practice ▷   **Graph.**   Check students' graphs.

1. the integers between
   −3 and 2   −2, −1, 0, 1

2. the integers between
   −1 and 0   no solution

Example 3

First find the solution.
Add 2 to each side.

Divide each side by 6.

Place a heavy dot at 3.

Graph $6y - 2 = 16$.
$$6y - 2 = 16$$
$$6y - 2 + 2 = 16 + 2$$
$$6y = 18$$
$$\frac{6y}{6} = \frac{18}{6}$$
$$y = 3$$
So, the solution is 3.

To graph an equation means to graph its solution(s).

practice ▷ **3.** Graph $4x + 3 = -9$. −3
Check students' graphs.

4. Graph $a - 3 = a + 5$. no solution

Example 4

Read: x is less than 3.
Substitute numbers for x.
Many numbers make $x < 3$ true.

Graph $x < 3$.

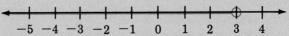

$$\begin{array}{c} \underline{x < 3} \\ 2 < 3 \\ \sqrt{2} < 3 \\ 0 < 3 \\ -1 < 3 \\ -2 < 3 \\ -2\frac{1}{2} < 3 \end{array}$$

All rational and irrational numbers less than 3 make $x < 3$ true.

All points that correspond to rational and irrational numbers less than 3 should be graphed. You can show this on a number line. We draw an arrow to show all such points.

The circle around 3 means 3 is not a solution.

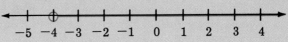

Example 5

Graph all points that correspond to numbers greater than −4.
−4 is not a solution.

Graph $y > -4$.

The rational numbers and the irrational numbers together form the *real numbers*. The graph of the real numbers is the real number line.

practice ▷ **5.** Graph $a > 2$.
all reals greater than 2   Check students' graphs.

6. Graph $x < -1$.   all reals less than −1

Example 6

Graph $x \leq 2$.
$x \leq 2$ is read x is less than 2 or x is equal to 2.

Any number that is less than 2 or equal to 2 makes $x \leq 2$ true. The graph includes 2 and all points in the negative direction from 2.

The dot at 2 means 2 is a solution.

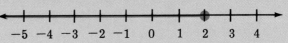

practice ▷   **7.**   Graph $y \leq 4$.   reals ≤ 4 **8.**   Graph $c \geq -2$.   reals ≥ -2
               Check students' graphs.

◇ ORAL EXERCISES ◇

Which symbol, $<$ or $>$, will make the inequality true?

1. $4 \underset{<}{\_\:?\:\_} 7$ **2.** $9 \underset{>}{\_\:?\:\_} 8$ **3.** $-1 \underset{>}{\_\:?\:\_} -3$ **4.** $-4 \underset{}{\_\:?\:\_} -2$ **5.** $0 \underset{<}{\_\:?\:\_} -3$
6. $-6 \underset{}{\_\:?\:\_} 2$ **7.** $3 \underset{}{\_\:?\:\_} -4$ **8.** $-2 \underset{}{\_\:?\:\_} 0$ **9.** $-1 \underset{}{\_\:?\:\_} 1$ **10.** $5 \underset{}{\_\:?\:\_} -5$

Read each inequality. (*Hint:* Read the variable first.) **Then replace**
x with a number to make the inequality true.   Answers may vary.

11. $x < 3$ **12.** $x > -1$ **13.** $x \leq -2$ **14.** $x \geq 4$ **15.** $x < -3$
16. $4 < x$ **17.** $-1 > x$ **18.** $2 \geq x$ **19.** $-3 \leq x$ **20.** $-6 \geq x$

◇ EXERCISES ◇

Graph.   Check students' graphs.

1.   the integers between 2 and 5   3, 4   **2.**   the integers between 0 and 6   1, 2, 3, 4, 5

3.   the integers between -5 and -1 **4.**   the integers between -4 and -2 -3
                        $-2, -3, -4$
5.   the integers between 3 and 4   no solution   **6.**   the integers between -2 and -1

7.   the integers between -3 and 4 $-2, -1, 0,$ **8.**   the integers between -6 and 2
                              1, 2, 3
★ **9.**   the positive even integers   2, 4, 6, . . . ★ **10.**   the positive odd integers   1, 3, 5, . . .

11. $4x - 5 = 3$ 2   **12.** $2y + 7 = 15$ 4   **13.** $3a - 4 = -10$ -2

14. $-5c + 6 = -4$ 2   **15.** $-6x - 9 = -3$ -1 **16.** $-x + 8 = 8$ 0

17. $7 = n + 3$ 4   **18.** $-4 = 3x - 13$ 3   **19.** $5 = -2z + 9$ -2

20. $x - 8 = x + 2$ no solution   **21.** $3r - 4 = 2r - 1$ 3   **22.** $2x + 6 = 2x - 4$

23. $4x - 1 = 2 + 5x$ -3 **24.** $2x - 1 = 5 + 3x$ -6 **25.** $3y - 7 = 4 + 3y$

26. $5a - 7 = 3a + 3$ 5   **27.** $4 - 5x = 6 - 5x$ no solution   **28.** $4c - 2 = 13 - c$ 3

★ **29.** $-2x + 2 = x - 8 - 3x$ ★ **30.** $5x + 7 + x = -5 + 6x$ no solution
      no solution

        6. no solution   **8.** $-5, -4, -3, -2, -1, 0, 1$ **22.** no solution   **25.** no solution

Graph.   Check students' graphs.

31. $x < 2$ reals < 2 **32.** $y > 3$ reals > 3 **33.** $a \leq -1$ reals ≤ -1 **34.** $c \geq 4$
35. $r \leq 0$ reals ≤ 0 **36.** $x \geq -3$ reals ≥ -3 **37.** $z < -2$ reals < -2 **38.** $n > -5$
39. $c > 1$ reals > 1 **40.** $y > 0$ reals > 0 **41.** $n \leq -4$ reals ≤ -4 **42.** $x > -1$
43. $-2 \geq y$ reals ≤ -2 **44.** $1 > r$ reals < 1 **45.** $a > -4$ reals > -4 **46.** $x \leq -2$

         34. reals ≥ 4 **38.** reals > -5 **42.** reals > -1 **46.** reals ≤ -2

Graph. (*Hint:* $\neq$ means is not equal to.)

★ **47.** $x \neq 2$ ★ **48.** $y \neq -3$ ★ **49.** $0 \neq n$ ★ **50.** $a \neq 1\frac{1}{2}$

   reals except 2   reals except -3 reals except 0   reals except $1\frac{1}{2}$

GRAPHING ON A NUMBER LINE   **395**

More on Graphing Inequalities

—◇ **OBJECTIVE** ◇—

To graph an inequality like
$4x - 9 > 2x + 5$

———— ◇ **RECALL** ◇ ————

Solve. $x - 3 \geq 4$

$$x - 3 \geq 4$$
$$x - 3 + 3 \geq 4 + 3$$
$$x + 0 \geq 7$$
$$x \geq 7$$

So, all real numbers ≥ 7 are solutions.

Solve. $3x < 15$

$$\frac{3x}{3} < \frac{15}{3}$$
$$1x < 5$$
$$x < 5$$

So, all real numbers < 5 are solutions.

You know how to graph an inequality like $x < 5$. To graph an inequality like $2x - 5 < 3$, first solve the inequality for x.

Example 1

Graph. $2x - 5 < 3$

Add 5 to each side.	$2x - 5 < 3$
The order is the same.	$2x - 5 + 5 < 3 + 5$ ← addition property for inequalities
Divide each side by 2.	$\frac{2x}{2} < \frac{8}{2}$ ← division property for inequalities
We divided by a positive number, so the order does not change.	$x < 4$

Any number less than 4 should make $2x - 5 < 3$ true.

Check two numbers that are less than 4.

Try 3.

$$\begin{array}{c|c} 2x - 5 & < 3 \\ \hline 2 \cdot 3 - 5 & 3 \\ 6 - 5 & \\ 1 & \\ & 1 < 3 \quad \text{True} \end{array}$$

Try -2.

$$\begin{array}{c|c} 2x - 5 & < 3 \\ \hline 2 \cdot (-2) - 5 & 3 \\ -4 - 5 & \\ -9 & \\ & -9 < 3 \quad \text{True} \end{array}$$

Check two numbers that are not less than 4.

Try 4.

$$\begin{array}{c|c} 2x - 5 & < 3 \\ \hline 2 \cdot 4 - 5 & 3 \\ 8 - 5 & \\ 3 & \\ & 3 < 3 \quad \text{False} \end{array}$$

Try 6.

$$\begin{array}{c|c} 2x - 5 & < 3 \\ \hline 2 \cdot 6 - 5 & 3 \\ 12 - 5 & \\ 7 & \\ & 7 < 3 \quad \text{False} \end{array}$$

Any number less than 4 makes $2x - 5 < 3$ true.

$$\begin{array}{ccccccccccc} \leftarrow\!\!+ & + & + & + & + & + & + & \oplus & + & + \!\!\rightarrow \\ -3 & -2 & -1 & 0 & 1 & 2 & 3 & 4 & 5 & 6 \end{array}$$

practice ▷ **Graph the inequality.** Check students' graphs.

1. $4x + 3 > -9$ reals > -3

2. $2y - 7 < -5$ reals < 1

Example 2

Graph.

$$-3a + 7 > 13$$

$$-3a + 7 > 13$$

Subtract 7 from each side.

$$-3a + 7 - 7 > 13 - 7$$

Same order

Divide each side by -3.

$$\frac{-3a}{-3} < \frac{6}{-3}$$

Reverse the order.

$$a < -2$$

Any number less than -2 makes $-3a + 7 > 13$ true.

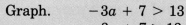

$$\begin{array}{ccccccccccc} + & + & + & \oplus & + & + & + & + & + & + & + \\ -5 & -4 & -3 & -2 & -1 & 0 & 1 & 2 & 3 & 4 \end{array}$$

practice ▷ **Graph the inequality.**   Check students' graphs.

3. $-2y - 7 > 3$ reals < -2 4. $-3a + 5 < 14$ reals > -3

Example 3

Graph.

$$5y - 6 \geq 4y - 1$$

$$5y - 6 \geq 4y - 1$$

Subtract $4y$ from each side.

$$5y - 4y - 6 \geq 4y - 4y - 1$$

Same order

$$y - 6 \geq -1$$

Add 6 to each side.

$$y - 6 + 6 \geq -1 + 6$$

Same order

$$y \geq 5$$

Any number greater than or equal to 5 makes $5y - 6 \geq 4y - 1$ true.

$$\begin{array}{cccccccccc} + & + & + & + & + & + & + & \bullet & + & + \\ -2 & -1 & 0 & 1 & 2 & 3 & 4 & 5 & 6 & 7 \end{array}$$

practice ▷ **Graph.**   Check students' graphs.

5. $4a + 1 \geq 3a - 2$ reals ≥ -5 6. $3x - 3 \leq 2x - 9$ reals ≤ -6

Example 4

Graph.

$$4x + 10 \geq 7x + 1$$

$$4x + 10 \geq 7x + 1$$

$$4x - 7x + 10 \geq 7x - 7x + 1$$

$$-3x + 10 \geq 1$$

$$-3x + 10 - 10 \geq 1 - 10$$

Divide each side by -3.

$$\frac{-3x}{-3} \leq \frac{-9}{-3}$$

Reverse the order.

$$x \leq 3$$

$$\begin{array}{cccccccccc} + & + & + & + & + & + & + & \bullet & + & + \\ -4 & -3 & -2 & -1 & 0 & 1 & 2 & 3 & 4 & 5 \end{array}$$

practice ▷ **Graph.**   Check students' graphs.

7. $2c - 7 \leq 6c + 13$ reals ≥ -5 8. $3y + 8 \geq 5y - 10$ reals ≤ 9

MORE ON GRAPHING INEQUALITIES

Graph. See TE Answer Section.

1. $2x - 5 > 5$
2. $3x + 7 < -2$
3. $4a + 6 \leq 10$
4. $5z - 10 > -20$
5. $-3y + 2 > -7$
6. $-6x - 8 \leq 10$
7. $2c - 3 \geq -3$
8. $-4a - 3 < -7$
9. $-x + 9 \leq 7$
10. $-c - 4 > 3$
11. $4 + 2n > 10$
12. $7 - 3x \leq -5$
13. $6 + x \geq 9$
14. $5 - r < 2$
15. $3z - 7 < 8$
16. $4a - 7 > -11$
17. $10y - 32 \leq 8$
18. $2 + 11c \geq 24$
19. $4x - 2 > 3x + 1$
20. $2y + 9 < y + 5$
21. $7a + 1 \leq 6a - 3$
22. $5x - 2 \geq 4x - 2$
23. $3c - 8 \leq 2c - 3$
24. $2a + 7 > a + 4$
25. $4x + 6 < x + 15$
26. $6y - 8 \geq 2y + 28$
27. $5a + 11 > 3a + 7$
28. $4x - 9 < 2x - 1$
29. $8n - 7 \leq 5n + 11$
30. $5z + 2 \geq 3z + 10$
31. $3y - 16 < 7y + 8$
32. $2a - 9 > 3a - 2$
33. $d - 3 \geq -4d + 22$
34. $8x - 13 \leq 21x - 26$
35. $-9y + 18 < 2y + 7$
36. $-4c + 22 > c - 3$
★ 37. $3x - 2x - 8 \geq 5x + 12$
★ 38. $7a + 2 + 4a < 3a - 14$
★ 39. $-8 - 6c > 2 + 8c - 4c$
★ 40. $-y \leq 7 - 8y + 4 - 4y$

★ 41. Five more than twice a number is less than 13. What numbers make this true?

★ 42. Four less than 3 times a number is greater than or equal to 11. What numbers make this true?

Calculator

Replace __?__ with $>$, $<$, or $=$. $(3 + 5)^2$ __?__ $3^2 + 5^2$ $>$

$(3 + 5)^2$	$3^2 + 5^2$
Press 3 ⊕ 5 ⊜	Press 3 ⊗ 3 ⊜
Display 8	Display 9 ← Write down and clear
Press ⊗ 8 ⊜	Press 5 ⊗ 5 ⊜ calculator.
Display 64	Display 25
	Press ⊕ 9 ⊜
	Display 34

So, $(3 + 5)^2 > 3^2 + 5^2$ since $64 > 34$.

Replace __?__ with $>$, $<$, or $=$.
1. $(1 + 1)^2$ __?__ $1^2 + 1^2$ $>$
2. $(7 + 9)^2$ __?__ $7^2 + 9^2$ $>$
3. $(8 + 15)^2$ __?__ $8^2 + 15^2$ $>$
4. $(31 + 24)^2$ __?__ $31^2 + 24^2$ $>$
5. For all whole numbers a and b, it appears that $(a + b)^2$ __?__ $a^2 + b^2$. $>$

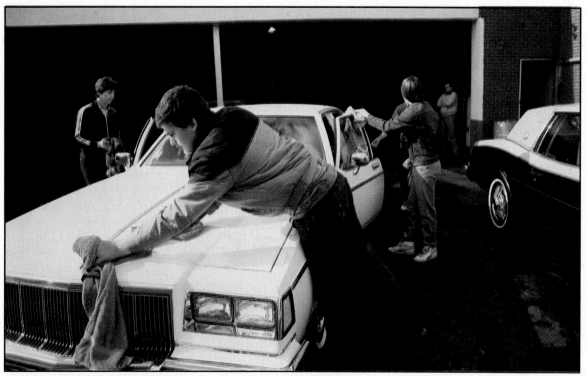

Problem Solving – Applications
Jobs for Teenagers

Example

Bill bought 12 gallons of gas at $1.15 per gallon, and he had a hot-wax treatment for $1.25. How much did he pay?

Identify what is to be found.

Bought: 12 gallons
Cost: $1.15 per gallon
Other cost: $1.25 for wax

Analyze the information.

Multiply 12 × $1.15 to find the cost of the gas. Add $1.25 for the wax.

Calculate.

$1.15   ← Multiply 12 × $1.15 to find the cost
× 12   of the gas.
─────
230
115
─────
$13.80
+ 1.25   ← Add $1.25 for the wax.
─────
$15.05   Total cost.

So, he paid $15.05.

practice ▷   Mary bought 9 gal of gas for $1.25/gal. What was her change from a 20-dollar bill? $8.75

Solve these problems.

1. Jim works at the Sparkle Car Wash from 3:00 P.M. until 6:00 P.M. Tuesday through Friday. On Saturday he works from 8:00 A.M. until 4:30 P.M. with a half-hour lunch break. He earns $3.75/h. How much does he earn per week? $75

2. Nan bought 15 gal of gas and had a polished wax job. If gas costs $1.14/gal, what was the total cost? $19.60

3. Mr. Stapleton had his car washed on Friday. He asked for a hot-wax treatment but bought no gas. How much did he pay? $4.75

4. Mrs. Schmidt had her car washed on a Thursday. She bought no gas but had a polished wax treatment. How much did Jim charge? $5.50

5. Ms. Helinski bought 15 gal of gas and had her car washed. If gas costs $1.29/gal, how much should Jim charge? $19.35

6. During the summer, Jan works full time at the Sparkle Car Wash. She earns $3.30 per hour plus time and a half for every hour over 40 hours a week. One week she worked 45 hours. How much did she earn? $156.75

7. In one hour on a Tuesday, 3 customers had their cars washed. One bought 5 gal of gas; another bought no gas but had a hot-wax treatment; the third bought 15 gal of gas. How much was taken in that hour if gas costs $1.20/gal? $30.75

8. Each of 5 customers bought 13 gal of gas and had their cars washed. How much money was taken in if gas costs $1.15/gal? $77.25

CAR WASH RATES

WITH NO GAS PURCHASE

TUES., WED. THURS.	FRI., SAT. SUN., HOLIDAYS
$3.00	$3.50

WITH GAS PURCHASE ANY DAY OF THE WEEK

NUMBER OF GALLONS OF GAS PURCHASED	COST OF CAR WASH
15	✱ FREE ✱
14	.25
13	.50
12	.75
11	1.00
10	1.25
9	1.50
8	1.75
7	2.00
6	2.25
5	2.50
4	2.75

- EXTRAS -

HOT WAX	$1.25
POLISHED WAX	$2.50

Graphing Points

To give the ordered pair for a
   point in a plane
To graph points in a plane

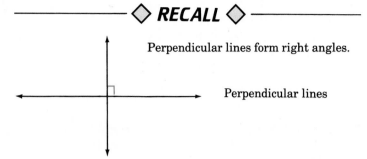

Perpendicular lines form right angles.

Perpendicular lines

Two perpendicular number lines determine a
coordinate plane. Each line is called an *axis*.

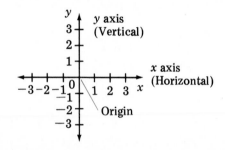

Example 1

Point *P* is in a coordinate plane.

The position of point *P* is described by
an ordered pair of numbers.

x-coordinate   *y*-coordinate

(1, 3)

Describe the location
of point *P*.

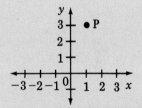

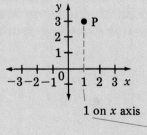

1 on *x* axis

3 on *y* axis

(1, 3)

The *x*-coordinate is always first.
The *y*-coordinate is always second.

So, the ordered pair (1, 3) describes point *P*.

**Give the ordered pair
for each point.**

1. *P* (−2, 3)   2. *Q* (−4, −2)
3. *R* (2, 0)   4. *S* (0, 0)
5. *T* (0, 2)   6. *U* (3, −2)

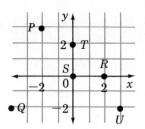

Example 2

Graph (3, 2). Then graph (2, −1).

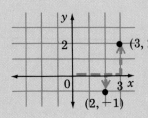

Start at the origin.
(3, 2)
Right 3   Up 2
(2, −1)
Right 2   Down 1

Example 3

Graph (−3, 1). Then graph (−4, −2).

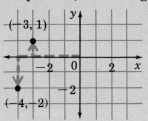

(−3, 1)
Left 3   Up 1
(−4, −2)
Left 4   Down 2

Summary

**To graph a point, the signs of the coordinates tell in
which directions to move from the origin.**

(3, 2)	(−3, 1)	(−4, −2)	(2, −1)
(+, +)	(−, +)	(−, −)	(+, −)
right up	left up	left down	right down

practice ▷   **Graph.**   Check students' graphs.

7. (1, 3)   8. (−2, 1)   9. (4, −2)   10. (−3, −3)

Example 4

Graph $(-3, 0)$.   Graph $(0, -1)$.

Start at the origin.

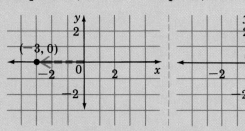

$(-3, 0)$ lies on the x axis.
$(0, -1)$ lies on the y axis.

$(-3, 0)$
left 3   neither up
nor down

$(0, -1)$
neither right   down 1
nor left

---

practice ▷   **Graph.**   Check students' graphs.

11. $(-4, 0)$ **12.** $(0, 2)$ **13.** $(1, 0)$ **14.** $(0, -3)$

◇ EXERCISES ◇

Give the ordered pair for each point.

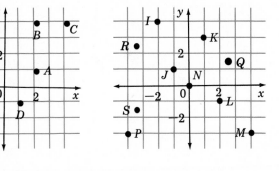

1. A (2, 1)   **2.** B (2, 4)   **3.** C (4, 4)

4. $D(1, -1)$ **5.** E (-2, -2)   **6.** F (-4, 3)

7. $G(-1, 2)$ **8.** H (-4, -2)   **9.** I (-2, 4)

10. $J(-1, 1)$ **11.** K (1, 3)   **12.** L (2, -1)

13. M (4, -3)   **14.** N (0, 0)   **15.** P (-4, -3)

★ **16.** Q $\left(2\frac{1}{2}, 1\frac{1}{2}\right)$ ★ **17.** R $\left(-3\frac{1}{2}, 2\frac{1}{2}\right)$ ★ **18.** S $\left(-3\frac{1}{2}, -1\frac{1}{2}\right)$

Graph each point. Use a different set of axes for every four
Check students' graphs.

19. $A(-1, 4)$ **20.** $B(-3, 0)$ **21.** $C(2, 3)$ **22.** $D(3, -2)$

23. $E(-2, -3)$ **24.** $F(-5, 1)$ **25.** $G(3, 0)$ **26.** $H(-2, -1)$

27. $I(0, 0)$ **28.** $J(4, -2)$ **29.** $K(-3, 3)$ **30.** $L(4, 1)$

31. $M(6, -3)$ **32.** $N(0, -1)$ **33.** $D(-4, -4)$ **34.** $Q(3, -5)$

★ **35.** $E\left(2\frac{1}{2}, 1\frac{1}{2}\right)$ ★ **36.** $F\left(-3\frac{1}{2}, 4\frac{1}{2}\right)$ ★ **37.** $G\left(-5\frac{1}{2}, 0\right)$ ★ **38.** $H\left(-1\frac{1}{2}, -4\frac{1}{2}\right)$

Horizontal and Vertical Lines

◇ **OBJECTIVE** ◇ ─── ─── ◇ **RECALL** ◇ ───

To determine whether a line is horizontal or vertical, given the coordinates of two points on the line

Give the coordinates of points A, B, C, and D.

$A(-2, 1)$
$B(3, 1)$

$C(2, 3)$
$D(2, -1)$

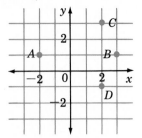

A *horizontal* line is parallel to the x-axis.
A *vertical* line is parallel to the y-axis.

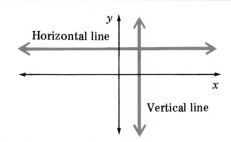

Horizontal line

Vertical line

Example 1

For the horizontal line, give the coordinates of points P, Q, and R. For the vertical line, give the coordinates of points S, T, and U.

$P(-2, 3)$, $Q(2, 3)$, $R(3, 3)$

same y-coordinate

$S(1, 2)$, $T(1, -2)$, $U(1, -3)$

same x-coordinate

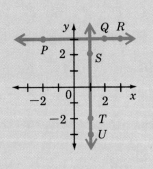

Example 1 suggests that

1. every point on a horizontal line has the same y-coordinate, and

2. every point on a vertical line has the same x-coordinate.

Example 2

Is $\overleftrightarrow{EF}$ a horizontal line or a vertical line?

$E(2, -1), F(2, 3)$	$E(-3, -4), F(1, -4)$
$E(2, -1) \quad\quad F(2, 3)$	$E(-3, -4) \quad\quad F(1, -4)$
same x-coordinate	same y-coordinate
So, $\overleftrightarrow{EF}$ is vertical.	So, $\overleftrightarrow{EF}$ is horizontal.

practice ▷ **Is $\overleftrightarrow{KL}$ a horizontal line or a vertical line?**

1. $K(2, -1), L(-3, -1)$ horizontal   **2.** $K(1, -4), L(1, 5)$ vertical

---

◇ EXERCISES ◇

Is $\overleftrightarrow{PQ}$ a horizontal line or a vertical line?

1. $P(2, 3), \quad Q(2, 1)$ vertical   **2.** $P(4, 3), \quad Q(1, 3)$ horizontal
3. $P(1, 4), \quad Q(1, -2)$ vertical   **4.** $P(-2, 1), Q(3, 1)$ horizontal
5. $P(1, -4), Q(-3, -4)$ horizontal   **6.** $P(4, 2), \quad Q(5, 2)$ horizontal
7. $P(3, -5), Q(3, -3)$ vertical   **8.** $P(-1, 6), Q(-1, -4)$ vertical
9. $P(5, -2), Q(-4, -2)$ horizontal   **10.** $P(-2, 5), Q(-2, -1)$ vertical

For what value of b will the line through R and S be horizontal or vertical as indicated?

★ **11.** $R(b, 3), \quad S(-3, -1)$; vertical -3 ★ **12.** $R(2, b), \quad S(5, -2)$; horizontal -2
★ **13.** $R(0, -1), S(1, b)$; horizontal -1 ★ **14.** $R(4, -4), S(b, 4)$; vertical 4

Challenge

Each pair of points is symmetrical with respect to the y-axis.
1. Graph and label four pairs of points that are symmetrical with respect to the x-axis.   Graphs will vary.
2. What is true of their x- and y-coordinates?

The x-coordinates are opposites; the y-coordinates are the same.

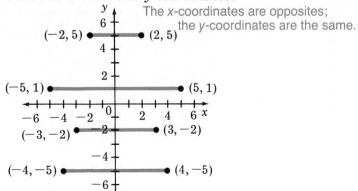

Graphing Equations and Inequalities

To graph an equation like
  $y = 2x - 1$
To graph an inequality like
  $y < -3x + 1$

Evaluate $3x - 5$ if x is 2; if x is -1.

$$3x - 5 \qquad\qquad 3x - 5$$
$$\downarrow \qquad\qquad\qquad \downarrow$$
$$3 \cdot 2 - 5 \qquad\qquad 3 \cdot (-1) - 5$$
$$6 - 5 \qquad\qquad\quad -3 - 5$$
$$1 \qquad\qquad\qquad\quad -8$$

Example 1

Find five ordered pairs that make the equation $y = x + 2$ true. Then graph the ordered pairs.

Select 5 values for x.	Evaluate $x + 2$ for each value of x.	For each value of x, $y = x + 2$.	Resulting ordered pairs
x	$x + 2$	y	(x, y)
0	$0 + 2$	2	$(0, 2)$
1	$1 + 2$	3	$(1, 3)$
2	$2 + 2$	4	$(2, 4)$
-1	$-1 + 2$	1	$(-1, 1)$
-2	$-2 + 2$	0	$(-2, 0)$

Graph the ordered pairs.

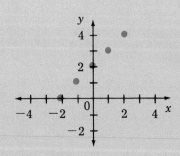

The five points appear to lie on a straight line.

---

practice ▷ **Graph five ordered pairs that make each equation true.**

1. $y = x - 1$ **2.** $y = -x + 3$ **3.** $y = -x - 2$

Answers may vary.

Example 2

Graph $y = -3x + 1$.

Find at least three ordered pairs that make the equation true.

x	$-3x + 1$	y	(x,y)
0	$-3 \cdot 0 + 1$	1	$(0, 1)$
2	$-3 \cdot 2 + 1$	-5	$(2, -5)$
-1	$-3 \cdot (-1) + 1$	4	$(-1, 4)$

Graph the ordered pairs.

All ordered pairs that make the equation true lie on the line.

Draw a straight line through them.

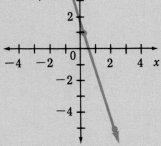

The graph of the line $y = -3x + 1$ is the *boundary line* for the graph of $y < -3x + 1$. Shading is used to show the region which is the graph. Shade *above* for $>$, *below for* $<$.

Example 3

Graph $y < -3x + 1$
Find ordered pairs that make $y < -3x + 1$ true.
Try $(-1, 4)$.   Try $(-3, -1)$.

Ordered pairs whose graphs lie on the *boundary line* of $y = -3x + 1$ *do not* make $y < -3x + 1$ true. Ordered pairs whose graphs lie below the line make $y < -3x + 1$ true.

y	$< -3x + 1$
4	$-3 \cdot (-1) + 1$
	$3 \qquad + 1$
	4

$4 < 4$ false

y	$< -3x + 1$
-1	$-3 \cdot (-3) + 1$
	$9 \qquad + 1$
	10

$-1 <$ 10 true

The dashed boundary line means the line is *not* part of the graph of $y < -3x + 1$.

Draw a dotted line for $y = -3x + 1$. Shade the region below the line.

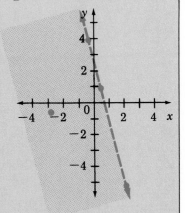

The shaded region is the graph of $y < -3x + 1$.

To *graph*

1. an equation like $y = -3x + 1$, graph ordered pairs that make the equation true and draw a straight line through the points.

2. an inequality like $y < -3x + 1$, find ordered pairs that make the inequality true and shade the region where the points lie.

◇ ORAL EXERCISES ◇

Evaluate for each value of x.

1.

x	$x + 3$
0	3
3	6
−1	2

2.

x	$-4x$
0	0
2	−8
−3	12

3.

x	$-x$
0	0
1	−1
−2	2

4.

x	$-x + 5$
0	5
2	3
−1	6

5.

x	$2x + 1$
0	1
1	3
−2	−3

6.

x	$3x - 2$
0	−2
2	4
−1	−5

7.

x	$-2x + 2$
0	2
3	−4
−2	6

8.

x	$-4x - 1$
0	−1
2	−9
−3	11

◇ EXERCISES ◇

Graph.   Check students' graphs.

1. $y = x + 1$
2. $y = x - 2$
3. $y = x + 3$
4. $y = x - 3$
5. $y = x$
6. $y = -x$
7. $y = -x + 2$
8. $y = -x - 1$
9. $y = 3x$
10. $y = -2x$
11. $y = 4x$
12. $y = -4x$
13. $y = 3x + 1$
14. $y = 2x + 1$
15. $y = -3x + 2$
16. $y = -2x + 2$
17. $y < 2x - 1$
18. $y < 3x - 2$
19. $y < 4x + 1$
20. $y < 5x - 3$
21. $y < 4x - 2$
22. $y < -5x + 2$
23. $y < -4x - 1$
24. $y < 5x + 4$
25. $y > -2x + 3$
26. $y > -4x + 3$
27. $y > 5x - 1$
28. $y > 6x - 5$
★ **29.** $x + y = 1$
★ **30.** $x - y = 3$
★ **31.** $4x + 2y = 6$
★ **32.** $6x - 3y = 9$
★ **33.** $y = \frac{1}{2}x$
★ **34.** $y = -\frac{1}{3}x$
★ **35.** $y = \frac{1}{2}x + 1$
★ **36.** $y = -\frac{1}{3}x - 2$
★ **37.** $y = \frac{2}{3}x$
★ **38.** $y = -\frac{4}{3}x$
★ **39.** $y = \frac{3}{4}x - 2$
★ **40.** $y = -\frac{5}{2}x + 1$

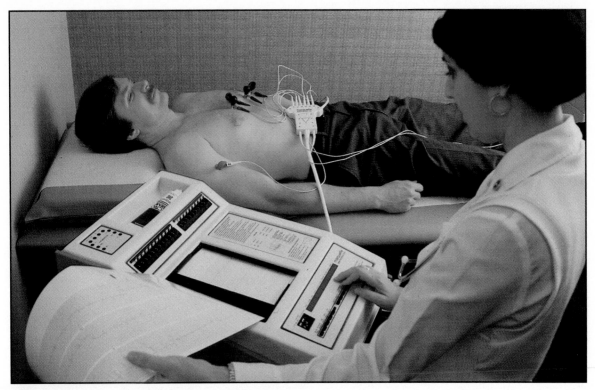

Problem Solving – Careers
Hospital Technicians

1. Ms. Ballin is an EKG technician. She operates equipment that monitors a patient's heart action. Her salary is $980 per month. What is her yearly salary? $11,760

2. Phyllis works 5 days a week. She gives 6 EKG's a day. How many EKG's does she give in 4 weeks? 120

3. Francis earns $210 a week as an EKG technician. He gets 2 weeks paid vacation. What is his yearly salary? $10,920

4. As a high school graduate, Mr. Sanders is entering a 3-month training program to become an EKG technician. As a trainee, his salary is $600 per month. After the training period, his salary will be $720 per month. How much more will he earn per year after training? $1,440

5. In 1974, there were about 10,000 EKG technicians working in the U.S.A. Due to increased reliance on EKG's in diagnosing heart conditions, the number of EKG technicians could increase by 25% by 1990. If it does, how many EKG technicians would be working in the U.S.A. by 1990? 12,500

Slope and Y-Intercept

To find the slope of a line, given
two points on the line
To find the slope and y-intercept
of a line, given its equation

Find three ordered pairs that make the equation
$y = 2x + 1$ true.

x	$2\ x + 1$	y	(x, y)
0	$2 \cdot 0 + 1$	1	$(0, 1)$
2	$2 \cdot 2 + 1$	5	$(2, 5)$
3	$2 \cdot 3 + 1$	7	$(3, 7)$

The steepness of a line is called the *slope* of the line.
The greater the steepness, the greater the slope.

$\overleftrightarrow{OA}$ is steeper than
$\overleftrightarrow{OB}$. So, the slope of
$\overleftrightarrow{OA}$ is greater
than the slope of $\overleftrightarrow{OB}$.

Example 1

Find the slope of $\overleftrightarrow{AB}$. Find the slope of $\overleftrightarrow{CD}$.

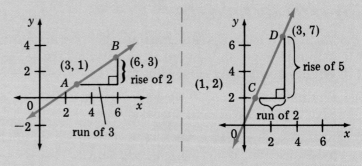

Rise is vertical distance.
Run is horizontal distance.

The *rise* is the difference of the
y-coordinates.
The *run* is the difference of the
x-coordinates.

Slope can be determined by dividing the amount of
rise by the amount of run.

y-coordinates

$A(3, 1), B(6, 3)$

x-coordinates

slope of $\overleftrightarrow{AB} = \dfrac{\text{rise}}{\text{run}}$

$= \dfrac{3 - 1}{6 - 3}$, or $\dfrac{2}{3}$

slope of $\overleftrightarrow{CD} = \dfrac{\text{rise}}{\text{run}}$

$= \dfrac{7 - 2}{3 - 1}$, or $\dfrac{5}{2}$

Slope of a non-vertical line $= \dfrac{\text{Difference of the } y\text{-coordinates}}{\text{Difference of the } x\text{-coordinates}}$

Example 2

Find the slope of $\overleftrightarrow{MN}$ for $M(4, 2)$ and $N(7, 3)$.

y-coordinates

$M(4, 2), N(7, 3)$

x-coordinates

$$\text{slope of } \overleftrightarrow{MN} = \frac{\text{Difference of the } y\text{-coordinates}}{\text{Difference of the } x\text{-coordinates}}$$

$$= \frac{3 - 2}{7 - 4}, \text{ or } \frac{1}{3}$$

practice ▷ **Find the slope of $\overleftrightarrow{AB}$ for the given pairs of points.**

1. $A(3, 1), B(6, 5)$ $\frac{4}{3}$ 2. $A(5, 7), B(3, 9)$ -1 3. $A(4, -3), B(2, -1)$ -1

The slope of a line can be determined directly from the equation of the line.

Example 3

Graph $y = 2x + 1$. Determine the slope of the line and the y-coordinate of the point of intersection of the line with the y-axis.

x	$2x + 1$	y	(x, y)
1	$2 \cdot 1 + 1$	3	$(1, 3)$
2	$2 \cdot 2 + 1$	5	$(2, 5)$
3	$2 \cdot 3 + 1$	7	$(3, 7)$

Slope $= \dfrac{7 - 3}{3 - 1} = \dfrac{4}{2}$, or 2.

The line intersects the y-axis at $(0, 1)$.

↑ *y intercept*

Graph the ordered pairs.

So, the slope is 2 and the y-intercept is 1.

The *y-intercept* is the y-coordinate of the point of intersection of a line with the y-axis.

Example 4

Graph $y = \frac{2}{3}x - 2$. Determine the slope and the y-intercept.

The line intersects the y-axis at $(0, -2)$.

x	$\frac{2}{3}x - 2$	y	(x, y)
0	$\frac{2}{3} \cdot 0 - 2$	-2	$(0, -2)$
3	$\frac{2}{3} \cdot 3 - 2$	0	$(3, 0)$
6	$\frac{2}{3} \cdot 6 - 2$	2	$(6, 2)$

slope $= \dfrac{-2 - 2}{0 - 6} = \dfrac{-4}{-6}$, or $\dfrac{2}{3}$.

So, the slope is $\frac{2}{3}$ and the y-intercept is -2.

Slope-intercept form of an equation $y = mx + b$ is an equation of a line.
m is the slope; b is the y-intercept.

Example 5

Find the slope and the y-intercept of the line whose equation is $y = -\frac{4}{5}x - 3$.

$$y = -\frac{4}{5}x - 3$$

$$y = mx + b$$

So, the slope is $-\frac{4}{5}$ and the y-intercept is -3.

---

practice ▷   **Give the slope and the y-intercept.**

 4. $y = 5x + 3$ 5; 3   **5.** $y = \frac{1}{3}x - 2$ $\frac{1}{3}$; -2 **6.** $y = -\frac{5}{7}x + 8$ $-\frac{5}{7}$;

◇ EXERCISES ◇

Find the slope of $\overleftrightarrow{AB}$ for the given pairs of points.

1. $A(4, 1), B(6, 3)$ 1   **2.** $A(2, 2), B(6, 5)$ $\frac{3}{4}$ **3.** $A(-1, 6), B(5, 9)$ $\frac{1}{2}$

4. $A(4, 3), B(-6, 4)$ $-\frac{1}{10}$ **5.** $A(7, 3), B(2, -8)$ $\frac{11}{5}$ **6.** $A(1, -6) B(7, 4)$ $\frac{5}{3}$

7. $A(5, 1), B(8, -3)$ $-\frac{4}{3}$ **8.** $A(4, 1), B(-6, 2)$ $-\frac{1}{10}$ **9.** $A(6, -2), B(7, -3)$ -1

Find the slope and the y-intercept.

10. $y = \frac{3}{2}x + 5$ $\frac{3}{2}$; 5   **11.** $y = \frac{1}{2}x - 4$ $\frac{1}{2}$; -4 **12.** $y = -\frac{2}{3}x + 1$ $-\frac{2}{3}$; 1

13. $y = \frac{4}{5}x + 7$ $\frac{4}{5}$; 7   **14.** $y = \frac{2}{5}x - 2$ $\frac{2}{5}$; -2 **15.** $y = -\frac{1}{3}x + 5$ $-\frac{1}{3}$; 5

The slope of a non-vertical line can be written as $\dfrac{y_2 - y_1}{x_2 - x_1}$, where y_2 and y_1 are the y-coordinates and x_2 and x_1 are the x-coordinates of two points on the line. That is, the two points on the line are (x_1, y_1) and (x_2, y_2).

Recall that for a horizontal line, the y-coordinates of its points are the same.

16. What is $\dfrac{y_2 - y_1}{x_2 - x_1}$ equal to if $y_1 = y_2$?   0   **17.** What is the slope of a horizontal line? 0

Recall that for a vertical line the x-coordinates are the same.

18. Does $\dfrac{y_2 - y_1}{x_2 - x_1}$ have a value if $x_1 = x_2$? Why or why not?   No, because the denominator is 0.

19. Why does the slope of a vertical line not exist?   Because $\dfrac{y_2 - y_1}{x_2 - x_1}$ does not exist for $x_1 = x_2$.

Find the slope and the y-intercept.

★ **20.** $2y - 3x = 8$ $\frac{3}{2}$; 4   ★ **21.** $x - 3y - 7 = 0$ $\frac{1}{3}$; $-2\frac{1}{3}$ ★ **22.** $4x + 2y = 6$ -2; 3

★ **23.** $3y + 5x = 6$ $-1\frac{2}{3}$; 2   ★ **24.** $2y - x = 6$ $\frac{1}{2}$; 3   ★ **25.** $5x - 3y = 5$ $1\frac{2}{3}$; $-1\frac{2}{3}$

      CHAPTER SIXTEEN

Solving Systems of Equations

──◇ **OBJECTIVES** ◇──   ──────◇ **RECALL** ◇──────

To determine whether
coordinates of a point satisfy
an equation

To solve a system of two
equations in two variables by
graphing

Graph $y = 2x - 1$.

x	$2x - 1$	y	(x, y)
0	$2 \cdot 0 - 1$	-1	$(0, -1)$
1	$2 \cdot 1 - 1$	1	$(1, 1)$
2	$2 \cdot 2 - 1$	3	$(2, 3)$

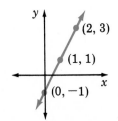

$y = -x + 3$		$y = 3x - 1$	
2	$-(1) + 3$	2	$3 \cdot 1 - 1$
	$-1 + 3$		$3 - 1$
	2		2
$2 = 2$	true	$2 = 2$	true

The graphs of $y = -x + 3$ and
$y = 3x - 1$ are shown. Point
$P(1, 2)$ lies on both lines. It is the
point of intersection of the two
lines. The coordinates of P satisfy
both equations.

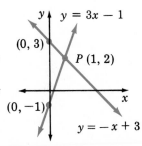

Example 1

Determine whether the coordinates of point $P(2, 3)$
satisfy both equations.
$$y = 3x - 3$$
$$y = -2x + 7$$
Substitute 2 for x and 3 for y in each equation, and
determine if the result is true.

$y = 3x - 3$		$y = -2x + 7$	
3	$3 \cdot 2 - 3$	3	$-2 \cdot 2 + 7$
	$6 - 3$		$-4 + 7$
	3		3
$3 = 3$	true	$3 = 3$	true

So, $(2, 3)$ satisfies both equations.

The coordinates of $P(2, 3)$ are solutions of both
equations. So, $(2, 3)$ is the solution of the system of
equations $y = 3x - 3$ and $y = -2x + 7$

practice ▷ **Determine whether the coordinates of $P(1, 2)$ satisfy both
equations.**

1. $y = -2x + 4$ yes
$y = 3x - 1$

2. $y = -x + 3$ no
$y = 2x - 1$

Example 2

Solve the system by graphing.
$$y = 2x + 5$$
$$y = -2x + 1$$

Graph each equation on the same set of axes.
Find at least three ordered
pairs that satisfy each equation.

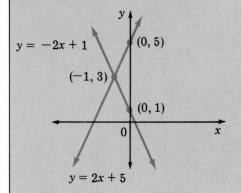

$y = -2x + 1$ (0, 5)

(−1, 3)

(0, 1)

0

$y = 2x + 5$

x	$2x + 5$	y	(x, y)
0	$2 \cdot 0 + 5$	5	(0, 5)
1	$2 \cdot 1 + 5$	7	(1, 7)
−1	$2 \cdot (-1) + 5$	3	(−1, 3)

x	$-2x + 1$	y	(x, y)
0	$-2 \cdot 0 + 1$	1	(0, 1)
1	$-2 \cdot 1 + 1$	−1	(1, −1)
2	$-2 \cdot 2 + 1$	−3	(2, −3)

The two lines intersect at $(-1, 3)$.
So, $(-1, 3)$ is the solution.

solving a system of equations by graphing
1. Graph each equation on the same set of axes.
2. Locate the coordinates of the point of intersection of the two lines.

practice ▷ **Solve each system by graphing.**   Check students' graphs.

3. $y = -2x + 4$ (1, 2)
   $y = 3x - 1$

4. $y = 2x - 5$ (2, −1)
   $y = -3x + 5$

◇ EXERCISES ◇

Determine whether the coordinates of $P(-1, 2)$ satisfy both equations.

1. $y = -x + 1$ yes
   $y = 2x + 4$

2. $y = 3x + 5$ no
   $y = -2x + 3$

3. $y = 2x + 4$ yes
   $y = -3x - 1$

4. $y = 3x - 7$ no
   $y = -x + 1$

5. $y = 4x$ no
   $y = 2x$

6. $y = -6x - 4$ yes
   $y = 9x + 11$

Solve each system by graphing.   Check students' graphs.

7. $y = -2x + 3$ (0, 3)
   $y = x + 3$

8. $y = x + 3$ (−1, 2)
   $y = 2x + 4$

9. $y = 3x - 5$ (2, 1)
   $y = -x + 3$

10. $y = 4x + 4$ (−2, −4)
    $y = 3x + 2$

11. $y = -2x$ (1, −2)
    $y = 3x - 5$

12. $y = -x + 3$ (4, −
    $y = -2x + 7$

COMPUTER ACTIVITIES

Recall that the slope of a line equals $\dfrac{\text{Difference of } y\text{-coordinates}}{\text{Difference of } x\text{-coordinates}}$.

You can write a program to find the slope of a line given the coordinates of two points. Before you can write the program you must know the formula for finding the slope.

Given the coordinates of any two points (x_1, y_1) and (x_2, y_2)

$$\text{slope} = \frac{y_2 - y_1}{x_2 - x_1}.$$ The formula in **BASIC:** S = (Y2 − Y1)/(X2 − X1).

Example Write a program to find the slope of a line given the coordinates of two points. Allow for three sets of data.

```
10  FOR I = 1 TO 3
20  INPUT "TYPE COORD OF FIRST POINT X1
    AND Y1 ";X1,Y1
30  INPUT "TYPE COORD OF SECOND POINT X2
    AND Y2 ";X2,Y2
40  LET S = (Y2 - Y1) / (X2 - X1)
50  PRINT "SLOPE IS ";S
60  NEXT I
70  END
```

This program will find the slope of any line that is NOT vertical. Consider the line through the points A(3, 4) and B(3, 11). This will be a vertical line; the x-coordinates are the same, 4.

The slope is $\dfrac{y_2 - y_1}{x_2 - x_1} = \dfrac{11 - 4}{3 - 3} = \dfrac{7}{0}$, which is *undefined*.

You can alter the program so that it will recognize vertical lines.
See the Computer Section beginning on page 420 for more information.

Exercises

1. Type and **RUN** the program above to find the slope of the line through the points (2, 5), (3, 7); (4, 6), (8, 8); (6, 1), (14, 7). $\quad 2, \frac{1}{2}, \frac{3}{4}$

2. **RUN** the program for the following three sets of points: (2, 4), (7, 9); (4, 5), (4, 12); (9, 1), (7,0).   1, Error message on second run. The slope is $\frac{7}{0}$, which is undefined. Why does the computer come up with an error message on the second **RUN**?

3. Use an **IF-THEN** statement to modify the program to avoid the error message above. (*Hint:* (Y2 − Y1)/(X2 − X1) is undefined if the denominator X2 − X1 is?)
   Add line 35 **IF X2 − X1 = 0 THEN PRINT "THIS IS A VERTICAL LINE."**

COMPUTER ACTIVITIES   **415**

Chapter Review

Graph. [393]

1. the integers between -2 and 3

2. $-2x - 8 = -6$
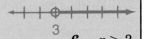

Graph. [393]

3. $x < -3$ **4.** $y \geq -1$ **5.** $c \leq 2$ **6.** $r > 3$

Graph. [396]

7. $2x - 3 > 7$ **8.** $-4y + 7 \leq -9$

9. $5a - 2 < 4a - 7$ **10.** $3c + 5 \geq 11 + 5c$

Give the ordered pair for each point. [401]

11. A $(2, 3)$ **12.** B $(0, -3)$
13. C $(-4, -4)$ **14.** D $(-2, 4)$
15. E $(4, 0)$ **16.** F $(3, -3)$

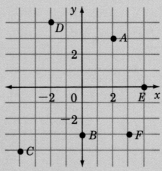

Graph each point. [401]   Check students' graphs.

17. $G(4, -1)$ **18.** $H(-3, -3)$
19. $J(-2, 0)$ **20.** $K(-4, 2)$

Is $\overleftrightarrow{RS}$ a horizontal line or a vertical line? [404]

21. $R(3, 1), S(3, -2)$ vertical   **22.** $R(5, -2), S(-1, -2)$ horizontal
23. $R(4, -4), S(3, -4)$ horizontal   **24.** $R(1, 0), S(-1, 0)$ horizontal

Graph. [406]   Check students' graphs.

25. $y = x + 2$ **26.** $y = -x + 1$
27. $y = 3x$ **28.** $y < -2x - 1$

Find the slope of $\overleftrightarrow{AB}$ for the given pairs of points. [410]

29. $A(3, 2), B(5, 3)$ $\frac{1}{2}$ **30.** $A(2, -3), B(6, 7)$ $\frac{5}{2}$

Find the slope and y-intercept. [410]   **Solve the system by graphing.** [413]

31. $y = \frac{2}{3}x + 5$ $\frac{2}{3}; 5$ **33.** $y = 2x - 5$ $(3, 1)$
                                                     $y = -2x + 7$

32. $y = -\frac{5}{6}x - 3$ $-\frac{5}{6}; -3$

Chapter Test

Graph.

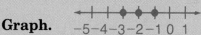

1. the integers between 0 and -4

2. $-3 = 3x - 12$

Graph.

3. $y > -2$

5. $c \geq 4$

4. $x \leq 0$

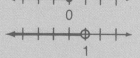

6. $m < 1$

Graph.

7. $3x - 6 > 9$

8. $-y + 2 \leq 7$

9. $4r - 2 \geq 5r + 1$

Give the ordered pair for each point.

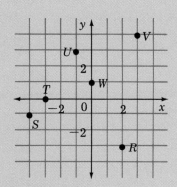

10. R $(2, -3)$ 11. S $(-4, -1)$

12. T $(-3, 0)$ 13. U $(-1, 3)$

14. V $(3, 4)$ 15. W $(0, 1)$

Graph each point. Check students' graphs.

16. $L(2, 4)$ 17. $M(0, -3)$

18. $N(-2, -3)$ 19. $P(-3, 2)$

Is $\overleftrightarrow{JK}$ a horizontal line or a vertical line?

20. $J(-1, 4), K(3, 4)$ horizontal

21. $J(-2, -1), K(-2, 0)$ vertical

Graph. Check students' graphs.

22. $y = x - 1$

23. $y = -2x$

24. $y < 3x + 1$

Find the slope of $\overleftrightarrow{AB}$ for the given points.

25. $A(3, 5), B(6, 9)$ $\frac{4}{3}$

26. $A(-1, 4), B(2, -7)$ $-\frac{11}{3}$

Find the slope and y-intercept.

27. $y = \frac{4}{5}x + 2$ $\frac{4}{5}; 2$

Solve the system by graphing.

28. $y = -x + 3$ $(1, 2)$
$\quad\ \ y = 3x - 1$

Cumulative Review

Perform the indicated operation.

1. $-22 + 17$ -5 2. $12 - (-16)$ 28 3. $-13(-6 + 9)$ -39 4. $-\dfrac{3}{4} \div \left(-\dfrac{5}{8}\right)$ $1\frac{1}{5}$

Simplify.

5. $3x^2 - 2 + 2x - 5x^2 - 7 - 5x$ $-2x^2 - 3x - 9$ 6. $(3a^2) \cdot (-4a^4)$ $-12a^6$ 7. $(a^2)^3 \cdot (-2a)^2$ $4a^8$

Write in mathematical terms.

8. 3 less than 3 times a number $3n - 3$ 9. three times $-x$ $-3x$ 10. r divided by -6 $\dfrac{r}{-6}$

Factor.

11. $-2x^3 - 6x^2 + 6x$ $-2x(x^2 + 3x - 3)$

Solve.

12. $-4x + 6 = -2$ 2 13. $-5x \geq 15$ $x \leq -3$

Solve the system by graphing.

14. $y = -x + 6$ Check students'
    $y = 5x - 6$ graphs. (2, 4)

Graph on a number line.

15. $x > -3$ reals > -3 16. $x \leq 4$ reals ≤ 4

Evaluate.

17. $(-3x)^2$ if $x = \dfrac{1}{9}$ $\frac{1}{9}$

18. $-5x^2y^2$ if $x = 2$ and $y = -\dfrac{1}{4}$ $-1\frac{1}{4}$

19. In a right triangle, one leg is 3 in. long and the hypotenuse is 7 in. long. How long is the other leg? $\sqrt{40}$ in.

20. In a triangle, two sides are 8 in. long each and the third side is 3 in. long. What kind of a triangle is it? isosceles

21. Name three kinds of angles. right, acute, obtuse, straight, reflex

22. m$\angle A = 25°$. What is its complement? Its supplement? 65°, 155°

23. What is the slope of $\overleftrightarrow{XY}$ with $X(1, 4)$ and $Y(5, 7)$? $\frac{3}{4}$

24. $A = \dfrac{bh}{2}$. Find A if $b = 7$ in. and $h = 9$ in. $31\frac{1}{2}$ in.²

25. The base of an isosceles triangle is 7 cm and the perimeter is 19 cm. Find the length of each leg. 6 cm

Solve these problems.

26. Martha bought 3 items at these prices: $3.78, $2.45, and $6.15. How much change will she receive from a $20 bill? $7.62

27. You worked $19\frac{3}{4}$ h Monday through Thursday. How many more hours must you work on Friday to work 26 h altogether? $6\frac{1}{4}$ h

Appendix

DISTANCE

OBJECTIVE

To use the distance formula

A bullet train averages 160 mi/h. To find the distance it covers in a given number of hours, use multiplication.

rate r	time t	distance d
160 mi/h	2 h	160 × 2 or 320 mi
160 mi/h	3 h	160 × 3 or 480 mi
160 mi/h	4 h	160 × 4 or ___?___
160 mi/h	8 h	___?___ or ___?___

Formula: $d = r \cdot t$

distance   rate   time

The formula $d = r \cdot t$ can be used to determine any one of the three quantities when the other two are known.

d = distance r = rate t = time

Example 1 The Adams family averaged 47 mi/h for a trip that lasted 12 h. How many miles did they travel?

$$d = r \cdot t$$
$$= 47 \cdot 12, \text{ or } 564 \text{ mi.}$$

So, the Adams family traveled 564 mi.

Practice **Use the distance formula $d = r \cdot t$.**

1. Find d if $r = 16$ mi/h and $t = 15$ h.   240 mi

2. A Boeing 757 jet averages 630 mi/h. What distance will it travel in $6\frac{1}{2}$ h?   4,095 mi

To solve the formula $d = rt$ for r, divide each side by t:

$$d = rt \qquad \frac{d}{t} = \frac{rt}{t} \qquad \frac{d}{t} = r$$

$$\text{rate} = \frac{\text{distance}}{\text{time}} \qquad r = \frac{d}{t}$$

Example 2   A boat traveled 18 mi in $1\frac{1}{2}$ h. What was the average speed of the boat?

$$r = \frac{d}{t} \qquad r = \frac{18}{1\frac{1}{2}}, \quad \text{or} \quad 12$$

So, the boat averaged 12 mi/h.

Practice   Use the formula $r = \frac{d}{t}$.

3. Find r if d = 220 mi and t = 3.5 h.   about 62.9 mi/h

4. A bus traveled 315 mi in 7 hours. What was the average speed (rate) of the bus?   45 mi/h

To solve the formula $d = rt$ for t, divide each side by r:

$$d = rt \qquad \frac{d}{r} = \frac{rt}{r} \qquad \frac{d}{r} = t$$

$$\text{time} = \frac{\text{distance}}{\text{rate}} \qquad t = \frac{d}{r}$$

Example 3   Clara bicycled for 13.5 mi averaging 9 mi/h. How long did it take her?

$$t = \frac{d}{r} \qquad t = \frac{13.5}{9}, \quad \text{or} \quad 1.5$$

So, it took Clara 1.5 h.

Practice   5. Find t if d = 350 mi and r = 50 mi/h.   7 h

6. Maria traveled 376 mi averaging 47 mi/h. How long did it take Maria to cover this distance?   8 h

Exercises

Use the distance formula.

1. Find d if r = 54 mi/h and t = $6\frac{1}{2}$ h.   351 mi

2. Find t if d = 375 mi and r = 25 mi/h.   15 h

3. Find r if d = 3,920 mi and t = 5.6 h.   700 mi/h

4. Pat drove from her town to the next town in 20 min. She drove at the rate of 36 mi/h. What distance did she drive?   12 mi

5. A jet traveled 2,170 mi in $3\frac{1}{2}$ h. What was the average rate of the jet?   620 mi/h

DISTANCE   **421**

PERIMETER

OBJECTIVE

To use perimeter formulas

A block in the shape of a rectangle
has the dimensions shown in the
picture at the right. What is the distance around
this block?

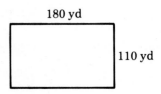

180 + 110 + 180 + 110 = 580

So, the distance around the block is 580 yd.

The distance around a polygon is called the *perimeter* of the polygon. The
perimeter of a polygon is found by adding the lengths of its sides.

Example 1   Find the perimeter of the equilateral
triangle ABC.

Each side of an equilateral triangle is
the same length.
So, $P = 3s$, where s is the length of
each side.

$$P = 2\frac{1}{2} \times 3$$

$$= 7\frac{1}{2}$$

So, the perimeter of an equilateral

triangle with each side $2\frac{1}{2}$ in. long is $7\frac{1}{2}$ in.

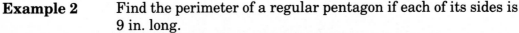

Example 2   Find the perimeter of a regular pentagon if each of its sides is
9 in. long.
$P = 5s$, where s is the length of
each side. Each side of a regular
polygon is the same length.
$P = 5 \cdot 9$, or 45 in.
So, the perimeter of a regular
pentagon with each side 9 in. long is
45 in.

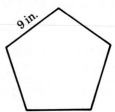

Practice

Find the perimeter of a square in which each side is $9\frac{1}{2}$ in. long. 38 in.

To derive the formula for the perimeter of a rectangle, add the lengths of its sides and simplify.

$$P = l + w + l + w$$
$$= (l + l) + (w + w)$$
$$= 2l + 2w, \text{ or } 2(l + w)$$

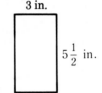

So, the perimeter of a rectangle is twice the sum of its length and width.

Example 3

Find the perimeter of the rectangle shown at the right. Use the formula $P = 2(l + w)$.

3 in.

$$P = 2(l + w)$$
$$= 2\left(5\frac{1}{2} + 3\right)$$
$$= 2 \times 8\frac{1}{2}, \text{ or } 17$$

$5\frac{1}{2}$ in.

So, the perimeter of the rectangle is 17 in.

Example 4

Find the perimeter of the right triangle shown.

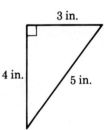

To find the perimeter, add the lengths of the sides.
$$P = 3 + 4 + 5, \text{ or } 12$$

So, the perimeter of the triangle is 12 in.

Exercises

Find the perimeters.

1. 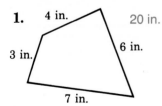 20 in.
4 in.
3 in.
6 in.
7 in.

2. 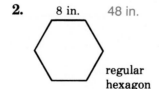 48 in.
8 in.
regular hexagon

3. 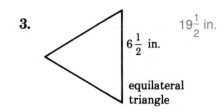 $19\frac{1}{2}$ in.
$6\frac{1}{2}$ in.
equilateral triangle

4. 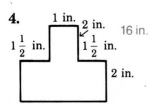 16 in.
1 in.
2 in.
$1\frac{1}{2}$ in.
$1\frac{1}{2}$ in.
2 in.

5. 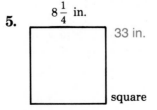 33 in.
$8\frac{1}{4}$ in.
square

6. 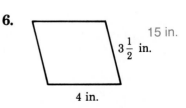 15 in.
$3\frac{1}{2}$ in.
4 in.

CIRCUMFERENCE

OBJECTIVES

To find the circumference of a circle
To find the radius or diameter of a circle given its circumference

A wheel with a diameter of 2 ft is rolled for one complete revolution. What distance will it cover?

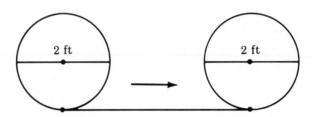

Performing this experiment will show that the wheel traveled about 6.3 ft.

The distance around the circle is called *circumference*.

So, the circumference of a circle with a diameter 2 ft long is about 6.3 ft.

Example 1   Perform the following experiment. Get three cylindrical cans of different sizes.

Step 1. Measure the diameter of each can.
Step 2. Measure the distance around (circumference).
Step 3. Divide the circumference by the length of a diameter; round the quotient to one decimal place.

Here are the results of one such experiment:

Diameter d	Circumference C	Quotient $\dfrac{c}{d}$
3 in.	9.4 in.	3.1
6 in.	18.8 in.	3.1
7 in.	22.0 in.	3.1

← Same quotient in each case

The quotient of the circumference and the length of a diameter in any circle is the same.

This quotient is the number π, which is approximately 3.14.

$$\pi \doteq 3.14$$

⌐ is approximately equal to

Since $\frac{c}{d} = \pi$, it follows that $C = \pi d$.

The circumference of any circle is equal to the product of π (about 3.14) and the length of a diameter.

$$\frac{\text{circumference}}{\text{diameter}} = \frac{C}{d} = \pi \qquad \pi \doteq 3.14$$

Example 2 Find the circumference. State the answer in terms of π.

$d = 3$ in. $r = 5$ in.
$C = \pi \cdot d$ $C = 2\pi r \qquad d = 2r$
  $= \pi \cdot 3$ $= 2 \cdot \pi \cdot 5$
  $= 3\pi$ in. $= 10\pi$ in.

Example 3 Find the length of a diameter of a circle whose circumference is 21 in. State the answer in terms of π.

$$C = \pi d, \text{ so } d = \frac{C}{\pi}.$$

$$d = \frac{21}{\pi} \text{ in.}$$

Practice **Find the circumference. State the answer in terms of π.**

1. $d = 5$ in. 5π in.   2. $r = 12$ in. 24π in.

3. Find the length of a diameter of a circle whose circumference is 34 in. State the answer in terms of π. $\frac{34}{\pi}$ in.

Example 4 Find the length of a radius of a circle whose circumference is 30 in. State the answer in terms of π.

$$C = 2\pi r, \text{ so } r = \frac{C}{2\pi}$$

$$r = \frac{30}{2\pi} \text{ in.} \quad \text{or} \quad \frac{15}{\pi} \text{ in.}$$

Example 5 Find the circumference of a circle to two decimal places if its radius is 8 in. long. Use 3.14 for π.

$$C = 2\pi r$$
$$\doteq 2 \times 3.14 \times 8$$
$$\doteq 50.24 \text{ in.}$$

CIRCUMFERENCE

Practice

1. Find the length of a radius of a circle whose circumference is 45 in. State the answer in terms of π. $\frac{45}{2\pi}$ in.

2. Find the circumference of a circle to two decimal places if its radius is 12 in. long. Use 3.14 for π.   75.36 in.

Exercises

Find the circumference. State the answer in terms of π.

1.
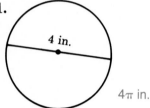
4 in.

4π in.

2.
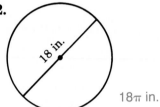
18 in.

18π in.

3.
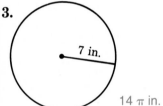
7 in.

14π in.

4.
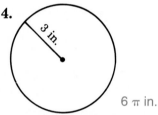
3 in.

6π in.

5. $d = 21$ in.
21π in.

6. $d = 32$ in.
32π in.

7. $r = 7$ ft
14π ft

8. $r = 9$ ft
18π ft

Solve.

9. A wheel has a diameter 2 ft long. How far will the wheel travel when it makes 3 complete turns?   about 19 ft

10. A wheel has a radius 27 in. long. How far will the wheel travel when it makes 10 complete turns?   about 141 ft

Find the perimeters to two decimal places. Use 3.14 for π.

11.
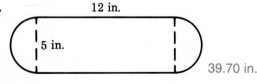
12 in.

5 in.

39.70 in.

12.
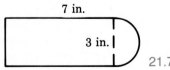
7 in.

3 in.

21.71 in.

GEOMETRY SUPPLEMENT

AREA

Objectives

To find the area of a square and a rectangle
To find the area of a parallelogram, a triangle, a trapezoid, and a circle

3 in.

Area is a number that tells how many square units are contained in a geometric figure. The area of this square is $3 \cdot 3$ or 9 in.^{2}

If the length of a side of a square is s, then the formula for the area of the square is $A = s^2$.

Example 1   Find the area of this rectangle.
The formula for the area of a rectangle is $A = lw$.

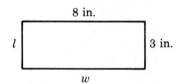
8 in.
l 3 in.
w

$A = lw$
$\quad = 8 \cdot 3$, or 24 in.^{2}

Example 2   Find the area of this parallelogram.
The formula for the area of a parallelogram is $A = bh$.

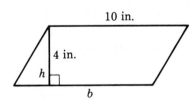
10 in.
4 in.
h
b

$A = bh$
$\quad = 10 \cdot 4$, or 40 in.^{2}

Practice   **Find the areas.**

1. square: $s = 12$ in.   144 in.^{2} 2.   rectangle: $l = 11$ in.; $w = 9$ in.
   99 in.^{2}
3. parallelogram: $b = 15$ in.; $h = 12$ in.   180 in.^{2}

A parallelogram can be divided into two congruent triangles by one of its diagonals. Therefore, the area of each triangle is onehalf of the area of the parallelogram. The height of the parallelogram is also the height (altitude) of each triangle.

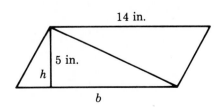
14 in.
5 in.
h
b

$A = \frac{1}{2}bh$

$\quad = \frac{1}{2} \cdot 14 \cdot 5$

$\quad = 7 \cdot 5$, or 35 in.^{2}

Practice **Find the areas.**

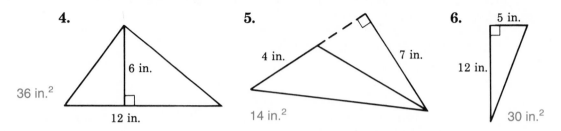

4.

6 in.

36 in.²

12 in.

5.

4 in.

7 in.

14 in.²

6. 5 in.

12 in.

30 in.²

A *trapezoid* is a quadrilateral (4-sided polygon) that has exactly one pair of parallel sides. The parallel sides are the bases of the trapezoid. In the figures below, two trapezoids, each the same size and shape, are placed together to form a new figure. The new figure formed is a parallelogram.

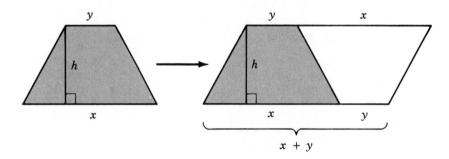

Observe.

1. The base of the parallelogram is the sum of the bases of the trapezoid ($x + y$).

2. The height of the parallelogram is the same as the height of the trapezoid (h).

3. The area of the parallelogram is twice the area of the trapezoid.

So, the area of the trapezoid is one half the area of the parallelogram:

$A_{\triangle} = \frac{1}{2}h\,(x + y)$, where h is the height and x and y are bases.

Example 3 Find the area of a trapezoid whose bases are 6 in. and 12 in. and whose height is 3 in.

$$A = \frac{1}{2}h\,(x + y)$$

$$= \frac{1}{2} \cdot 3 \cdot (6 + 12), \text{ or } 27 \text{ in.}^2$$

To derive the formula for the area of a circle, imagine that a circle is divided into a large number of wedges. The wedges are then arranged to form a parallelogram as is shown in the picture.

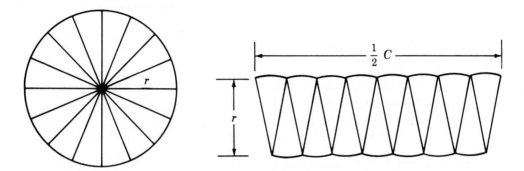

The base of the parallelogram is one half of the circumference of the circle. The height is the length of a radius of the circle.

So, the area of the parallelogram is $\frac{1}{2}Cr$.

$$A_\square = \frac{1}{2}Cr = \frac{1}{2} \cdot 2\pi r \cdot r = \pi r^2$$

But the area of the circle is equal to the area of the parallelogram.

So, $A_\bigcirc = \pi r^2$ where r is the length of a radius of the circle.

Example 4   Find the area of the circle shown in the picture to two decimal places.

$A = \pi r^2$
$\quad = \pi \cdot 4^2, \quad \text{or} \quad 16\pi \text{ in.}^2$
$\pi \doteq 3.14, \quad \text{so} \quad A \doteq 16 \times 3.14, \quad \text{or} \quad 50.24 \text{ in.}^2$

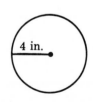

Exercises

1. Find the area of a trapezoid with bases 8 in. and 14 in. and height 9 in.   99 in.²

2. Find the area of a circle with a radius of 9 in. to two decimal places.   254.34 in.²

Find the areas.

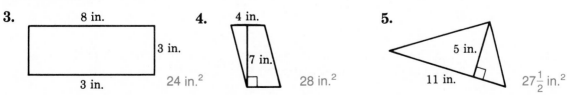

3. 8 in. / 3 in. / 3 in.   24 in.²

4. 4 in. / 7 in.   28 in.²

5. 5 in. / 11 in. $27\frac{1}{2}$ in.²

SURFACE AREA

OBJECTIVES

To find surface areas of rectangular solids, cylinders, pyramids, and cones

To find the surface area of a rectangular solid unfold it. It will form 6 rectangles as is shown in the picture below.

Example 1   Find the surface area of the rectangular solid shown below.

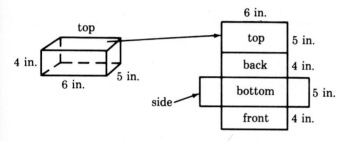

Surface area:
Area of top: $6 \cdot 5$, or 30 in.^{2}
Area of back: $6 \cdot 4$, or 24 in.^{2}
Area of bottom: $6 \cdot 5$, or 30 in.^{2}
Area of side: $5 \cdot 4$, or 20 in.^{2}
Area of side: $5 \cdot 4$, or 20 in.^{2}
Area of front: $6 \cdot 4$, or 24 in.^{2}
Surface area: 148 in.^{2}

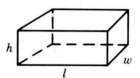

In general,
surface area $= 2lh + 2lw + 2hw.$

The surface area of a cylinder can also be found by unfolding the cylinder.

Example 2   Find the surface area of the cylinder shown below.

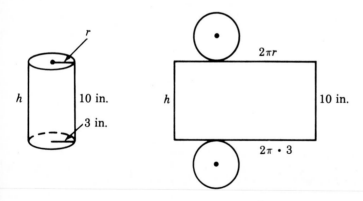

Note that the vertical portion of the cylinder opens to make a rectangle. One side of the rectangle is the circumference of each base ($2\pi r$). The other side is the height of the cylinder (h).

So, surface area of the cylinder is $2\pi r^2 + 2\pi rh$
$$= 2\pi \cdot 3^2 + 2\pi \cdot 3 \cdot 10, \text{ or } 78\pi \text{ in.}^2$$

Practice   1.   Find the surface area of a rectangular solid with length of the base 7 in., height 6 in., and width 12 in.   504 in.^{2}

To find the surface area of this pyramid, observe the following.

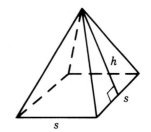

(1) The base is a square.
(2) Each of the four faces is a triangle.
(3) Each triangle has the same area.

If the length of each side of the base is s, then the area of the base is s^2.

If the altitude of each triangle is h, then its area is $\frac{1}{2}sh$.

For the four triangles: $4 \cdot \frac{1}{2} \cdot sh = 2sh$.

So, the surface area of a pyramid is

$$A = s^2 + 2sh.$$

Practice   **Find the surface area.**

2.

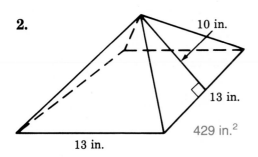

10 in.

13 in.

429 in.^{2}

13 in.

3.

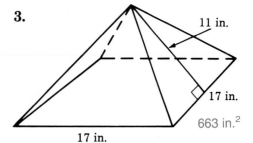

11 in.

17 in.

663 in.^{2}

17 in.

To find the surface area of a cone, observe the following.

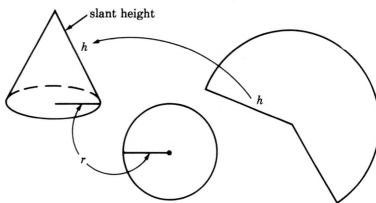

slant height

When unfolded,
(1) the base is a circle;
(2) the curved surface is a portion of a circle.

The area of the circle is πr^2.
The area of the portion of a circle is πrh.
So, the surface area of a cone is $\pi r^2 + \pi rh$, where r is the length of a radius of the base and h is the slant height.

Practice   Find the surface area. Use $\pi \doteq 3.14$ and compute the answer to one decimal place.

4.

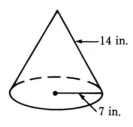

—14 in.

7 in.

$\text{Area}_{\text{base}} = \pi r^2 = \underline{\ ?\ }$ 49π in.^{2} or 153.9 in.^{2}

$\text{Area}_{\text{curved surface}} = \pi r h = \underline{\ ?\ }$ 98π in.^{2} or 307.7 in.^{2}

$\text{Surface area} = \underline{\ ?\ }$ 147π in.^{2} or 461.6 in.^{2}

Exercises

Find the surface areas.

1.   384 in.^{2}

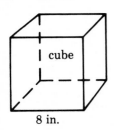

cube

8 in.

2.

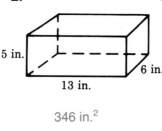

5 in.

13 in.

6 in.

346 in.^{2}

3. 56π ft^2 or 175.8 ft^2

12 ft

2 ft

4.

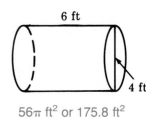

6 ft

4 ft

56π ft^2 or 175.8 ft^2

5.

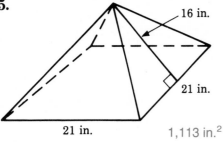

16 in.

21 in.

21 in.

1,113 in.^{2}

6. 225π in.^{2} or 706.5 in.^{2}

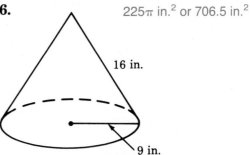

16 in.

9 in.

7.   A gallon of paint will cover 320 ft^2. How many gallons of paint are needed to paint the walls and ceiling of a room (no windows) that is 24 ft long, 12 ft wide, and 8 ft high?   2.7 gal

VOLUME

OBJECTIVE

To find volumes of rectangular solids, cylinders, pyramids, and cones

Volume is a number that tells how many cubic units are contained in a solid.

The rectangular solid shown at the right has a length of 8 units, a width of 3 units, and a height of 4 units. Note that one layer at the bottom has 8 × 3, or 24 units. There are 4 such layers. So, altogether there are 8 × 3, or 24 units. This suggests a formula for the volume of a rectangular solid.

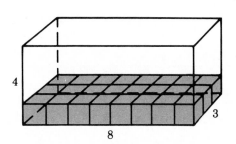

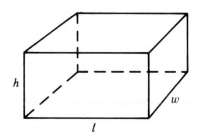

$$V = lwh$$
$$l = \text{length}, w = \text{width}, h = \text{height}$$

Example 1   Find the volume of this rectangular solid (prism).

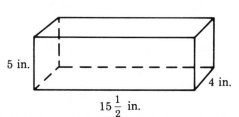

$$V = lwh$$
$$= 15\tfrac{1}{2} \times 4 \times 5 \text{ or } 310 \text{ in.}^3 \text{ (cubic inches)}$$

So, the volume is 310 in.^{3}

Note that for the solid (prism) at the top of the page, 8 × 3, or 24 square units is the area of the base of the prism. So, the volume can be viewed as being the product of the area of the base and height.

$$V = B \cdot h$$

↑ ↑

area height
of
base

Example 2   Find the volume of the cylinder shown below.

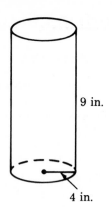

9 in.

4 in.

$$\text{Area of base } B = \pi r^2$$
$$= \pi \cdot 4^2, \text{ or } 16\pi \text{ in.}^2$$
$$\text{Volume} = Bh$$
$$= 16\pi \cdot 9, \text{ or } 144\pi \text{ in.}^3$$

Formula for the volume of a cylinder:
$$V = \pi r^2 h$$

Practice   **Find the volumes.**

1. Rectangular prism: $l = 17$ in.; $w = 6$ in.; $h = 7$ in.   714 in.³

2. Cylinder: radius of base $= 3$ in.; $h = 14$ in.   126π in.³

3. Compute the volume in Exercise 2 to the nearest tenth. Use 3.14 for π.
    395.4 in.³

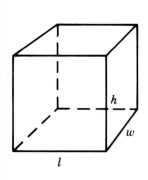

h

w

l

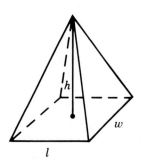

h

w

l

Experiment: This rectangular prism and pyramid have the same base and height. If the pyramid is filled with water and emptied into the prism, then 3 fillings of the pyramid will fill up the prism. It follows that the volume of the pyramid is one third of the volume of the corresponding prism.

$$V = \frac{1}{3}lwh$$

Example 3   Find the volume of a pyramid whose base is a square with a side 7 in. long and height 12 in.

$$\text{Volume} = \frac{1}{3} \cdot 7 \cdot 7 \cdot 12, \text{ or } 196 \text{ in.}^3$$

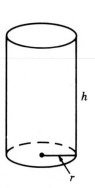

h

r

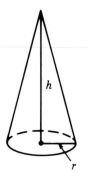

h

r

This cylinder and cone have the same size base and the same height. An experiment like the one with the rectangular prism and corresponding pyramid will show that 3 cones fit into the cylinder. It follows that the volume of the cone is one third of the volume of the corresponding cylinder.

$$V_{\text{cone}} = \frac{1}{3}\pi r^2 h$$

Practice   **Find the volumes.**

4. Cone: Radius of base $= 6$ in.; height $= 13$ in. 156π in.^{3}

5. Cone: Radius of base $= 10$ in.; height $= 15$ in. 500π in.^{3}

Exercises

1. Find the volume of a rectangular solid with length of the base 6 in., height 9 in., and width 7 in.   378 in.^{3}

2. Find the volume of a cylinder with the radius of the base 8 in. and height 16 in. $1{,}024\pi$ in.^{3}

3. Find the volume of a pyramid with the base a square with each side 8 in. long and height 9 in.   192 in.^{3}

4. Find the volume of a cone with the radius of base 3 ft and height 6 ft. Compute the volume to the nearest tenth using $\pi \doteq 3.14$.   56.5 ft^3

5.

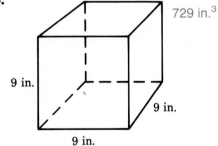

729 in.^{3}

9 in.

9 in.

9 in.

6.

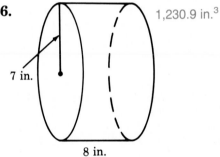

1,230.9 in.^{3}

7 in.

8 in.

7.

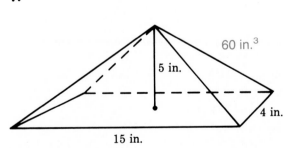

60 in.^{3}

5 in.

4 in.

15 in.

8.

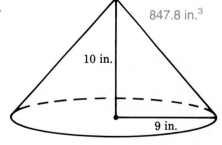

847.8 in.^{3}

10 in.

9 in.

9. If one cubic inch of some substance weights 12 oz, what is the weight of a bar of this substance that is 15 in. long, 7 in. wide, and 4 in. high?   5,040 oz

Computer Section _____

INPUT

OBJECTIVE
To write programs using the **INPUT** statement

Suppose you want to calculate $7 + 2y$ for any value of y.

First assign some variable (letter) to the value of $7 + 2y$, say S.

Write a formula. $\qquad\qquad\qquad\qquad\qquad$ $S = 7 + 2 \cdot Y$

Now express the formula in **BASIC:** $\qquad$ **LET S = 7 + 2 $*$ Y**

You are now ready to write a program to evaluate $7 + 2 * Y$ for any value of Y.

Example 1 Write a program to evaluate $7 + 2 * Y$ for any value of Y.
Run the program to find the value of $7 + 2 * Y$ for $Y = 3$.
Remember to press **RETURN** at the end of each line.

Tells the computer that you will be assigning a value to Y. The computer will ask the user for a value of Y by printing a "?".	10 INPUT Y
This formula tells how to find S.	20 LET S = 7 + 2 $*$ Y
	30 PRINT S
Tells the computer that the program is **END**ed.	40 END
Type **RUN.** (No line number needed.)	]RUN
The computer is asking for a value for Y.	?
The value for Y.	Type **3.** Press **RETURN.**
The computer displays the results.	**13**

You can now use the program in **EXAMPLE 1** to evaluate $7 + 2 * Y$ for any other value of Y by **RUN**ning the program for each new value of Y.

Example 2 The formula for the perimeter of a rectangle is $P = 2l + 2w$. Write a program to find the perimeter of any rectangle. Then type and **RUN** the program to find P for $l = 8$ and $w = 6$.

First write the formula in **BASIC**.

$P = 2 \cdot l + 2 \cdot w$ becomes **P = 2 $*$ L + 2 $*$ W.**

Computer will expect a value for *L*.	`10   INPUT L`
Computer will expect a value for W.	`20   INPUT W`
Formula in **BASIC**.	`30   LET P = 2 * L + 2 * W`
	`40   PRINT P`
	`50   END`

Now **RUN** the program to find the value of P for L = 8 and W = 6.

Press **RETURN.**

Type **RUN.**

Press **RETURN.**

The computer asks you for a value for L.	**?** is displayed on screen.
The value of L.	You type **8.**
	Press **RETURN.**
The computer now wants a value for W.	**?** is displayed again.
The value of W.	You type **6.**
	Press **RETURN.**
The value of P for L = 8, W = 6	**28** is displayed.

Example 3 The following is a copy of a program. Use your knowledge of algebra to fill in the blank in the screen display.

Formula for finding T.	`10   INPUT A`
	`20   LET T = (4 * A + 10) / (2 * A - 2)`
	`30   PRINT T`
	`40   END`
	`]RUN`
	`?2`
Programmer types in 2 as value of A.	☐
What will be displayed on the screen?	

The program gives the value of T in the formula. T = 9

Exercises

Fill in the missing blanks. Then type and RUN the program to evaluate for the given variable.

1. Evaluate $3A + 4$: $A = 7$

   ```
   10 INPUT A
   20 LET S =  3 * A + 4
   30 PRINT  S
   40 END  25
   ```

2. Evaluate $\dfrac{2X + 6}{4}$: $X = 9$

   ```
   10 INPUT  X
   20 LET T = (2 * X + 6) / 4
   30 PRINT  T
   40 END  6
   ```

3. Evaluate x^2: $X = 19$

   ```
   10 INPUT  X
   20 LET M =  X ∧ 2
   30  PRINT M
   40  END 361
   ```

4. Evaluate $4X + 9Y$: $X = 5$, $Y = 2$

   ```
   10 INPUT  X
   20 INPUT  Y
   30 LET G =  4 * X + 9 * Y
   40  PRINT G
   50  END 38
   ```

5. Evaluate $\dfrac{A + 4}{B}$: $A = 6$, $B = 2$

   ```
   10 INPUT  A
   20 INPUT  B
   30 LET K = (A + 4) / B
   40  PRINT K
   50  END 5
   ```

6. Evaluate $\dfrac{A + B + C}{3}$: $A = 6$, $B = 9$, $C = 12$

   ```
   10 INPUT  A
   20 INPUT  B
   30  INPUT C
   40 LET  T = (A + B +
   50  PRINT T
   60  END 9
   ```

Use your knowledge of algebra to predict the final screen display.

7.
```
10   INPUT A
20   LET S = 3 * A + 4
30   PRINT S
40   END

]RUN
?5
☐    19
```

8.
```
10   INPUT M
20   LET G = 24 - 2 * M
30   PRINT G
40   END

]RUN
?5
☐    14
```

Evaluate. Then write and RUN a program for each evaluation.

9. $5A + 9$ for $A = 6$
   39

10. X^3 for $X = 8$
    512

11. $\dfrac{3B + 6}{A}$ for $B = 2$, $A = 4$
    3

12. The formula for the area of a square is $A = s^2$ where s is the length of a side. Write a program to find the area of any square. **RUN** the program to find A for $s = 6$.
    See TE Answer Section.

13. The formula for the braking distance in meters for a car traveling at a given rate in kilometers per hour is $B = 0.006s^2$. Write a program to find the distance for any speed. **RUN** the program to find B for $s = 45$ km/h. See TE Answer Section.

14. For the program in **Exercise 12** above, find out what happens to the area when a side is doubled. Is the area doubled? (*Hint:* **RUN** for $s = 12, 24, 48$).
    No, the area is multiplied by 4. It equals 48.6 m.

LOOPS

OBJECTIVE
To write programs using **FOR-NEXT** loops

The formula for finding a person's salary if you know the wage and the hours worked is S = W · H. Suppose you want to write a program to find S for different values of W and H. Once the program is written and **RUN** you can get the result by merely typing in the values of W and H. However, for each new set of values of W and H you must retype **RUN**. Programmers use a technique that avoids the repetition of typing **RUN** over and over. The technique involves using a *loop* as shown below.

Example 1 Write a program to find the value of S in the formula S = W · H. Allow for executing the program for three different sets of values of W and H. Then **RUN** the program to find S for: W = $5.95, H = 42; W = $7.00, H = 39; W = $5.75, H = 40.

First write the formula in **BASIC**.

S = W · H becomes **S = W * H.**

Use I or any letter other than S, W, or H.	┌►10 FOR I = 1 TO 3
	L ╎ 20 INPUT W
	O ╎ 30 INPUT H
The program calculates S 3 times.	O ╎ 40 LET S = W * H
The computer loops back 3 times for 3	P ╎ 50 PRINT S
different sets of values for W and H.	└ 60 NEXT I
	70 END
	]RUN
Type 5.95 for the first wage.	?5.95
Type 42 for the number of hours.	?42
The value of S	249.9
The computer asks for second set of	?7
values for W and H.	?39
The value of S for W = 7 and H = 39.	273
Type in the third set of values.	?5.75
	?40
The value of S for the third set of values of W and H corresponds to I = 3.	230

FOR I =1 TO 3 instructed the computer to calculate S for 3 sets of values of W and H and then stop or **END**.

In the program found in Example 1, the statements between lines 10 and 60 are processed 3 times because the **FOR** statement in line 10 tells the computer to do so. Each time the computer reaches line 60, it goes back to line 10 for the next value of I. After using I = 3, the computer goes back with a value of I = 4. It checks and finds that I cannot be bigger than 3. The computer then drops out of the **FOR-NEXT** loop and goes to line 70 **END**. In any **FOR-NEXT** loop, there must be agreement between the letter used in the **FOR** statement and the letter used in the **NEXT** statement.

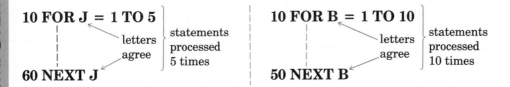

10 FOR J = 1 TO 5 — letters agree — statements processed 5 times — **60 NEXT J**

10 FOR B = 1 TO 10 — letters agree — statements processed 10 times — **50 NEXT B**

Example 2   A rocket is launched with an initial velocity of 128 ft/s. The formula for the height of the rocket at the end of T seconds is $H = 128T - 16T^2$. Write a program to find the value of H for any 2 values of T. Then **RUN** the program to find H for T = 2, T = 5.

First, write the formula in **BASIC**.

$H = 128T - 16T^2$ becomes **H = 128 * T − 16 * T ∧ 2.**

Use K or any letter other than T or H.	
| | ```
10   FOR K = 1 TO 2
20   INPUT T
30   LET H = 128 * T - 16 * T ^ 2
40   PRINT H
50   NEXT K
60   END
``` |

| Computer asks for first value of T; you type 2. Screen displays the answer; height of 192. Computer loops back to line 10, asks for second and last value of T; you type 5. Computer **PRINTS** the answer, 240. | ```
]RUN
?2
192
?5
240
``` |

Since the computer is told that there are only 2 values of T **(FOR K = 1 TO 2)**, the looping process is now complete. The computer goes to the next line of the program: 60 **END**. The program is **END**ed.

Exercises

Complete the program by filling in the missing blanks.

1. Write a program to find the value of Y for 6 different values of X.

   $Y = 4X^3 + 5X$

   10 FOR I = 1 to <u>6</u>
   20 INPUT X
   30 LET Y = <u>4 * X ∧ 3 + 5 * X</u>
   40 PRINT Y
   50 NEXT <u>I</u>
   60 END

2. Write a program to find the value of A for 8 different sets of values of B and H.

   $A = \dfrac{BH}{2}$

   10 FOR K = <u>1 to 8</u>
   20 INPUT <u>B</u>
   30 INPUT H
   40 LET <u>A = B * H / 2</u>
   50 PRINT <u>A</u>
   60 <u>NEXT K</u>
   70 <u>END</u>

Use the following program to answer Exercises 3 through 5.

```
10   FOR M = 1 TO 4
20   INPUT X
30   LET Y = 2 * X ∧ 3
40   PRINT Y
50   NEXT M
60   END
```

3. What is the significance of line 10? It tells the computer to loop 4 times.

4. What will cause the program to go to line 60? when the looping is completed

5. Predict, without **RUN**ning the program, the output for the following values of X: X = 2, X = 4, X = 5, X = 7. Then **RUN** the program to see if the computer agrees with your algebraic interpretation. 16, 128, 250, 686

6. The formula for simple interest is I = PRT.
   Write a program to evaluate I for any 2 different sets of values of P, R, and T. **RUN** the program for P = $2000, R = 12%, T = 4 years, and for P = $4500, R = 23%, T = 3 years. (*Hints:* Type values of P without "$". Type 12% as .12, 23% as .23. Do not use commas in numbers like 2000 or 4500.)

   $960, $3,105   See TE Answer Section for program.

7. $Y = 4X^2 + 3X$.
   Write a program to evaluate Y for any 5 different values of X.
   Compute the values of Y for the following values of X:
   5, 4, 6, 7, 2. Then **RUN** the program. Check whether or not the computer agrees with your computations. 115, 76, 162, 217, 22
   See TE Answer Section for program.

MORE ON PRINT AND INPUT

OBJECTIVES

To use the print command with expressions in quotation marks
To write programs using *enhanced* **INPUT**

Example 1   Type and **RUN** the following program. What is the difference in results among the 3 **PRINT** statements?

<table>
<tr><td>Line 20 instructs the computer to print exactly what is in quotation marks, the letter P, not the value of P.
Line 30 results in the PRINTing of the value of P, 5.
Line 40 combines the two types of PRINT command.</td><td>

```
10   LET P = 5
20   PRINT "P"
30   PRINT P
40   PRINT "P="P
50   END

]RUN
P
5
P=5
```

</td></tr>
</table>

Example 2   What will be the output of the following program when **RUN**?

<table>
<tr><td></td><td>

```
10   LET L = 8 * 3
20   PRINT "L="L
30   END
```

</td></tr>
<tr><td>When the program is RUN, line 20 will produce this output.
The value of L is 24 since 8 * 3 = 24.</td><td>

```
]RUN
L=24
```

</td></tr>
</table>

Recall that when a program using **INPUT** is **RUN** a "**?**" appears on the screen. If someone else were **RUN**ning the program the meaning of the "**?**" might not be clear. In this lesson you will learn how to *enhance* the **INPUT** statement. It will be clear to anyone **RUN**ning your program what is expected when the "**?**" appears. This is illustrated in **Example 3**.

Example 3 Write a program to find the product of any two numbers. Allow for **RUN**ning the program for any two pairs of numbers. When **RUN,** the screen should display an actual instruction, not merely a "?". The output should also explain what the answer represents. Then **RUN** the program to find the products 5 ∗ 8 and 4 ∗ 6.

First write a formula in **BASIC.**
Product means multiply.
Let P = A ∗ B, where A and B are any two numbers.

Tells the operator what the program will do	`10  PRINT "PROGRAM FINDS PRODUCTS."`
	`20  FOR I = 1 TO 2`
INPUT is split into two separate statements.	`30  INPUT "TYPE IN A NUMBER ";A`
	`40  INPUT "TYPE IN A SECOND NUMBER ";B`
	`50  LET P = A * B`
	`60  PRINT "THEIR PRODUCT IS "P`
Allows one line of space before next question	`70  PRINT`
	`80  NEXT I`
	`90  END`

Leave 1 space after **IS** so the computer will print a space before the answer.

```
]RUN
PROGRAM FINDS PRODUCTS.
```
Type 5.   `TYPE IN A NUMBER 5`
Type 8.   `TYPE IN A SECOND NUMBER 8`
Screen displays result.   `THEIR PRODUCT IS 40`
This corresponds to I = 2.   `TYPE IN A NUMBER 4`
```
TYPE IN A SECOND NUMBER 6
THEIR PRODUCT IS 24
```

Notice the structure and role of the *enhanced* **INPUT** statement in lines 30 and 40 of the program.

30 INPUT "TYPE IN A NUMBER "; A

             space

    Instruction in   A semicolon is
    quotation marks   needed.

There is a space between number and " to avoid screen display **TYPE IN A NUMBER 5.** The result is that when the program is **RUN,** the **INPUT** is clarified or *enhanced* by an actual instruction, **TYPE IN A NUMBER.**

COMPUTER SECTION

Exercises

What will be the output of each program when it is RUN?

1. 10 LET W = 4
   20 PRINT "W"
   30 PRINT W
   40 PRINT "W="W
   50 END

   W
   4
   W = 4

2. 10 LET M = 3 * 14
   20 PRINT "M="M
   30 END

   M = 42

3. 10 LET S = 12 / 4
   20 PRINT "S="S
   30 END

   S = 3

4. When the following program is **RUN** what will be the output: if B = 8?; if B = 5?

   10 PRINT "PROGRAM EVALUATES 6+4B FOR ANY TWO DIFFERENT VALUES OF B"
   20 FOR I = 1 TO 2
   30 INPUT "TYPE IN A VALUE FOR B ";B
   40 LET Y = 6 + 4 * B
   50 PRINT "THE VALUE OF Y IS ";Y
   60 NEXT I
   70 END

   PROGRAM EVALUATES 6 + 4B FOR ANY TWO DIFFERENT VALUES OF B.
   TYPE IN A VALUE FOR B 8(5).
   THE VALUE OF Y IS 38(26).

Find the error in each statement. Then rewrite it correctly.

5. **10 INPUT TYPE IN A VALUE FOR A; A**   10 INPUT "TYPE IN A VALUE FOR A "; A

6. **10 INPUT "WHAT IS THE VALUE OF B?" B**   10 INPUT "WHAT IS THE VALUE OF B?"

7. **10 INPUT "WHAT IS THE VALUE OF M?; M**   10 INPUT "WHAT IS THE VALUE OF M"

8. **10 PRINT THE ANSWER IS W**   10 PRINT "THE ANSWER IS W"

9. **10 PRINT "THE ANSWER IS W"** (Numerival value of W is desired output.)
   10 PRINT "THE ANSWER IS " W

For each of the following:

(a) Write a program using *enhanced* **INPUT** and the new **PRINT** technique of this lesson;
(b) predict the output algebraically before **RUN**ning the program;
(c) type and **RUN** the program for the given value(s) of the variable(s).

10. Evaluate $7X + 4$ if: $X = 8$; $X = 3$; $X = 23$ 60, 25, 165

11. Evaluate $17 - 2M$ if: $M = 4$; $M = 5$ 9, 7

12. Evaluate $2L + 2W$ if: $L = 8$; $W = 6$; $L = 9$, $W = 6$ 28; 30

13. Evaluate $\dfrac{A}{2A - 4}$ if: $A = 6$; $A = 8$; $A = 20$ 0.75, 0.6\overline{6}, 0.5

14. Evaluate $3B^5$ if: $B = 2$; $B = 6$; $B = 5$; $B = 7$ 96; 23,328; 9,375; 50,421

15. Find the quotient of any two non-zero numbers. **RUN** the program for the following pairs of numbers: 12 and 6; 45 and 15; 48 and 96; 22 and 88.
    2, 3, 0.5, 0.25 See TE Answer Section for program.

THE IF...THEN STATEMENT

OBJECTIVE
To write programs using the **IF . . . THEN** statement

Consider the following department store advertisement.

To find the cost of 6 tapes you must make a decision.
Do you multiply by $7.00 or $4.00?
Since 6 is more than 5, you must use $4.00. The cost is $4.00 × 6 or $24.00.

A computer program for finding the total cost of the tapes must tell the
computer how to "decide" whether to multiply by 7 or 4.

Example   Write a program to find the cost of the tapes described above. Allow
for 3 calculations. Type and **RUN** the program to find the cost of 2
tapes, 8 tapes, 5 tapes.

The computer will have to "decide" which of two formulas to use.

Total = 7 times Number bought   OR   Total = 4 times Number bought

$$T = 7 * N \qquad\qquad\qquad T = 4 * N$$

Note that in the program below the symbol for *is greater than* is >.

```
                              10   FOR K = 1 TO 3
                              20   INPUT "TYPE IN NUMBER OF TAPES ";N
 The computer checks if N > 5.  30   IF N > 5 THEN 60
  If so, it skips down to line 60.  40   LET T = 7 * N
     If N ≤ 5 it goes to line 40.  50   GOTO 70
                              60   LET T = 4 * N
                              70   PRINT "TOTAL COST IS "T
                              80   PRINT : PRINT
                              90   NEXT K
                              100  END
```

The **RUN** for this program is shown on page 430.

THE IF . . . THEN STATEMENT

445

You type in 2 since 2 < 5. Line 40 is used: 7 * 2 = 14.	]RUN TYPE IN NUMBER OF TAPES 2 TOTAL COST IS 14
Line 80 **PRINT:PRINT** produces 2 lines of separation between results. This corresponds to K = 2 and 8 > 5. Line 60 is used: 4 * 8 = 32.	TYPE IN NUMBER OF TAPES 8 TOTAL COST IS 32
Third and last calculation 5 is not greater than 5. So, line 40 is used: 7 * 5 = 35.	TYPE IN NUMBER OF TAPES 5 TOTAL COST IS 35

In general, the **IF . . . THEN** statement has the following form.

the word **IF** . . . the word **THEN**

30 IF N > 5 THEN 60

line number

any line number other than that of the statement

Two arithmetic expressions joined by
a decision symbol such as > comparing N with 5.

BASIC Decision Symbols

Symbol	Meaning
=	is equal to
>	is greater than
> =	is greater than or equal to
<	is less than
< =	is less than or equal to
< >	is not equal to

Exercises

Identify each of the following IF . . . THEN statements as correct or incorrect. If it is incorrect, copy and correct it.

Incorrect; after then there must be a different number.

correct   Incorrect; should have a decision symbol.

1. 10 IF Y > 0 THEN 90   **2.**   20 IF A + M THEN 100   **3.**   40 IF G < 6 THEN 40

4. Write a program to find the cost of pens as advertised: "Special sale on pens: $1.50 each; only $1.25 each if you buy more than 6." Allow for three calculations. Then type and **RUN** the program to find the cost of 2 pens; 8 pens; 6 pens.   See TE Answer Section for program and results.

Use the following program to answer Exercises 5–10.

```
10   FOR J = 1 TO 4
20   INPUT "WHAT IS THE VALUE OF X? ";X
30   IF X < 4 THEN 60
40   LET Y = 3 * X
50   GOTO 70
60   LET Y = 12 / X
70   PRINT "THE VALUE OF Y IS "Y
80   PRINT : PRINT : PRINT
90   NEXT J
100  END
```

5. What is the significance of line 10?   It tells the computer to loop 4 times.

6. Predict without **RUN**ning the program the value of Y, if X = 6.   18

7. Predict without **RUN**ning the program the value of Y, if X = 3.   4

8. At what point in the program will the computer carry out line 100?
    after it has been run 4 times.

9. Type and **RUN** the program for X = 2, 4, 8, and 12.   6, 3, 24, 36

10. You can delete a line from a program by typing only the number of the line. Omit or delete line 50 by typing 50. Now **RUN** the program for the same values of X: 2, 4, 8, and 12. What happens? Why?   The results would be 6, 3, 1.5, and 1 because every run would use line 60 to compute the answer.

11. Bill is a salesman. If the total of his sales for each of two weeks is greater than $750, his commission is $50. Otherwise he is paid no commission. Complete the following program for determining whether or not Bill is paid a commission.

    10 FOR K = 1 TO 3
    20 INPUT "TYPE IN SALES FOR THE FIRST WEEK "; F
    30 INPUT   "TYPE IN SALES FOR THE SECOND WEEK"; S
    40 LET T = F +   S
    50 IF   T > 750   **THEN**   80
    60 PRINT "NO COMMISSION"
    70 GO TO   90
    80 PRINT   "COMMISSION IS $50"
    90 NEXT   K
    100 END

Type and **RUN** the program for sales of $475 and $319; sales of $295 and $125; sales of $335 and $435. $50, no commission, $50

12. A teacher wants to write a program to **PRINT** the average of 4 grades. If the average is less than 70, the computer should also **PRINT "SEND A WARNING NOTICE"**. Write a program allowing for two sets of grades. Type and **RUN** the program to average: 72, 60, 80, 60; 80, 90, 90, 80.
    See TE Answer Section for program and results.

THE IF . . . THEN STATEMENT   447

SEQUENTIAL INPUT

OBJECTIVE

To write programs involving input of numbers that are in sequence

Example 1   Write a program to find the square of any 6 numbers.

First write a formula in **BASIC**.

LET S = Number squared

LET S = N^2

LET S = N $\wedge$ 2

The computer will loop back 6 times.	10 FOR I = 1 TO 6
Note that I and N are not the same.	20 INPUT "TYPE NUMBER BEING SQUARED ";N
	30 LET S = N ^ 2
This identifies the output.	40 PRINT "SQUARE IS "S
	50 NEXT I
	60 END

Now suppose you want to write a program to square all numbers in *sequence* from 1 to 100; that is, $1^2, 2^2, 3^2, 4^2, 5^2, 6^2, 7^2, \ldots, 99^2, 100^2$.

When the program is **RUN** you would have to enter 100 values for **N**.
There is a simpler way to write the program when the *input* is *sequential*.

Example 2   Write and **RUN** a program to square the numbers from 1 to 5 using the *sequential input* method.

This provides a heading for the output.	10 PRINT "NUMBER","SQUARE OF NUMBER"
INPUT statement not needed.	20 FOR N = 1 TO 5
It is N itself being squared.	30 LET S = N ^ 2
Thus, when N = 1, 1 is squared, and	40 PRINT N,S
so on.	50 NEXT N
	60 END

The output of **RUN** is shown on the next page.

```
]RUN
NUMBER          SQUARE OF NUMBER
1               1
2               4
3               9
4               16
5               25
```

Note the use of the comma in line 40 of the program: **PRINT N, S.**
This resulted in the printing of two columns; the number in one column
and the square of the number in the second column.

The numbers 3, 7, 11, 15, 19, 23, . . ., and so on are in sequence. Successive
numbers of the sequence differ by 4. The number after 11 is **STEP**ped up by 4,
or increased by 4. The numbers—30, 25, 20, 15, 10—are also in sequence.
Successive numbers are decreased by 5. The number following 25 is **STEP**ped
down (decreased) by 5. The use of **STEP** in sequential looping is illustrated in
the next Example.

Example 3   Write a program to evaluate 2A + 5 for the following values of A:
3, 5, 7, 9, 11, 13, 15.

The values of A are in sequence; successive numbers are
STEPped up (increased) by 2.

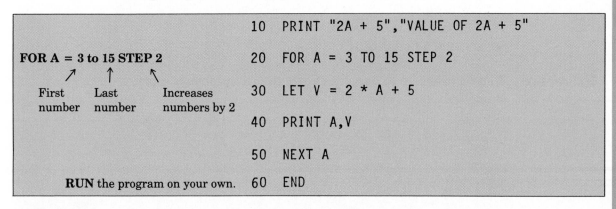

```
10   PRINT "2A + 5","VALUE OF 2A + 5"

20   FOR A = 3 TO 15 STEP 2

30   LET V = 2 * A + 5

40   PRINT A,V

50   NEXT A

60   END
```

FOR A = 3 to 15 STEP 2

First number / Last number / Increases numbers by 2

RUN the program on your own.

Exercises

Write and RUN a program to find each of the following.

1. The square of the numbers 4, 7, 10, 13, 16.
2. The value of X^3, if X = 1, 2, 3, 4, 5, 6, 7, 8, 9.
3. The value of 4B + 3, if B = 6, 8, 10, 12, 14, 16.
4. The value of $X^2 + 3$, if X = 1, 7, 13, 19, 25, 31, 37.
5. The squares of the numbers 14, 12, 10, 8, 6, 4, 2. (*Hint:* the numbers are
   successively **STEP**ped down by 2; **STEP** − 2.)
   See TE Answer Section for programs and results.

INT

OBJECTIVE

To use **INT** to determine if one number is a factor of another

The **INT**eger part of the number 8.679543 is 8.
In **BASIC,** the instruction **PRINT** the **INT**eger part of 8.679543 is written as **PRINT INT(8.679543).**

Example 1 Find each: **INT(3.45649)** **INT(3.87569)** **INT(3)**

INT(3.45649) = 3 **INT(3.87569)** = 3 **INT(3.0)** = 3

Example 2 Use **INT** to determine whether 4 is a factor of 15.

Does 4 divide into 15 evenly?

Does 15/4 = **INT(15/4)**?

$$\begin{array}{r} 3.75 \\ 4\overline{)15.00} \end{array}$$

Does 3.75 = **INT(3.75)**?

No. $3.75 \neq 3$

So, 4 is not a factor of 15.

Example 3 Write a program to determine whether a number is a factor of 528.

```
10  INPUT "TYPE IN A POSSIBLE FACTOR OF 528 ";F
20  IF 528 / F =  INT (528 / F) THEN 50
30  PRINT F" IS NOT A FACTOR OF 528."
40  GOTO 60
50  PRINT F" IS A FACTOR OF 528."
60  END
```

Exercises

1. Find each of the following: **INT(32.18976)**; **INT(10 * 0.3455654 + 1)**.
   32; 3

2. Modify the program above with a loop so that you can test whether any 5 different numbers are factors of 528. Then **RUN** the program to determine which of the following numbers are factors of 528: 4, 17, 24, 96, 176.
   Add line 5 **FOR I = 1 TO 5** and line 55 **NEXT I.** yes, no, yes, no, yes

RND

Objective
To represent random numbers

When tossing a die (one of a pair of dice), the die may come up a 5 or a 6 or a 3, or any one of 6 possible numbers. A computer can be programmed to simulate this by choosing *random* numbers from 1 to 6.

Most computers use **RND(1)** to generate random numbers between 0 and 1. So, the command **10 PRINT RND(1)** would result in the printing of numbers such as .779343355 or .103117626.

Example 1   Show that **INT(6 * RND(1) + 1)** produces random numbers from 1 through 6. Use the two values of **RND(1)** above.

INT(6 * RND(1) + 1)	**INT(6 * RND(1) + 1)**
INT(6 * .779343355 + 1)	**INT(6 * .103117626 + 1)**
INT(4.67606013 + 1)	**INT(.618705756 + 1)**
INT(5.6760606013)	**INT(1.618705756)**
5	1

So, **INT(6 * RND(1) + 1)** generates numbers from 1 through 6 such as 1 and 5.

Example 2   Write an expression to produce random numbers from:

1 through 4	1 through 23
INT(4 * RND(1) + 1)	**INT(23 * RND(1) + 1)**

Exercises

1. Which of the following could be a result of printing **RND(1)**?   c

   a. 4.678567857
   b. 5.678678905
   c. 0.9987765554

2. Write a representation for random numbers from 1 through 49.
   INT(49 * RND(1) + 1)

Extra Practice

PRACTICE ON SOLVING ADDITION AND SUBTRACTION EQUATIONS: CHAPTER 1

Evaluate for the given values of the variables.

1. $x + 15$; $x = 5$ 20
2. $35 - a$; $a = 18$ 17
3. $a + b - c$; $a = 15, b = 20, c = 5$ 30
4. $a + b + c$; $a = 3, b = 6, c = 8$ 17
5. $c - a + b$; $a = 3, b = 9, c = 1$ 7
6. $a - b + c$; $a = 10, b = 6, c = 7$ 11

Rewrite to make computation easier. Then compute.

7. $4 + 72 + 96$ 172
8. $17 + 13 + 183$ 213
9. $3 + 96 + 297$ 396
10. $15 + 10 + 85$ 110
11. $45 + 65 + 55$ 165
12. $346 + 49 + 54$ 449

Which, if any, of the values is a solution of the open sentence?

13. $a + 3 = 10$; 5, 13   none
14. $42 = x - 4$; 37, 46, 47   46
15. $a + 12 = 4 + 9$; 0, 1, 25   1
16. $y - 18 = 32 + 8$; 22, 42, 58   58
17. $x + 12 = 12$; 0, 1, 12   0
18. $b - 8 = 14 - 8$; 6, 10, 12   none

Solve each equation. Check the solution.

19. $x + 5 = 10$ 5
20. $x - 6 = 15$ 21
21. $x + 15 = 26$ 11
22. $14 = a - 12$ 26
23. $a + 15 = 17$ 2
24. $b + 6 = 19$ 13
25. $y - 10 = 15$ 25
26. $m + 12 = 20$ 8
27. $a + 32 = 40$ 8
28. $x - 12 = 18$ 30
29. $y + 16 = 36$ 20
30. $y - 17 = 3$ 20

PROBLEM SOLVING

31. If the cost of a pair of shoes is decreased by $8, the price is $38. Find the cost. $46

32. The sum of Karen's salary and a $40 commission is $430. Find her salary. $390

33. 18 is the same as a number decreased by 10. What is the number?   28

34. 15 more than Merv's age is 38. How old is Merv?   23

Find P.

35. $P = a + b + c$ P = 36 in.

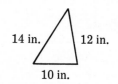

14 in. 12 in. 10 in.

36. $P = l + w + l + w$ P = 24 cm

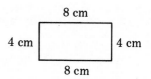

8 cm 4 cm 4 cm 8 cm

PRACTICE ON SOLVING MULTIPLICATION AND DIVISION EQUATIONS: CHAPTER 2

Evaluate for the given values of the variables.

1. $8a; a = 4$ 32
2. $\frac{12}{x}; x = 6$ 2
3. $\frac{6t}{4}; t = 6$ 9
4. $\frac{144}{b}; b = 16$ 9
5. $15a; a = 15$ 225
6. $\frac{12x}{5}; x = 10$ 24
7. $6x + 5; x = 3$ 23
8. $\frac{a}{2} + 5h + 2; a = 4, b = 2$ 14
9. $4 + 5m + 2n; m = 4, n = 5$ 34
10. $4xy + 5; x = 8, y = 5$ 165
11. $10ab + a - b; a = 6, b = 4$ 242
12. $40ab - 20ab; a = 2, b = 3$ 120
13. $25 - 4x + 5xy; x = 4, y = 12$ 249
14. $16 + 8m + 2mn; m = 5, n = 10$ 156

Rewrite to make computation easier. Then compute.

15. $5 \cdot 32 \cdot 20$ 3,200
16. $2 \cdot 9 \cdot 50$ 900
17. $4 \cdot 8 \cdot 5 \cdot 5$ 800
18. $50 \cdot 70 \cdot 2$ 7,000
19. $2 \cdot 14 \cdot 5 \cdot 10$ 1,400
20. $8 \cdot 9 \cdot 25$ 1,800

Solve each equation. Check the solution.

21. $8x = 40$ 5
22. $72 = 9m$ 8
23. $12 = \frac{x}{5}$ 60
24. $36 = 4y$ 9
25. $6a = 12$ 2
26. $4x = 16$ 4
27. $\frac{a}{10} = 7$ 70
28. $\frac{x}{32} = 3$ 96
29. $15 = \frac{y}{6}$ 90
30. $45 = \frac{x}{2}$ 90
31. $3n = 90$ 30
32. $\frac{a}{10} = 12$ 120

PROBLEM SOLVING

33. 8 times a number is 32. Find the number.   4
34. 12 times a number is 60. Find the number.   5
35. The $90 selling price of a coat is twice the cost. Find the cost. $45
36. Jamie's age of 6 is the same as her aunt's age divided by 5. How old is her aunt?   30
37. Cathy's salary of $800 per month is the same as her mother's salary divided by 3. Find her mother's salary. $2,400
38. Rico's 36 base hits this season is twice his number of hits last season. How many base hits did he have last year?   18 hits
39. The total cost of a restaurant bill divided among 5 people is $4 each. Find the total cost. $20
40. Three times O'Neil's salary is $315. What is his salary? $105
41. $A = l \cdot w$. Find A if $l = 8$ cm and $w = 6$ cm. $A = 48$ cm^2
42. $P = 2l + 2w$. Find P if $l = 12$ cm and $w = 4$ cm.   32 cm

PRACTICE ON SIMPLIFYING EXPRESSIONS AND SOLVING EQUATIONS: CHAPTER 3

Simplify.

1. $4 \cdot 3m$ 12m
2. $8m \cdot 7$ 56m
3. $3(4x + 3)$ 12x + 9
4. $12(2a - 1)$ 24a − 12
5. $5a + 6a$ 11a   m + 2n + 3
6. $6x - 3 + 4x$ 10x − 3
7. $3a + 2a - 5$ 5a − 5
8. $3m + 2n + 3 - 2m$
9. $5(3a - 3)$ 15a − 15
10. $5 + 2a + 4a$ 5 + 6a
11. $8x + 5y + x + y$ 9x + 6y
12. $14 + 5x + 2y - 12 + x + y$ 6x + 3y + 2

Simplify. Then evaluate for these values of the variables.

13. $4x + 9 + 3x; x = 5$ 44
14. $7a + 3 + 16a; a = 10$ 233
15. $10 + a + 6a; a = 8$ 66
16. $2x + 3y - 10; x = 5, y = 10$ 30
17. $15m + 10m - 3m + 2; m = 5$ 112
18. $9b + 6a - 3a + 10b; a = 2, b = 3$ 63
19. $5x + 3x + 8y - 2y; x = 4, y = 5$ 62
20. $12 + 3a + 5 + 2b + 6a;$
   $a = 6, b = 7$ 85

Solve each equation. Check the solution.

21. $3n + 6 = 15$ 3
22. $2x - 7 = 9$ 8
23. $12 = \frac{x}{5} - 3$ 75

24. $21 = 7x + 7$ 2
25. $36 = 8n + 4$ 4
26. $\frac{a}{6} + 7 = 13$ 36

27. $10y - 9 = 81$ 9
28. $\frac{y}{12} + 7 = 11$ 48
29. $11m - 4 = 73$ 7

30. $32 = 9a - 22$ 6
31. $16 = 5a - 19$ 7
32. $46 = 5y + 11$ 7

Find the value.

33. 3^2 9
24. 2^4 16
35. 1^7 1
36. 6^2 36
37. 5^3 125
38. 4^3 64
39. 3^5 243
40. 7^3 34.
41. 2^6 64
42. 3^3 27
43. 5^4 625
44. 4^5 1,0

Write using exponents.

45. $5 \cdot 5 \cdot 5$ 5^3
46. $a \cdot a \cdot a \cdot a$ a^4
47. $3 \cdot 3 \cdot 4 \cdot 4 \cdot 4 \cdot 4$ $3^2 \cdot 4^4$
48. $m \cdot m \cdot n \cdot n \cdot n$ $m^2 n^3$
49. $6 \cdot 6 \cdot x \cdot x \cdot x \cdot x \cdot x$ $6^2 x^5$
50. $4 \cdot 4 \cdot 4 \cdot m \cdot m$ $4^3 m^2$
51. $x \cdot x \cdot x \cdot y \cdot y \cdot y \cdot y$ $x^3 y^4$
52. $2 \cdot 2 \cdot 2 \cdot b \cdot b \cdot b \cdot b$ $2^3 b^4$

PROBLEM SOLVING

53. 3 less than 5 times a number is 32. Find the number.   7

54. 18 is the same as twice a number decreased by 6. Find the number.   12

55. 45 is the same as 3 times a number increased by 6. Find the number.   13

56. 12 more than a number divided by 6 is 23. Find the number.   66

PRACTICE ON MULTIPLYING AND DIVIDING FRACTIONS: CHAPTER 4

Multiply.

1. $5 \cdot \frac{1}{3}$ $1\frac{2}{3}$ **2.** $6 \cdot \frac{1}{5}$ $1\frac{1}{5}$ **3.** $4 \cdot \frac{2}{7}$ $1\frac{1}{7}$ $4\frac{2}{3}$ **4.** $\frac{2}{3} \cdot 7$

5. $3 \cdot 2\frac{4}{5}$ $8\frac{2}{5}$ **6.** $2\frac{2}{3} \cdot 2$ $5\frac{1}{3}$ **7.** $10 \cdot \frac{3}{7}$ $4\frac{2}{7}$ $5\frac{3}{5}$ **8.** $4 \cdot 1\frac{2}{5}$

9. $\frac{2}{3} \cdot \frac{4}{5}$ $\frac{8}{15}$ **10.** $\frac{3}{2} \cdot \frac{5}{4}$ $1\frac{7}{8}$ **11.** $\frac{1}{5} \cdot 3\frac{1}{2}$ $\frac{7}{10}$ $4\frac{2}{15}$ **12.** $6\frac{1}{5} \cdot \frac{2}{3}$

Simplify.

13. $\frac{4}{10}$ $\frac{2}{5}$ **14.** $\frac{10}{15}$ $\frac{2}{3}$ **15.** $\frac{12}{28}$ $\frac{3}{7}$ **16.** $\frac{36}{15}$ $2\frac{2}{5}$

17. $\frac{5}{12} \cdot \frac{4}{15}$ $\frac{1}{9}$ **18.** $\frac{8}{3} \cdot 1\frac{7}{8}$ 5 **19.** $8 \cdot \frac{5}{24}$ $1\frac{2}{3}$ 6 **20.** $3\frac{3}{7} \cdot \frac{14}{8}$

21. $\frac{2}{3}$ of 15 10 **22.** $\frac{3}{5}$ of 25 15 **23.** $2\frac{4}{5} \cdot 5\frac{5}{7}$ 16 6 **24.** $1\frac{3}{4} \cdot 3\frac{3}{7}$

Divide.

25. $\frac{3}{4} \div \frac{12}{5}$ $\frac{5}{16}$ **26.** $5\frac{1}{3} \div 4$ $1\frac{1}{3}$ **27.** $20 \div 3\frac{1}{5}$ $6\frac{1}{4}$ $\frac{1}{3}$ **28.** $\frac{7}{4} \div 5\frac{1}{4}$

29. $16 \div 1\frac{1}{3}$ 12 **30.** $\frac{4}{9} \div \frac{14}{15}$ $\frac{10}{21}$ **31.** $3\frac{1}{5} \div 8$ $\frac{2}{5}$ $\frac{12}{25}$ **32.** $\frac{3}{10} \div \frac{15}{24}$

33. $1\frac{1}{2} \div 2\frac{2}{3}$ $\frac{9}{16}$ **34.** $\frac{4}{5} \div 1\frac{1}{4}$ $\frac{16}{25}$ **35.** $1\frac{7}{8} \div \frac{5}{9}$ $3\frac{3}{8}$ $8\frac{1}{3}$ **36.** $3\frac{1}{3} \div \frac{2}{5}$

37. $2\frac{2}{9} \div 3\frac{1}{5}$ $\frac{25}{36}$ **38.** $2\frac{4}{7} \div 4\frac{2}{7}$ $\frac{3}{5}$ **39.** $1\frac{1}{6} \div 2\frac{5}{8}$ $\frac{4}{9}$ $\frac{7}{10}$ **40.** $4\frac{2}{3} \div 6\frac{2}{3}$

Solve.

41. $\frac{2}{3}x = 30$ 45 **42.** $\frac{4}{5}x = 24$ 30 $1\frac{1}{5}$ **43.** $\frac{7}{6}x = \frac{21}{15}$

44. $\frac{1}{2}x = \frac{8}{7}$ $2\frac{2}{7}$ **45.** $\frac{5}{3}x = \frac{25}{12}$ $1\frac{1}{4}$ 1 **46.** $\frac{4}{9}x = \frac{12}{27}$

47. $\frac{3}{4}x = \frac{1}{2}$ $\frac{2}{3}$ **48.** $\frac{7}{8}x = \frac{3}{4}$ $\frac{6}{7}$ $1\frac{5}{9}$ **49.** $\frac{3}{7}x = \frac{2}{3}$

50. $8 = \frac{3}{2}x$ $5\frac{1}{3}$ **51.** $\frac{4}{7} = \frac{5}{14}x$ $1\frac{3}{5}$ 66 **52.** $36 = \frac{6}{11}x$

PROBLEM SOLVING

Find A or V in terms of π, where possible.

53. $A = \frac{1}{2}bh; b = \frac{10}{3}, h = 12$ $A = 20$ **54.** $A = \frac{1}{2}bh; b = \frac{9}{5}, h = 5$ $A = 4\frac{1}{2}$

55. $A = \pi r^2; r = 1\frac{2}{3}$ $A = 2\frac{7}{9}\pi$ **56.** $A = \pi r^2, r = \frac{3}{5}$ $A = \frac{9}{25}\pi$

57. $V = \frac{1}{3}\pi r^2 h; r = 1\frac{1}{5}, h = 4$ $V = 1\frac{23}{25}\pi$ **58.** $V = \frac{1}{3}\pi r^2 h; r = 2\frac{2}{5}, h = \frac{1}{3}$ $V = \frac{16}{25}\pi$

59. $V = \frac{4}{3}\pi r^3; r = 3\frac{1}{4}$ $V = 45\frac{37}{48}\pi$ **60.** $V = \frac{4}{3}\pi r^3; r = 2\frac{1}{3}$ $V = 16\frac{76}{81}\pi$

PRACTICE ON ADDING AND SUBTRACTING FRACTIONS: CHAPTER 5

Compute. Simplify if possible.

1. $\frac{4}{5} + \frac{3}{5}$ $1\frac{2}{5}$

2. $\frac{3}{7} + \frac{2}{7}$ $\frac{5}{7}$

3. $\frac{5}{9} - \frac{4}{9}$ $\frac{1}{9}$

4. $\frac{7}{11} - \frac{2}{11}$ $\frac{5}{11}$

5. $\frac{4}{5} + \frac{2}{15}$ $\frac{14}{15}$

6. $\frac{4}{3} - \frac{4}{9}$ $\frac{8}{9}$

7. $\frac{2}{3} - \frac{1}{2}$ $\frac{1}{6}$

8. $\frac{5}{6} + \frac{7}{9}$ $1\frac{11}{18}$

9. $\begin{array}{r} \frac{5}{6} \\ - \frac{2}{3} \\ \hline \end{array}$ $\frac{1}{6}$

10. $\begin{array}{r} \frac{5}{6} \\ + \frac{2}{5} \\ \hline \end{array}$ $1\frac{7}{30}$

11. $\begin{array}{r} \frac{15}{4} \\ - \frac{3}{2} \\ \hline \end{array}$ $2\frac{1}{4}$

12. $\begin{array}{r} \frac{1}{7} \\ + \frac{3}{14} \\ \hline \end{array}$ $\frac{5}{14}$

13. $\begin{array}{r} \frac{3}{7} \\ + \frac{5}{3} \\ \hline \end{array}$ $2\frac{2}{21}$

14. $\begin{array}{r} \frac{7}{10} \\ - \frac{3}{5} \\ \hline \end{array}$ $\frac{1}{10}$

15. $\begin{array}{r} \frac{7}{8} \\ + \frac{2}{3} \\ \hline \end{array}$ $1\frac{13}{24}$

16. $\begin{array}{r} \frac{5}{9} \\ - \frac{2}{5} \\ \hline \end{array}$ $\frac{7}{45}$

17. $\frac{2}{9} + \frac{2}{3} + \frac{3}{4}$ $1\frac{23}{36}$

18. $\frac{4}{5} + \frac{4}{3} + \frac{1}{2}$ $2\frac{19}{30}$

19. $\frac{6}{5} + \frac{4}{3} + \frac{3}{4}$ $3\frac{17}{60}$

20. $\begin{array}{r} 5\frac{3}{5} \\ + 2\frac{1}{10} \\ \hline \end{array}$ $7\frac{7}{10}$

21. $\begin{array}{r} 7\frac{3}{4} \\ - 2\frac{1}{6} \\ \hline \end{array}$ $5\frac{7}{12}$

22. $\begin{array}{r} 5\frac{3}{8} \\ + 7 \\ \hline \end{array}$ $12\frac{3}{8}$

23. $\begin{array}{r} 17 \\ - 6\frac{2}{5} \\ \hline \end{array}$ 10

24. $4\frac{2}{3} + 5\frac{1}{2}$ $10\frac{1}{6}$

25. $5\frac{3}{4} + 4\frac{2}{5} + 3\frac{1}{2}$ $13\frac{13}{20}$

26. $7\frac{1}{2} + 6\frac{1}{5} + 4\frac{3}{5}$ 1

27. $3\frac{2}{5} - 1\frac{2}{3}$ $1\frac{11}{15}$

28. $7\frac{3}{4} - 4\frac{2}{3}$ $3\frac{1}{12}$

29. $16\frac{5}{8} - 8\frac{3}{16}$ $8\frac{7}{16}$

Solve. Simplify if possible.

30. $x + \frac{2}{3} = \frac{4}{5}$ $\frac{2}{15}$

31. $x - \frac{1}{2} = \frac{5}{6}$ $1\frac{1}{3}$

32. $x + \frac{2}{5} = \frac{1}{2}$ $\frac{1}{10}$

33. $x - 4\frac{1}{2} = 5\frac{1}{3}$ $9\frac{5}{6}$

34. $x + 3\frac{1}{4} = 7\frac{2}{3}$ $4\frac{5}{12}$

35. $x - 6\frac{1}{2} = 6\frac{3}{5}$ 13

PROBLEM SOLVING

36. Harry cut a $6\frac{3}{4}$ ft board from a sheet of plywood 10 ft long. How long was the remaining piece? $3\frac{1}{4}$ ft

37. Amy weighs $110\frac{1}{2}$ lb. How many pounds must she gain in order to weigh $115\frac{3}{5}$ lb? $5\frac{1}{10}$ lb

38. Susan needs $5\frac{7}{8}$ yd of wool for a skirt, $1\frac{1}{4}$ yd for a vest, and $3\frac{1}{2}$ yd for a jacket. How many yards of wool does she need for the outfit? $10\frac{5}{8}$ yd

39. Gene jogs $5\frac{1}{2}$ mi on Monday, $6\frac{1}{4}$ mi on Wednesday, and $8\frac{1}{3}$ mi on Saturday. How many miles does he jog? $20\frac{1}{12}$ mi

PRACTICE ON READING TABLES AND READING AND MAKING GRAPHS: CHAPTER 6

Answer questions 1–4 using the table.

BALANCE REMAINING OF $1,000 LOAN OVER 5 YEARS IF YOU ARE MAKING THE REQUIRED ANNUAL PAYMENTS.

Rate of Interest	1 Year	2 Years	3 Years	4 Years	5 Years
11%	$841	$664	$446	$246	Paid off
$11\frac{1}{2}\%$	$843	$667	$470	$248	Paid off
12%	$845	$670	$473	$250	Paid off
$12\frac{1}{2}\%$	$846	$673	$476	$253	Paid off
13%	$848	$675	$479	$255	Paid off

1. You borrowed $1,000 at 11%. How much money will you still owe after 3 yr? $446

2. You borrowed $1,000 at 13%. How much money will you still owe after 2 yr? $675

3. You borrowed $1,000 at $12\frac{1}{2}\%$. How much money will you still owe after 4 yr? $253

4. You borrowed $1,000 at 12%. When would this loan be paid off?   after 5 yr

5. Construct a pictograph using the information below.   Check students' graphs for
   35,000 Fans, Game 1; 40,000 Fans, Game 2;   Exercises 5–8.
   38,000 Fans, Game 3; 45,000 Fans, Game 4

6. Make a vertical bar graph to show these baseball scores: Game 1, 3 runs; Game 2, 12 runs; Game 3, 7 runs; Game 4, 1 run.

7. Make a horizontal bar graph to show the average monthly temperature for January in Chicago, 49°; Denver, 30°; Miami, 67°; Honolulu, 72°.

8. Make a line graph to show these test scores. First test, 95; Second test, 100; Third test, 88; Fourth test, 92; Fifth test, 98.

PROBLEM SOLVING
What is the number of births for each of the months?

9. September 1984   60

10. October 1982   40

11. November 1982   70

12. December 1984   70

13. Which of the years had the most births for Sept. through Dec.?   1984

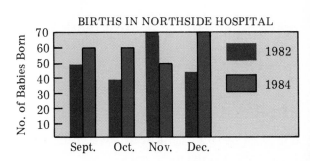

PRACTICE ON DECIMAL COMPUTATION: CHAPTER 7

Round to the nearest tenth and to the nearest hundredth.

1. 0.736 0.7 0.74 **2.** 7.659 7.7 7.66 **3.** 19.576 19.6 19.58 **4.** 16.435 16.4 16.44 **5.** 9.016 9.0 9.0²

6. 97.455 97.5 97.46 **7.** 0.150 0.2 0.15 **8.** 173.497 173.5 173.50 **9.** 4.055 4.1 4.06 **10.** 10.986 11 10

Change to decimals. Round to the nearest tenth and to the nearest hundredth.

11. $\frac{658}{100}$ 6.6 6.58 **12.** $\frac{6}{10}$ 0.6 0.60 **13.** $\frac{86}{1000}$ 0.1 0.09 **14.** $\frac{3}{5}$ 0.6 0.60 **15.** $\frac{2}{3}$ 0.7 0.67

16. $\frac{5}{11}$ 0.5 0.45 **17.** $\frac{6}{7}$ 0.9 0.86 **18.** $\frac{3}{4}$ 0.8 0.75 **19.** $\frac{5}{12}$ 0.4 0.42 **20.** $\frac{7}{9}$ 0.8 0.78

Add. **21.** 48.53 **22.** 110.824 **23.** 19.851 **24.** 7.906 **25.** 178.111

21. $34.65 + 4.32 + 9.56$ **22.** $8.07 + 93.654 + 9.1$ **23.** $5.1 + 6.83 + 7.921$

24. $0.35 + 0.256 + 7.3$ **25.** $43.1 + 72.01 + 63.001$ **26.** $41.1 + 38.23 + 0.6$

27. $100.41 + 73.256 + 0.3$ **28.** $9.101 + 12.35 + 0.721$ **29.** $0.75 + 6.25 + 1.378$

26. 79.93 **27.** 173.966 **28.** 22.172 **29.** 8.378

Subtract.

30. 73.47 − 41.68   31.79

31. 96.563 − 7.358   89.205

32. 127.065 − 72.198   54.867

33. Subtract 4.23 from 67.1. 62.87 **34.** Subtract 4.37 from 73. 68.63

Multiply.

35. $(81.3)(7.45)$ 605.685 **36.** $(43.75)(6.15)$ 269.0625 **37.** $(132.1)(0.73)$ 96.433

38. 81.3 × 7.5   609.75

39. 596 × 0.35   208.6

40. 68.5 × 7.39   506.215

Divide. Round to the nearest hundredth.

41. $0.03\overline{)4.987}$ 166.23 **42.** $3.5\overline{)9.721}$ 2.78 **43.** $0.65\overline{)72.956}$ 112.24

44. $5.65\overline{)93.498}$ 16.55 **45.** $0.71\overline{)4.1987}$ 5.91 **46.** $2.42\overline{)91.865}$ 37.96

Solve. Round to the nearest tenth.

47. $0.2x = 0.48$ 2.4 **48.** $0.06x = 9.6$ 160 **49.** $3.2x = 4.87$ 1.5

50. $5.01x = 6.843$ 1.4 **51.** $1.2x = 48.3$ 40.3 **52.** $0.05x = 9.56$ 191.2

PROBLEM SOLVING

53. Carlos ran 28.5 km in 2.5 h. Find his average speed. 11.4 km/h

54. Kathy saves $16.35 each week. How much does she save in 36 weeks? $588.6(

PRACTICE ON SOLVING PROBLEMS INVOLVING METRIC UNITS: CHAPTER 8

Change as indicated.

1. 48 mm to cm   4.8
2. 9.38 km to m   9,380
3. 8 cm to m   0.08
4. 12 hm to m   1,200
5. 6 dm to m   0.6
6. 0.008 m to mm   8
7. 8 dm to m   0.8
8. 75 m to hm   0.75
9. 42 km to m   42,000
10. 15 cm to m   0.15
11. 40 mm to cm   4
12. 1,500 m to km   1.5

EXTRA PRACTICE

Make true sentences.

15.   3,700   24. 7,300

13. 1 m = __?__ cm   100
14. 8.7 m = __?__ cm   870
15. 3.7 km = __?__ m
16. 8 cm = __?__ mm   80
17. 12 km = __?__ m   12,000
18. 4 cm = __?__ m   0.04
19. 4.6 m² = __?__ cm²   46,000
20. 5 m² = __?__ cm²   50,000
21. 40,000 m² = __?__ ha   4
22. 8 L = __?__ mL   8,000
23. 15 mL = __?__ L   0.015
24. 7.3 L = __?__ mL
25. 30 mL = __?__ L   0.03
26. 12 L = __?__ mL   12,000
27. 25 mL = __?__ L
28. 6.2 g = __?__ gm   6,200
29. 4,500 g = __?__ kg   4.5
30. 15 kg = __?__ t   0.015

27.   0.025

Change Celsius to Fahrenheit. Compute to the nearest degree.

34.   189°F

31. C = 25°   77°F
32. C = 41°   106°F
33. C = 9°   48°F
34. C = 87°
35. C = 32°   90°F
36. C = 15°   59°F
37. C = 61°   142°F
38. C = 99°
39. C = 55°   131°F
40. C = 72°   162°F
41. C = 48°   118°F
42. C = 21°

38.   210°F   42.   70°F

Change Fahrenheit to Celsius. Compute to the nearest degree.

46.   23°C

43. F = 100°   38°C
44. F = 40°   4°C
45. F = 52°   11°C
46. F = 73°
47. F = 77°   25°C
48. F = 51°   11°C
49. F = 85°   29°C
50. F = 33°
51. F = 64°   18°C
52. F = 49°   9°C
53. F = 96°   36°C
54. F = 200°

50.   1°C   54.   93°C

PROBLEM SOLVING

55. A square measures 6.8 cm on each side. Find the area in mm².   4,624 mm²

56. A rectangular field measures 210 m by 90 m. Find the area in hm².   1.89 hm²

57. A family drank 3.5 L of milk. How many milliliters is this?   3,500 mL

58. A recipe calls for 0.75 L of lemonade. How many milliliters is this?   750 mL

59. A car weighs 950 kg. How many metric tons is this?   0.95 t

60. Lorin must take 6 mg of medicine each day. How many grams will he take in a year? (*Hint:* There are 365 days in a year.)   2.19 g

61. John drove his car for 75 km. How many meters is this?   75,000 m

62. A pole is 15 m long. How many centimeters is this?   1,500 cm

EXTRA PRACTICE

459

MIXED PRACTICE: CHAPTERS 1–8

Evaluate for the given values of the variables.
1. $y + 7; y = 6$ 13
2. $9n; n = 10$ 90
3. $16a + 3 + 7a; a = 4$ 9⁣
4. $15ab + a + b + 20; a = 5, b = 4$ 329
5. $17 + 4xy + 8y; x = 5, y = 3$ 101

Which, if any, of the values is a solution of the open sentence?
6. $n + 5 = 12; 7, 17$ 7
7. $a - 9 = 16 - 2; 5, 14, 24$ none

Solve each equation. Check the solution.
8. $y - 3 = 19$ 22
9. $42 + x = 50$ 8
10. $9y = 72$ 8
11. $47 = \frac{x}{3}$ 141
12. $4m + 9 = 53$ 11
13. $41 = 5n - 4$ 9
14. $\frac{2}{3}x = 20$ 30
15. $\frac{3}{9}x = \frac{27}{2}$ 40.5
16. $\frac{3}{5}x = 15$ 25
17. $x - \frac{1}{2} = \frac{2}{3}$ $1\frac{1}{6}$
18. $x + \frac{3}{5} = \frac{1}{2}$ $-\frac{1}{10}$
19. $x + 2\frac{1}{2} = 4\frac{3}{5}$ $2\frac{1}{10}$

Find the value.
20. 4^2 16
21. 5^3 125
22. 1^8 1
23. 3^5 243

Simplify.
24. $\frac{3}{5} \cdot \frac{15}{21}$ $\frac{3}{7}$
25. $3\frac{1}{2} \cdot \frac{16}{9}$ $6\frac{2}{9}$
26. $\frac{3}{5} + \frac{8}{3}$ $3\frac{4}{15}$
27. $7\frac{3}{5} - 2\frac{3}{4}$ $4\frac{17}{20}$
28. $\frac{2}{5}$ of 25   10
29. $\frac{2}{3} \div \frac{7}{12}$ $1\frac{1}{7}$
30. $3\frac{1}{2} \div 1\frac{7}{8}$ $1\frac{13}{15}$
31. $16 \div 1\frac{1}{3}$ 12

Round to the nearest tenth and to the nearest hundredth.
32. 0.437   0.4, 0.44
33. 9.758   9.8, 9.76
34. 12.019   12.0, 12.02
35. 63.455   63. ⁣
36. Add $36.45 + 0.316 + 147.4$ 184.166
37. Subtract 3.87 from 43.96   40.09
38. Divide $48.6\overline{)493.45}$ 10.15
39. Multiply $(19.6)(123.45)$ 2,419.62

Change as indicated.
40. 63 mm to cm   6.3
41. 4.63 km to m   4,630
42. 10 cm to m   0.1
43. 9 dm to m   0.9
44. 43 m to hm   0.43
45. 13 hm to m   1,300

Change to Celsius or Fahrenheit. Find to the nearest degree.
46. $C = 30°$ 86°F
47. $F = 90°$ 32°C
48. $C = 80°$ 176°F
49. $F = 10°$ –

50. 22 is the same as a number increased by 8. What is the number?   14
51. 9 times a number decreased by 8 is 37. Find the number.   5

52. $A = \frac{1}{2}bh$. Find A if $b = \frac{6}{7}$, and $h = 14$.   6
53. Juanita saves $21.50 each month. How much is saved after 8 months? $17⁣

54. Make a vertical bar graph to show the attendance for the following games:
   Game 1, 1500; Game 2, 1800; Game 3, 1650; Game 4, 1200   Check students' graphs.

PRACTICE ON PERCENT: CHAPTER 9

Change to decimals.

1. 50% 0.5
2. 7% 0.07
3. 400% 4.00
4. 32.4% 0.324
5. 0.7% 0.007
6. 19.7% 0.197
7. 425% 4.25
8. 3.6% 0.036
9. 0.0007% 0.000007
10. $54\frac{1}{3}$% 0.5433

Change to percents.

11. 0.07 7%
12. 0.654 65.4%
13. 18 1,800%
14. $\frac{3}{5}$ 60%
15. $0.12\frac{1}{2}$ $12\frac{1}{2}$%
16. $\frac{5}{8}$ $62\frac{1}{2}$%
17. $0.45\frac{3}{5}$ $45\frac{3}{5}$%
18. 0.02 2%
19. 45.6 4,560%
20. 9.32 932%

Compute.

21. 15% of 743 is what number? 111.45
22. 8.7% of 48.64 is what number? 4.23
23. $6\frac{1}{4}$% of 156 is what number? 9.75
24. 70% of 496.50 is what number? 347.55
25. 36.4% of 945.34 is what number? 344.10
26. $5\frac{3}{4}$% of 90 is what number? 5.175

Compute.

27. What % of 90 is 54? 60%
28. What % of 120 is 48? 40%
29. What % of 155 is 73? about 47%
30. What % of 750 is 150? 20%
31. What % of 745 is 42.6? about 5.7%
32. What % of 12.8 is 5.7? about 44.5%

Compute.

33. 24 is 60% of what number? 40
34. 90 is 50% of what number? 180
35. 18.6 is 20% of what number? 93
36. 43.2 is 37.5% of what number? 115.2
37. 120 is 70% of what number? 171.43
38. 16.5 is 92.3% of what number? about 17.9

PROBLEM SOLVING

39. The Tigers played 30 games. They won 70% of the games. How many games did they win? 21

40. The Wildcats played 20 games. They lost 4 games. What % of the games played did they lose? 20%

41. Jane got a $1,200 raise. The rate was 8% of her salary. What was her salary? $15,000

42. Darlene earns 12% commission on all sales. This week her sales were $950. Find her commission. $114

43. Find the cost of the following purchases at a 30% discount: $32, sweater; $48.50, shoes; $125.40, coat.
$22.40; $33.95; $87.78

44. Find the cost of the following purchases at a "$\frac{1}{4}$ off" sale: $34.50, toaster; $125.30, fan; $64.25 radio.
$25.88; $93.98; $48.19

PRACTICE ON WORKING WITH STATISTICS AND PROBABILITY: CHAPTER 10

Find the range. Find the mean.

1. 18, 12, 20, 16, 14, 10   r = 10; m = 15
2. 98, 100, 96, 98, 94, 96   r = 6; m = 97
3. 253, 248, 295, 309, 290   r = 61; m = 279
4. 1000, 1010, 995, 990, 1200   r = 210; m =
5. 50, 54, 48, 46, 43, 56, 60   r = 17; m = 51
6. 10, 9, 7, 8, 8, 6, 10, 9, 7, 8, 9, 7   r = 4; m =

Find the median.

7. 100, 98, 98, 97, 96, 95, 94   97
8. 18, 20, 19, 22, 26, 24, 21   21
9. 12, 10, 19, 18, 16, 14   15
10. 100, 98, 96, 96, 95, 98, 94, 96   96
11. 50, 48, 46, 48, 52, 54, 50   50
12. 12, 10, 8, 10, 9, 8, 8, 10, 11, 12, 10   10

Find the mode(s).

13. 98, 96, 100, 98, 94, 93, 92   98
14. 12, 10, 9, 10, 12, 8, 6, 10, 11   10
15. 4, 5, 9, 8, 6, 4, 5, 3   4 and 5
16. 100, 98, 98, 100, 96, 94, 93, 92   98 and 100
17. 56, 54, 50, 53, 52, 51   none
18. 250, 248, 246, 248, 245, 244   248

Use these scores for Exercises 19 and 20.

98	96	98	100	100	97	90	94
96	98	93	88	86	88	94	100

19. Make a frequency table.
    Check students' tables.
20. Find the mean.   94.75

Answer the questions about the spinner.

21. How many possible outcomes?   12
22. How many even outcomes?   7
23. How many odd outcomes?   5
24. P(2)? $\frac{1}{3}$ **25.** P(3)? $\frac{1}{6}$ **26.** P(4)? $\frac{1}{4}$
27. Spin the arrow 100 times. About how many times should it stop on an even number?   58

PROBLEM SOLVING

28. A deck of cards contains 52 cards. There are 4 jacks in the deck. Find P(jack). $\frac{1}{13}$

29. A bag contains 80 red marbles and 100 white marbles. Find P(red). $\frac{4}{9}$ Find P(white). $\frac{5}{9}$

30. There are 5 candidates for treasurer and 4 for secretary. In how many ways can these offices be filled?   20

31. How many three-digit numbers can be formed from the digits 2, 4, and 5, if each digit can be repeated?   27

PRACTICE ON SOLVING AND APPLYING RATIOS AND PROPORTIONS: CHAPTER 11

A bag contains 14 pennies, 10 dimes, 7 quarters, 4 half-dollars, and 6 silver dollars. Find each ratio. Simplify.

1. dimes to quarters $\frac{10}{7} = 1\frac{3}{7}$
2. half-dollars to pennies $\frac{4}{14} = \frac{2}{7}$
3. pennies to dimes $\frac{14}{10} = 1\frac{2}{5}$
4. silver dollars to pennies $\frac{6}{14} = \frac{3}{7}$
5. quarters to half-dollars $\frac{7}{4} = 1\frac{3}{4}$
6. dimes to silver dollars $\frac{10}{6} = 1\frac{2}{3}$

Solve each proportion.

7. $\frac{4}{5} = \frac{n}{10}$ $n = 8$

8. $\frac{4}{3} = \frac{x}{6}$ $x = 8$

9. $\frac{a}{2} = \frac{5}{10}$ $a = 1$

10. $\frac{7}{x} = \frac{3}{6}$ $x = 14$

11. $\frac{n}{9} = \frac{3}{27}$ $n = 1$

12. $\frac{12}{3} = \frac{2}{y}$ $y = \frac{1}{2}$

13. $x{:}5 = 3{:}15$ $x = 1$

14. $12{:}y = 24{:}36$ $y = 18$

15. $3{:}5 = 18{:}n$ $n = 30$

16. $\frac{n}{4} = \frac{18}{9}$ $n = 8$

17. $\frac{8}{7} = \frac{24}{a}$ $a = 21$

18. $\frac{8}{1} = \frac{n}{3}$ $n = 24$

Which are true proportions?

19. $\frac{4}{9} = \frac{24}{54}$ yes

20. $\frac{9}{2} = \frac{45}{10}$ yes

21. $\frac{3}{15} = \frac{15}{60}$ no

22. $\frac{3}{10} = \frac{18}{60}$ yes

23. $\frac{28}{35} = \frac{4}{5}$ yes

24. $\frac{10}{30} = \frac{1}{3}$ yes

PROBLEM SOLVING

Solve each word problem. Use a proportion.

25. William drove 180 km in 2 hours. How long will it take him to drive 300 km?   3 h 20 min

26. Tammy made $7.50 for 2 hours of cutting grass. How much did she earn for 3 hours of work? $11.25

27. Hubie bought 3 m of cloth for $18.75. How many meters of cloth can he buy for $31.25?   5 m

28. It takes 8 cups of flour to bake 3 cakes. How many cups of flour are needed to bake 5 cakes? $13\frac{1}{3}$ cups

29. Helen bought 4 pairs of socks for $9.00. How many pairs of socks can she buy for $13.50?   6

30. It takes 10 eggs to make 4 quiches. How many eggs does it take to make 6 quiches?   15

31. Tanya earned $2,600 by working part-time. Make a circle graph to show how she spent her earnings.   Check students' graphs.

Savings	$1,200	Charity	$ 100
School supplies	$ 400	Entertainment	$ 400
Clothing	$ 500		

EXTRA PRACTICE

PRACTICE ON ADDING AND SUBTRACTING INTEGERS: CHAPTER 12

Compare. Use > or <.

1. $7 \equiv 12$ <
2. $15 \equiv 14$ >
3. $-3 \equiv 1$ <
4. $0 \equiv -7$ >
5. $-9 \equiv -16$ >
6. $-10 \equiv 10$ <
7. $19 \equiv -16$ >
8. $9 \equiv -9$ >
9. $0 \equiv 10$ >
10. $6 \equiv -19$ >
11. $-21 \equiv 35$ <
12. $-41 \equiv -50$ >

Add.

13. $9 + 2$ 11
14. $12 + (-2)$ 10
15. $-4 + (-6)$ -10
16. $-5 + 8$ 3
17. $15 + (-1)$ 14
18. $-16 + 9$ -7
19. $7 + (-15)$ -8
20. $16 + (-3)$ 13
21. $-10 + (-16)$ -16
22. $-21 + (-3)$ -24
23. $-15 + 9$ -6
24. $-3 + (-7)$ -10

Copy and complete.

25. $16 + (-16) = ?$ 0
26. $12 + ? = 0$ -12
27. $-9 + ? = 0$ 9
28. $19 + ? = 19$ 0
29. $-15 + 15 = ?$ 0
30. $20 + ? = 0$ -20

Subtract.

31. $15 - 3$ 12
32. $10 - 12$ -2
33. $-6 - 5$ -11
34. $-3 - (-5)$ 2
35. $-10 - 9$ -19
36. $-20 - (-10)$ -10
37. $16 - 4$ 12
38. $12 - (-3)$ 15
39. $-30 - (-15)$ -15
40. $-17 - 3$ -20
41. $-40 - (-15)$ -25
42. $-12 - 12$ $-$

Solve and check.

43. $x + (-6) = -3$ 3
44. $5 + x = -4$ -9
45. $x - 3 = 7$ 10
46. $x - (-2) = 8$
47. $12 + x = -2$ -14
48. $x - (-7) = -3$ -10
49. $9 + x = -3$ -12
50. $x + (-3) = 12$
51. $x - (-10) = 10$ 0
52. $x - 3 = -7$ -4
53. $x - (-2) = -8$ -10
54. $x - 9 = 1$ 10

PROBLEM SOLVING

55. At 7:00 A.M., the temperature was 45°F. It rose 10°F by noon and rose 23°F between noon and 5:00 P.M. What was the temperature at 5:00 P.M.?   78°F

56. At 6:00 A.M., the temperature was 8°C. It dropped 2°C by 10:00 A.M. and rose 4°C by 2:00 P.M. What was the temperature at 2:00 P.M.?   10°C

57. Fred won 200 points, lost 75 points, lost 15 points, and won 32 points. How many points did he have?   142

58. A football team gained 12 yards, lost 3 yards, and lost 5 yards. What was the net result?   gain of 4 yd

59. In a game, Jo lost 40 points, lost 10 points, won 60 points, and won 150 points. How many points did she have?   160

60. Sam had $95.50 in his checking account. He wrote a check for $8.75, deposited $15.35, wrote a check for $9.50 and another check for $10.95. What was his final balance? $81.65

464

PRACTICE ON MULTIPLYING AND DIVIDING INTEGERS: CHAPTER 13

Multiply.

1. $15 \cdot 18$ 270   **2.** $20 \cdot (-3)$ -60 **3.** $-10 \cdot 35$ -350 **4.** $-43 \cdot 0$ 0

5. $-32 \cdot (-5)$ 160   **6.** $-73 \cdot 42$ $-3,066$ **7.** $-34 \cdot (-45)$ 1,530   **8.** $18 \cdot (-75)$ $-1,350$

9. $0 \cdot 98$ 0   **10.** $-36 \cdot (-68)$ **11.** $9 \cdot (-48)$ -432 **12.** $-53 \cdot (-38)$
                              2,448   2,014

Multiply. Use the distributive property.

                                                                   -52

13. $-5 \cdot (-3 + 6)$ -15 **14.** $-6 \cdot (-2 - 4)$ 36   **15.** $-13 \cdot (10 - 6)$

16. $8 \cdot (6 - 9)$ -24 **17.** $-10 \cdot (15 - 3)$ -120 **18.** $12 \cdot (-3 + 9)$ 72

19. $-11 \cdot (-5 - 10)$ 165   **20.** $20 \cdot (-2 + 9)$ 140   **21.** $-14 \cdot (-2 - 10)$
                                                                                 168

Divide, if possible.

22. $-45 \div 1$ -45 **23.** $65 \div (-1)$ -65 **24.** $-44 \div 11$ -4 **25.** $-33 \div (-3)$ 11

26. $0 \div 43$ 0   **27.** $-400 \div 40$ -10 **28.** $-200 \div (-10)$ 20 **29.** $96 \div (-12)$ -8

30. $-500 \div (-25)$ **31.** $-36 \div (-9)$ 4 **32.** $90 \div (-9)$ -10**33.** $-72 \div 0$
     20   not possible

Divide.

34. $\dfrac{2}{3} \div \dfrac{7}{5}$ $\dfrac{10}{21}$ **35.** $-\dfrac{3}{4} \div \left(-\dfrac{2}{5}\right)$ $1\dfrac{7}{8}$ **36.** $\dfrac{5}{6} \div \left(-\dfrac{3}{7}\right)$ $-1\dfrac{17}{18}$ **37.** $-\dfrac{9}{5} \div \dfrac{7}{4}$ $-1\dfrac{1}{35}$

38. $1\dfrac{1}{2} \div \left(-\dfrac{3}{8}\right)$ -4 **39.** $-\dfrac{3}{4} \div 2\dfrac{1}{4}$ $-\dfrac{1}{3}$ **40.** $-\dfrac{3}{8} \div \left(-5\dfrac{1}{7}\right)$ $\dfrac{7}{96}$**41.** $2\dfrac{2}{5} \div 1\dfrac{5}{9}$ $1\dfrac{19}{35}$

42. $-2\dfrac{1}{5} \div 4\dfrac{1}{2}$ $-\dfrac{22}{45}$ **43.** $-1\dfrac{1}{3} \div \left(-2\dfrac{1}{5}\right)$ **44.** $\dfrac{8}{3} \div \left(-5\dfrac{1}{2}\right)$ $-\dfrac{16}{33}$ **45.** $-6\dfrac{1}{3} \div 5\dfrac{1}{2}$
                                        $\dfrac{20}{33}$ $-1\dfrac{5}{11}$

Simplify.

46. $3a + (-2a)$ a **47.** $6x - (-3x)$ $9x$ **48.** $-5 \cdot 4x$ $-20x$

49. $-6 \cdot (-5m)$ $30m$ **50.** $4(2a - 5)$ $8a - 20$ **51.** $-(3 + 2y)$

52. $-(6 - 5)$ $-18a + 15$ **53.** $-2a - (-3a)$ a **54.** $-7(4x + 3)$

55. $-2(5a - 3)$ $-10a + 6$ **56.** $-(3 - 2y)$ $-3 + 2y$ **57.** $-4(6a - 3)$

58. $-6n + (-5n)$ $-11n$ **59.** $-(-3 + 2y)$ $3 - 2y$ **60.** $-4a - 7a$ $-11a$

                                                51. $-3 - 2y$ **54.** $-28x - 2y$
                                                                   57. $-24a + 12$

Solve and check.

61. $-6x = 36$ -6 **62.** $-5y = -45$ 9 **63.** $4a = -48$ -12 **64.** $-11x = -33$ 3

65. $7n = -21$ -3 **66.** $-10y = 120$$-12$ **67.** $-9x = -9$ 1   **68.** $5a = -65$ -13

69. $\dfrac{a}{-3} = 9$ -27 **70.** $\dfrac{m}{4} = -10$ -40**71.** $\dfrac{x}{-9} = -3$ 27   **72.** $\dfrac{y}{-2} = 12$ -24

73. $\dfrac{n}{-5} = 10$ -50 **74.** $\dfrac{x}{-9} = -20$ 180 **75.** $\dfrac{y}{10} = -10$ -100 **76.** $\dfrac{a}{-1} = -6$ 6
                                                          -3 -5 -1

77. $-2x + 5 = 9$ -2**78.** $4y - 3 = -15$ -3 **79.** $-3n - 1 = 14$ -5 **80.** $-6a - 7 = -1$ -1

8.1 $\dfrac{c}{4} - 5 = -3$ 8 **82.** $\dfrac{a}{-3} + 2 = -1$ 9 **83.** $\dfrac{x}{-2} - 4 = -2$ -4 **84.** $\dfrac{y}{5} + 6 = 3$ -15

PRACTICE ON SOLVING EQUATIONS AND INEQUALITIES: CHAPTER 14

Solve.

1. $-3y - 15 = 5y + 25$ -5
2. $6a - 3 = 13 - 2a$ 2
3. $-7x - 3 = -4x + 3$ -2
4. $2(x - 5) = 12$ 11
5. $-3(a - 2) = -6$ 4
6. $-5(2 - 2y) = 2(4y - 6)$ -1
7. $n - 3 > 7$ $n > 10$
8. $a - 4 \le -6$ $a \le -2$
9. $3h \ge -15$ $h \ge -5$
10. $-2y < -14$ $y > 7$
11. $\frac{1}{2}x \le -2$ $x \le -4$
12. $-\frac{y}{5} > -3$ $y < 15$

Evaluate.

13. a^5 if $a = -2$ -32
14. $-4x^6$ if $x = -1$ -4
15. $3a^2b$ if $a = 4$ and $b = -1$ -48
16. $8a^2b^3$ if $a = -3$ and $b = 8$ $36{,}864$

Simplify.   26. $x^2 + x - 5$ 27. $-c^2 - 12c + 12$ 28. $13y^2 + y + 17$

17. $a^5 \cdot a^3$ a^8
18. $x \cdot x^6$ x^7
19. $(y^2)^3$ y^6
20. $(x^5)^2$ x^{10}
21. $(xy)^3$ x^3y^3
22. $(a^2b)^3$ a^6b^3
23. $(-3a^2)^2$ $9a^4$
24. $(-4ab^2)^3$ $-64a^3b^6$
25. $(2x^2y^3)^3$ $8x^6y^9$
26. $3x^2 - 4x - 1 - 2x^2 + 5x - 4$
27. $-4c + 6c^2 + 5 - 8c - 7c^2 + 7$
28. $5y^2 + 7y + 14 - 6y + 8y^2 + 3$
29. $6a^6 + 5 - 3a^4 + a^6 + 5a^4 + 9$
30. $-3d + 6d^2 + 7d^4 - 3 + 5d + 8d^2 - d^4$ $6d^4 + 14d^2 + 2d - 3$

Multiply.   29. $7a^6 + 2a^4 + 14$ 32. $-3x^3 + 6x^2 - 15x$

33. $2y^3 + 6y^2 + 4y$

31. $a(-3a^2 + a)$ $-3a^3 + a^2$
32. $-3x(x^2 - 2x + 5)$
33. $-2y(-y^2 - 3y - 2)$
34. $-6x(5 - 3x - 2x^2)$
35. $2m(m - 3m^2 + 1)$
36. $5(-3 + 6a - 3a^2)$
34. $12x^3 + 18x^2 - 30x$
35. $-6m^3 + 2m^2 + 2m$
36. $-15a^2 + 30a - 15$

Factor.

37. $3x - 9$ $3(x - 3)$
38. $3a^3 - 12a^2 + 18a$ $3a(a^2 - 4a + 6)$
39. $6a^3 + a$ $a(6a^2 + 1)$
40. $4x^2 - 8x + 24$ $4(x^2 - 2x + 6)$

PROBLEM SOLVING

$1\frac{1}{2}$

41. Twice a number, decreased by 8, is equal to 14. What is the number?   11

42. Six times a number, increased by 10, is equal to 19. What is the number?

43. Four times a number, decreased by 7, is equal to 21. Find the number.   7

44. Three times a number, increased by 2, is equal to 4. Find the number. $\frac{2}{3}$

45. Ten times a number, increased by 5, is equal to 35. Find the number.   3

46. Five times a number, decreased by 15, is equal to 15. Find the number.   6

47. Twice a number, increased by 16, is equal to 48. Find the number.   16

48. Nine times a number, decreased by 10, is 71. Find the number.   9

PRACTICE ON CLASSIFYING GEOMETRIC FIGURES AND ON FINDING SQUARE ROOTS: CHAPTER 15

Complementary or Supplementary?

1. 50°, 40° C
2. 130°, 50° S
3. 90°, 90° S
4. 60°, 30° C

In the figure at the right, line m ∥ line n. Name two pairs of congruent angles for each of the following angles.
Answers may vary.   Sample answers given.

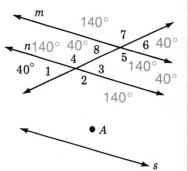

5. vertical ∠1 and ∠3
6. alternate interior ∠4 and ∠5
7. corresponding ∠6 and ∠3
8. congruent ∠2 and ∠4
9. Determine the measures of angles labeled 2 through 8.
10. Construct a line perpendicular to line s that passes through point A. Check students' answers.

How many sides does each of the following have?

11. triangle 3
12. pentagon 5
13. hexagon 6
14. quadrilateral 4

Draw each of the following. Check students' graphs.

15. square
16. triangle
17. pentagon
18. hexagon

Classify the triangles according to the measures of their angles.

19. 43°, 52°, 85° acute
20. 80°, 80°, 20° acute
21. 60°, 30°, 90° right

Find the measure of the third angle, given the measures of two angles of a triangle.

22. 38°, 72° 70°
23. 40°, 120° 20°
24. 10°, 165° 5°

Give two square roots of each number.

25. 36 ±6
26. 81 ±9
27. 100 ±10
28. 25 ±5
29. 144 ±12

Approximate to the nearest tenth. Use the table on page 473.

30. $\sqrt{5}$ 2.2
31. $\sqrt{32}$ 5.7
32. $\sqrt{18}$ 4.2
33. $\sqrt{3}$ 1.7
34. $\sqrt{98}$ 9.9

For each right triangle, find the missing length to the nearest tenth.

35. $a = 3, b = 4$ $c = 5$
36. $a = 5, c = 13$ $b = 12$
37. $b = 5, c = 7$ $a = 4.9$

PROBLEM SOLVING

38. One of two supplementary angles is 3 times the size of the other. What are the measures of the two angles? 45°, 135°

39. The lengths of two sides of a right triangle are 5 in. and 8 in. Find the length of the hypotenuse. 9.4 in.

PRACTICE ON GRAPHING INTEGERS, EQUATIONS, AND INEQUALITIES: CHAPTER 16

EXTRA PRACTICE

Graph.   Check students' graphs.

1. the integer between -1 and 4
2. the integers greater than 2
3. the integers less than 1
4. the integers greater than -1

5. $2x - 5 = -3$
6. $3b + 6 = 15$
7. $-5n + 2 = -3$
8. $6y + 2 = 12 - y$
9. $-3a - 4 = 2a + 11$
10. $4x - 5 = 3x + 5$
11. $x > -2$
12. $x \le 3$
13. $x \ge 4$
14. $y \le 0$
15. $y > 1$
16. $y < 2$
17. $2x - 2 > 6$
18. $3y + 5 \ge -4$
19. $5x - 1 < 4$
20. $4a - 9 \ge 16 - a$
21. $7x - 3 > 9 - 5x$
22. $y - 6 \le 5y - 2$

Give the ordered pair for each point.

23. A (3, 1)
24. B (0, 3)
25. C $(-4, 2)$
26. D (5, 0)
27. E $(2, -2)$
28. F $(-3, -2)$
29. G $(0, -3)$
30. H $(-2, 0)$

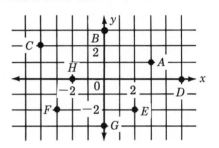

Graph each point.   Check students' graphs.

31. $A(3, 2)$
32. $M(-2, 4)$
33. $F(2, -1)$
34. $Y(0, 3)$
35. $B(-3, -4)$
36. $C(-4, 0)$

Graph.   Check students' graphs.

37. $y = x - 1$
38. $y = -x + 2$
39. $y = x + 3$
40. $y = 2x$
41. $y = 2x + 1$
42. $y = 3x - 2$

Find the slope of $\overleftrightarrow{AB}$ for the given pairs of points.

43. $A(2, 3); B(5, 7)$ $\frac{4}{3}$
44. $A(-4, 9); B(2, -3)$ -2
45. $A(0, 5); B(4, 3)$ $-\frac{1}{2}$
46. $A(8, 3); B(6, 1)$ 1

Find the slope and the y-intercept of the line with the given equation.

47. $y = 4x - 2$ $4, -2$
48. $y = 2x + 9$ $2, 9$
49. $y = \frac{3}{5}x - 1$ $\frac{3}{5}, -1$

50. $y = \frac{5}{3}x + 1$ $\frac{5}{3}, 1$
51. $y = \frac{1}{4}x + \frac{3}{4}$ $\frac{1}{4}, \frac{3}{4}$
52. $y = -2x$ $-2, 0$

Solve each system of equations by graphing.   Check students' graphs.

53. $y = x - 3$ $(-2, -5)$
     $y = 3x + 1$
54. $y = 2x + 12$ $(-2, 8)$
     $y = -4x$
55. $y = -5x$ $(1, -5)$
     $y = -x - 4$

EXTRA PRACTICE

MIXED PRACTICE: CHAPTERS 9–16

Change to decimals.

1. 60% 0.60 **2.** 8% 0.08 **3.** 500% 5.0 **4.** 89.6% 0.896 **5.** 0.0065%
0.000065

Change to percents.

6. 0.09 9% **7.** 0.537 53.7% **8.** $\frac{2}{5}$ 40% **9.** $0.15\frac{1}{2}$ 15.5% **10.** 48.35
4,835%

Compute.

11. 12% of 648 is what number? 77.76   **12.** 6.5% of 54 is what number? 3.51
13. What % of 80 is 56? 70%   **14.** What % of 890 is 356? 40%
15. 43.2 is 80% of what number? 54   **16.** 270 is 60% of what number? 450
17. Find the range, mean, median, and mode.
12, 10, 8, 12, 6, 8, 14 8, 10, 10, 8, and 12

Solve and check.

72

18. $\frac{3}{5} = \frac{n}{10}$ 6 **19.** $\frac{8}{x} = \frac{3}{6}$ 16 **20.** $\frac{4}{9} = \frac{8}{y}$ 18 **21.** $\frac{c}{8} - 4 = 5$
22. $x - 5 = 7$ 12 **23.** $8 + x = 16$ 8 **24.** $x + (-3) = 6$ 9 **25.** $-3a = 15$
26. $-12 = -2n + 6$ 9 **27.** $-2n + 9 = 7$ 1 **28.** $-3(a - 1) = 6$ −1
29. $-2y + 12 = 3y - 3$ 3 **30.** $5a - 2 = 13 + 2a$ 5 **31.** $-6(3 - 2y) = 18$
32. $a + 3 \geq 7$ $a \geq 4$ **33.** $-2x < -18$ $x > 9$ **34.** $3h \leq -21$ $h \leq -7$
25. −5 **31.** 3

Solve for x.

35. $x + a = b$ $x = b - a$ **36.** $x - q = p$ $x = p + q$ **37.** $-x + c = -d$
$x = c + d$

Simplify.   **55.** $-2x^3 + 6x^2 - 10x$

38. $16 + (-2)$ 14 **39.** $-12 + 6$ −6 **40.** $-20 + (-3)$ −23 **41.** $18 - 3$ 15
42. $-20 - (-6)$ −14 **43.** $-8 - 8$ −16 **44.** $-15 \cdot (-3)$ 45 **45.** $60 \cdot (-2)$
46. $-18 \div 2$ −9 **47.** $-45 \div -9$ 5 **48.** $\frac{2}{3} \div \frac{4}{9}$ $\frac{3}{2}$ **49.** $-\frac{3}{5} \div \frac{21}{10}$
50. $a^6 \cdot a^3$ a^9 **51.** $(x^2y)^2$ x^4y^2 **52.** $(-2xy^3)^4$ $16x^4y^{12}$ **53.** $(-2a^2)(9a^3)$
54. $n(-3n^2 + n)$ $-3n^3 + n^2$ **55.** $-2x(x^2 - 3x + 5)$ **56.** $3y(-y + 2y^2 - 3)$
57. $-(m^2 - 2m + 6)$ $-m^2 + 2m - 6$ **58.** $-4a^2 + 3a - 2 - (2a - 3a^2 + 1)$
$-a^2 + a - 3$ **56.** $6y^3 - 3y^2 - 9y$

Evaluate. $a = 3$, $b = 4$, $x = -2$, and $y = -1$ **45.** −120 **49.** $-\frac{2}{7}$ **53.** $-18a^5$

59. $3x^2y$ −12 **60.** $-4x^3y^2$ 32 **61.** $-(3b)^2$ −144 **62.** $(-2a)^2$ 36

Graph.   Check students' graphs.

63. $A(-4, 2)$ **64.** $B(3, -4)$ **65.** $C(-1, -3)$
66. $y = 3x - 1$ **67.** $y = x - 4$ **68.** $y = 2x$

Factor.

69. $4a^2 - 12a + 36$ $4(a^2 - 3a + 9)$

70. $10x^3 - 15x^2 + 25x$ $5x(2x^2 - 3x + 5)$

**Find the measure of the third angle, given the measures
of two angles of a triangle.**

71. $42°, 68°$ $70°$

72. $50°, 100°$ $30°$

73. $170°, 5°$ $5°$

74. A bag contains 120 white marbles and 60 red marbles. Find P(white). Find P(red).

$\frac{2}{3}, \frac{1}{3}$

A baseball team won 10 games and lost 4. Find each ratio.

75. Wins to total $\frac{10}{14}$ or $\frac{5}{7}$

76. Wins to losses $\frac{10}{4}$ or $\frac{5}{2}$

77. Losses to total $\frac{4}{14}$ or $\frac{2}{7}$

Find each ratio. Leave answers in fractional form.

78. $\sin A$ $\frac{3}{5}$

79. $\cos B$ $\frac{3}{5}$

80. $\tan A$ $\frac{3}{4}$

81. $\tan B$ $\frac{4}{3}$

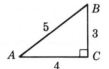

Write in mathematical terms.

82. 8 decreased by 6 $8 - 6$

83. 3 more than 6 times a number $6n + 3$

**For each right triangle, find the missing length to the nearest tenth.
a and b are legs, c is the hypotenuse.**

84. $a = 12, b = 7$ 13.9

85. $b = 8, c = 15$ 12.7

86. Classify each of the triangles in the
picture according to its sides. Tell
why you classified each as you did.
Check students' answers.

a. equilateral

b. isosceles

c. right

Approximate to the nearest tenth. Use the table on page 473.

87. $\sqrt{53}$ 7.3

88. $\sqrt{14}$ 3.7

89. $\sqrt{3}$ 1.7

90. $\sqrt{75}$ 8.7

Find the slope of $\overleftrightarrow{AB}$ for the given pairs of points.

91. $A(3, 4); B(6, 8)$ $\frac{4}{3}$

92. $A(-2, 3); B(7, 8)$ $\frac{5}{9}$

Find the slope and the y-intercept of the line with the given equation.

93. $y = \frac{2}{3}x + 3$ $\frac{2}{3}, 3$

94. $y = -\frac{4}{5}x - 5$ $-\frac{4}{5}, -5$

Solve each system by graphing. Check students' graphs.

95. $y = 3x - 1$ $(1, 2)$
$y = -2x + 4$

96. $y = -x + 4$ $(1, 3)$
$y = 2x + 1$

97. Measure each angle and classify as
acute, right, or obtuse.

a. obtuse A

b. right B

c. acute C

98. One of two complementary angles is
four times the other. Find the
measures of the two angles. $18°, 72°$

99. One of two supplementary angles is
five times the other. Find the
measures of the two angles. $30°, 150°$

Answers to Practice Exercises

Pages 1–2

1. 7 **2.** 10 **3.** 1 **4.** 11 **5.** 14
6. 95 **7.** 17 **8.** 52 **9.** 46
10. 80 **11.** 38 **12.** terms: 11, p;
variable: p; constant: 11 **13.** terms: t,
4, u; variables: t, u; constant: 4
14. terms: x, 8, y; variables: x, y; constant:
8 **15.** terms: x, 9, y, 4; variables: x, y;
constants: 9, 4

Pages 3–4

1. associative **2.** commutative
3. identity **4.** $49 + 1 + 28 = 78$
5. $93 + 7 + 62 = 162$ **6.** $198 + 2 +$
$19 = 219$

Pages 5–7

1. not open; false **2.** open **3.** not
open; true
4. 9 **5.** 21 **6.** 0,1
7. 0, 1, 2

Pages 8–10

1. 9 **2.** 8 **3.** x **4.** 8 **5.** 21
6. 10 **7.** 6 **8.** 20 **9.** 15
10. 19 **11.** 10 **12.** 19

Pages 11–12

1. $6 + 7$ **2.** $5 - 3$ **3.** $x + 11$
4. $x + 4$ **5.** $6 + y$ **6.** $b - 5$
7. $t + 8$ **8.** $x - 13$ **9.** $x - 5$

Pages 13–15

1. 152 lb **2.** $75 **3.** 15 yr
4. 14 yr

Pages 18–19

1. $P = 36$ in. **2.** $P = 68$ m
3. $S = \$60$ **4.** $C = \$54$
5. $W = T - L$ **6.** $I = d + p$

Pages 26–27

1. 35 **2.** 108 **3.** 416 **4.** 322
5. 2 **6.** 4 **7.** 34 **8.** 4 **9.** 6
10. 12 **11.** 10 **12.** 3

Pages 28–29

1. associative **2.** commutative
3. identity **4.** $25 \cdot 4 \cdot 17 = 1{,}700$
5. $2 \cdot 50 \cdot 19 = 1{,}900$
6. $5 \cdot 20 \cdot 33 = 3{,}300$

Pages 30–32

1. 30 **2.** 2 **3.** 47 **4.** 4 **5.** 35
6. 25 **7.** 56 **8.** 29 **9.** 18
10. 26

Pages 33–34

1. 24 **2.** 10 **3.** 35 **4.** 69
5. 36 **6.** 26 **7.** 13 **8.** 56
9. 37

Pages 36–38

1. 8 **2.** k **3.** 25 **4.** x **5.** 5
6. 3 **7.** 2 **8.** 8 **9.** 56 **10.** 16
11. 45 **12.** 18 **13.** 5 **14.** 40
15. 5 **16.** 63

Pages 39–41

1. $2 \cdot 9$ **2.** $\dfrac{p}{q}$ **3.** $13w$ **4.** 3
5. 20 yr   **6.** 40   **7.** 28

Pages 46–47

1. 322 in.^{2} **2.** 91 ft^2 **3.** 128 cm^2
4. 24 cm   **5.** 64 ft

Pages 53–55

1. $9 \cdot 7 + 9 \cdot 5 = 108$ **2.** $5 \cdot 4 - 3 \cdot 4 = 8$ **3.** $6 \cdot 7 + 6 \cdot 9 = 96$
4. $8(6 + 11)$ **5.** $(13 - 10)3$
6. $14(12 + 9)$ **7.** $16x + 24$
8. $14b - 35$ **9.** $20x - 30$

Pages 56–57

1. $9t$ **2.** not possible
3. $13m + 4n + 7$ **4.** $7a + 4$
5. $8p + 8$ **6.** $4a + 9$

Pages 58–60

1. 19   **2.** 29   **3.** 45   **4.** 22
5. 77   **6.** 42   **7.** 49   **8.** 66

Pages 61–62

1. 3   **2.** 5   **3.** 3   **4.** 15   **5.** 48
6. 30

Pages 63–64

1. $4x + 3$ **2.** $2x - 9$ **3.** $\dfrac{n}{6} - 7$
4. $\dfrac{x}{7} + 6$

Pages 65–68

1. 10   **2.** 32   **3.** \$30   **4.** 24

Pages 69–71

1. 16   **2.** 9   **3.** 125   **4.** 64
5. 8   **6.** 1   **7.** 216   **8.** 5^3 **9.** 8^4
10. 9^2 **11.** 7^3 **12.** 6^4 **13.** 5^4
14. $7^3 \cdot 8^2$ **15.** $a^5 \cdot b^2$ **16.** 8

17. 64   **18.** 243   **19.** 128   **20.** 7
21. 405

Pages 72–74

1. 36   **2.** 64   **3.** 169   **4.** 361
5. 16π **6.** 9π **7.** 100π **8.** 121π
9. 200π **10.** 90π **11.** 400π
12. 147π **13.** 54   **14.** 216
15. 5,400   **16.** 600

Pages 79–81

1. $\dfrac{1}{4}$ **2.** $\dfrac{1}{9}$ **3.** $\dfrac{3}{8}$ **4.** $\dfrac{4}{7}$ **5.** no

6. no   **7.** yes   **8.** $\dfrac{3}{5}$ **9.** $\dfrac{4}{9}$ **10.** $\dfrac{5}{6}$

11. 1   **12.** 1

Pages 82–84

1. 4   **2.** 4   **3.** $4\dfrac{1}{6}$ **4.** $5\dfrac{1}{3}$

5. $1\dfrac{5}{7}$ **6.** $\dfrac{19}{5}$ **7.** $\dfrac{17}{3}$ **8.** $\dfrac{22}{5}$

9. $\dfrac{19}{3}$ **10.** $\dfrac{29}{6}$ **11.** $\dfrac{39}{4}$ **12.** $6\dfrac{2}{3}$

13. $5\dfrac{1}{3}$ **14.** $2\dfrac{2}{5}$ **15.** $7\dfrac{1}{2}$ **16.** $3\dfrac{3}{4}$

Pages 85–87

1. $1\dfrac{1}{9}$ **2.** $2\dfrac{2}{9}$ **3.** $\dfrac{15}{28}$ **4.** $1\dfrac{1}{6}$

5. $1\dfrac{7}{8}$ **6.** $\dfrac{5}{6}$ **7.** $\dfrac{10}{27}$ **8.** $\dfrac{9}{28}$

9. $\dfrac{12}{25}$ **10.** $\dfrac{8}{9}$

Pages 88–89

1. $3 \cdot 2$ **2.** $2 \cdot 5$ **3.** $3 \cdot 7$
4. $5 \cdot 7$ **5.** $2^2 \cdot 5$ **6.** prime
7. $2 \cdot 3^2$ **8.** 2^3 **9.** $2^3 \cdot 3$
10. $2^3 \cdot 5$ **11.** 2^5 **12.** $2 \cdot 3^3$

Pages 90–91

1. 4   **2.** 3   **3.** 1   **4.** 2   **5.** 10
6. 9   **7.** 4   **8.** 6

Pages 92–94

1. $\frac{3}{5}$ 2. $\frac{9}{14}$ 3. $\frac{2}{3}$ 4. $\frac{14}{25}$ 5. $\frac{2}{3}$

6. $\frac{2}{5}$ 7. $1\frac{1}{3}$ 8. $1\frac{4}{5}$ 9. $1\frac{7}{8}$

10. $1\frac{1}{2}$ 11. $1\frac{3}{4}$ 12. $1\frac{1}{5}$ 13. $\frac{3}{5}$

14. $\frac{2}{3}$ 15. $\frac{3}{4}$ 16. $\frac{4}{15}$ 17. $\frac{16}{27}$

Pages 95–97

1. $\frac{1}{3}$ 2. $\frac{2}{5}$ 3. $\frac{8}{9}$ 4. $\frac{1}{4}$ 5. 3

6. 2   7. $7\frac{1}{2}$ 8. 12   9. $10\frac{1}{2}$

10. $6\frac{2}{3}$ 11. 12   12. $13\frac{1}{3}$

Pages 99–100

1. $\frac{3}{5}$ 2. $\frac{1}{2}$ 3. $\frac{1}{3}$ 4. $\frac{1}{2}$ 5. $\frac{3}{8}$

6. 4   7. $2\frac{1}{2}$ 8. $\frac{1}{4}$

Pages 102–103

1. 20   2. 15   3. 14   4. 36

5. $3\frac{1}{3}$ 6. $8\frac{1}{3}$ 7. $1\frac{5}{7}$ 8. $1\frac{1}{2}$

Pages 104–105

1. $6\frac{2}{3}$ 2. $38\frac{1}{2}$ 3. $1\frac{7}{9}\pi$ 4. $13\frac{1}{2}\pi$

Pages 111–113

1. $\frac{6}{7}$ 2. $\frac{7}{11}$ 3. $\frac{4}{5}$ 4. $\frac{4}{9}$ 5. $\frac{2}{3}$

6. $\frac{2}{3}$ 7. $\frac{1}{2}$ 8. $\frac{1}{3}$ 9. $1\frac{1}{2}$ 10. $1\frac{1}{3}$

11. $2\frac{2}{3}$ 12. $1\frac{1}{5}$

Pages 114–115

1. 12   2. 30   3. 60   4. 30

Pages 116–118

1. $\frac{1}{2}$ 2. $\frac{5}{6}$ 3. $\frac{4}{15}$ 4. $\frac{1}{5}$ 5. $1\frac{1}{2}$

6. $1\frac{1}{3}$ 7. $1\frac{1}{5}$ 8. $1\frac{2}{3}$ 9. $\frac{8}{15}$

10. $\frac{16}{21}$ 11. $\frac{5}{14}$ 12. $\frac{1}{10}$

Pages 119–121

1. $1\frac{1}{18}$ 2. $1\frac{1}{20}$ 3. $\frac{5}{12}$ 4. $1\frac{5}{18}$

5. $\frac{7}{12}$ 6. $\frac{13}{20}$ 7. $\frac{7}{18}$ 8. $\frac{11}{15}$

9. $1\frac{2}{3}$ 10. $1\frac{5}{12}$ 11. $1\frac{1}{5}$

Pages 122–124

1. $8\frac{1}{5}$ 2. $13\frac{1}{2}$ 3. $5\frac{1}{2}$ 4. $7\frac{3}{4}$

5. $5\frac{5}{12}$ 6. $9\frac{4}{15}$ 7. $4\frac{1}{12}$ 8. $2\frac{5}{8}$

9. $11\frac{3}{4}$ 10. $8\frac{7}{8}$ 11. $3\frac{3}{5}$ 12. $7\frac{9}{10}$

13. $11\frac{2}{5}$ 14. $9\frac{1}{5}$ 15. $13\frac{3}{8}$

Pages 126–128

1. 4   2. 9   3. 7   4. $3\frac{1}{2}$ 5. $1\frac{1}{2}$

6. $2\frac{3}{4}$ 7. $2\frac{9}{10}$ 8. $3\frac{4}{5}$ 9. $4\frac{1}{4}$

10. $\frac{2}{3}$ 11. $4\frac{1}{2}$

Pages 129–130

1. $\frac{10}{11}$ 2. $\frac{4}{5}$ 3. $\frac{2}{5}$ 4. $\frac{2}{3}$ 5. $\frac{7}{12}$

Pages 132–134

1. $4\frac{1}{2}$ 2. $\frac{2}{5}$ 3. $2\frac{2}{3}$ 4. $3\frac{1}{3}$

5. $2\frac{1}{2}$ 6. $4\frac{1}{2}$ 7. $2\frac{1}{2}$ 8. $9\frac{1}{3}$

9. 20   10. $4\frac{1}{2}$ 11. 12   12. $11\frac{2}{3}$

Pages 140–142

1. $4,973   2. $6,270

3.

Investment Growth					
Rate	Year				
	0	1	2	3	4
5%	$1,000	$1,051	$1,105	$1,161	$1,221
6%	$1,000	$1,061	$1,127	$1,197	$1,271

Pages 143–144

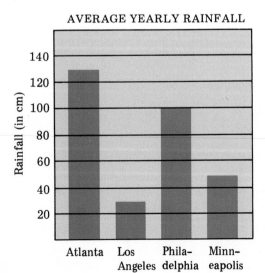

ATTENDANCE AT SOCCER GAMES

Key: Each represents 1,000 fans.

Pages 147–149

1. Lee: 95; Connie: 85; Roger: 70
2. grade 7: 350; grade 8: 300; grade 9: 500

AVERAGE YEARLY RAINFALL

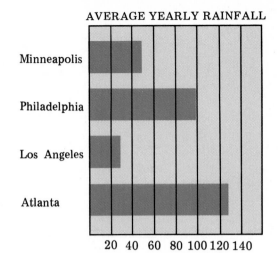

AVERAGE YEARLY RAINFALL

Pages 150–152

1. increased   2. stayed the same
3. decreased
4.

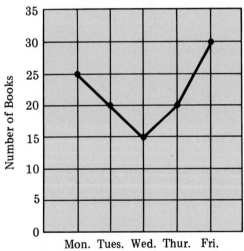

BOOKS SOLD EACH DAY

Pages 154–156

1. Marita   2. Sam   3. hogs: 1940;
lambs: 1930   4. $27 for 50 kg

ANSWERS TO PRACTICE EXERCISES

Pages 157–159

1. 5 cm   **2.** 21   **3.** 6 cm   **4.** 12 cm

Pages 165–167

1. 6.8   **2.** 0.000403   **3.** seventeen and four tenths   **4.** six hundred fifty-nine ten-thousandths   **5.** 0.7
6. 0.65   **7.** 0.046   **8.** 0.09   **9.** 0.6; 0.63   **10.** 0.7; 0.66   **11.** 0.7; 0.68
12. 24.2; 24.18

Pages 168–170

1. 0.6   **2.** 0.25   **4.** 0.625   **4.** 0.15
5. 0.7   **6.** 0.8   **7.** 0.8   **8.** 0.6
9. 0.57   **10.** 0.56   **11.** 0.92   **12.** 0.17

Pages 171–172

1. 241.48   **2.** 97.132   **3.** 13.22
4. 7.766   **5.** 21.85   **6.** 8.84

Pages 173–174

1. 149.52   **2.** 0.06324   **3.** 0.018468
4. 4.512   **5.** 0.63

Pages 177–179

1. 92.1   **2.** 3,423.1   **3.** 715
4. 3.8; 3.83   **5.** 0.6; 0.61   **6.** 0.8; 0.83

Pages 180–181

1. 28.7   **2.** 438.5   **3.** 9,876
4. 5,325   **5.** 96,257   **6.** 1,041.56
7. 2   **8.** 3   **9.** 2.34×10^2
10. 4.9×10^4 **11.** 3×10^7

Pages 182–183

1. 2.9   **2.** 8.61   **3.** 3.4   **4.** 2.8
5. 22.475   **6.** 0.5   **7.** 3.8   **8.** 0.3
9. 1.3

Pages 190–191

1. 3 cm   **2.** 6 cm   **3.** 37 mm   **4.** 42 mm

Pages 192–194

1. 10   **2.** 0.01   **3.** 10   **4.** 0.001

Pages 195–197

1. 1,250 cm   **2.** 7,000 m   **3.** 0.432 m
4. 1.5 m   **5.** 0.529 km   **6.** 0.082 km

Pages 200–201

1. 2.88   **2.** 3,600   **3.** 59,000
4. 1.589   **5.** 0.046   **6.** 2.19

Pages 202–204

1. mL   **2.** L   **3.** 8,200   **4.** 30,000
5. 0.62   **6.** 3.8   **7.** 600   **8.** 2,300

Pages 205–207

1. g   **2.** mg   **3.** 5,600   **4.** 3,000
5. 0.08   **6.** 0.4   **7.** 0.0045   **8.** 63,000

Pages 209–210

1. F = 50°   **2.** F = 77°
3. F = 64.4°   **4.** C = 26.$\overline{6}$°
5. C = 15.$\overline{5}$°   **6.** C = 10°

Pages 216–218

1. 0.73   **2.** 0.62   **3.** 0.35   **4.** 0.56
5. 0.85   **6.** 0.426   **7.** $0.67\frac{1}{3}$
8. 2.25   **9.** 2.00   **10.** 0.713
11. 0.04   **12.** 0.004   **13.** 0.043
14. 0.05   **15.** 0.007   **16.** 7%
17. 46%   **18.** 66.7%   **19.** $35\frac{1}{2}$%
20. 800%   **21.** 40%   **22.** 50%
23. 75%   **24.** $33\frac{1}{3}$%   **25.** $28\frac{4}{7}$%
26. $22\frac{2}{9}$%

Pages 219–221

1. 51.1   **2.** 17.28   **3.** 1.64016
4. 46.655   **5.** 6.9375   **6.** 1.0835
7. 10   **8.** 17

Pages 222–223

1. 50%   **2.** $83\frac{1}{3}$%   **3.** 8%

Pages 224–225

1. 90   **2.** 20   **3.** 57   **4.** 233
5. $280

Pages 226–228

1. $11.13   **2.** $234.25   **3.** $322.50

Pages 229–231

1. $15.26   **2.** $103.96   **3.** $43.84
4. $4.64

Pages 232–233

$43.89

Pages 240–242

1. r = 12; m = 91.5   **2.** r = 6; m =
46.6   **3.** r = 8; m = 31.6   **4.** 7
5. 8.5   **6.** 90   **7.** 9   **8.** none
9. 10 and 9

Pages 243–245

1. 8.65   **2.** 12.56

Pages 248–249

1. $\frac{1}{4}$ **2.** $\frac{1}{4}$ **3.** 40 times **4.** 30 times

Pages 250–251

1. $\frac{1}{12}$ **2.** $\frac{1}{12}$ **3.** $\frac{1}{12}$ **4.** $\frac{1}{2} \cdot \frac{1}{6} = \frac{1}{12}$

Pages 252–253

1. 20   **2.** 16

Pages 260–262

1. $\frac{13}{3}$ or 13:3   **2.** $\frac{3}{13}$ or 3:13   **3.** $\frac{13}{16}$ or
13:16 **4.** $\frac{1}{7}$ **5.** $\frac{3}{7}$ **6.** $\frac{4}{7}$ **7.** $\frac{7}{7}$ or 1

8. $\frac{n}{7}$ **9.** $\frac{50}{1}$ or   50:1 **10.** $\frac{1}{35}$ or 1:35

Pages 263–265
1. yes   **2.** yes   **3.** no   **4.** yes
5. 20   **6.** 30   **7.** 90   **8.** 20
9. 4   **10.** 9   **11.** 9   **12.** 8

Pages 266–268
1. yes   **2.** yes   **3.** no   **4.** no
5. $54   **6.** 7 days

Pages 270–272
1. Transportation: $174; Food: $366;
Savings: $120; Housing: $300; Clothing:
$180; Other: $60

Fractional Part	Measure of Central Angle	Percent Planted
$\frac{50}{200}$ or $\frac{1}{4}$	$\frac{1}{4}(360°) = 90°$	25%
$\frac{25}{200}$ or $\frac{1}{8}$	$\frac{1}{8}(360°) = 45°$	$12\frac{1}{2}$%
$\frac{100}{200}$ or $\frac{1}{2}$	$\frac{1}{2}(360°) = 180°$	50%
$\frac{25}{200}$ or $\frac{1}{8}$	$\frac{1}{8}(360°) = 45°$	$12\frac{1}{2}$%
1	360°	100%

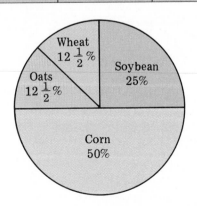

Pages 274–276

1. 60 km **2.** 37.5 km **3.** 2.25 m
4. 4.5 m **5.** 4.5 m

Pages 277–279

1. $x = 9; y = 4$ **2.** 8 m

Pages 280–282

1. 4 **2.** 3 **3.** 5 **4.** $\frac{12}{5}$ **5.** $\frac{12}{13}$

6. $\frac{5}{13}$ **7.** 1.732 **8.** 0.866
9. 0.500

Pages 288–290

1. 2 **2.** −5 **3.** 0 **4.** −1
5. −2 **6.** 6 **7.** −6 **8.** −3
9. 3 before; 5 after **10.** 0 before; 2 after **11.** −4 before; −2 after
12. −6 before; −4 after **13.** −1 before; 1 after **14.** < **15.** >
16. < **17.** < **18.** 6 **19.** 7
20. −5

Pages 291–293

1. 8 **2.** 7 **3.** −3 **4.** −6
5. −2 **6.** 3 **7.** 1 **8.** −4
9. 0 **10.** 0 **11.** 0 **12.** 0

Pages 295–297

1. 4 **2.** −5 **3.** −3 **4.** −6
5. −2 **6.** 8 **7.** 5 **8.** 7

Pages 299–300

1. −3 **2.** 8 **3.** −7 **4.** 10
5. −21 **6.** 0 **7.** 0 **8.** 25
9. −12 **10.** 25 **11.** −17 **12.** 33

Pages 301–302

1. 3 **2.** 9 **3.** −5 **4.** −6
5. −13 **6.** −13 **7.** −21
8. −27 **9.** 3 **10.** ⁻9 **11.** 2
12. 5 **13.** 10 **14.** 15 **15.** 8
16. 9

Pages 303–304

1. a gain of 5 yd **2.** no gain or loss
2. 1,700 m **4.** 2,400 m

Pages 305–306

1. −24 **2.** −5 **3.** −8 **4.** 3
5. −2 **6.** −8 **7.** 18 **8.** −6
9. −8

Pages 312–313

1. 54 **2.** −63 **3.** −48 **4.** −36
5. −504 **6.** −1,653 **7.** −6,566
8. −874 **9.** 0 **10.** 0 **11.** 0
12. 0

Pages 314–315

1. 15 **2.** 28 **3.** 27 **4.** 81
5. 3,807 **6.** 2,898 **7.** 4,140
8. 16,600

Pages 316–318

1. 36 **2.** −525 **3.** −18 **4.** −98
5. $-12 \cdot (-5) = 60$ **6.** $8 \cdot (-6) = -48$ **7.** $-7 \cdot 9 = -63$ **8.** $10 \cdot (-11) = -110$ **9.** 60 **10.** 180
11. −600 **12.** 160 **13.** 160
14. −30 **15.** −30 **16.** −8

Pages 320–322

1. 7 **2.** 7 **3.** 8 **4.** 9 **5.** -5
6. -5 **7.** -8 **8.** -8 **9.** 44
10. -36 **11.** 85 **12.** -99 **13.** 0
14. 0 **15.** not possible **16.** not possible

Pages 323–324

1. $\frac{3}{4}$ **2.** $8\frac{1}{2}$ **3.** $-8\frac{1}{2}$ **4.** $-5\frac{1}{3}$
5. $-2\frac{2}{5}$ **6.** $1\frac{1}{2}$ **7.** $\frac{2}{3}$ **8.** $-1\frac{5}{7}$

Pages 326–327

1. $-4x$ **2.** $-2x$ **3.** $-8x$ **4.** 0
5. $4x$ **6.** $-4x$ **7.** $11x$ **8.** $-5x$
9. $21x$ **10.** $-20y$ **11.** $12y$
12. $-18x$ **13.** $6x - 3$ **14.** $-2y + 8$
15. $-3 - 4x$ **16.** $-5 + 3y$

Pages 328–329

1. -9 **2.** -7 **3.** 7 **4.** -20
5. 12 **6.** -48 **7.** -3 **8.** -1
9. -24

Pages 335–336

1. -3 **2.** -1 **3.** 5 **4.** 2

Pages 337–338

1. 4 **2.** 10 **3.** -13 **4.** 3

Pages 339–341

1. $n + 7 = 15$ **2.** $5x - 3 = 7$
3. $n - 6 = 11$; \$17 **4.** $3n + 5 = 23$; 6

Pages 344–346

1. $-8 \le -1$ **2.** $2 \ge -1$ **3.** $-4 < 20$
4. $4 > -20$ **5.** $-2 \ge -3$
6. $2 \le 3$

Pages 347–349

1. $x > -7$ **2.** $y \le -8$ **3.** $n \ge -4$
4. $x < -4$ **5.** $y < -5$ **6.** $y \ge 8$

Pages 351–352

1. 9 **2.** 24 **3.** 32 **4.** 64
5. -64 **6.** 1,296

Pages 353–355

1. b^4 **2.** $-12y^7$ **3.** $35r^9$ **4.** z^{10}
5. x^{12} **6.** c^8 **7.** a^4b^4 **8.** $16y^2$
9. $-8s^3$

Pages 356–357

1. trinomial **2.** monomial
3. binomial **4.** $4y^2 + 5y - 2$
5. $-3a^4 + 6a^2 - 2a - 1$

Pages 358–359

1. $a^4 - 5a^2$ **2.** $-12c^4 - 18c^3 + 6c^2$
3. $-5x^2 - 4$ **4.** $-a^2 - 7a + 4$
5. $2c^2 - 3c - 3$

Pages 360–362

1. $5(c - 2)$ **2.** $7(3 + z)$ **3.** $6(2 - a)$
4. $6(x^2 - 3)$ **5.** $2(y^2 - 3y + 5)$
6. $2(2n^2 + 3n - 6)$ **7.** $y(y + 4)$
8. $c^2(c - 2)$ **9.** $n^2(n^2 + 3)$
10. $3y(y - 4)$ **11.** $5y(2y^2 + 3y - 1)$
12. $3y(y^3 - 2y + 3)$

Pages 368–369

1. **2.** **3.**

4. B, C **5.** E, G, H

Pages 370–371

1. 90°; right **2.** 130°; obtuse
3. 75°; acute **4.** complementary
5. supplementary **6.** supplementary
7. complementary

Pages 373–374

1. vertical angles: $\angle 1$ and $\angle 4$, $\angle 2$ and $\angle 3$, $\angle 5$ and $\angle 8$, $\angle 6$ and $\angle 7$; alternate interior angles: $\angle 3$ and $\angle 6$, $\angle 4$ and $\angle 5$; corresponding angles: $\angle 1$ and $\angle 5$, $\angle 2$ and $\angle 6$, $\angle 3$ and $\angle 7$, $\angle 4$ and $\angle 8$

Pages 377–378

1. right **2.** acute **3.** obtuse

Pages 382–384

1. 48 is not a perfect square. **2.** 25 is a perfect square. **3.** 81 is a perfect square. **4.** 66 is not a perfect square.
5. 2.8 **6.** 8.2 **7.** 5.7 **8.** 8.7
9. $s \doteq 7.3$ m

Pages 385–387

1. 8.2 cm **2.** 9.2 m **3.** 7.1 cm **4.** 9.4 m

Pages 393–395

1. The integers between -3 and 2.

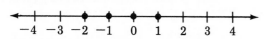

2. The integers between -1 and 0.

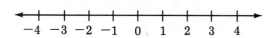

3. -3 **4.** no solution **5.** all reals > 2
6. all reals < -1 **7.** all reals ≤ 4
8. all reals ≥ -2

Pages 396–398

1. all reals > -3 **2.** all reals < 1
3. all reals < -2 **4.** all reals > -3
5. all reals ≥ -3 **6.** all reals ≤ -6
7. all reals ≥ -5 **8.** all reals ≤ 9

Pages 401–403

1. $(-2, 3)$ **2.** $(-4, -2)$ **3.** $(2, 0)$
4. $(0, 0)$ **5.** $(0, 2)$ **6.** $(3, -2)$
7. –10.

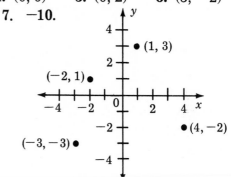

11. –14.

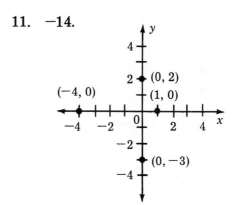

Pages 404–405
1. horizontal **2.** vertical

Pages 406–408
Answers may vary. Five ordered pairs are given for each.

1. $y = x - 1$

x	$x - 1$	y	(x,y)
0	$0 - 1$	-1	$(0,-1)$
1	$1 - 1$	0	$(1,0)$
2	$2 - 1$	1	$(2,1)$
-1	$-1 - 1$	-2	$(-1,-2)$
-2	$-2 - 1$	-3	$(-2,-3)$

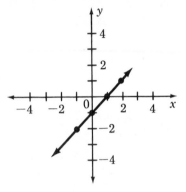

2. $y = -x + 3$

x	$-x + 3$	y	(x,y)
0	$0 + 3$	3	$(0,3)$
1	$-1 + 3$	2	$(1,2)$
2	$-2 + 3$	1	$(2,1)$
3	$-3 + 3$	0	$(3,0)$
4	$-4 + 3$	-1	$(4,-1)$

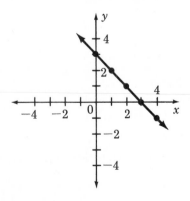

3. $y = -x - 2$

x	$-x - 2$	y	(x,y)
0	$-0 - 2$	-2	$(0,-2)$
1	$-1 - 2$	-3	$(1,-3)$
-1	$-(-1) - 2$	-1	$(-1,-1)$
-2	$-(-2) - 2$	0	$(-2,0)$
-3	$-(-3) - 2$	1	$(-3,1)$

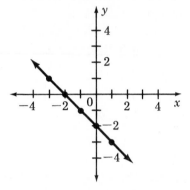

4. $y = -3x$

x	$-3x$	y	(x,y)
0	$-3(0)$	0	$(0,0)$
1	$-3(1)$	-3	$(1,-3)$
-1	$-3(-1)$	3	$(-1,3)$

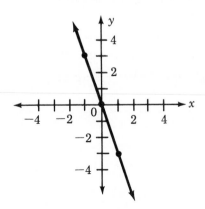

5. $y < 4x - 3$

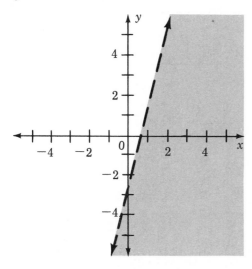

6. $y > 3x + 4$

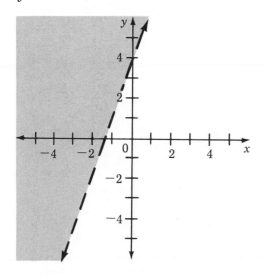

Pages 410–412

1. $\frac{4}{3}$ **2.** -1 **3.** -1 **4.** The slope **is 5. The** y-intercept is 3.   **5.** The slope is $\frac{1}{3}$. The y-intercept is -2.   **6.** The slope is $\frac{-5}{7}$. The y-intercept is 8.

Pages 413–414

1. yes   **2.** no
3.

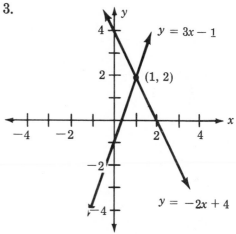

4.

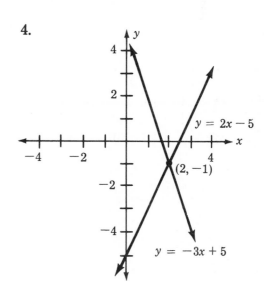

Answers to Odd-numbered Extra Practice Exercises

Chapter 1

1. 20   **3.** 30   **5.** 7   **7.** 172
9. 396   **11.** 165   **13.** none   **15.** 1
17. 0   **19.** 5   **21.** 11   **23.** 2
25. 25   **27.** 8   **29.** 20   **31.** $46
33. 28   **35.** $P = 36$ in.

Chapter 2

1. 32   **3.** 9   **5.** 225   **7.** 23
9. 34   **11.** 242   **13.** 249
15. 3,200   **17.** 800   **19.** 1,400
21. 5   **23.** 60   **25.** 2   **27.** 70
29. 90   **31.** 30   **33.** 4   **35.** $45
37. $2,400   **39.** $20   **41.** $A = 48$ cm^2

Chapter 3

1. $12\,m$ **3.** $12x + 9$ **5.** $11a$
7. $5a - 5$ **9.** $15a - 15$ **11.** $9x + 6y$ **13.** 44   **15.** 66   **17.** 112
19. 2   **21.** 3   **23.** 75   **25.** 4
27. 9   **29.** 7   **31.** 7   **33.** 9
35. 1   **37.** 125   **39.** 243   **41.** 64
43. 625   **45.** 5^3 **47.** 4^4
49. $6^2 x^5$ **51.** $x^3 y^4$ **53.** 7   **55.** 13

Chapter 4

1. $1\frac{2}{3}$ **3.** $1\frac{1}{7}$ **5.** $8\frac{2}{5}$ **7.** $4\frac{2}{7}$

9. $\frac{8}{15}$ **11.** $\frac{7}{10}$ **13.** $\frac{2}{5}$ **15.** $\frac{3}{7}$

17. $\frac{1}{9}$ **19.** $1\frac{2}{3}$ **21.** 10   **23.** 16

25. $\frac{5}{16}$ **27.** $6\frac{1}{4}$ **29.** 12   **31.** $\frac{2}{5}$

33. $\frac{9}{16}$ **35.** $3\frac{3}{8}$ **37.** $\frac{25}{36}$ **39.** $\frac{4}{9}$

41. 45   **43.** $1\frac{1}{5}$ **45.** $1\frac{1}{4}$ **47.** $\frac{2}{3}$

49. $1\frac{5}{9}$ **51.** $1\frac{3}{5}$ **53.** $A = 20$

55. $A = 2\frac{7}{9}\pi$ **57.** $V = 1\frac{23}{25}\pi$

59. $V = 45\frac{37}{48}\pi$

Chapter 5

1. $1\frac{2}{5}$ **3.** $\frac{1}{9}$ **5.** $\frac{14}{15}$ **7.** $\frac{1}{6}$ **9.** $\frac{1}{6}$

11. $2\frac{1}{4}$ **13.** $2\frac{2}{21}$ **15.** $1\frac{13}{24}$

17. $1\frac{23}{36}$ **19.** $3\frac{17}{60}$ **21.** $5\frac{7}{12}$

23. $10\frac{3}{5}$ **25.** $13\frac{13}{20}$ **27.** $1\frac{11}{15}$

29. $8\frac{7}{16}$ **31.** $1\frac{1}{3}$ **33.** $9\frac{5}{6}$

35. $13\frac{1}{10}$ **37.** $5\frac{1}{10}$ lb   **39.** $20\frac{1}{12}$ mi

Chapter 6

1. $446   **3.** $253

5. ATTENDANCE AT FOOTBALL GAMES

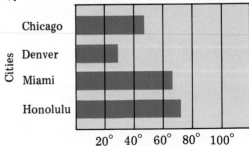

Key: Each 👤 represents 5,000 fans.

7. AVERAGE MONTHLY TEMPERATURE

9. 60   **11.** 70   **13.** 1984

Chapter 7

1. 0.7; 0.74 **3.** 19.6; 19.58 **5.** 9.0; 9.02 **7.** 0.2; 0.15 **9.** 4.1; 4.06 **11.** 6.6; 6.58 **13.** 0.1; 0.09 **15.** 0.7; 0.67 **17.** 0.9; 0.86 **19.** 0.4; 0.42 **21.** 48.53 **23.** 19.851 **25.** 178.11 **27.** 173.966 **29.** 8.378 **31.** 89.205 **33.** 62.87 **35.** 605.685 **37.** 96.433 **39.** 208.6 **41.** 166.23 **43.** 112.24 **45.** 5.91 **47.** 2.4 **49.** 1.5 **51.** 40.3 **53.** 11.4 km/h

Chapter 8

1. 4.8 **3.** 0.08 **5.** 0.6 **7.** 0.8 **9.** 42,000 **11.** 4 **13.** 100 **15.** 3,700 **17.** 12,000 **19.** 46,000 **21.** 4 **23.** 0.015 **25.** 0.03 **27.** 0.025 **29.** 4.5 **31.** 77°F **33.** 48°F **35.** 90°F **37.** 142°F **39.** 131°F **41.** 118°F **43.** 38°C **45.** 11°C **47.** 25°C **49.** 29°C **51.** 18°C **53.** 36°C **55.** 4,624 mm² **57.** 3,500 mL **59.** 0.95 t **61.** 75,000 m

Mixed Practice for Chapters 1–8

1. 13 **3.** 95 **5.** 101 **7.** none **9.** 8 **11.** 141 **13.** 9 **15.** 40.5 **17.** $1\frac{1}{6}$ **19.** $2\frac{1}{10}$ **21.** 125 **23.** 243 **25.** $6\frac{2}{9}$ **27.** $4\frac{17}{20}$ **29.** $1\frac{1}{7}$ **31.** 12 **33.** 9.8; 9.76 **35.** 63.5; 63.46 **37.** 40.09 **39.** 2,419.62 **41.** 4,630 **43.** 0.9 **45.** 1,300 **47.** 32°C **49.** −12°C **51.** 5 **53.** $172

Chapter 9

1. 0.5 **3.** 4.00 **5.** 0.007 **7.** 4.25 **9.** 0.000007 **11.** 7% **13.** 1,800% **15.** $12\frac{1}{2}\%$ **17.** $45\frac{3}{5}\%$ **19.** 4,560% **21.** 111.45 **23.** 9.75 **25.** 344.10 **27.** 60% **29.** about 47% **31.** about 5.7% **33.** 40 **35.** 93 **37.** 171.43 **39.** 21 **41.** $15,000 **43.** sweater: $22.40; shoes: $33.95; coat: $87.78

Chapter 10

1. $r = 10$; $m = 15$ **3.** $r = 61$; m $= 279$ **5.** $r = 17$; $m = 51$ **7.** 97 **9.** 15 **11.** 50 **13.** 98 **15.** 4 **17.** none **19.**

Score(s)	Tally	Frequency(f)	Sum(f · s)
100					3	300
98					3	294
97			1	97
96				2	192
94				2	188
93			1	93
90			1	90
88				2	176
86			1	86
Total		16	1,516

21. 12 **23.** 5 **25.** $\frac{1}{6}$ **27.** 58 **29.** $P(\text{red}) = \frac{4}{9}$; $P(\text{white}) = \frac{5}{9}$ **31.** 27

Chapter 11

1. $\frac{10}{7} = 1\frac{3}{7}$ **3.** $\frac{14}{10} = 1\frac{2}{5}$ **5.** $\frac{7}{4} = 1\frac{3}{4}$ **7.** 8 **9.** 1 **11.** 1 **13.** 1 **15.** 30 **17.** 21 **19.** yes **21.** no **23.** yes **25.** 3 h 20 min **27.** 5 m **29.** 6

31.

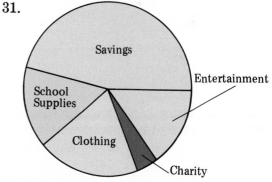

Chapter 12

1. < **3.** < **5.** > **7.** > **9.** < **11.** < **13.** 11 **15.** −10 **17.** 14 **19.** −8 **21.** −16 **23.** −6 **25.** 0 **27.** 9 **29.** 0 **31.** 12 **33.** −11 **35.** −19 **37.** 12 **39.** −15 **41.** −25 **43.** 3 **45.** 10 **47.** −14 **49.** −12 **51.** 0 **53.** −10 **55.** 78°F **57.** 142 **59.** 160

Chapter 13

1. 270   **3.** -350 **5.** 160
7. 1,530   **9.** 0   **11.** -432
13. -15 **15.** -52 **17.** -120
19. 165   **21.** 168   **23.** -65
25. 11   **27.** -10 **29.** -8 **31.** 4
33. not possible   **35.** $1\frac{7}{8}$ **37.** $-1\frac{1}{35}$
39. $-\frac{1}{3}$ **41.** $1\frac{19}{35}$ **43.** $\frac{20}{33}$
45. $-1\frac{5}{33}$ **47.** $9x$ **49.** $30m$
51. $-3 - 2y$ **53.** a **55.** $-10a + 6$
57. $-24a + 12$ **59.** $3 - 2y$
61. -6 **63.** -12 **65.** -3 **67.** 1
69. -27 **71.** 27   **73.** -50
75. -100 **77.** -2 **79.** -5
81. 8   **83.** -4

Chapter 14

1. -5 **3.** -2 **5.** 4   **7.** $n > 10$
9. $h \geq -5$ **11.** $x \leq -4$ **13.** -32
15. -48 **17.** a^8 **19.** y^6 **21.** x^3y^3
23. $9a^4$ **25.** $8x^6y^9$ **27.** $-c^2 - 12c + 12$ **29.** $7a^6 + 2a^4 + 14$
31. $-3a^3 + a^2$ **33.** $2y^3 + 6y^2 + 4y$
35. $-6m^3 + 2m^2 + 2m$ **37.** $3(x - 3)$
39. $a(6a^2 + 1)$ **41.** 11   **43.** 7
45. 3   **47.** 16

Chapter 15

1. complementary   **3.** supplementary
For examples **5** and **7,** answers may vary.
Examples are given.   **5.** $\angle 1$ and $\angle 3$
7. $\angle 6$ and $\angle 3$ **9.** $\angle 2$: 140°; $\angle 3$: 40°;
$\angle 4$: 140°; $\angle 5$: 140°; $\angle 6$: 40°; $\angle 7$: 140°;
$\angle 8$: 40°   **11.** 3   **13.** 6
15.   **17.**

19. acute   **21.** right   **23.** 20°
25. ± 6 **27.** ± 10 **29.** ± 12
31. 5.7   **33.** 1.7   **35.** $c = 5$
37. $a = 4.9$ **39.** 9.4 in.

Chapter 16

1. the integers between -1 and 4
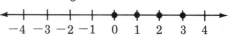

3. the integers less than 1
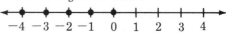

5. $x = 1$

7. $n = 1$

9. $a = -3$

11. $x > 2$

13. $x \geq 4$

15. $y > 1$

17. $x > 4$

19. $x < 1$

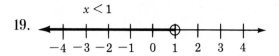

21. $x > 1$

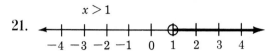

23. $(3, 1)$ **25.** $(-4, 2)$ **27.** $(2, -2)$
29. $(0, -3)$

31., 33., 35.

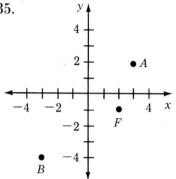

37.

x	$x - 1$	y	(x, y)
0	$0 - 1$	-1	$(0, -1)$
1	$1 - 1$	0	$(1, 0)$
2	$2 - 1$	1	$(2, 1)$

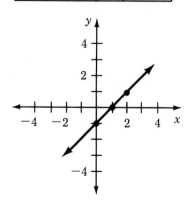

39.

x	$x + 3$	y	(x, y)
-1	$-1 + 3$	2	$(-1, 2)$
0	$0 + 3$	3	$(0, 3)$
1	$1 + 3$	4	$(1, 4)$

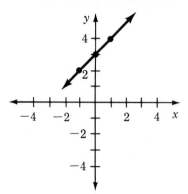

40.

x	$2x + 1$	y	(x, y)
-1	$2(-1) + 1$	-1	$(-1, -1)$
0	$2(0) + 1$	1	$(0, 1)$
1	$2(1) + 1$	3	$(1, 3)$

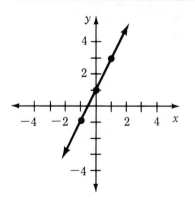

43. $\frac{4}{3}$ **45.** $-\frac{1}{2}$ **47.** 4; -2 **49.** $\frac{3}{5}$;

-1 **51.** $\frac{1}{4}$; $\frac{3}{4}$ **53.** $(-2, -5)$

55. $(1, -5)$

Mixed Practice for Chapter 9–16

1. 0.60 **3.** 5.0 **5.** 0.000065
7. 53.7% **9.** 15.5% **11.** 77.76
13. 70% **15.** 54 **17.** 8, 10, 10, 8
and 12 **19.** 16 **21.** 72 **23.** 8
25. -5 **27.** 1 **29.** 3 **31.** 3
33. $x > 9$ **35.** $x = b - a$ **37.** $x = c + d$ **39.** -6 **41.** 15 **43.** -16

45. -120 **47.** 5 **49.** $-\frac{2}{7}$

51. $x^4 y^2$ **53.** $-18a^5$ **55.** $-2x^3 + 6x^2 - 10x$ **57.** $-m^2 + 2m - 6$
59. -12 **61.** -144

63., 65.

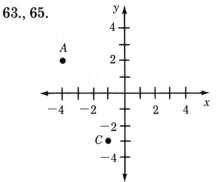

67.

x	$x - 4$	y	(x, y)
2	$2 - 4$	2	$(2, -2)$
1	$1 - 4$	3	$(1, -3)$
0	$0 - 4$	4	$(0, -4)$

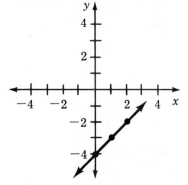

69. $4(a^2 - 3a + 9)$ **71.** 70° **73.** 5°

75. $\frac{10}{14}$ or $\frac{5}{7}$ **77.** $\frac{4}{14}$ or $\frac{2}{7}$ **79.** $\frac{3}{5}$

81. $\frac{4}{3}$ **83.** $6n + 3$ **85.** 12.7

87. 7.3 **89.** 1.7 **91.** $\frac{4}{3}$ **93.** $\frac{2}{3}$, 3

95. $(1, 2)$ **97. a.** obtuse **b.** right
c. acute **99.** 30°, 150°

Table of Measures

	METRIC	CUSTOMARY

Length

METRIC

1 kilometer (km) = 1,000 meters
1 hectometer (hm) = 100 meters
1 dekameter (dam) = 10 meters
1 meter (m)
1 decimeter (dm) = 0.1 meter
1 centimeter (cm) = 0.01 meter
1 millimeter (mm) = 0.001 meter

CUSTOMARY

1 foot (ft) = 12 inches (in.)

$1 \text{ yard (yd)} = \begin{cases} 3 \text{ feet} \\ 36 \text{ inches} \end{cases}$

$1 \text{ mile (mi)} = \begin{cases} 5,280 \text{ feet} \\ 1,760 \text{ yards} \end{cases}$

Mass/Weight

METRIC

1 metric ton (t) = 1,000 kilograms
1 kilogram (kg) = 1,000 grams
1 hectogram (hg) = 100 grams
1 dekagram (dag) = 10 grams
1 gram (g)
1 decigram (dg) = 0.1 gram
1 centigram (cg) = 0.01 gram
1 milligram (mg) = 0.001 gram

CUSTOMARY

1 pound (lb) = 16 ounces (oz)
1 ton (T) = 2,000 pounds

Capacity

METRIC

1 kiloliter (kL) = 1,000 liters
1 hectoliter (hL) = 100 liters
1 dekaliter (daL) = 10 liters
1 liter (L)
1 deciliter (dL) = 0.1 liter
1 centiliter (cL) = 0.01 liter
1 milliliter (mL) = 0.001 liter

CUSTOMARY

1 cup (c) = 8 fluid ounces (fl oz)
1 pint (pt) = 2 cups
1 quart (qt) = 2 pints
1 gallon (gal) = 4 quarts

1 liter = 1,000 cubic centimeters (cm^3)
1 milliliter = 1 cubic centimeter

TIME

1 minute (min) = 60 seconds (s)
1 hour (h) = 60 minutes
1 day (d) = 24 hours
1 week = 7 days

$1 \text{ year (y)} = \begin{cases} 12 \text{ months} \\ 365 \text{ days} \end{cases}$

1 decade = 10 years
1 century = 100 years

$<$	is less than
$>$	is greater than
3^4	the fourth power of 3
x^3	the third power of x, or x cubed
$\doteq$	is approximately equal to
$\neq$	is not equal to
$0.2\overline{2}$	repeating decimal
$\overline{AB}$	line segment AB
$\cong$	is congruent to
$\overrightarrow{AB}$	ray AB
$\overleftrightarrow{TR}$	line containing points T and R
$\angle XYZ$	angle XYZ
$\triangle ABC$	triangle ABC
$m\angle A$	measure of angle A
π	pi (about 3.14)
$\overleftrightarrow{KL} \div \overleftrightarrow{MN}$	line KL is parallel to line MN
$\overleftrightarrow{AB} \perp \overleftrightarrow{CD}$	line AB is perpendicular to line CD
$\approx$	is similar to
$\sqrt{25}$	the square root of 25
20%	20 percent
$+6$	positive 6
-4	negative 4
$|-5|$	absolute value of negative 5
BASIC	Beginners All-Purpose Symbolic Instruction Code

Formulas

Perimeter

rectangle	$P = 2l + 2w$
triangle	$P = a + b + c$
square	$P = 4s$

Circumference

circle	$C = \pi d = 2\pi r$

Area

rectangle	$A = lw$
square	$A = s^2$
parallelogram	$A = bh$
triangle	$A = \frac{1}{2}bh$
circle	$A = \pi r^2$

Surface area

rectangular prism	$A = 2lw + 2lh + 2wh$
cube	$A = 6s^2$

Volume

rectangular prism	$V = lwh$
cube	$V = s^3$
cylinder	$V = \pi r^2 h$
rectangular pyramid	$V = \frac{1}{3}Bh$
cone	$V = \frac{1}{3}\pi r^2 h$

Table of Squares and Square Roots

No.	Square	Square Root	No.	Square	Square Root	No.	Square	Square Root
1	1	1.000	35	1.225	5.916	68	4,624	8.246
2	4	1.414	36	1.296	6.000	69	4,761	8.307
3	9	1.732	37	1,369	6.083	70	4,900	8.357
4	16	2.000	38	1,444	6.164	71	5,041	8.426
5	25	2.236	39	1,521	6.245	72	5,184	8.485
6	36	2.449	40	1,600	6.325	73	5,329	8.544
7	49	2.646	41	1,681	6.403	74	5,476	8.602
8	64	2.828	42	1,764	6.481	75	5,625	8.660
9	81	3.000	43	1,849	6.557	76	5,776	8.718
10	100	3.162	44	1,936	6.633	77	5,929	8.775
11	121	3.317	45	2,025	6.708	78	6,084	8.832
12	144	3.464	46	2,116	6.782	79	6,241	8.888
13	169	3.606	47	2,209	6.856	80	6,400	8.944
14	196	3.742	48	2,304	6.928	81	6,561	9.000
15	225	3.875	49	2,401	7.000	82	6,724	9.055
16	256	4.000	50	2,500	7.071	83	6,889	9.110
17	289	4.123	51	2,601	7.141	84	7,056	9.165
18	324	4.243	52	2,704	7.211	85	7,225	9.220
19	361	4.359	53	2,809	7.280	86	7,396	9.274
20	400	4.472	54	2,916	7.348	87	7,569	9.327
21	441	4.583	55	3,025	7.416	88	7,744	9.381
22	484	4.690	56	3,136	7.483	89	7,921	9.434
23	529	4.796	57	3,249	7.550	90	8,100	9.487
24	576	4.899	58	3,364	7.616	91	8,281	9.539
25	625	5.000	59	3,481	7.681	92	8,464	9.592
26	676	5.099	60	3,600	7.746	93	8,649	9.644
27	729	5.196	61	3,721	7.810	94	8,836	9.695
28	784	5.292	62	3,844	7.874	95	9,025	9.747
29	841	5.385	63	3,969	7.937	96	9,216	9.798
30	900	5.477	64	4,096	8.000	97	9,409	9.849
31	961	5.568	65	4,225	8.062	98	9,604	9.899
32	1,024	5.657	66	4,356	8.124	99	9,801	9.950
33	1,089	5.745	67	4,489	8.185	100	10,000	10.000
34	1,156	5.831						

Glossary

The explanations given in this glossary are intended to be brief descriptions of the terms listed. They are not necessarily definitions.

Acute angle An angle whose measure is less than 90°.

Acute triangle A triangle in which each angle measures less than 90°.

Addition property for equations If $a = b$ is true, then $a + c = b + c$ is also true for all numbers a, b, and c.

Addition property for inequalities If $a < b$, then $a + c < b + c$ for all numbers a, b, and c.

Adjacent angles Two angles with a common vertex, a common side, and no common interior points.

Alternate interior angles Two inside angles on the same side of the transversal that cuts two lines.

Altitude (of a triangle) A segment that originates at a vertex of the triangle and is perpendicular to the line containing the opposite side.

Associative property of addition For all numbers a, b, and c, $(a + b) + c = a + (b + c)$.

Associative property of multiplication For all numbers a, b, and c, $(a \cdot b) \cdot c = a \cdot (b \cdot c)$.

Average The number found from adding several numbers and then dividing the sum by the number of numbers added.

◆

Bar graph A method of comparing quantities by the use of solid bars.

Binomial A polynomial with two terms.

Capacity To measure the capacity of an object, we measure how much it holds.

Central angle An angle whose vertex is at the center of a circle.

Circle Closed curve in a plane (flat surface) such that every point on the circle is the same distance from the center.

Circumference The measure of the distance around a circle.

Coefficient The multiplier of a variable. *Example* In *4x*, 4 is the coefficient.

Collinear points Points that are contained in one line.

Commission Amount of extra money salespeople are given for selling something. The more they sell, the more they make if they are paid a "commission."

Commutative property of addition $a + b = b + a$

Commutative property of multiplication $a \cdot b = b \cdot a$

Complementary angles Two angles the sum of whose measure is 90°.

Complete factorization A number shown as a product of prime numbers only: $36 = 3 \cdot 3 \cdot 2 \cdot 2$.

Congruent segments Segments that are of the same length.

Coordinate A number assigned to a point on the number line.

Coordinate plane Two perpendicular number lines in a plane make up a coordinate plane, or a coordinate system. Each point in a coordinate plane corresponds to an ordered pair of numbers, and vice versa.

Corresponding angles Two angles on the same side of a transversal, one on the inside, the other on the outside.

Cosine (cos) The cosine of an angle in a right triangle is the ratio of the length of the side adjacent to that angle to the length of the hypotenuse.

Counting principle One event can happen in x ways. Another event can happen in y ways. The total number of ways that both events can happen is $x \cdot y$ ways.

Cube A three-dimensional solid with all faces squares.

Decagon A polygon with ten sides.

Decimal Numbers with decimal points used to indicate place value.

Degree Celsius (°C) A metric unit of temperature.

Denominator In the fraction $\frac{3}{7}$, the number below the fraction bar is the denominator.

Dependent events Two events are dependent if the outcome of one event affects the outcome of the other.

Diameter A line segment whose endpoints are on the circle and which contains the center of the circle.

Discount A reduction applied to the price of an article.

Distributive property of multiplication over addition For all numbers a, b, and c, $a \cdot (b + c) = a \cdot b + a \cdot c$.

Divisible A number is divisible by another number if the remainder is zero, when the first number is divided by the second number.

Divisor In $7\overline{)28}$, 7 is the divisor, the number by which you are dividing.

Equation A mathematical sentence in which the $=$ symbol is used. $3x = 15$ and $x + 6 = 14$ are examples of equations.

Equation of a line $y = mx + b$ is an equation of a line. m is the slope of the line, and b is the y-intercept.

Even number Any number divisible by 2 is an even number.

Exponent In 2^4, 4 is the exponent. It tells how many times 2 is used as a factor in the product $2 \cdot 2 \cdot 2 \cdot 2$.

Extremes (in a proportion) The extremes in a proportion are the first and the fourth terms.

Example $2{:}3 = 4{:}6$

or $\dfrac{2}{3} = \dfrac{4}{6}$ 2 and 6 are the extremes.

Factor A number is a factor of a given number if the given number is divisible by the number.

$5 \cdot 2 = 10$

5 and 2 are factors of 10.

Formula An equation that relates some quantities.

$A = \pi r^2$ is a formula.

Frequency The number of times that a number appears in a set of data.

Gram (g) A metric unit of weight, 0.001 of a kilogram.

Greatest common factor (GCF) The largest number by which each of a given set of two or more numbers is divisible. For example, 4 is the GCF of 8, 16, 20.

Hexagon A polygon with six sides.

Hypotenuse The side opposite the right angle in a right triangle.

Independent events Two events are independent if the outcome of one does not affect the outcome of the other.

Inequality A mathematics sentence that has the symbol $>$ or $<$ in it.

Integer A directed whole number, positive or negative or zero.

Irrational number A number named by a nonterminating, nonrepeating decimal. For example: 1.424224222 . . ., $\sqrt{2}$, π.

Least common denominator The same as the least common multiple.

Least common multiple (LCM) The smallest number of which each of two or more given numbers is a factor. The LCM of 3, 5, 2, is 30. 30 is the smallest number that has 3, 5, and 2 as factors.

Line segment Segment $\overline{AB}$ is the set of points on a line including A and B and all the points in between.

Liter (L) A metric unit of capacity, 1,000 mL.

Mean The mean is the average. To find the mean, add the numbers and divide the sum by the number of numbers.

Means (in a proportion) The means in a proportion are the second and the third terms.

Example 2:3 = 4:6
or $\frac{2}{3} = \frac{4}{6}$ 3 and 4 are the means.

Median In a set of numbers arranged from least to greatest, the median is the middle number. If there is an even number of items, the median is the mean of the two middle numbers.

Meter A metric unit of length that is equal to 100 centimeters.

Metric system A system of measurement based on the number 10.

Mixed number A number that consists of the sum of an integer and a fraction. For example: $4\frac{2}{3}$.

Mode The number occurring more frequently than any other number. A set may have more than one mode.

Monomial A polynomial with one term.

Multiple Any product of the given number and a whole number.

Non-collinear points Three or more points that are not contained in the same line.

Number line A line in which the numbers and the points have been matched one-to-one.

Numerator In a fraction, the number above the fraction line is the numerator. For example, in the fraction $\frac{5}{7}$, 5 is the numerator.

Obtuse angle An angle whose measure is between 90° and 180°.

Obtuse triangle A triangle that has one obtuse angle.

Octagon A polygon with eight sides.

Odd number Any number not divisible by 2 is an odd number.

Ordered pair $(2, -3)$ is an ordered pair of numbers. Each ordered pair of numbers corresponds to exactly one point in a coordinate plane, and vice versa.

Origin The point for zero on a number line. The point for (0, 0) in a coordinate plane.

Parallel lines Lines that lie in the same plane and do not intersect.

Parallelogram A quadrilateral with two pairs of opposite sides parallel.

Percent Indicates how many out of a hundred.

Symbol for percent: %

Example 42% means 42 out of a hundred.

Perfect square A number whose principal square root is a whole number.

Perimeter The distance around a polygon.

Perpendicular lines Lines intersecting at right angles.

Pi Pi or π is the number, approximately equal to 3.14, used in finding the circumference or area of a circle.

π is the ratio between the circumference and twice the radius.

Pictographs A graph that uses pictures to show data.

Plane A two-dimensional flat surface.

Polygon A geometric figure with three or more sides.

Polynomial A polynomial is an expression containing one or more terms.

Examples $3x, 5a^2 - 2$

$4y^2 - 3y + 1$

Prime number A whole number greater than 1 having only two factors, 1 and itself. For example, 7 is a prime. The only factors of 7 are 7 and 1.

Principal square root The positive square root of a number. For example, 5 is the principal square root of 25. $\sqrt{25} = 5$

Probability The probability of an event is the number of favorable outcomes divided by the total number of all possible outcomes.

Property of 1 for multiplication The product of any number and 1 is that number. $x \cdot 1 = x$

Proportion A proportion is an equation that states that two ratios are equal.

$\frac{a}{b} = \frac{c}{d}$ or $a{:}b = c{:}d$

Protractor A device for measuring angles.

Pythagorean theorem In any right triangle, the square of the length of the hypotenuse equals the sum of the squares of the lengths of the other two sides. If $\triangle ABC$ is a right triangle with c the hypotenuse, then $a^2 + b^2 = c^2$.

Pythagorean triple Any three numbers a, b, c such that $a^2 + b^2 = c^2$.

Quadrilateral A polygon with four sides.

Radius A segment from a point on a circle to the center of the circle.

Range The difference between the highest and lowest numbers in a set.

Ratio A ratio is a comparison of two numbers by division. A fraction is a ratio. For example, $\frac{a}{b}$.

Rational number A number that can be named by a fraction. For example, a fraction or a mixed number or a terminating or repeating decimal.

Ray A part of a straight line that has a beginning point and continues infinitely in one direction.

Real numbers The real numbers include all of the rational and irrational numbers.

Reciprocal Two numbers are reciprocals if their product is 1. 3 and $\frac{1}{3}$ are reciprocals since $3 \times \frac{1}{3} = 1$.

Rectangle A four-sided polygon with all right angles.

Regular polygon A polygon with all sides of the same length and all angles of the same measure.

Repeating decimal A decimal that repeats a digit or group of digits forever.

Example $0.4444\ldots = 0.4\overline{4}$

Rhombus A quadrilateral with four sides of the same length.

Right angle An angle whose measure is 90°.

Right triangle A triangle that contains one right angle.

Scientific notation A number written as the product of a number from 1 to 10 and a power of 10.

Example $2.39 \times 10^3 = 2,390$

Similar triangles Triangles with the same shape but not necessarily the same size.

Sine (sin) The sine of an angle in a right triangle is the ratio of the length of the side opposite the angle to the length of the hypotenuse.

Slope of a line The slope of a line can be shown by this ratio:

$$\text{slope} = \frac{\text{Difference of } y\text{-coordinates}}{\text{Difference of } x\text{-coordinates}}$$

Sphere A round three-dimensional figure shaped like a baseball or basketball. All points on a sphere are the same distance from the center.

Square A quadrilateral in which all sides have the same measure and all angles are right angles.

Square root If $x \cdot x = n$, then x is a square root of n.

Straight angle An angle with a measure of 180°.

Supplementary angles Two angles the sum of whose measures is 180°.

System of equations Two equations with two variables form a system of equations.

Example $y = 2x - 5$
$y = -3x + 5$

Tangent (tan) The tangent of an angle in a right triangle is the ratio of the length of the side opposite the angle to the length of the side adjacent to the angle.

Transversal A line that intersects two or more lines.

Trapezoid A quadrilateral with exactly one pair of parallel sides.

Triangle A polygon with three sides.

Trinomial A polynomial with three terms.

Variable A letter that can be replaced by numbers.

Vertical angles Angles formed by two intersecting lines that are opposite each other; they have the same measure.

Volume The measure of the amount of space occupied by a figure in three dimensions.

Weight A number that tells how heavy an object is.

Whole number The numbers 0, 1, 2, 3, 4, . . . are whole numbers.

x-axis The horizontal number line in a coordinate plane.

x-coordinate The first number in an ordered pair of numbers.

y-axis The vertical number line in a coordinate plane.

y-coordinate The second number in an ordered pair of numbers.

y-intercept The y-intercept is the y-coordinate of the point of intersection of a line with the y-axis.

Index